TRANSACTIONS OF THE INTERNATIONAL ASTRONOMICAL UNION VOLUME XXVIB

PROCEEDINGS OF THE TWENTY SIXTH GENERAL ASSEMBLY PRAGUE 2006

COVER ILLUSTRATION

THE ASTRONOMICAL TOWER OF THE KLEMENTINUM COMPLEX IN PRAGUE'S OLD TOWN, CZECH REPUBLIC

Klementinum is the oldest Jesuit college in Czech Lands founded immediately after Jesuits came to Prague in 1556. In 1622 Klementinum merged its activities and libraries with the Charles University, Prague (founded 1348) and formed Charles-Ferdinand University in 1654. After the Jesuit order was abolished in 1773, Klementinum continued to house the University and focused on philosophical and mathematical studies, astronomy and theology.

The Astronomical Tower was erected together with the completion of the Baroque library and of the Mirror Chapel in 1722. The tower is crowned with a statue of an Atlas holding the celestial sphere on his shoulders.

Josef Stepling (1716–'78), a study director at the Philosophical Faculty, equipped the tower with astronomical instruments and thus transformed it into an astronomical observatory for both scientists and students. The computation and designs necessary to construct these instruments were provided by an excellent mechanician, Jan Klein (1684–1762). The instruments he produced were among the best of what his epoch could offer.

Antonín Strnad (1746–'99) was appointed the observatory adjunct in 1774. He improved significantly systematic meteorological observations performed in Klementinum since 1752. This series of daily temperature measurements of a high quality, later enriched by other meteorological data and continuing without any interruption up to now, is used as a benchmark for Central European climatologic research.

The Astronomical Tower was used by astronomers to keep time in Prague (Tempus Pragense). From 1842, a man waving a flag from the top of the tower signaled mean Prague noon. This signal was followed by a cannon shot from 1891. This tradition was maintained till 1926. This all came to an end in 1928, when the astronomical observations were moved to the new Ondřejov Observatory and all scientific research, except the meteorological recording, came to an end in 1939.

INTERNATIONAL ASTRONOMICAL UNION

UNION ASTRONOMIQUE INTERNATIONALE

International Astronomical Union

PROCEEDINGS OF THE TWENTY SIXTH GENERAL ASSEMBLY PRAGUE 2006

TRANSACTIONS OF THE INTERNATIONAL ASTRONOMICAL UNION VOLUME XXVIB

Edited by

KARAEL A. VAN DER HUCHT

General Secretary of the Union

CAMBRIDGE UNIVERSITY PRESS
The Edinburgh Building, Cambridge CB2 8RU, United Kingdom
32 Avenue of the Americas, New York, NY 10013-2473, USA
477 Williamstown Road, Port Melbourne, VIC 3207, Australia
Ruiz de Alarcón 13, 28014 Madrid, Spain
Dock House, The Waterfront, Cape Town 8001, South Africa

First published 2008

Printed in the United Kingdom at the University Press, Cambridge

Typeset in System LaTeX 2ε

A catalogue record for this book is available from the British Library

Library of Congress Cataloguing in Publication data

ISBN 9780 521 856065 hardback
ISSN 1743-9213

Président de l'Union Astronomique Internationale

RONALD D. EKERS

PRESIDENT OF THE INTERNATIONAL ASTRONOMICAL UNION

2003–2006

Table of Contents

SECOND SESSION

CLOSING CEREMONY

III – RESOLUTIONS OF THE GENERAL ASSEMBLY

IV – REPORT OF THE EXECUTIVE COMMITTEE 2003 - 2006

VI – STATUTES, BYE-LAWS AND WORKING RULES.

SYDNEY, AUSTRALIA, 2006

VII – RULES FOR IAU SCIENTIFIC MEETINGS

VIII – EXECUTIVE COMMITTEE WORKING GROUPS

IX – DIVISIONS, COMMISSIONS, WORKING GROUPS, AND PROGRAM GROUPS 2006 - 2009

X – NATIONAL MEMBERSHIP

XI – INDIVIDUAL MEMBERSHIP

Preface

The IAU XXVI General Assembly, hosted by the National Committee for Astronomy of the Czech Republic under the auspices of the Czech Academy of Sciences, took place in the Prague Congress Center, 14–25 August 2006, and welcomed a total of 2412 registered participants from 73 countries, including 540 students and 115 seniors, as well as 208 accompanying guests.

The National Organizing Committee, co-chaired by Professors Jan Palouš and Jan Vondrák, with support of a 27-member Board and a Local Organizing Committee chaired by Professor Cyril Ron, arranged for a very well-organized, constructive and informative General Assembly.

Financial support for a limited number of the participants was provided by the IAU, while co-sponsoring by ESO and NASA is gratefully acknowledged.

The Assembly featured a rich scientific programme, organized by Professor Oddbjørn Engvold, IAU General Secretary 2003–2006, and comprising six Symposia, 17 Joint Discussions and seven Special Sessions. During the scientific programme about 650 papers were presented and more than 1550 posters displayed. The proceedings of the six IAU GA Symposia: No. 235, F. Combes & J. Palouš (eds.); No. 236, A. Milani, B. Valsecchi & D. Vokrouhlický (eds.); No. 237, B. G. Elmegreen & J. Palouš (eds.); No. 238, V. Karas & G. Matt (eds.); No. 239, F. Kupka, I. W. Roxburgh & K. L. Chan (eds.); and No. 240, W. I. Hartkopf, E. F. Guinan & P. Harmanec (eds.), have been published in 2007 in the regular *IAU Symposium Proceedings Series* by Cambridge University Press. The proceedings of the Invited Discourses, including the discourse presented during the Inaugural Ceremony on 'Astronomy in Prague, from the Past to the Present' by Dr. Alena Adravová; of the 17 Joint Discussions; as well as of six Special Sessions have been published in the *Highlights of Astronomy*, Volume 14, by K.A. van der Hucht (ed., 2007, CUP).

Plenary events included the First and Second Session of the General Assembly, four Invited Discourses, the Gruber Cosmology Prize 2006 lecture by John C. Mather, and a lunch debate on planet definition.

The EC Working Group on *Women in Astronomy* and Commission 46 on *Astronomy Education and Development* organized, respectively, a well attended lunch meeting on 'Women in Astronomy' (301 participants) and a 'Young Astronomers Lunch Debate' (221 participants). Commission 46 was also in charge of the 'Young Astronomers Consulting Service', open during the Assembly and assisted by 82 consultants.

In addition to its scientific programme, the IAU XXVI General Assembly hosted the regular Business Meetings of the EC, the 12 Divisions, 40 Commissions and 75 Working Groups. The present volume, the *Transactions of the International Astronomical Union, Volume XXVIB*, records the organizational and administrative business of the IAU XXVI General Assembly and the status of the Membership of the IAU.

It is my pleasure to thank those IAU Division and Commission presidents and Working & Program Group chairpersons who have provided reports of their Business Meetings, and the staff of the IAU Secretariat, Mme Vivien Reuter and Mme Maiténa Mitschler, for their professional support in assembling this volume.

Last but not least, I wish to acknowledge the formidable accomplishment of the Prague NOC/LOC, chaired by Professors Jan Palouš, Jan Vondrák and Cyril Ron, in organizing a memorable IAU XXVI General Assembly.

Karel A. van der Hucht
IAU General Secretary
Paris, Utrecht, 30 September 2008

Transactions IAU, Volume XXVIB
Proc. IAU XXVI General Assembly, August 2006
Karel A. van der Hucht, ed.

doi:10.1017/S1743921308023612

CHAPTER I

TWENTY SIXTH GENERAL ASSEMBLY

INAUGURAL CEREMONY

15 August 2006, 14:00 hr
Kongresové centrum Praha, Congress Hall

Master of Ceremonies:
Jan Palouš, chairman of the National Organizing Committee

Opening address by Dr. Ronald D. Ekers, President, International Astronomical Union

Mr. Ladislav Mrgavec, from the office of the President of the Czech Republic, Mr. Miloš Gregar, from the Lord Major's department, Professor Václav Pačes, President of the Academy of Sciences, national representatives, our hosts, IAU members and invited participants:

We have come to meet here in Prague and for some of you, you have come again to meet here in Prague. We re-unite again, of course to meet old friends, but also here in Prague to make connections with this wonderful city and its rich astronomical heritage. This meeting is about new and exciting developments and discoveries in our Universe. It is about an IAU which is very actively creating for itself a future relevant for the 21-st century.

But here in Prague, one is inevitably drawn into the history, as well as looking to the future. So I cannot resist drawing a little from the history of Prague, by using it, I think also as an inspiration for the future. 400 years ago, just before Galileo had sighted the moons of Jupiter, when astronomers had perhaps one of the most interesting and serious debates ever in the history of astronomy, on the Copernican system, whether the Earth was the center or not. You may think the questions you ask now are big questions, but just think of what it was like here in Prague 400 years ago and the questions they were addressing. In looking at what was happening 400 years ago and especially with Tycho Brahe and Johannes Kepler here in the city of Prague, the analogies flow so smoothly, that I thought I would just run with this story for a bit in my talk.

Tycho had the fabulous Uraniborg Observatory in Denmark, a well funded fabulous observatory, comparable in terms of fraction of GNP to any of the great modern observatories. However, there were changes. His funding agency was his patron, who had become Emperor Rudolf II. They had different ideas on what should be funded. I would say Tycho perhaps failed almost completely to adapt to the changes that the funding agency was looking for and his funds were withdrawn. He had no more money to stay in Denmark and he abandoned the observatory and came here to the scientific sanctuary in Prague. Now Tycho had truly realized that, and I'll quote his words, *"an Astronomer must be cosmopolitan, because ignorant statesmen cannot be expected to value their services"*. Not bad for 400 years ago. And so what he had done, was make sure that his instruments and his equipment were designed to be transportable, so he was able to move them here to Prague and set them up in this city.

At almost the same time Johannes Kepler faced very different problems. In his hometown of Graz he was a Lutheran in a Catholic area of Austria, and because of religious persecution he also moved to Prague. Which, again, afforded him scientific sanctuary. Now the moral of this story is not that everybody who feels pressured by their government or persecuted for their beliefs should move to Prague. I think even that might test the enormous skills of our host organizers.

But there is a serious message here, that I want you to remember. That is: our community of science and astronomers is an international community. We can move with enormous freedom around the world and work in different countries, and the national barriers are not strong limitations to what we do. This is a wonderful privilege that we have in this time and in our area of research.

Now to continue the story, Tycho and Kepler made a perfect team. Tycho had accumulated some of the best collections of observations in the history of astronomy, and it would prove that Kepler would be an excellent interpreter of these observations. But either alone could not have made happen what happened. Tycho did not have the vision or the theoretical background, and of course Kepler didn't have the data. But how these two teamed up is in itself one of the great stories in the history of astronomy in Prague. For Tycho carefully guarded his notebooks, he didn't share his observations. There was the dream of whatever 400 years ago was the equivalent of the Nobel Prize masterpiece. That was his focus. Tycho was a Danish aristocrat and Kepler was a mere peasant from Germany. These were barriers which were too strong to form a collaboration in the way we might form it today.

What happened, as you may know the story, was bizarre. Tycho was attending a dinner, hosted by Petr Vok Ursinus Baron Rozmberk here in Prague. Tycho, as was normal I gather, drank to his usual excess but, in the presence of Baron Rozmberk, was too polite to leave the table to relieve himself when he should have done so. As a result, he got a bladder infection and died shortly afterwards. Kepler then took over the observational data base and made the interpretations. So for you observers, if you are being wined and dined by theoreticians here in Prague, be careful.

So that was 400 years ago. Let's take a shorter period of time, a tenth of that. Thirty-nine years ago we had a General Assembly here in Prague, a small fraction of the history of Prague, but in fact about half of the lifetime of the IAU. I thought it might be interesting in fact to have a very quick show of hands for those of you here in the audience and I know that how many are, many were here in Prague in 1967. Where are you? ...Look at that, isn't that amazing.

Just briefly, looking at what happened in that General Assembly, 39 years ago, there are some incredible things and no details. You know many of the details, QSOs had been discovered a few years before. There were workshops, meetings and obviously enormous activity, just as there are today. That problem has remained exciting throughout the last 40 years. Two years before the meeting, Arno Penzias and Robert Woodrow Wilson discovered the microwave background, one of the greatest discoveries and later to receive a Nobel Prize, just before that meeting in Prague. Probably it was too hot a topic, because I don't see much indication that it was discussed in Prague, but of course in this meeting in Prague, in the next ceremony, we will be awarding the Gruber Prize in

Cosmology to somebody whose work happened as a consequence of the discovery of the microwave background.

Perhaps – and I didn't realize it until I was looking at my history carefully – to me the most remarkable thing that was happening during the General Assembly in Prague, in 1967, was that the astronomers in Cambridge had observed pulsars two weeks before the General Assembly and were deciding what to do with this incredible discovery. Jocelyn Bell Burnell is here and I hope people will talk to her afterwards about what was going on at that time. But to the best of my knowledge, at that time they were thought to be little green men and the issue was what would you tell the government and how would you do it. That was while the discussions were happening in Prague. We have of course major, major pulsar discussions at this meeting.

Just one or two things, before I close, about the IAU. The IAU really works in the background, to provide the lubrication for the wheels of the international machinery of astronomy and international science. When I accepted the position as IAU President, I knew relatively little about what the organization does and I imagine it is much the same for most of you here. But this is the way it should be. The IAU needs to facilitate astronomy and can do it in the background. But there is a huge team of Commissions and a network of Working Groups that are all busy behind the scenes. And the results of some of these activities are revealed and come out during these meetings and result in sometimes complex resolutions, but also sometimes other resolutions. More of that in a moment. This is also a meeting the shape of which is set very much by the Division Presidents. The Division Presidents represent the areas of astronomy in the Union, and we now have a new structure in which they are directly involved in the Union activities. I also found it of incredible interest that the first proposal for this kind of structure in detail came from Luboš Perek at the Prague 1967 General Assembly when he took over as General Secretary and he will be speaking to you shortly. It took the 40 years, or a good fraction of them, before such as change was in fact acceptable to the community, but now we have implemented it, primarily since the Kyoto 1997 General Assembly, and it is an important part of the Union.

Now I mentioned some things that happen in the background and result in resolutions. There are events which will be unfolding at this General Assembly, which will be unexpected and are certainly not the usual way things happen. I wanted to say a word or two about what will happen in the next week-and-a-half, and some of you may not be aware of what is going on. Because of international treaties, because of the IAU's mandate to get agreement on nomenclature, it is part of our job to make sure that we can communicate by having well-defined terms.

The definition of a planet has been sadly left not well-defined for hundreds of years. However, in order to proceed with IAU resolutions relating to the definition of a planet we have had to be very circumspect and to operate almost *in camera*. Because of the intense public interest in this topic and the intense pressure from the press, we have set up some groups representing not just astronomers, but historians and people from outreach, science writers and educators, to discuss this issue. Tomorrow, in the GA newspaper you will read about the results of these discussions. Everything is being embargoed until tomorrow. This is certainly not the way the IAU normally works, but I hope you will accept that in this case, with the enormous outside pressure, we felt it was the best course of action. So, by tomorrow everybody will be fully informed, and you have to be fully informed. Because in this next week-and-a-half, you will have an opportunity to debate what is being proposed, to give your input, to think about it. It is a complex issue, think carefully. And in the Second Session of this IAU General Assembly we will be making a vote on an IAU resolution related to the planets.

So my final comment is: after this IAU XXVIth General Assembly in Prague I think it would be wonderful if, instead of the Prague Spring, it could be remembered as the Prague Planet Protocol.

I now open this XXVIth General Assembly of the International Astronomical Union.

Address by Dr. Miloš Gregar, Councilor, City of Prague

Mr. President, ladies and gentlemen:

I am delighted and honored to be able to welcome your congress in our City of Prague, the heart of Europe. Especially, I welcome the representatives of the International Astronomical Union. The city of Prague is proud that Prague is hosting your meeting already for the second time. The last General Assembly in Prague took place 39 years ago. Therefore, Prague has become the third city, along with Sydney and Rome, where this congress comes for the second time.

Ladies and gentlemen, Prague has many attributes: Golden Prague, Prague – Mother of Cities, Prague – Heart of Europe, Prague – Rome of the North, Prague – Hundred Towers. None of the attributes belongs to Prague so indubitably as the one of "Prague Astronomical". In this respect, Prague's *genius loci* could hardly find a rival in the world.

Prague is literally an "astronomical" city. Its original, up to this day preserved town-planning disposition, which the emperor Charles IV gave to the city almost seven centuries ago, was drawn up according to the strict astronomical rules. Even the most unique UNESCO-protected architectural landmarks, as the Charles Bridge, the Old Town Bridge Tower and others are for experts something like small astronomy textbooks. One of the most unique technical sights in the world, the 600 years old Prague Astronomical Clock, has been indicating about twenty astronomical entries with remarkable precision up to the present day. We could hardly find any city in the world connected as much to the history of astronomy as Prague. At every corner you can find places remaining almost untouched by the flow of time, you can stand in the niches where Giordano Bruno, Tycho Brahe, Johannes Kepler, Tadeáš Hájek of Hájek, Joost Bürgi, David Gans, Jehuda Löw, Christian Doppler, Ernst Mach or Albert Einstein used to live and work on their inventions. Their legacy has been connected to Prague forever.

Prague not only protects, but also tries to further develop their legacy. The remembrance of these titans is essential for today's astronomy. We can say that we are probably successful in our efforts, since the modern Prague Observatory belongs among the most visited educational institutions in our country. Your congress is a great example of Prague's astronomical reputation.

Ladies and gentlemen, I believe that despite of your challenging program you will be able to visit the places where your predecessors used to reside with us, and that you will find the time for both relaxation and entertainment.

I wish your IAU General Assembly every success. Thank you for your attention.

Address by Prof. Václav Pačes, President, Academy of Sciences of the Czech Republic

Ladies and gentlemen:

It is a pleasure and honor for me to address you on behalf of the Academy of Sciences of the Czech Republic. The Academy of Sciences in this country is a system of research institutes ranging from mathematics up to humanities and social sciences, with the emphasis on natural sciences.

I am proud that the Institute of Astronomy of the Academy undertook the task to co-organize this IAU XXVI General Assembly. The Institute of Astronomy is a medium sized institute that is devoted to solar physics and stellar astrophysics, to black holes, to galaxies, to astrometry and to the small bodies of the solar system.

It operates the 2-m telescope that was put in operation 39 years ago at the occasion of your IAU XIII General Assembly, which, as it was already mentioned, was held in Prague. Prague is, I think, a very good place for your Assembly because it always was a city of science and especially, as mentioned here, it is a city where, for instance, Johannes Kepler and Tycho Brahe lived and worked.

It was in the time of Emperor Rudolf II, who reigned over the whole Habsburg empire from the Prague castle. Rudolf II was a collector of arts and he was supporter of art and science. Of course at that time, the 17th century, science was different from today's science. Chemistry was mixed with alchemy, astronomy was mixed with astrology.

It was the time when the Prague Rabbi Loew created the artificial servant with the name Golem. There is a nice old Czech movie on these times. Rudolf II in this movie, invited Tycho Brahe to the Prague Castle to let him explain how the planets move and he did so in the company of the famous alchemist, Magister Kelly, and several of the noblemen at the Prague Castle. But these people – in the movie – wanted to get the Golem, and they couldn't do it without removing Rudolf II, so they decided to poison him. They put poison in his glass of wine at the very moment when Tycho Brahe started to show them how the planets moved, and I thought that you might be interested in seeing how your predecessor Tycho Brahe was explaining the movement of planets in the solar system, and so I took a piece of this movie and would like to project it to you, so please ... [*showing an excerpt from the movie "The Emperor and the Golem", 1951*].

Thank you for your attention.

Address by Prof. Jana Musilová, Vice-Rector for Research, Masaryk University, Brno

Dear participants of this so highly anticipated event – astronomers from all over the world, dear distinguished guests, ladies and gentlemen:

I am very glad and honoured by the opportunity of addressing you on behalf of the Rector of Masaryk University, Professor Petr Fiala.

Astronomy and astrophysics have significantly contributed to our understanding of the picture of the world. This is in fact not different from other fields of physics. However, as the Sun, the Moon, the stars in the sky, and outer space altogehter have always fascinated everybody – adults or children, scientists or poets – astronomy has in its hands a rare chance to get as many people as possible acquainted with physics cognition, to support the general knowledge of its importance. We are very aware of it at Masaryk University.

No famous astronomer has worked in Brno, neither in past nor in presence – neglecting for a while that a number of them, such as Josef Mohr, Luboš Perek, Jirí Grygar or Luboš Kohoutek, who became famous in Prague or abroad, had started their careers in Brno. The university observatory owns only a 60-cm telescope. Nevertheless, some world astronomer personalities maybe appear there, because of the enormous interest of students in astrophysics fields of study at Masaryk University.

Therefore, I wish your discussions in your sessions be successful, strengthen the already existing scientific contacts and establish new ones. Thank you for your attention.

Address by Prof. Jan Bednář, Vice-Rector, Charles University, Prague

Dear President of the International Astronomical Union, dear participants of the General Assembly, distinguished guests, ladies and gentlemen:

Astronomy and astrophysics belong to the deepest and most common roots of natural sciences, as one of the main sources of human civilization, development and advance. They study not only planetary and stellar systems, interstellar matter and similar physical problems, but they offer essential and fruitful contributions to the understanding of the whole Universe, including deep and sophisticated philosophical interpretations.

Astronomy has provided during its history a great number of initialization and interpretations for mathematics, mechanics, classical relativistics and quantum physics.

We can all say that astronomy has deeply contributed to the human community progress during long history of our human world. It is surely pleasant and inspirational for the organizers of this General Assembly and also for me, representing Charles University in Prague, that astronomical studies have had a long and fruitful tradition, also in the local territory of the Czech Republic, especially in Prague in the field of the Charles University. I hope that this will be presented in detail in the lectures during other ceremonies.

I would like to wish to the General Assembly great scientific success, presentations of new substantial scientific results, new hypotheses and theories and generally many inspirational ideas. I also wish you a very pleasant stay in Prague.

Thank you very much for your attention.

Address by Prof. Václav Havlíček, Rector, Czech Technical University, Prague

Mr. President, ladies and gentlemen, dear colleagues and friends:

It is a great honour for me to greet the participants of the XXVIth General Assembly of the International Astronomical Union on behalf of the Czech Technical University, the oldest university of technology in Central Europe.

The Engineering School of the Czech Estates was founded on the basis of an edict, issued by the holy Roman Emperor Joseph I in 1707, and classes began in 1780. The studies were at first directed toward military engineering, but the orientation soon moved toward engineering for civilian purposes. In the early 19th century the studies were thoroughly re-organized by Frank-Joseph Gerstner and the modern Prague Polytechnic opened in 1806.

Engineering education at the University has always been based on mathematics and physics and there have been strong connections with astronomy. Many of our professors of mathematics or physics have done research in astronomy too. I would like to mention especially Professor Christian Doppler, who taught mathematics at Prague Polytechnic from 1835 to 1847. In 1842 he published the paper "Über das farbige Licht der Doppelsterne", writing about the shift of the light spectrum of double stars. In this paper he first formulated Doppler's principle, which is still being used in astronomy and other technical applications today.

The Czech Technical University nowadays consists of seven faculties. One of them is the Faculty of Nuclear Science and Physical Engineering. Research in nuclear physics, closely connected with astronomy, provides answers to basic questions of the existence of the Universe and is an important practical application of astronomical research results.

Last but not least, I wish you interesting papers, good discussions, important results and a pleasant stay in Prague. Thank you for your attention.

Address by Emeritus Professor Luboš Perek, Astronomical Institute of the Academy of Sciences of the Czech Republic, Prague; chairman NOC of the XIIIth IAU General Assembly in Prague, 1967; General Secretary of the International Astronomical Union 1967-1970

Ladies and Gentlemen, welcome again after 39 years:

At the XIIIth General Assembly of the IAU, held thirty-nine years ago, I had the privilege to invite the audience to meet again soon in Prague. Thirty-nine years is a short time in astronomy, but in human life it means two generations.

Many things have changed in that time. All branches of astronomy made substantial advances, thanks to space research, to computer technology, and, in the first place, thanks to a larger number of human brains working in the field. It is impossible to give an account of all new discoveries and of new understanding of old problems. Please be referred to 200 volumes of IAU Colloquium proceedings and 200 volumes of IAU Symposium proceedings, which appeared in those 39 years.

There are things, however, which have not changed. Among them is the individual membership in the IAU, an important support of personal contacts across space and time. As regards space, we greet astronomers from 67 countries and expect participants from eight additional countries to join us today or tomorrow.

As regards time, connecting past with the present, we have in Prague four former Presidents of the IAU. The youngest, in terms of service, is Franco Pacini, whose name is closely connected with rotating neutron stars. He was preceded by Lodewijk Woltjer, a supporter of the Very Large Telescope at Mount Paranal. Yoshihide Kozai stands for lunisolar perturbations of satellite orbits. The oldest in service is Adriaan Blaauw. He put all runaway stars into their place in an improved cosmic distance scale. More than half a century ago, I had the honor and pleasure to share an office with Adriaan at the Leiden Observatory, where the atmosphere consisted not of air or oxygen, but of pure astronomy.

Seven former General Secretaries, who devoted part of their lives to the IAU, are among us, starting with my predecessor, Jean-Claude Pecker, my life-long friend, who attended more IAU congresses than anybody else, or almost anybody else. My successor, Kees de Jager, made the Sun his permanent residence. Further, Jean-Pierre Swings, supporter of Mars exploration, Derek McNally, fighter against adverse environmental impacts, Johannes Andersen, director of the Nordic Optical Telescope, and Hans Rickman, observer of the comet impact on Jupiter. These seven musketeers have been recently joined by Jacqueline Bergeron. Therefore, the total count is now eight former General Secretaries.

Names of all former presidents of commissions, professors, and colleagues who connect the past with the present are too many to be listed here and now. They are all welcome, as well as all those who will become friends and colleagues at this General Assembly.

Ladies and gentlemen, next time, please, do not wait thirty-nine years. You are welcome any time. Thank you.

Presentation of the Peter Gruber Foundation Cosmology Prize Award 2006 and the Peter Gruber Foundation Fellowships 2006

Introduction by Patricia Murphy Gruber

Welcome to the Peter Gruber Foundation Cosmology Prize Ceremony presentation. We are happy to present this award in partnership with the IAU.

With the vision of my husband, Peter Gruber, we established this Prize in the year 2000 and it was the first of our international Prizes. We have established five Prizes: in Cosmology, Justice, Genetics, Neuro-science and Women's Rights. Each award recognizes discoveries and achievements that produce fundamental expansion of human knowledge.

I would like to acknowledge the vision and leadership of my husband Peter Gruber. It has been his ideas and the income from his successful career that entirely funds the work of the Peter Gruber Foundation.

Introduction by Peter Gruber

First of all, I just want to welcome all of you to this wonderful event. I am bedazzled by the spectacle of being here in Prague and seeing this wonderful city and its unfolding. I want to make this a very brief speech, if I can. There is an old expression that says "I throw my glass so that others will throw their jade".

Many years ago, and then I will stop, many years ago, when I was a little boy about eleven years old, I was in boarding school, and Brother Darcy, who was one of the teachers, loved two things. He loved astronomy and he loved his garden. And those people who helped him in his garden, got up at night to go to see the stars.

And that is how I became involved and very much interested in this, in the whole issue of astronomy. I just wanted to share that with you and tell you that it has been a wonderful trip, and I am enjoying it enormously. I just want to thank you so much for being here. Thank you.

Introduction by Patricia Murphy Gruber, continued

So Cosmology was our very first Prize and as Peter said, originated from when he was eleven years old being able to look through a telescope.

We award this Cosmology Prize annually, recognizing ground-breaking theoretical, analytical or conceptual discoveries. Past recipients are Allan Sandage and Jim Peebles (2000), who is here with us today; British Astronomer Royal, Martin Rees (2001); Rashid Sunyaev (2003), who is also here with us today; Alan Guth and Andrei Linde (2004); and, last year, James Gunn (2005).

Before announcing the 2006 laureate, I would like first to present our Fellowships for young astronomers. These fellowships were established with the IAU, with the aim of encouraging promising young investigators in cosmology. They are selected by the IAU from numerous applications. The 2006 Fellows are two: Inma Martinez-Valpuesta and Hum Chand.

Inma cannot be with us today. She was born in Spain and she is currently studying at the Centre for Astrophysics Research at the University of Hertfordshire. She is investigating the evolution of stellar bars in disk galaxies.

Hum is from India and is studying at the inter-university Centre for Astronomy and Astrophysics at Pune. He and his colleagues have found that the fine structure constant is not increasing with time, it really is constant. This is a good thing for the world of physics. I am told it means you do not have to rewrite your theories.

And now for the 2006 Cosmology Prize of the Foundation. This Prize carries a gold medal and an unrestricted cash prize of 250.000 US dollars. A distinguished international board of advisers has guided our selection of the 2006 laureate. These advisers are nominated by the IAU and other scientific unions to ensure the scientific integrity of the Prize. The members, this year, are James Peebles, Ron Ekers, Jocelyn Bell Burnell, Roger Penrose, Peter Galison, Simon White, and Jacqueline Bergeron. Owen Gingerich and Virginia Trimble advise the Foundation on cosmology matters. It is gratifying to acknowledge the serious dedication, the knowledge and the enthusiasm that the advisers bring to the judging process.

Now it is time to announce the Prize. The 2006 Gruber Prize for Cosmology is awarded for ground-breaking studies, looking back over 13 billion years to the early Universe and confirming that our Universe was born in a hot Big Bang.

The 2006 Prize is awarded to Dr. John C. Mather and the *COBE* team. The Prize is shared by Dr. John Mather and the *COBE* science working group, representing the hundreds of people who contributed to *COBE*.

Before asking John Mather to accept the award, I'd like to ask two of the advisers to speak briefly about the achievements of Dr. Mather and the *COBE* team. Firstly, Prof. Virginia Trimble from the University of California at Irvine and Las Cumbres Observatory, she will outline John Mather's personal achievements.

Laudation by Virginia L. Trimble

Dr. John Cromwell Mather is, in a sense, a typical American, in another sense not. He just celebrated a significant birthday, and so is on the absolute cutting edge of the baby boom and the generation whose lives were transformed by Sputnik and its successors.

On the other hand, in a country of transients, he lives now only one state line away from where he was born, in Roanoke, Virginia. John grew up on a farm, in Sussex County, New Jersey, where his father was working in a dairy cattle experimental station of Rutgers. His mother was an elementary schoolteacher, and it is clear they taught him a good deal more science than was available in the local school, visited every other week by a Bookmobile. My school had lots of books, we were visited once a year by a cow, so that we could see what it looked like.

Sputnik went up when John was eleven, and by high school he was an enthusiastic consumer of summer school college classes in mathematics and physics. He was also even then persistent in the face of experimental failures. His first home-built short wave radio receiver didn't work, so he got some more parts and built another one.

When the discovery of the CMB was announced in 1965, Mather was a freshman majoring in physics at Swarthmore. Having read books by George Gamow, he was not surprised, though the discoverers were. In 1968, bachelor's degree in hand, John was awarded both NSF and Woodrow Wilson Fellowships. He accepted the NSF Fellowship that was worth about US$ 400 more per year (a lot of money in those days), and went off to the University of California, Berkeley.

And we now discover that John Mather is very far on the hedgehog end of the fox-hedgehog spectrum of scientists, for, when he settled down to a thesis project in 1970, it was to work with Paul Richards on CMB measurements. (Rocket data had shortly before apparently suggested a large excess of short wavelength emission, no!) They, together with David Woody and Michael Werner (with some encouragement from Charles Townes) built a suitable widget and hauled it up 12,000 feet, almost to the top of White Mountain. John is the first author on the 1971 ApJ Letter that reports that they saw no spectral features or excesses.

Next came a balloon experiment. Incidentally, physicists say 'experiment' where astronomers would say 'observation'. Mather and Woody drove that one in a University truck from Berkeley to Palestine, Texas, where the flight was a success but the instrument failed. Back to Berkeley, where Woody rebuilt the instrument. Later flights worked as planned and still saw no excess or feature emission.

Meanwhile, Mather wrote up a thesis and headed off to New York to be a 'proper' radio astronomer, working with Pat Thaddeus on observations of interstellar molecules. But the CMB was calling, and two years later, he was at Goddard Space Flight Center, on track to become the lead scientist on what we now call *COBE*.

Let us fast-forward to a Saturday afternoon at the January 1990 meeting of the American Astronomical Society, a session called Cosmic Background Radiation. The *COBE* authors had been very brave, because the abstracts had been submitted before the 18 November launch date. Nancy Boggess, who is also here, introduced the session and the satellite. We heard from Mike Hauser and from George Smoot and then from John Mather. He actually said rather little, and put up the measured spectrum, a more perfect black body than the calibrating sources, and remarked that it represented only a tiny fraction of the data. The audience spontaneously rose and broke into applause. The session chair growled into the microphone: ... *What are you clapping for? The Universe?*

And I will hand over the microphone to Jim Peebles, to make just a little more clear what we were clapping for!

Laudation by P. James E. Peebles

Mr. President, friends and colleagues:

It is a distinct pleasure to join you in celebrating a deep advance in physical science.

I remind you that in the early 1970s we knew space is filled with microwave radiation. We had the idea that this radiation is a remnant from the Early Universe. But we felt reasonably sure that if this were so, then the radiation would have the distinctive thermal blackbody spectrum. We knew that the long-wavelength part of the spectrum is indeed close to blackbody. We had indications that there is an excess over blackbody near the peak. That apparent anomaly persisted for some 20 years, during which time we theorists were forced to consider the possibility that our universe is a good deal more complicated than we would have liked to have imagined.

We rose to the occasion and we discussed pertubations to the radiation spectrum by decaying dark matter or magnetized superconducting annihilating cosmic strings, for example. This is beautiful physics, and it still may be part of our physical world, but we know that it is not dominant in the energy inventory. The *COBE*-FIRAS experiment, with principal investigator John Mather, showed that our Universe on this scale of things is wonderfully simple, and furthermore that the microwave background is almost uniquely interpretable as a fossil from a time when our Universe was very different from now. What a wonderful advance.

The second experiment on *COBE*, DMR, with George Smoot as PI, aimed to detect the disturbance from an exactly homogeneous distribution of this radiation that must be present, given that our Universe is not exactly homogeneous. When *COBE* was planned, in 1974, we knew the physics of this disturbance; in fact we had too many theories of what might have happened to drive the formation of galaxies and concentrations of galaxies in the expanding Universe, in the process disturbing the spatial distribution of the thermal radiation. The deep contribution by DMR was to show the community the direction that it has taken since then to get now a wonderfully tested cosmology that describes the expansion of the Universe and the evolution of structure within it. This theory rests on the general relativity Einstein introduced nearly a century ago. General relativity has stood the test of time, and an enormous extrapolation from the Solar System, where it was first tested, to the Universe. This is a wonderful result, and another great contribution from *COBE*.

The third experiment on *COBE*, DERBY, with Michael Hauser as PI, addressed another question with a long history. From the time people seriously considered the possibility that the Universe is homogeneous in the large scale average, a fascinating issue has been what has become of the starlight from the generations of stars. One of the goals for the DERBY experiment was to address an aspect of that question that Mike Hauser and I had been debating in the 1970s: what is the present energy density of this accumulated radiation, and what is its distribution in wavelength? DERBY set us on the right track: we have now a more or less a consistent story relating what is observed to what has been happening in galaxies and their AGNs. This is another important advance.

I asked John yesterday: "Weren't you a little nervous putting three eggs – the three experiments on *COBE* – in one basket?" He was young, a post-doc, when he set out in 1974 to become project-scientist for a mission of considerable size and importance. It was a brave adventure and a marvelous success.

A final word. The awarding of major prizes is a wonderful thing, not so much for the individual, although that is fun, but to the community to show our respect, and to the broader community to show that people are doing things in this field that are worth respecting. But during the many years of the tradition of major prizes in science, the way science is done has evolved: great advances depend on crucial contributions from increasingly large numbers of people. I don't think we have broken any very unique ground in the way this award has been arranged, but it does at least take a step toward recognition of how science has evolved. It is a pleasure to consider that this award to the *COBE* mission team struck a good and proper balance.

Thank you.

Words of thanks by John C. Mather

I would like very much to thank the Gruber Foundation for recognizing the work of our team, from the bottom of my heart and on behalf of the whole *COBE* team.

You said out loud that great discoveries come from teams and not just from one or two or three people. I have just heard my talents described in a delightfully embarrassing way, but I have to tell you I didn't do this project myself at all. I think, as individuals we all know our history, we know how we came together to do something beautiful. But I don't know if we really appreciate the flow of history and faith, the forces that come together for a moment to do something magnificent. Standing here in Prague, a beautiful city, maybe a thousand years old, with a very exciting and interesting and difficult history, one can really feel that and sense it.

We astronomers have been blessed by these forces. How could Hans Lipperhey have known 400 years ago, what he would do to astronomy? Or how could Thomas Jefferson and Benjamin Franklin, two of the first scientists in the New World, have guessed that their future tax payers would support fundamental science for the good of all mankind, or that the benefits would be so immense? Or that the discipline that it takes to put spaceships on the moon, using sliderules to design them, would lead so far.

Now to be more specific, I would like to thank some people, especially my family who gave me great opportunity and showed me about science. My father was a scientist, my mother's father was a scientist. I thank my dear wife Jane, who encouraged me when the going was tough, which it was often, and understood what it meant that I was terrified that I might make a big mistake while helping to lead a team of a total of 1500 people.

I thank my thesis-adviser Paul Richards, who is here today. He had actually designed the prototype instrument. The prototype spectrometer should measure the cosmic background radiation spectrum, while I was his graduate student. I thank my fellow student David Woody, who actually made the apparatus work, after I left Berkeley and proved that it could be done. I thank my post-doctoral adviser Pat Thaddeus, who saw that this idea actually could fly in space and helped me organize the team that we grew to write a proposal to NASA, back in 1974.

Then NASA Headquarters, in the person of Nancy Boggess, saw that this proposal was good and chose six people: four from our team and two from two competing teams and said that we should define a new *COBE* mission. Nancy had leadership in this new field of infrared astronomy for NASA Headquarters. She actually backed four major observatories that are coming along. The first was the *InfraRed Astronomical Satellite*, then there was the *COBE*. Now the *Spitzer Space Telescope* which is flying now, and what is coming next is the Stratospheric Observatory for Infrared Astronomy, the SOFIA. It is almost done. So, she did that even though the technology was very hard to develop. I am sure that without Nancy's leadership our project would never have got started.

On our science team we had an outstanding chairman, Rainer Weiss, who actually spent some of his childhood right here in Prague. My fellow principal investigators and our deputies, George Smoot and his deputy Chuck Bennett, Mike Houser and his deputy Tom Castle, my deputy Rick Shaver, they were brilliant to work with.

But beyond all of these individual outstanding people, I have to express the appreciation of the science team for the professional engineers, who actually took these challenges and made the *COBE* work. The expert engineers worked side by side with the scientists. Reaching the limits and finding ways around them and they worked like crazy. There are many stories of the ways that the *COBE* almost failed, but was saved because somebody was working late at night or in weekends and noticed something was not quite right. If you really want to know what happened, you have to read the book.

So, on behalf of all of us on the *COBE* team: our grateful thanks and appreciation to the Gruber Foundation. We have a glorious future ahead of us.

Thank you.

Dr. Alena Hadravová

Astronomy in Prague: from the past to the present

Research Center for the History of Sciences and Humanities,
Academy of Sciences, Prague, Czech Republic

Published in:
Hadravová, A. & Hadrava, P. 2007, in: K. A. van der Hucht (ed.),
Highlights of Astronomy, Vol. 14 (Cambridge: CUP), p. 3

Cultural performances during the Inaugural Ceremony were presented by:

The Lesser Town Singers
The Children Traditional Ensemble Rosénka
The Brass Ensemble of the Prague Castle Guard Orchestra

Transactions IAU, Volume XXVIB
Proc. IAU XXVI General Assembly, August 2006
Karel A. van der Hucht, ed.

doi:10.1017/S1743921308023624

CHAPTER II

TWENTY SIXTH GENERAL ASSEMBLY

First Session: 15 August 2006, 15:45 - 16:50 hr
Second Session: 24 August 2006, 14:00 -16:45 hr
Closing Ceremony: 24 August 2006, 16:45 -17:15 hr

Kongresové centrum Praha, Congress Hall

Prof. Ronald D. Ekers, President, in the chair

FIRST SESSION

1. General Assembly. Opening and welcome

Following the Inaugural Ceremony, the President welcomed the participants to the first plenary working session of the IAU XXVII General Assembly, notably the representatives of the National Members.

2. Representatives of the National Members and members of the Nominating Committee

National Member	**Category of Adherence**	**Votes (a)**	**National Representative**	**Nominating Committee member**
Argentina	II	0	Olga I. Pintado	Olga I. Pintado
Armenia	I	0	Areg M. Mickaelian	
Australia	IV	1	Brian J. Boyle	Michael A. Dopita
Austria	I	1	Michel Breger Jörg Pfleiderer	Michael Breger Jörg Pfleiderer
Belgium	IV	1	Conny Aerts	Conny Aerts
Bolivia	*Interim*	*0*	-	-
Brazil	II	1	Daniela Lazzaro	Horacio A. Dottori
Bulgaria	I	0	-	-
Canada	V	1	James E. Hesser	Gregory G. Fahlman
Chile	I	1	José Maza	Mónica Rubio
China Nanjing	V	1	Shuang Nan Zhang	Zhao Yongheng
China Taipei	I	1	Wen Ping Chen Hsian-Kuang Chang	Wen Ping Chen Hsian-Kuang Chang
Croatia	I	1	-	-
Cuba	*Interim*	*0*	Oscar Alvarez Pomares	Oscar Alvarez Pomares
Czech Republic	II	1	Jan Palouš	Jan Vondrák
Denmark	III	1	Johannes Andersen	Johannes Andersen
Egypt	III	0	Ahmed A. Hady	Ahmed A. Hady

Estonia	I	1	Laurits Leedjärv	Laurits Leedjärv
Finland	II	1	Ilkka V. Tuominen	Ilkka V. Tuominen
France	VII	1	Françoise Combes	Françoise Genova
Germany	VII	1	Günther Hasinger	Jürgen H.M.M. Schmitt
Greece	III	1	Paul G. Laskarides	Paul G. Laskarides
Hungary	II	1	Bela Szeidl	Bela Szeidl
Iceland	I	1	-	-
India	IV	1	S. Sirajul Hasan	Umesh C. Joshi
Indonesia	I	1	-	-
Iran	I	1	Jamshid Ghanbari	Ahmad Kiasat Pour
Ireland	I	1	Paul Callanan	Paul Callanan
Israel	II	1	Elia M. Leibowitz	Elia M. Leibowitz
Italy	V	1	Isabella M. Gioia	Isabella M. Gioia
			Ginevra Trinchieri	
Japan	VII	1	Norio Kaifu	Sadanori Okamura
Korea RP	I	1	Young-Woon Kang	
Latvia	I	1		
Lebanon	*Interim*	*1*	-	-
Lithuania	I	0	Gražina Tautvaišiene	Gražina Tautvaišiene
Malaysia	*Interim*	*1*	-	-
Mexico	II	1	Christine Allen	Rafael Costero
			Alberto Carramiñana	Alberto Carramiñana
Mongolia	*Interim*	*1*	-	-
Morocco	*Interim*	*0*	-	-
Netherlands	IV	1	Pieter C. van der Kruit	J. Mathijs van der Hulst
				Jan Lub
New Zealand	I	1	Willam J. Baggaley	Willam J. Baggaley
Nigeria	I	1	-	-
Norway	I	1	Kaare Aksnes	Kaare Aksnes
Peru	*Interim*	*0*	-	-
Philippines	*Interim*	*1*	Cynthia P. Celebre	Cynthia P. Celebre
Poland	III	1	Stepien Kazimierz	Stepien Kazimierz
			Rafal Moderski	Rafal Moderski
Portugal	II	1	José Pereira Osório	José Pereira Osório
Romania	I	1	Magda G. Stavinschi	Magda G. Stavinschi
Russian Federation	V	1	Dmitrij V. Bisikalo	Rustam D. Dagkesamansky
Saudi Arabia	I	1		
Serbia, Rep. of	I	1	Zoran Knežević	Zoran Knežević
			Olga M. Atanacković-Vukmanović	Olga M. Atanacković-Vukmanovi
Slovakia	I	1	Vladimir Porubčan	Vladimir Porubčan
South Africa	III	1	Patricia A. Whitelock	Patricia A. Whitelock
Spain	IV	1	Rafael Bachiller	Rafael Bachiller
Sweden	III	1	Nikolai E. Piskunov	Nikolai E. Piskunov
			Dainis Dravins	Dainis Dravins
Switzerland	III	1	Matthias Liebendörfer	Matthias Liebendörfer
			Daniel Pfenniger	Daniel Pfenniger
Tajikistan	I	0	Pulat B. Babadjanov	Pulat B. Babadjanov
Thailand	I	1	Boonrucksar Soonthornthum	Boonrucksar Soonthornthum
Turkey	I	0	M. Ali Alpar	M. Ali Alpar
UK	VII	1	Michael Rowan-Robinson	Ian D. Howarth
Ukraine	II	0	Iryna B. Vavilova	Iryna B. Vavilova
			Yaroslav S. Yatskiv	Yaroslav S. Yatskiv
USA	IX	1	John P. Huchra	John A. Graham
Vatican City State	I	1	Christopher J. Corbally	José G. Funes
Venezuela	I	1	Gustavo R. Bruzual	Gustavo R. Bruzual

3. Appointment of Official Tellers

Haili Hu (Netherlands), Styliani Kafka (Greece/Chile), Jana Kašparová (Czech Republic), Luca Sbordone (Italy), and Richard Wünsch (Czech Republic) agreed to serve as Official Tellers, under the supervision of Virginia L. Trimble (USA), for the First and Second Session. Their appointment was approved by acclamation.

4. Admission of new National Members to the Union

Proposals for admission to the Union had been received from three countries: Lebanon (requesting interim status), Mongolia (requesting interim status), and Thailand (requesting full status).

4.1. *Presentations by new National Members*

The proposal from Thailand was officially presented in an address by Dr. Suchinda Chotipanich, Deputy Permanent Secretary of the Thai Ministry of Science and Technology. Lebanon and Mongolia did not present themselves.

4.2. *Vote on admission of new National Members*

By unanimous vote of the present 36 National Representatives, the General Assembly admitted the following National Members of the Union:
- Lebanon, interim status, the National Council for Scientific Research being the Adhering Organization;
- Mongolia, interim status, the Mongolian Academy of Sciences being the Adhering Organization; and
- Thailand, full status, the National Astronomical Research Institute being the Adhering Organization.

5. Votes on proposed revisions of the Statutes and Bye-Laws

A proposal for revisions of the Statutes and Bye-Laws, based on requests from the membership and on consideration by the Executive Committee, had been published in IAU *Information Bulletin* No. 98 (June 2006), section 6.2, pp. 53 - 61. The proposal had been submitted in good time to the Adhering Organizations. IAU Vice-President Robert Williams introduced the revisions to the General Assembly. Each revision was voted on individually and approved in voting ratios (in-favor:abstention:opposed) 33:3:0, 33:2:1, 34:0:2, 36:0:0. The Statutes, Bye-Laws, and Working Rules are presented in chapter VI of these *Transactions*.

6. Appointment of the Finance Committee

Full National Member	**Category of Adherence**	**Dues units**	**Votes (b)**	**Representative**
Argentina	II	2	0	Olga I. Pintado
Armenia	I	1	0	Areg M. Mickaelian
Australia	IV	6	5	Brian J. Boyle Matthew Colless
Austria	I	1	2	Michel Breger Jörg Pfleiderer
Belgium	IV	6	5	Conny Aerts
Brazil	II	2	3	Horatio A. Dottori
Bulgaria	I	1	0	-
Canada	V	10	6	James E. Hesser Gregory F. Fahlman

Chile	I	1	2	Mónica Rubio
China Nanjing	V	10	6	Xiangqun Cui
China Taipei	I	1	2	Wen Ping Chen
				Hsian-Kuang Chang
Croatia	I	1	2	-
Czech Republic	II	2	3	Petr Heinzel
Denmark	III	4	4	Birgitta Nordström
Egypt	III	4	0	Ahmed A. Hady
Estonia	I	1	2	Laurits Leedjärv
Finland	II	2	3	Ilkka V. Tuominen
France	VII	20	8	Suzanne V. Debarbat
Germany	VII	20	8	Rheinhard Schlickeiser
Greece	III	4	4	Paul G. Laskarides
Hungary	II	2	3	Lajos G. Balazs
Iceland	I	1	2	-
India	IV	6	5	Umesh C. Joshi
Indonesia	I	1	2	-
Iran	I	1	2	Sadollah G. Nasiri
Ireland	I	1	2	Paul Callanan
Israel	II	2	3	Elia M. Leibowitz
Italy	V	10	6	Isabella M. Gioia
				Ginevra Trinchieri
Japan	VII	20	8	Toshio Fukushima
Korea RP	I	1	2	
Latvia	I	1	2	-
Lithuania	I	1	0	Gražina Tautvaišiene
Mexico	II	2	3	Alfonso Serrano
Netherlands	IV	6	5	Jan Lub
New Zealand	I	1	2	Willam J. Baggaley
Nigeria	I	1	2	
Norway	I	1	2	Jan-Erik Solheim
Poland	III	4	4	Stepien Kazimierz
				Rafal Moderski
Portugal	II	2	3	José Pereira Osório
Romania	I	1	2	Magda G. Stavinschi
Russian Federation	V	10	6	Nikolaj Samus
Saudi Arabia	I	1	2	-
Serbia, Rep. of	I	1	2	Zoran Knežević
				Olga M. Atanacković-Vukmanović
Slovakia	I	1	2	Vladimir Porubčan
South Africa	III	4	4	John W. Menzies
Spain	IV	6	5	Rafael Bachiller
Sweden	III	4	4	Nikolai E. Piskunov
				Dainis Dravins
Switzerland	III	4	4	Matthias Liebendörfer
				Daniel Pfenniger
Tajikistan	I	1	0	Pulat B. Babadjanov
Thailand	I	1	2	Boonrucksar Soonthornthum
Turkey	I	1	0	M. Ali Alpar
UK	VII	20	8	Paul G. Murdin
Ukraine	II	2	0	Iryna B. Vavilova
				Alexander A. Konovalenko
USA	IX	35	10	Robert W. Milkey
Vatican City State	I	1	2	Richard P. Boyle
Venezuela	I	1	2	-

7. Appointment of the Nominating Committee

The President appointed the members of the Nominating Committee, as listed in section 2 of this report.

8. Report of the Executive Committee 2003 - 2006

The General Secretary briefly presented the Report of the Executive Committee 2003-2006. The full version of this report is published in chapter IV of these *Transactions*.

9. Report on the work of the Special Nominating Committee

The President reported that the Special Nominating Committee proposed the following slate of IAU members for Officers and for Executive Committee members for the triennium 2006 - 2009:

President	Catherine J. Cesarsky	(France)
President-Elect	Robert Williams	(MD, USA)
General Secretary	Karel A. van der Hucht	(Netherlands)
Assistant General Secretary	Ian F. Corbett	(UK)
Vice-President	Beatriz Barbuy	(Brazil)
Vice-President	Cheng Fang	(China)
Vice-President	Martha P. Haynes	(NY, USA)
Vice-President	George K. Miley	(Netherlands)
Vice-President	Giancarlo Setti	(Italy)
Vice-President	Brian Warner	(South Africa)
Advisor, Past President	Ronald D. Ekers	(Australia)
Advisor, Past General Secretary	Oddbjørn Engvold	(Norway)

The General Assembly took note of these nominations.

10. Proposals to host the IAU XXVIII General Assembly in 2012

The President reported that proposals to host the IAU XXVIII General Assembly in 2012 had been received from candidate organizing committees in Beijing (China), Hamburg (Germany), Paris (France), and Thessaloniki (Greece), and that the selection of the host by the Executive Committee would be announced in Session 2 of this General Assembly.

SECOND SESSION

11. Individual Membership

By acclamation the General Assembly welcomed 925 new Individual Members, who had been admitted to the IAU upon recommendation of the Nominating Committee.

The General Secretary reported the names of 172 Individual Members who passed away since IAU GA XXV in 2003.

As of 15 August 2006 the total number of IAU Individual Members was 9785, distributed over 87 countries world-wide.

12. Resolutions

12.1. *Report of the Resolutions Committee*

The members of the Resolutions Committee 2003-2006 were Christopher J. Corbally (chair, USA), Jocely S. Bell Burnell (UK), Matthew Colless (Australia) Georges Meylan (Switzerland), Silvia Torres-Peimbert (Mexico), Rachel L. Webster (Australia), and Robert Williams (USA).

The Committee communicated by e-mail prior to the General Assembly and held several informal meetings between the two sessions of the General Assembly. The Committee received no Resolutions from the Adhering Organizations and no Resolutions were received which had financial implications for the Union. Initially, four scientific resolutions of Type B were received and considered. One merited some exchange with the proposers, but all were quite swiftly recommended, without modifications, to the Executive Committee. These Resolutions 1, 2, 3, and 4, were posted on the IAU web, in IAU *Information Bulletin* No. 98 (June 2006) and in the 3rd issue of *Dissertatio cum Nuncio Sidereo III*, the General Assembly newspaper.

Minor revisions were made to Resolutions 1 and 3 as a result of interactions within Division I working sessions. The revised versions were given to all present at the General Assembly.

A few days before the General Assembly the Resolutions Committee received from the Executive Committee a draft Resolution on 'The Definition of a Planet'. The Resolutions Committee presented initial comments on the Resolution to the EC and worked diligently with members of the EC and its Planet Definition Committee [Owen Gingerich (chair, USA), Richard P. Binzel (USA), André Brahic (France), Catherine J. Cesarky (IAU, ex officio), Dava Sobel (USA), Jun-ichi Watanabe (Japan), and Iwan P. Williams] to incorporate the reactions and suggestions of the IAU members, especially those of Division III, into an acceptable formulation of the Resolution. It was found best to break the draft Resolution into two separate resolutions, Resolution 5A/B (on the definition of 'planet') and Resolution 6A/B (on the category of Pluto).

Acknowledgment is gratefully given to the essential roles of Jocelyn Bell-Burnell, Rachel Webster, and Robert Williams in the re-crafting of the draft Resolution.

Christopher J. Corbally, chairman of the Resolutions Committee, 29 November 2006

12.2. *Presentation of Resolutions of type B by Proposers*

The chair of the Resolutions Committee, Christopher J. Corbally, summarized the recommendations of the Committee concerning the resolutions of Type B, which had been submitted to the Committee in due time before the General Assembly and had been posted before the Second Session of the General Assembly. Subsequently each of the resolutions was presented by a representative of the proposers.

The full text of the final Resolutions is presented in chapter III of these *Transactions.*

12.3. *Vote on Resolutions*

Resolutions 1, 2, 3 and 4 were carried by a clear majority of the General Assembly.

The text of the Resolutions 5A/B and 6A/B was published in the 9th issue of the *Dissertatio cum Nuncio Sidereo III*, with some small amendments added on the spot.
Resolution 5A (on definition of a planet) was carried by a majority of the General Assembly.
Resolution 5B (to insert the word 'classical' before the word 'planet' in Resolution 5A) did not pass, with only 91 votes in favour.
Resolution 6A (on definition of Pluto-class objects) was carried by a majority of the General Assembly, with 237 members in favour, 157 members opposed, and 30 members abstaining.
Resolution 6B (on 'plutonian objects' as the category name for Pluto-class objects) did not pass, with 183 members in favour and 186 members opposed.

12.4. *Appointment of the Resolutions Committee 2006 - 2009*

By acclamation, the General Assembly appointed the following Resolutions Committee for the triennium 2006 - 2009: Jocelyn Susan Bell Burnell (chair, UK), Michel Dennefeld (France), Brian Warner (South Africa), and Rachel L. Webster (Australia).

13. Place and date of the IAU XXVIII General Assembly in 2012

From a slate of four candidate host countries the Executive Committee selected by vote the invitation from China. Therefore, upon invitation from the Chinese Astronomical Society, the IAU XXVIII General Assembly will be held in Beijing, China, 20 - 31 August 2012. The General Assembly accepted the invitation by acclamation.

14. Division and Commission matters

14.1. *Adjustments to the Divisional structure*

Proposed changes in the Division, Commission and Working Groups Structure were accepted and are presented in chapter IX, §§ 2 and 3, of these *Transactions.* The new structure of all Divisions, Commissions and Organizing Committees is presented in chapter IX of these *Transactions.*

14.2. *Election of Division Presidents, Vice-Presidents and Organizing Committees 2006 - 2009*

Each Division held internal elections for President, Vice-President and Organizing Committee for the triennium 2006 - 2009. The elections were approved by the General Assembly.

Division	President	Vice-President
I	Jan Vondrák (Czech Republic)	Dennis D. McCarthy (DC, USA)
II	Donald B. Melrose (Australia)	Valentin Martínez Pillet (Spain)
III	Edward L.G. Bowell (AZ, USA)	Karen J. Meech (HI, USA)
IV	Monique Spite (France)	Christopher J. Corbally (Vatican City State)
V	Alvaro Giménez (Spain)	Steven D. Kawaler (IA, USA)
VI	Thomas J. Millar (Northern Ireland)	You-Hua Chu (IL, USA)
VII	Ortwin Gerhard (Germany)	Despina Hatzidimitriou (Greece)
VIII	Sadanori Okamura (Japan)	Elaine M. Sadler (Australia)
IX	Rolf-Peter Kudritzki (HI,USA)	Andreas Quirrenbach (Germany)
X	Rendong Nan (China)	Russell A.Taylor (Canada)
XI	Günther Hasinger (Germany)	Christine Jones (MA, USA)
XII	Malcolm G. Smith (Chile)	Françoise Genova (France)

14.3. *Election of Commission Presidents, Vice-Presidents and Organizng Committees 2006 - 2009*

Each Commission held internal elections for President, Vice-President and Organizing Committee for the triennium 2006 - 2009. The elections were approved by the General Assembly.

Commission	President	Vice-President	under Division
4	Toshio Fukushima (Japan)	George H. Kaplan (USA)	I
5	Raymond P. Norris (Australia)	Masatoshi Ohishi(Japan)	XII
6	Alan C. Gilmore (New Zealand)	Nicolay Samus (Russia)	XII
7	Joseph A. Burns (USA)	Zoran Knežević (Serbia)	I
8	Irina I. Kumkova (Russia)	Dafydd Wyn Evans (UK)	I
10	James A. Klimchuk (USA)	Lidia van Driel-Gesztelyi (France)	II
12	Valentin Martinez Pillet (Spain)	Alexander Kosovichev (USA)	II
14	Steven R. Federman (USA)	Glenn M. Wahlgren (USA)	XII
15	Walter F. Huebner (USA)	Alberto Cellino (Italy)	III
16	Régis Courtin (France)	Melissa A. McGrath (USA)	III
19	Aleksander Brzezinski (Poland)	Chopo Ma (USA)	I
20	Julio A. Fernndez (Uruguay)	Makoto Yoshikawa (Japan)	III
21	Adolf N. Witt (USA)	Jayant Murthy (India)	III
22	Pavel Spurný (Czech Republic)	Jun-ichi Watanabe (Japan)	III
25	Peter Martinez (South Africa)	Eugene F. Milone (Canada)	IX
26	Christine Allen (Mexico)	Jose A.D. Docobo (Spain)	IV
27	Steven D. Kawaler (USA)	Gerald Handler (Austria)	V
28	Françoise Combes (France)	Roger L. Davies (UK)	VIII
29	Mudumba Parthasarathy (India)	Nikolai E. Piskunov (Sweden)	IV
30	Stephane Udry (Switzerland)	Guillermo Torres (USA)	IX
31	Pascale Defraigne (Belgium)	Richard N. Manchester (Australia)	I
33	Ortwin Gerhard (Germany)	Rosemary F. Wyse (USA)	VII
34	Thomas J. Millar (N. Ireland)	You-Hua Chu (USA)	VI
35	Francesca d'Antona (Italy)	Corinne Charbonnel (Switzerland)	IV
36	John D. Landstreet (Canada)	Martin Asplund (Australia)	V
37	Despina Hatzidimitriou (Greece)	Charles J. Lada (USA)	VII
40	Rendong Nan (China)	Russell A.Taylor (Canada)	X
41	Il-Seong Nha (South Korea)	Clive L.N. Ruggles (UK)	XII
42	Slavek Rucinski (Canada)	Ignasi Ribas (Spain)	V
44	Günther Hasinger (Germany)	Christine Jones (USA)	XI
45	Sunetra Giridhar (India)	Richard O. Gray (USA)	IV
46	Magda G. Stavinschi (Romania)	Rosa M. Ros (Spain)	XII
47	Rachel L. Webster (Australia)	Thanu Padmanabhan (India)	VIII
49	Jean-Louis Bougeret (France)	Rudolf von Steiger (Swiss)	II
50	Richard J. Wainscoat (USA)	. . .	XII
51	Alan P. Boss (USA)	William M. Irvine (USA)	III

52	Sergei A. Klioner (Germany)	Gérard Petit (France)	I
53	Michel Mayor (Switzerland)	Alan P. Boss (USA)	III
54	Guy S. Perrin (France)	Stephen T. Ridgway (USA)	IX
55	Ian E. Robson (UK)	Dennis Crabtree (Canada)	XII

The General Assembly took note of these elections. The complete structure of the IAU Divisions, Commissions, Working Groups and Programme Groups is presented in chapter IX of these *Transactions*.

15. Financial matters

IAU Statutes, clause VII.13d reads: The General Assembly appoints a Finance Committee, consisting of one representative of each National Member having the right to vote on budgetary matters according to clause 14.a. , to advise it on the approval of the budget and accounts of the Union. The General Assembly also appoints a Finance Sub-Committee to advise the Executive Committee on its behalf on budgetary matters between General Assemblies.

IAU Statutes, clause XII.23.a. reads: The Finance Committee examines the accounts of the Union from the point of view of responsible expenditure within the intent of the previous General Assembly, as interpreted by the Executive Committee. It also considers whether the proposed budget is adequate to implement the policy of the General Assembly. It submits reports on these matters to the General Assembly before its decisions concerning the approval of the accounts and of the budget.

15.1. *Report of the Finance Committee on the IAU accounts of the triennium 2004 - 2006, and the budget 2007 - 2009*

SUMMARY

The IAU is financially sound and the General Fund has reached an adequate level to serve as a stable reserve for the Union. The accounting procedures in place provide an accurate record of the financial transactions of the Union and the annual independent audit provides an important check on this.

The budget for the coming triennium is adequate for the programs currently planned and the risks, while not negligible, are manageable.

REPORT

15.1.1. *Introduction*

The Finance Committee (FC) has reviewed the preliminary report prepared by the Finance Sub-Committee (FSC) for the period 2003 - 2005 and the recommendations on the budget for 2007 - 2009.

The FSC was appointed at the 2003 General Assembly in Sydney and has operated as a standing committee in the intervening years, advising the General Secretary and reviewing accounts and budgetary proposal as they became available. It was composed of Robert W. Milkey (USA, chair), Toshio Fukushima (Japan), Paul G. Murdin (UK), Brigitta Nordström (Denmark), John W. O'Byrne (Australia), and Tónu Viik (Estonia). Fukushima (the former chair), Milkey, and Viik were continuing from the previous period.

All amounts quoted below are rounded to the last digit, and stated in thousands of Swiss Francs (CHF), abbreviated KCHF. The IAU operates on a three-year financial cycle with budgets established for a triennium which ends in a General Assembly year. At the time of this report the triennium 2004 - 2006 has not ended, so we examine the current accounts on a year by year basis. We include 2003, the final year of the triennium 2000 - 2003, because that was not completed at the time of the last report.

15.1.2. *Accounts for the years 2003 - 2005*

Accounts for the years 2003, 2004 and 2005 were examined individually together with reports from the General Secretary and the reports from the independent auditors. We are pleased to report that the funds were expended responsibly and in accordance with the budgets adopted by the 25th General Assembly, as modified by Executive Committee actions.

The accounts compare to the budgets as follows (in KCHF):

	2003		2004		2005	
	budget	actual	budget	actual	budget	actual
revenues	913	1022	919	1068	947	1243
expenditures	1031	1067	865	965	890	857
net	−118	−44	55	103	57	386

The total of special contributions in the years 2003 - 2005 was 168 KCHF and an additional 114 KCHF was received from ICSU for a special project on Near Earth Objects. This type of revenue is not budgeted, but is of great value to the IAU when received.

The most notable departure from budget planning is the large excess revenue in 2005. This is due to a very successful collection of contributions (dues) from Adhering Organizations and significant recovery of shortfalls from prior years, including the previous triennium. In the period 2000 - 2002 the member contributions were reported as below budget by 317 KCHF, while in the period 2003-2005 these revenues were 202 KCHF above budget. We commend the secretariat for efforts in this area and recommend continuing attention to this matter.

The 2004 actual revenues and expenditures each include an amount of slightly over 60 KCHF which was collected at Sydney for proceedings and transmitted to IAU in 2004. Subsequently this was passed on to the publisher. While this transaction is shown in the accounts for that year, it was not included in the budget projections.

The net benefit to the IAU General Fund (reserves) over this period was 446 KCHF, bringing the value of the General Fund to 1134 KCHF. This satisfies a recommendation by the 2003 Finance Committee that the General Fund be held at approximately one-year's operating budget for the Union. For reference, the average annual expense proposed in the budget for 2007 - 2009 is 988 KCHF. After the planned deficit of 113 KCHF in 2006, the reserves will still exceed this amount.

Reports from the audit firm of MBM Conseil for the years 2003 through 2005 were submitted to and examined by the Finance Committee. These reports verify the revenues, expenditures, and net assets for the period in question. The auditors also examined the accounting procedures used by the IAU.

15.1.3. *Budget for 2007 - 2009*

15.1.3.1. *Income*

The income is forecast on the basis of a 3% annual growth in the contribution unit to offset inflation. This is consistent with recent history and current expectations for inflation factors in both the U.S.A. and Europe as reported by the U.S. Department of Labor and by Eurostat. The same inflation factors have been used for both revenue and expenses.

Income from Adhering Organizations has been forecast using the current level of contribution for each National Member and the assumption that all National Members will pay without delay. This category accounts for 97% of the budgeted revenue for IAU and thus it is extremely important that remittances be made in a timely manner. While two new levels of adherence have been created, it is not yet clear whether negotiations with member nations, particularly the United States, will lead to increased revenues in the coming triennium.

The second largest category of budgeted revenue is that from IAU Publications. Due to the recent change in publisher this has been somewhat difficult to forecast, but forecasts become more reliable as each year goes by. It is hoped that this may even rise above the budget expectations.

'Special Contributions' is another important category of revenue but one which is not included in the budgetary forecasts. This is an appropriate budgeting process, because these cannot be assured. However, as each such contribution is received, the Executive Committee and the General Secretary can make the appropriate modification to the expenditure plan to pursue the activities for which these are intended.

Programs associated with the *International Year of Astronomy* will probably require additional fundraising in the coming triennium.

15.1.3.2. *Expenses*

The proposed expenses are consistent with the status quo for IAU programs.

The operating costs of the Secretariat have risen in the past couple of years as the increased web capability was implemented and the membership database was updated. The cost for support of the electronic infrastructure is included in this category, in effect moved in part from the Publications category. We still view the costs of administration as low for an organization of this size and would not be surprised if these were to increase further in the near future.

The budget forecasts a major saving in publications costs over prior periods. This is a desirable result of the move to electronic distribution for the IAU *Information Bulletin*, etc. The IAU is to be commended for this development. The distribution of budgeted expenses 2007-2009 by category in the proposed budget is: Scientific Activities 36% Educational Activities 10%, Publications 4%, Other Unions 4%, Governance 11%, Administration 38%.

A major element of support for activities of the IAU does not appear anywhere in the budget because it is the in-kind contribution of the home institution of the General Secretary. It is safe to say that without this very substantial contribution of effort by a senior scientist, the IAU would find it extremely difficult to accomplish its programs. For a period of six years, first as Assistant General Secretary, and then as General Secretary, an individual devotes a major portion, if not all, of his or her time to the affairs of the IAU. A further substantial in-kind contribution is made by the provision of the office accommodation by the Institute d'Astrophysique in Paris. We would like to express our gratitude for this support.

15.1.3.3. *Risks to budget*

We identify several risks to the budget. If any of these develop, this will endanger the achievement of the program within the established budget.

15.1.3.3.1. *Inflation* — Should inflation significantly outpace the assumed 3%, and especially should this affect labor costs in France where the Secretariat is located, then program expenses will have to be adjusted to meet these costs.

15.1.3.3.2 *Currency Exchange Fluctuations* — Two scenarios:

- The Swiss Franc drops with respect to the Euro, the currency in which most expenses are incurred. This seems unlikely since these two currencies have been extremely stable with respect to each other in recent years.
- Other currencies fall with respect to the Swiss Franc making payment by adhering organizations increasingly difficult.

15.1.3.3.3. *Adhering Organizations are late paying* — This would require IAU to draw on reserves to meet current expenses.

15.1.3.3.4. *Airfares increase* — Airfares could increase substantially due to fuel costs or other factors, thereby increasing the costs of both scientific program support (grants) and governance (Executive Committee).

15.1.3.3.5. *Labor Costs in the Secretariat* — We feel that the Secretariat is minimally staffed for the program needs of the IAU, and staff turnover may require some re-organization of the office during the coming triennium. The IAU has grown and the demands on the General Secretary and Secretariat staff have increased commensurately. Despite the dedication and hard work of the staff these demands may still require further staff or contract services above those foreseen in the budget to keep pace with the needs and expectations of the membership.

Should any of these be realized, we recommend that the Executive Committee and the General Secretary take immediate steps to adjust the expenditure budget, so as to avoid any unacceptable impact to the bottom line.

15.1.4. *General Fund*

The General Fund is the IAU's buffer against financial difficulties and as such is essential in guaranteeing the long-term health of the organization. This fund has now reached the level recommended by the previous Finance Committee and it is incumbent on the IAU to maintain this level and plan for growth to preserve against inflationary erosion. While maintaining this fund, the IAU may use any excess earnings for program support.

15.1.5. *Recommendations*

Overall our recommendations are based on our perception that the IAU is severely limited in its resources. To enable the IAU to expand its programs and fill the role potentially available for it in the domain of world astronomy additional income is necessary. A large proportion of current revenues must necessarily be used just to maintain the basic infrastructure for an international organization of the size of the IAU. Program growth will require additional resources.

15.1.5.1. Revenue Collections: continue program for aggressively collecting any payments that are in arrears. Negotiate, where possible, for increased contributions from member organizations.

15.1.5.2. Continue annual reviews of accounts by the Finance Sub Committee.

15.1.5.3. Re-examine investment policy to determine if higher earnings can be obtained at acceptable risks. If additional revenues, even of modest amounts, could be obtained it could have a noticeable impact on the programs.

15.1.5.4. Seek additional sources for revenue, including fund raising and development of new products for which the IAU 'brand' would have some meaning.

15.1.6. *Finance Sub-Committee for 2006 - 2009*

Based on the standing Working Rules for the FSC, the FC proposes:
Continuing Members: Paul G. Murdin (UK, chair), Brigitta Nordström (Denmark), John W. O'Byrne (Australia).

New Members: Kevin B. Marvel (USA), Cyril Ron (Czech Republic), Cui Xiangqun (China Nanjing).

Robert W. Milkey, chair FSC, IAU XXVI General Assembly, August 24, 2006

15.2. *Vote on proposed budget 2007 - 2009*

The General Assembly approved the proposed budget 2007 - 2009 (Appendix I to this chapter).

15.3. *Appointment of the Finance Sub-Committee 2006 - 2009*

Following the recommendation of the Finance Committee, the General Assembly approved a Finance Sub-Committee for the period 2003 - 2009 with members: Paul G. Murdin (UK, chair), Kevin B. Marvel (USA), Brigitta Nordström (Denmark), John W. O'Byrne (Australia), Cyril Ron (Czech Republic), and Cui Xiangqun (China Nanjing).

16. Appointment of the Special Nominating Committee 2006 - 2009

Following the vote of the Division Presidents, the following composition of the Special Nominating Committee for the triennium 2006 - 2009 was unanimously approved by the General Assembly. *Members:* Catherine J. Cesarsky (chair, France), Alan H. Batten (Canada), Philip Allan Charles (South-Africa), Ronald D. Ekers (Australia), Julieta Fierro (Mexico), Eva K. Grebel (Germany), Sadanori Okamura (Japan), and Graina Tautvaiiene (Lithuania). *Advisers:* Karel A. van der Hucht (General Secretary, Netherlands and Ian F. Corbett (Assistant General Secretary, UK).

17. Election of the Executive Committee Members 2006 - 2009

Upon nomination by the Special Nominating Committee and approved by the Executive Committee, the General Assembly elected:

President: Catherine J. Cesarsky (France); President-Elect: Robert Williams (MD, USA); General Secretary: Karel A. van der Hucht (Netherlands); Assistant General Secretary: Ian F. Corbett (UK); and Vice-Presidents (2006-2012): Martha P. Haynes (NY, USA), George K. Miley (Netherlands), and Giancarlo Setti (Italy).

Accordingly, the composition of the Executive Committee for the triennium 2006 - 2009 is as follows:

President	Catherine J. Cesarsky	(France)
President-Elect	Robert Williams	(MD, USA)
General Secretary	Karel A. van der Hucht	(Netherlands)
Assistant General Secretary	Ian F. Corbett	(UK)
Vice-President (2003-2009)	Beatriz Barbuy	(Brazil)
Vice-President (2003-2009)	Cheng Fang	(China)
Vice-President (2006-2012)	Martha P. Haynes	(NY, USA)
Vice-President (2006-2012)	George K. Miley	(Netherlands)
Vice-President (2006-2012)	Giancarlo Setti	(Italy)
Vice-President (2003-2009)	Brian Warner	(South Africa)
Advisor, Past President	Ronald D. Ekers	(Australia)
Advisor, Past General Secretary	Oddbjørn Engvold	(Norway)

This being the last agenda item of Second Session of the General Assembly, President Ronald D. Ekers invited the new Executive Committee to the stage, and the General Assembly proceeded to the Closing Ceremony.

CLOSING CEREMONY

18. Invitation to the IAU XXVII General Assembly in Brazil in 2009. Address by Prof. Daniela Lazzaro

It is my honor and pleasure, on behalf of the Brazilian Ministry of Science and Technology, the Brazilian Astronomical Society and the National Organizing Committee, to invite you to Brazil to participate in the IAU XXVII General Assembly to be held in Rio de Janeiro, 3 - 14 August 2009.

This beautiful city, located on the coast of Brazil, is not only well known for its unique tourist attractions, but is also the city where astronomical research officially began in the country. It was back in 1827 when Emperor Don Pedro I created the Observatório Imperial, later renamed Observatório Nacional. This is considered the first scientific institution in Brazil. In its long and not always easy history, the Observatório Nacional has received important visitors, among which Albert Einstein. Indeed it was a great event, in particular because, as many of you surely know, it was in Brazil, more specifically in the small town of Sobral in the north-east region of Brazil, that five years earlier, on 29 May 1919, the theory of General Relativity proposed by Einsten was observationally confirmed by measuring the bending of star light during a solar eclipse.

Nearly hundred years later, at the beginning of the 20-th century, was created in São Paulo what is nowadays the largest astronomical institute in Brazil. Initially it was a department of the Meteorological and Astronomical Service of the State, located at the São Paulo Observatory. Only later on it was renamed Instituto Astronômico e Geofísico, being an institute of the Escola Politécnica of São Paulo. Exactly sixty years ago it became part of the University of São Paulo and very recently it moved to its present location on the university campus. Nowadays, astronomical research is developed at several universities and institutes all over the Brazil, in the states of Rio Grande do Sul, Minas Gerais, Rio Grande do Norte, Santa Catarina, Bahia, Pernambuco and several others.

Astronomical research in Brazil received a large impulse in the 1970's with the operational debut of several telescopes, including the 1.6 m National Optical Telescope, the radio-telescope at Itapetinga, and the activities of the Space Research Institute. Presently we are entering into a new phase, thanks to our participation in the Gemini consortium and the SOAR consortium.

In 1974 the Brazilian Astronomical Society was founded, which presently has around 400 members. Since its creation it has organized annual meetings which offer the opportunity to astronomers, especially students and post-docs, to present the results of their research. The Society has been giving support to symposia and workshops. Seven IAU Symposia and two Latin American Regional IAU Meetings were held in Brazil. Now the time has come for an IAU General Assembly. This will occur in Rio de Janeiro.

So, be sure not to miss the IAU XXVII General Assembly: great science and a great time is waiting for you! The members of the Brazilian Astronomical Society and all Brazilian astronomers as well as the "cariocas" are looking forward to welcome you warmly in Rio. Thank you!

Daniela Lazzaro, Chair of the National Organizing Committee, Prague, 2006

19. Address by the incoming General Secretary, Karel A. van der Hucht

Dear colleagues, friends,

The secret of the success of an AGS in surveying the preparations for the IAU Scientific Meetings and its Proceedings Series is communication with the organizers of these meetings and with the editors of their proceedings.

As IAU AGS in the past three years, communication was necessarily my middle name. Communication is time consuming and can occasionally be disappointing, but mostly communication is rewarding, gratifying and constructive. Sharing information should be considered neither a waste of time nor a threat, but rahter a necessity and a blessing.

That is why the IAU EC has created the concept of the IAU Editorial Board, starting its activities in early 2004. The IAU Editorial Board serves as a communication platform to support all IAU Symposium and Colloquium Proceedings' Editors. Here they can exchange experience and ask for advice, whenever necessary, in their efforts to ensure that (*a*) all papers published in the IAU Proceedings are of the highest possible standard; and (*b*) the IAU Proceedings are published within six months after the event.

For each calendar year, the IAU EB comprises as Working Members the Chief Editors of the IAU Symposia of that year, assisted by four Advisers and by the AGS as chair. The Working Members who are encouraged to share editing experiences with each other within the EB. Through seeing and watching each other perform, the EB provides a sense of community. In this way, the goal of having the IAU Proceedings published within six months after the event has almost been achieved in 2004 and 2005, the first two years that we have been working together with our new Publisher, Cambridge University Press. In 2004 - 2005 the average production time was 7.5 months. We hope that, thanks to the hard and efficient work of the Editors of all nine Symposia held in 2006, six of them during this GA, we will get closer to the desired six-months production time, while maintaining the highest possible scientific standard.

Are the IAU Symposia still of relevance in an age when almost every day, somewhere on this planet, a new astronomical meeting takes place? We think they are. Because (*a*) the IAU still receives 2 - 3 times more proposals for IAU Symposia than the nine that can be financially accommodated by the IAU in any one year; (*b*) the IAU Symposia cover broad topics and attract broad communities, including talented astronomers from less privileged countries supported by IAU Grants.

The eight IAU Symposia and Colloquia in 2004 attracted 1229 participants, of which 303 participants from 48 countries received IAU grants. The nine IAU Symposia and Colloquia in 2005 attracted 1648 participants, of which 253 participants from 39 countries received IAU grants. Thus together, the 17 IAU Symposia and IAU Colloquia held in 2004 and 2005 attracted 2877 participants, more than in any single IAU General Assembly. Approximately 20% of the participants received IAU travel grants.

In 2005, the IAU also accommodated two Regional IAU Meetings (RIMs), an Asian-Pacific RIM, and a Latin-American RIM. Together, they attracted 559 participants, of which 65 participants received IAU grants.

The triennium 2003-2005 has been quite busy in terms of publishing activities: 38 Proceedings' volumes of IAU Symposia, Colloquia and Regional IAU meetings saw the light.

As for all above mentioned issues of IAU Symposia and Regional IAU Meetings, my successor Ian F. Corbett will see to it, with your help, that we maintain and improve our level of performance.

Communication will have to be my middle name also as your IAU General Secretary. You may remind me of that anytime. The next three years will be as important for the IAU as any other three years. My communication with you will be through the frequently updated IAU web site; the *Highlights* and *Transactions* of this GA; the bi-annual *Information Bulletin*; - the e-Newsletter; and, as a matter of course, the contacts with the Presidents of our 12 Divisions and 40 Commissions, and the chairpersons of the 75 Working/Program Groups whom you have voted into office, as well as with National Members.

As a special issue for the coming three years, the IAU has taken upon itself to be the facilitating organization of the *International Year of Astronomy 2009*. The main concept of this program is Communication with the Public: sharing with the community at large what we do, why we do it, and explaining why it is of relevance for all human beings.

Thank you, Oddbjørn Engvold, for showing me the way in the past three years. Thank you, Ian Corbett, for taking over the responsibilities as AGS. Thank you, IAU members, for giving me your vote of confidence. It will be my pleasure to serve you and to communicate with you for another three years until our great community gets together again, in Rio de Janeiro, 2009, for the IAU XXVII General Assembly.

20. Address by the retiring President, Ronald D. Ekers

It has certainly been a great honor for me to serve you as President of the IAU for the last triennium.

Those of you who know me may realize that I sometimes felt frustration at the rate of progress the IAU and how its members are adapting to this quite rapidly changing world. But I am immensely pleased with the way the new structure of the IAU is working, with the way in which the scientific matters of the Union are now being handled through the Division Presidents as representatives of the scientific part of the Union. And, of course, I also note that international organizations must work by providing consensus between all of the different nations and groups of astronomers, in order to proceed essentially as an international organization.

I have really enjoyed this, and I want to thank the incredible team I had working with me for the last three years. I thank all of you for your support.

21. Address by the incoming President, Catherine J. Cesarsky

We started this conference, in magnificent Prague, with a lovely rendition of "What poor astronomers are they", from John Dowland [1563 - 1626, *Dissertatio cum Nuncio Sidereo III*, no. 3, p. 1]. I would like to close it with my motto: "What lucky astronomers we are": we lead purposeful and interesting lives.

Look at this General Assembly. We have just had two weeks of passionate discussions. From black holes, dark energy, NEOs, to the meaning of the width of a spectral line or of a glitch in the radio emission of a pulsar, detection limits for extra solar planets, confusion limits in the infrared, the definition of a planet, every celestial object, every concept, every observation, every prediction got scrutinized, debated, refined . . . until next time.

Among astronomers, there are no national barriers; we all share the Universe. And thus, the Union – the IAU –, is not a vain word; it embodies one small and distinctive part of humanity, completely absorbed in the study of the heavens. I really love the feeling of togetherness in intellectual adventure and path finding, enhanced by our continuous arguing and squabbling. This is why I am not so much proud, as deeply happy that you have elected me President of the IAU.

In the next three years, we will not have only the fortune of astronomers to celebrate. We will be commemorating, in 2009, the 400th anniversary of Galileo's first observations with a telescope, which brought about a fundamental change in our perception of the universe. We have decided through a resolution passed at the last General Assembly, at the initiative of Franco Pacini, that 2009 would be the International Year of Astronomy. UNESCO has endorsed our resolution and we hope that soon the UN will follow through. This offers an ideal opportunity to highlight astronomy's role in enriching all human cultures, to promote astronomy in the developing nations, to inform the public about our latest discoveries, and to emphasize the essential role of astronomy in science education. Individual countries will be undertaking their own initiatives, considering their own national needs, while the IAU will act as catalyst and coordinator of 2009 IYA on the global scale. We plan to liaise with, and involve, as many as possible of the ongoing outreach and education efforts throughout the world, including those organized by amateur astronomers. One interesting example is a programme geared at small children primarily in developing countries, 'Universe Awareness'. There will also be international events, events at the General Assembly in Rio de Janeiro, as well as an opening and a closing event.

2009 is also the year of the 90th anniversary of the IAU, and on 22 July 2009 will come about the longest duration total solar eclipse of the 21st century. More celebrations!

We welcome your involvement and your ideas. Let us all together make this exceptional year an astounding success.

22. Closure of the IAU XXVI General Assembly

President Ronald D. Ekers called to the stage the members of the Czech NOC/LOC of the IAU XXVI General Assembly and thanked them for a wonderful experience in a wonderful Prague.

On behalf of the participating guests, Mr. Harvey Green from Australia thanked the NOC/LOC for all they did to make the participating guests feel comfortable during their stay in Prague.

The President declared the General Assembly closed.

A cultural performance at the start of the Second Session was presented by the Vocal Quartet Affetto.

APPENDIX I

Proposed IAU budget 2007 - 2009 (CHF)

INCOME	**2007**	**2008**	**2009**	**2007 - '09**
unit of contribution	3685	3800	3900	
adjustment for inflation	3%	3%	3%	
number of units	262	262	262	
Adhering Organizations	965 470	995 600	1 021 800	2 982 870
royalties	20 000	20 000	20 000	60 000
bank interest	10 000	10 000	10 000	30 000
total INCOME	*995 470*	*1 025 600*	*1 518 000*	*3 072 870*

EXPENDITURE	**2007**	**2008**	**2009**	**2007 - '09**
SCIENTIFIC ACTIVITIES				
General Assemblies				
Travel Grants	-	-	235 000	235 000
operations	4 000	8 000	50 000	62 000
sub-sub-total GAs	*4 000*	*8 000*	*285 000*	*297 000*
Meetings				
Symposia	225 000	225 000	225 000	675 000
Regional Meetings	30 000	30 000	-	60 000
co-sponsored meetings	-	-	-	-
sub-sub-total Meetings	*255 000*	*255 000*	*225 000*	*735 000*
Working Groups				
EC Working Groups	5 000	5 000	5 000	15 000
Commission Working Groups	5 000	5 000	5 000	15 000
Minor Planet Center	12 000	12 000	12 000	36 000
CB for Astronomical Telegrams	4 000	4 000	4 000	12 000
Meteor Data center	2 000	2 000	2 000	6 000
sub-sub-total Comm/WGs	*28 000*	*28 000*	*28 000*	*84 000*
sub-total SCIENTIFIC ACTIVITIES	*287 000*	*291 000*	*538 000*	*1 116 000*
EDUCATIONAL ACTIVITIES				
PG-ISYA	45 000	45 000	-	90 000
PG-TAD	45 000	45 000	45 000	135 000
PG-EA	15 000	15 000	15 000	45 000
other activities	15 000	15 000	15 000	45 000
sub-total EDUCATIONAL ACTIVITIES	*120 000*	*120 000*	*75 000*	*315 000*
DELEGATES TO OTHER UNIONS	*12 000*	*12 000*	*12 000*	*36 000*
DUES TO OTHER UNIONS				
ICSU	7 500	7 500	7 500	22 500
ERS/FAGS	10 000	10 000	10 000	30 000
IUCAF	7 500	7 500	7 500	22 500
sub-total DUES TO OTHER UNIONS	*25 000*	*25 000*	*25 000*	*75 000*

EXECUTIVE COMMITTEE				
Officers' Meetings	12 000	12 000	12 000	36 000
Executive Committee Meetings	45 000	80 000	80 000	205 000
General Secretary	30 000	30 000	30 000	90 000
Assistant General Secretary	2 000	2 000	2 000	6 000
EC WG IYA2009	2 000	2 000	8 000	12 000
sub-total EXECUTIVE COMMITTEE	*91 000*	*126 000*	*132 000*	*349 000*
PUBLICATIONS	*10 000*	*10 000*	*10 000*	*30 000*
ADMINISTRATION				
salaries and charges	286 790	286 790	286 790	860 370
training courses	5 000	5 000	5 000	15 000
general office expenses	82 000	85 000	90 000	257 000
auditor	2 500	2 500	2 500	7 500
bank charges	4 000	4 000	4 000	12 000
sub-total ADMINISTRATION	*380 290*	*383 290*	*388 290*	*1 151 870*
total EXPENDITURE	*995 470*	*1 025 600*	*1 051 800*	*3 072 870*
net result	*70 180*	*58 310*	*−128 490*	-

Transactions IAU, Volume XXVIB
Proc. IAU XXVI General Assembly, August 2006
Karel A. van der Hucht, ed.

doi:10.1017/S1743921308023636

CHAPTER III

RESOLUTIONS OF THE GENERAL ASSEMBLY

1. Introduction

The members of the Resolutions Committee 2003-2006 were Christopher J. Corbally (chair, USA), Jocely S. Bell Burnell (UK), Matthew Colless (Australia) Georges Meylan (Switzerland), Silvia Torres-Peimbert (Mexico), Rachel L. Webster (Australia), and Robert Williams (USA).

The report of the Resolutions Committee is presented in Chapter II, section 12.1 of these *Transactions*.

2. Proposed Resolutions

The Resolutions Committee received the following Resolutions of Type B:

Resolution 1.
Adoption of the P03 Precession Theory and Definition of the Ecliptic.
Proposed by: James L. Hilton, Division I WG on " Precession and the Ecliptic".
Supported by: Division I.

Resolution 2.
Supplement to the IAU 2000 Resolutions on Reference Systems
Proposed by: Nicole Capitaine, Division I WG on "Nomencalture for Fundamental Astronomy". *Supported by:* Division I.

Resolution 3.
Re-definition of Barycentric Dynamical Time, TBD
Proposed by: Nicole Capitaine, Division I WG on "Nomencalture for Fundamental Astronomy". *Supported by:* Division I.

Resolution 4.
Endorsement of the Washington Charter for Communicating Astronomy with the Public
Proposed by: Ian Robson, Division XII WG on "Communicating Astronomy with the Public". *Supported by:* Division XII.

Resolution 5.
Definition of a Planet in the Solar System
Proposed by: EC Planet Definition Committee. *Supported by:* Division III.

Resolution 6.
Pluto
Proposed by: EC Planet Definition Committee. *Supported by:* Division III.

Transactions IAU, Volume XXVIB
Proc. IAU XXVI General Assembly, August 2006
Karel A. van der Hucht, ed.

doi:10.1017/S1743921308023648

APPROVED RESOLUTIONS / RÉSOLUTIONS APPROUVÉES

RESOLUTION B.1

Adoption of the P03 Precession Theory and Definition of the Ecliptic

The XXVIth International Astronomical Union General Assembly,

Noting

1. the need for a precession theory consistent with dynamical theory,

2. that, while the precession portion of the IAU 2000A precession-nutation model, recommended for use beginning on 1 January 2003 by resolution B1.6 of the XXIVth IAU General Assembly, is based on improved precession rates with respect to the IAU 1976 precession, it is not consistent with dynamical theory, and

3. that resolution B1.6 of the XXIVth General Assembly also encourages the development of new expressions for precession consistent with the IAU 2000A precession-nutation model, and

Recognizing

1. that the gravitational attraction of the planets make a significant contribution to the motion of the Earth's equator, making the terms *lunisolar precession* and *planetary precession* misleading,

2. the need for a definition of the ecliptic for both astronomical and civil purposes, and

3. that in the past, the ecliptic has been defined both with respect to an observer situated in inertial space (inertial definition) and an observer co-moving with the ecliptic (rotating definition),

Accepts

The conclusion of the IAU Division I Working Group on Precession and the Ecliptic published in Hilton *et al.* 2006, *Celest.Mech.* (**94**, 351), and

Recommends

1. that the terms *lunisolar precession* and *planetary precession* be replaced by *precession of the equator* and *precession of the ecliptic*, respectively,

2. that, beginning on 1 January 2009, the precession component of the IAU 2000A precession nutation model be replaced by the P03 precession theory, of Capitaine *et al.* (2003, *A&A*, **412,** 567-586) for the precession of the equator (Eqs. 37) and the precession of the ecliptic (Eqs. 38); the same paper provides the polynomial developments for the P03 primary angles and a number of derived quantities for use in both the equinox based and CIO based paradigms,

3. that the choice of precession parameters be left to the user, and

4. that the ecliptic pole should be explicitly defined by the mean orbital angular momentum vector of the Earth-Moon barycenter in the Barycentric Celestial Reference System (BCRS), and this definition should be explicitly stated to avoid confusion with other, older definitions.

Notes

1. *Formulas for constructing the precession matrix using various parameterizations are given in Eqs. 1, 6, 7, 11, 12 and 22 of Hilton et al. (2006). The recommended polynomial developments for the various parameters are given in Table 1 of the same paper, including the P03 expressions set out in expressions (37) to (41) of Capitaine et al. (2003) and Tables 3-5 of Capitaine et al. (2005).*

2. *The time rate of change in the dynamical form factor in P03 is* $dJ_2/dt = -0.3001\times10^{-9}$ *century*$^{-1}$

References

Capitaine, N., Wallace, P. T., & Chapront, J. 2003, *A&A*, **412**, 567
Capitaine, N., Wallace, P. T., & Chapront, J. 2005, *A&A*, **432**, 355
Hilton, J. L., Capitaine, N., Chapront, J., Ferrandiz, J. M., Fienga, A., Fukushima, T., Getino, J., Mathews, P., Simon, J.-L., Soffel, M., Vondrak, J., Wallace, P., & Williams, J. 2006, *Celest. Mech., 94, 351*

RÉSOLUTION B.1

Adoption de la théorie P03 de la précession et définition de l'écliptique

La XXVIème Assemblée générale de l'Union astronomique internationale,

Notant

1. le besoin d'une théorie de la précession cohérente avec une théorie dynamique,

2. que la partie précession du modèle de précession-nutation UAI 2000A, dont l'adoption a été recommandée par la Résolution B1.6 à compter du 1er Janvier 2003, bien qu'étant basée sur des valeurs de vitesses de précession améliorées par rapport à celles du modèle de précession UAI 1976, n'est pas cohérente avec une théorie dynamique,

3. que la Résolution B1.6 de la XXIVème Assemblée générale de l'UAI encourage également le développement de nouvelles expressions de la précession compatibles avec le modèle UAI 2000A de précession-nutation et,

Reconnaissant

1. que le potentiel gravitationnel des planètes apporte une contribution significative au mouvement de l'équateur terrestre, rendant ambiguës les expressions *précession luni-solaire* et *précession planétaire*,

2. le besoin d'une définition de l'écliptique dans des buts à la fois astronomiques et civils,

3. que, dans le passé, l'écliptique a été défini par rapport à un observateur situé dans un repère inertiel (définition inertielle) et par rapport à un observateur situé dans une repère tournant avec l'écliptique (définition rotationnelle),

Accepte

les conclusions du groupe de travail de la Division 1 de l'UAI sur la précession et l'écliptique, publiées par Hilton et al. (2006, *Celest. Mech.* **94**, 351) et

Recommande

1. que les expressions *précession luni-solaire* et *précession planétaire* soient remplacées respectivement par les expressions *précession de l'équateur* et *précession de l'écliptique*,

2. que, à compter du 1er Janvier 2009, la partie précession du modèle de précession-nutation UAI 2000A soit remplacée par la théorie de la précession P03 de Capitaine et al. (2003, A&A 412, 567-586) pour la précession de l'équateur (Eqs. 37) et la précession de l'écliptique (Eqs. 38); la même publication donne les développements polynomiaux pour les quantités primaires et un certain nombre de quantités dérivées pour usage à la fois dans le paradigme basé sur l'équinoxe et dans celui basé sur la CIO,

3. que le choix des paramètres de précession soit laissé à l'utilisateur, et,

4. que le pôle de l'écliptique soit explicitement défini par le vecteur moment cinétique orbital moyen du barycentre du système Terre-Lune dans le Système de référence céleste barycentrique (BCRS), et que cette définition soit explicitement donnée pour éviter toute confusion avec des définitions antérieures.

Note 1

Les formules nécessaires à la construction de la matrice de précession utilisant diverses paramétrisations sont données par les Eqs 1, 6, 7, 11, 12 et 22 de Hilton et al. (2006). Les développements polynomiaux recommandés pour les divers paramètres sont donnés dans la Table 1 du même article, incluant les expressions P03 données par les équations (37) à (41) de Capitaine et al. (2003) et les Tables 3-5 de Capitaine et al. (2005).

Note 2

La variation temporelle du facteur de forme J_2 est dans P03: $\mathrm{dJ_2/dt} = -0.3001 \times 10^{-9}$ siècle^{-1}

Références

Capitaine, N., Wallace, P. T. & Chapront, J., 2003, *A&A*, **412,** 567

Capitaine, N., Wallace, P. T. & Chapront, J., 2005, *A&A*, **432**, 355

Hilton, J. L., Capitaine, N., Chapront, J., Ferrandiz, J. M., Fienga, A., Fukushima, T., Getino, J., Mathews, P., Simon, J.-L., Soffel, M., Vondrák, J., Wallace, P., & Willams, J., 2006, *Celest. Mech.* **94**, 351

RESOLUTION B.2

Supplement to the IAU 2000 resolutions on reference systems

Recommendation 1

Harmonizing the name of the pole and origin to "intermediate"

The XXVIth International Astronomical Union General Assembly,

Noting

1. the adoption of resolutions IAU B1.1 through B1.9 by the IAU General Assembly of 2000,

2. that the International Earth Rotation and Reference Systems Service (IERS) and the Standards Of Fundamental Astronomy (SOFA) activity have made available the models, procedures, data and software to implement these resolutions operationally, and that the Almanac Offices have begun to implement them beginning with their 2006 editions, and

3. the recommendations of the IAU Working Group on "Nomenclature for Fundamental Astronomy" (IAU Transactions XXVIA, 2005), and

Recognizing

1. that using the designation "intermediate" to refer to both the pole and the origin of the new systems linked to the Celestial Intermediate Pole and the Celestial or Terrestrial Ephemeris origins, defined in Resolutions B1.7 and B1.8, respectively would improve the consistency of the nomenclature, and

2. that the name "Conventional International Origin" with the potentially conflicting acronym CIO is no longer commonly used to refer to the reference pole for measuring polar motion as it was in the past by the International Latitude Service,

Recommends

1. that, the designation "intermediate" be used to describe the moving celestial and terrestrial reference systems defined in the 2000 IAU Resolutions and the various related entities, and

2. that the terminology "Celestial Intermediate Origin" (CIO) and "Terrestrial Intermediate Origin" (TIO) be used in place of the previously introduced "Celestial Ephemeris Origin" (CEO) and "Terrestrial Ephemeris Origin" (TEO), and

3. that authors carefully define acronyms used to designate entities of astronomical reference systems to avoid possible confusion.

Recommendation 2

Default orientation of the Barycentric Celestial Reference System (BCRS) and Geocentric Celestial Reference System (GCRS)

The XXVIth International Astronomical Union General Assembly,

Noting

1. the adoption of resolutions IAU B1.1 through B1.9 by the IAU General Assembly of 2000,

2. that the International Earth Rotation and Reference Systems Service (IERS) and the Standards Of Fundamental Astronomy (SOFA) activity have made available the models, procedures, data and software to implement these resolutions operationally, and that the Almanac Offices have begun to implement them beginning with their 2006 editions,

3. that, in particular, the systems of space-time coordinates defined by IAU 2000 Resolution B1.3 for (a) the solar system (called the Barycentric Celestial Reference System, BCRS) and (b) the Earth (called the Geocentric Celestial Reference System, GCRS) have begun to come into use,

4. the recommendations of the IAU Working Group on "Nomenclature for Fundamental Astronomy" (IAU Transactions XXVIA, 2005), and

5. a recommendation from the IAU Working Group on "Relativity in Celestial Mechanics, Astrometry and Metrology",

Recognizing

1. that the BCRS definition does not determine the orientation of the spatial coordinates,

2. that the natural choice of orientation for typical applications is that of the ICRS, and

3. that the GCRS is defined such that its spatial coordinates are kinematically non-rotating with respect to those of the BCRS,

Recommends

that the BCRS definition is completed with the following: "For all practical applications, unless otherwise stated, the BCRS is assumed to be oriented according to the ICRS axes. The orientation of the GCRS is derived from the ICRS-oriented BCRS."

RÉSOLUTION B.2

Supplément aux résolutions 2000 de l'UAI sur les systèmes de référence

Recommendation 1

Harmonisation en "intermédiaire" de la dénomination relative au pôle et à l'origine

La XXVIème Assemblée générale de l'Union astronomique internationale,

Notant

1. l'adoption des Résolutions B1.1 à B1.9 de l'UAI par l'Assemblée générale 2000 de l'UAI,
2. que le Service international de la rotation de la Terre et des systèmes de référence (IERS) et l'activité des Standards de l'astronomie fondamentale (SOFA) ont rendu disponibles les modèles, les procédés, les données et les logiciels nécessaires pour implémenter les résolutions de façon opérationnelle et que les services d'éphémérides ont commencé à les mettre en œuvre à partir de leurs éditions 2006, et
3. les recommandations du groupe de travail de l'UAI sur la "Nomenclature pour l'astronomie fondamentale" (IAU Transactions XXVIA, 2005), et

Reconnaissant

1. qu'utiliser la désignation "intermédiaire" pour se rapporter à la fois au pôle et à l'origine des nouveaux systèmes liés au Pôle céleste intermédiaire et aux origines céleste ou terrestre des éphémérides, définies dans les Résolutions B1.7 et B1.8, respectivement, améliorerait la cohérence de la nomenclature,
2. que le nom "Origine conventionnelle internationale" avec l'acronyme CIO, source potentielle de conflit, n'est plus utilisé pour désigner le pôle de référence pour la mesure du mouvement du pôle comme c'était le cas dans le passé par le Service International des Latitudes,

Recommande

1. que la désignation "intermédiaire" soit utilisée pour décrire les systèmes de référence céleste et terrestre définis dans les Résolutions 2000 de l'UAI et les diverses entités correspondantes,

2. que la terminologie "Origine céleste intermédiaire" (CIO) et "Origine terrestre intermédiaire" (TIO) soit utilisée à la place de celles qui ont été introduites précédemment de "Origine Céleste des Ephémérides" (CEO) et "Origine Terrestre des Ephémérides" (TEO), et

3. que les auteurs définissent soigneusement les acronymes utilisés pour désigner les entités relatives aux systèmes de référence astronomiques pour éviter toute confusion possible.

Recommendation 2

Orientation par défaut du Système de référence céleste barycentrique (BCRS) et du Système de référence céleste géocentrique (GCRS)

La XXVIème Assemblée générale de l'Union astronomique internationale,

Notant

1. l'adoption des Résolutions B1.1 à B1.9 de l'UAI par l'Assemblée générale 2000 de l'UAI,

2. que le Service international de la rotation de la Terre et des systèmes de référence (IERS) et l'activité des Standards de l'astronomie fondamentale (SOFA) ont rendu disponibles les modèles, les procédés, les données et les logiciels nécessaires pour implémenter les résolutions de façon opérationnelle et que les services d'éphémérides ont commencé à les mettre en œuvre à partir de leurs éditions 2006, et

3. qu'en particulier, les systèmes de coordonnées d'espace-temps définis par la Résolution B1.3 pour (a) le système solaire (appelé le Système de référence céleste barycentrique, BCRS) et (b) la Terre (appelé le Système de référence céleste géocentrique, GCRS), ont commencé à entrer en usage,

4. les recommandations du groupe de travail de l'UAI sur la "Nomenclature pour l'astronomie fondamentale" (IAU Transactions XXVIA, 2005), et

5. une recommandation du groupe de travail de l'UAI sur la "Relativité en mécanique céleste, astrométrie et métrologie",

Reconnaissant

1. que la définition du BCRS ne détermine pas l'orientation des axes des coordonnées spatiales,

2. que le choix naturel d'orientation pour des applications usuelles est celui de l'ICRS,

3. que le GCRS est défini de telle sorte que ses axes de coordonnées spatiales sont cinématiquement non tournants par rapport à ceux du BCRS,

Recommande

que la définition du BCRS soit complétée par ce qui suit: "Pour toutes les applications pratiques, sauf indications contraires, le BCRS est supposé être orienté selon les axes de l'ICRS. L'orientation du GCRS est déduite de l'orientation du BCRS par l'ICRS."

RESOLUTION B.3

Re-definition of Barycentric Dynamical Time, TDB

The XXVIth International Astronomical Union General Assembly,

Noting

1. that IAU Recommendation 5 of Commissions 4, 8 and 31 (1976) introduced, as a replacement for Ephemeris Time (ET), a family of dynamical time scales for barycentric ephemerides and a unique time scale for apparent geocentric ephemerides, that IAU Resolution 5 of Commissions 4, 19 and 31 (1979) designated these time scales as Barycentric Dynamical Time (TDB) and Terrestrial Dynamical Time (TDT) respectively, the latter subsequently renamed Terrestrial Time (TT), in IAU Resolution A4, 1991,

2. that the difference between TDB and TDT was stipulated to comprise only periodic terms, and

3. that Recommendations III and V of IAU Resolution A4 (1991) (i) introduced the coordinate time scale Barycentric Coordinate Time (TCB) to supersede TDB, (ii) recognized that TDB was a linear transformation of TCB, and (iii) acknowledged that, where discontinuity with previous work was deemed to be undesirable, TDB could be used, and

Recognizing

1. that TCB is the coordinate time scale for use in the Barycentric Celestial Reference System,

2. the possibility of multiple realizations of TDB as defined currently,

3. the practical utility of an unambiguously defined coordinate time scale that has a linear relationship with TCB chosen so that at the geocenter the difference between this coordinate time scale and Terrestrial Time (TT) remains small for an extended time span,

4. the desirability for consistency with the Teph time scales used in the Jet Propulsion Laboratory (JPL) solar-system ephemerides and existing TDB implementations such as that of Fairhead & Bretagnon (*A&A* **229**, 240, 1990), and

5. the 2006 recommendations of the IAU Working Group on "Nomenclature for Fundamental Astronomy" (IAU Transactions XXVIB, 2006),

Recommends

that, in situations calling for the use of a coordinate time scale that is linearly related to Barycentric Coordinate Time (TCB) and, at the geocenter, remains close to Terrestrial Time (TT) for an extended time span, TDB be defined as the following linear transformation of TCB:

$$\mathrm{TDB} = \mathrm{TCB} - \mathrm{L}_B \times (\ \mathrm{JD}_{\mathrm{TCB}} - \mathrm{T}_0\ ) \times 86400 + \mathrm{TDB}_0,$$

where $\mathrm{T}_0 = 2443144.5003725$,
and $\mathrm{L}_\mathrm{B} = 1.550519768 \times 10^{-8}$ and $\mathrm{TDB}_0 = -6.55 \times 10^{-5}$ s are defining constants.

Notes

1. $\mathrm{JD}_{\mathrm{TCB}}$ is the TCB Julian date. Its value is $\mathrm{T}_0 = 2443144.5003725$ for the event 1977 January 1 00h 00m 00s TAI at the geocenter, and it increases by one for each 86400s of TCB.

2. The fixed value that this definition assigns to L_B is a current estimate of $\mathrm{L}_C + \mathrm{L}_G - \mathrm{L}_C \times \mathrm{L}_G$, where L_G is given in IAU Resolution B1.9 (2000) and L_C has been determined (Irwin & Fukushima, 1999, *A&A* **348**, 642) using the JPL ephemeris DE405. When using the JPL Planetary Ephemeris DE405, the defining L_B value effectively eliminates a linear drift between TDB and TT, evaluated at the geocenter. When realizing TCB using other ephemerides, the difference between TDB and TT, evaluated at the geocenter, may include some linear drift, not expected to exceed 1 ns per year.

3. The difference between TDB and TT, evaluated at the surface of the Earth, remains under 2 ms for several millennia around the present epoch.

4. The independent time argument of the JPL ephemeris DE405, which is called Teph (Standish, *A&A*, **336**, 381, 1998), is for practical purposes the same as TDB defined in this Resolution.

5. The constant term TDB_0 is chosen to provide reasonable consistency with the widely used TDB − TT formula of Fairhead & Bretagnon (1990).
   n.b. The presence of TDB_0 means that TDB is not synchronized with TT, TCG and TCB at 1977 Jan 1.0 TAI at the geocenter.

6. For solar system ephemerides development the use of TCB is encouraged.

RÉSOLUTION B.3

Redéfinition du Temps dynamique barycentrique, TDB

La XXVIème Assemblée générale de l'Union astronomique internationale,

Notant

1. que la Recommandation 5 (1976) des Commissions 4, 8 et 31 de l'UAI a introduit, en remplacement du temps des éphémérides (TE), une famille d'échelles de temps dynamique pour les éphémérides barycentriques et une unique échelle de temps pour les éphémérides apparentes géocentriques,

2. que la Résolution 5 (1979) des Commissions 4, 19 et 31 de l'UAI a désigné ces échelles de temps par Temps dynamique barycentrique (TDB) et Temps dynamique terrestre (TDT) respectivement, cette dernière échelle de temps ayant été par la suite renommée Temps terrestre (TT) par la Résolution A4, 1991,

3. que la différence entre TDB et TDT a été spécifiée comme ne comprenant que des termes périodiques, et

4. que les Recommandations III et V de la Résolution A4 (1991) de l'UAI (i) ont introduit l'échelle de temps-coordonnée barycentrique (TCB) pour remplacer TDB, (ii) ont reconnu que TDB était une fonction linéaire de TCB, et (iii) ont admis que, lorsqu'une discontinuité avec les travaux antérieurs était jugée indésirable, TDB pouvait être utilisé, et

Reconnaissant

1. que TCB est l'échelle de temps-coordonnée à utiliser dans le Système de référence céleste barycentrique,

2. la possibilité de réalisations multiples de TDB tel qu'il est défini actuellement,

3. l'utilité pratique d'une échelle de temps définie de façon non ambiguë par une relation linéaire avec TCB, choisie de façon à ce que, au géocentre, la différence entre cette échelle de temps-coordonnée et le Temps terrestre (TT) reste faible pendant un long intervalle de temps,

4. l'avantage d'une cohérence avec les échelles de temps Teph utilisées pour les éphémérides du système solaire du Jet Propulsion Laboratory (JPL) et les réalisations de TDB telles que celle de Fairhead & Bretagnon (*A&A* **229**, 240, 1990), et

5. les recommandations 2006 du Groupe de travail de l'UAI sur la "Nomenclature pour l'astronomie fondamentale" (IAU Transactions XXVIB, 2006),

Recommande

que, dans des situations qui demandent l'utilisation d'une échelle de temps-coordonnée qui soit reliée linéairement au Temps-coordonnée barycentrique (TCB) et reste, au géocentre, proche du Temps terrestre (TT) pendant un long intervalle de temps, TDB soit défini par la transformation linéaire suivante de TCB:

$$\mathrm{TDB} = \mathrm{TCB} - \mathrm{L}_B \times (\mathrm{JD}_{TCB} - \mathrm{T}_0) \times 86400 + \mathrm{TDB}_0,$$

où $\mathrm{T}_0 = 2443144.5003725$ et
où $\mathrm{L_B} = 1.550519768 \times 10^{-8}$ et $\mathrm{TDB}_0 = -\ 6.55 \times 10^{-5}$ s sont des constantes de définition.

Notes

1. $\mathrm{JD_{TCB}}$ est la date Julienne TCB. Sa valeur est $\mathrm{T}_0 = 2443144.5003725$ pour l'évènement 1977 Janvier 1 00h 00m 00s TAI, au géocentre, et il augmente de 1 par 86400 s de TCB.

2. La valeur fixe que cette définition assigne à L_B est une estimation actuelle de $\mathrm{L}_C + \mathrm{L}_G - \mathrm{L}_C \times \mathrm{L}_G$, où L_G est donné dans la Résolution B1.9 de l'UAI (2000) et L_C a été déterminé (Irwin & Fukushima, 1999, *A&A* **348**, 642) en utilisant les éphémérides DE405 du JPL. Quand on utilise les éphémérides planétaires DE405 du JPL, la valeur de définition L_B élimine très efficacement une dérive linéaire entre TDB et TT, évaluée au géocentre. Lorsque l'on réalise TCB en utilisant d'autres éphémérides, la différence entre TDB et TT, évaluée au géocentre, peut inclure une dérive linéaire, qui ne devrait pas dépasser 1 ns par an.

3. La différence entre TDB et TT, évaluée à la surface de la Terre, reste en dessous de 2 ms durant plusieurs millénaires autour de l'époque actuelle.

4. L'argument temporel utilisé pour les éphémérides DE405, qui est appelé Teph (Standish, *A&A*, **336**, 381, 1998), est, pour des applications pratiques, le même que TDB tel qu'il est défini dans cette Résolution.

5. Le terme constant TDB_0 est choisi de façon à assurer une cohérence satisfaisante avec la formule de Fairhead & Bretagnon (1990), qui est largement utilisée pour TDB − TT.
   n.b. La présence de TDB_0 signifie que TDB n'est pas synchronisé avec TT, TCG et TCB pour 1977 Jan 1.0 TAI, au géocentre.

6. L'usage de TCB est encouragé pour le développement des éphémérides dans le système solaire.

RESOLUTION B.4

Endorsement of the Washington Charter for Communicating Astronomy with the Public

The Washington Charter was one of the outcomes of the 2ndInternational Conference on Communicating Astronomy with the Public, held in Washington DC in October 2003. Council endorsed the Washington Charter in March 2004. Nineteen other societies, organizations and facilities have endorsed the Charter, including the BAA and PPARC.
At the Communicating Astronomy with the Public 2005 meeting in Garching last June, a revised version of the Charter was proposed. This softened the language and also tidied up some of the phraseology. This was endorsed by the attendees and accepted by the IAU Working Group. The revised version is appended.
The IAU General Assembly is requested to confirm endorsement of the Revised Washington Charter.

The Washington Charter for Communicating Astronomy with the Public

As our world grows ever more complex and the pace of scientific discovery and technological change quickens, the global community of professional astronomers needs to communicate more effectively with the public. Astronomy enriches our culture, nourishes a scientific outlook in society, and addresses important questions about humanity's place in the universe. It contributes to areas of immediate practicality, including industry, medicine, and security, and it introduces young people to quantitative reasoning and attracts them to scientific and technical careers. Sharing what we learn about the universe is an investment in our fellow citizens, our institutions, and our future. Individuals and organizations that conduct astronomical research - especially those receiving public funding for this research - have a responsibility to communicate their results and efforts with the public for the benefit of all.

Recommendations

For Funding Agencies:

Encourage and support public outreach and communication in projects and grant programs.
Develop infrastructure and linkages to assist with the organization and dissemination of outreach results. Emphasize the importance of such efforts to project and research managers.
Recognize public outreach and communication plans and efforts through proposal selection criteria and decisions and annual performance awards.
Encourage international collaboration on public outreach and communication activities.

For Professional Astronomical Societies:

Endorse standards for public outreach and communication.
Assemble best practices, formats, and tools to aid effective public outreach and communication.

Promote professional respect and recognition of public outreach and communication.
Make public outreach and communication a visible and integral part of the activities and operations of the respective societies.
Encourage greater linkages with successful ongoing efforts of amateur astronomy groups and others.

For Universities, Laboratories, Research Organizations and Other Institutions:

Acknowledge the importance of public outreach and communication.
Recognize public outreach and communication efforts when making decisions on hiring, tenure, compensation and awards.
Provide institutional support to enable and assist with public outreach and communication efforts. Collaborate with funding agencies and other organizations to help ensure that public outreach and communication efforts have the greatest possible impact.
Make available formal public outreach and communication training for researchers.
Offer communication training in academic courses of study for the next generation of researchers.

For Individual Researchers:

Support efforts to communicate the results and benefits of astronomical research to the public, convey the importance of public outreach and communication to team members.
Instill this sense of responsibility in the next generation of researchers

RESOLUTION B.5

Definition of a Planet in the Solar System

Contemporary observations are changing our understanding of planetary systems, and it is important that our nomenclature for objects reflect our current understanding. This applies, in particular, to the designation 'planets'. The word 'planet' originally described 'wanderers' that were known only as moving lights in the sky. Recent discoveries lead us to create a new definition, which we can make using currently available scientific information.
The IAU therefore resolves that planets and other bodies, except satellites, in our Solar System be defined into three distinct categories in the following way:

(1) A planet[1] is a celestial body that (*a*) is in orbit around the Sun, (*b*) has sufficient mass for its self-gravity to overcome rigid body forces so that it assumes a hydrostatic equilibrium (nearly round) shape, and (*c*) has cleared the neighbourhood around its orbit.

(2) A "dwarf planet" is a celestial body that (*a*) is in orbit around the Sun, (*b*) has sufficient mass for its self-gravity to overcome rigid body forces so that it assumes a hydrostatic equilibrium (nearly round) shape,[2] (*c*) has not cleared the neighbourhood around its orbit, and (*d*) is not a satellite.

(3) All other objects,[3] except satellites, orbiting the Sun shall be referred to collectively as "Small Solar System Bodies."

---

1. The eight planets are: Mercury, Venus, Earth, Mars, Jupiter, Saturn, Uranus, and Neptune.

2. An IAU process will be established to assign borderline objects into either dwarf planet and other categories.

3. These currently include most of the Solar System asteroids, most Trans-Neptunian Objects (TNOs), comets and other small bodies.

RESOLUTION B.6

The IAU further resolves:

Pluto is a "dwarf planet" by the above definition and is recognized as the prototype of a new category of Trans-Neptunian Objects[1].

---

1. An IAU process will be established to select a name for this category.

Transactions IAU, Volume XXVIB
Proc. IAU XXVI General Assembly, August 2006
Karel A. van der Hucht, ed.

doi:10.1017/S174392130802365X

CHAPTER IV

REPORT OF THE EXECUTIVE COMMITTEE

1. Executive Committee 2003 - 2006

1.1. *Composition of the Executive Committee*

During the triennium 2003 - 2006, the Executive Committee was composed as follows:

Ronald D.	Ekers	President
Catherine J.	Cesarsky	President-Elect
Oddbjørn	Engvold	General Secretary
Karel A.	van der Hucht	Assistant General Secretary
Beatriz	Barbuy	Vice-President
Chen	Fang	Vice-President
Kenneth A.	Pounds	Vice-President
Silvia	Torres-Peimbert	Vice-President
Brian	Warner	Vice-President
Robert	Williams	Vice-President
Franco	Pacini	Adviser
Hans	Rickman	Adviser

1.2. *Meetings of the Executive Committee*

The IAU Executive Committee met as follows during the triennium:
- 78th Meeting, 25-26 July 2003, during the IAU XXV General Assembly, Sydney, Australia
- 79th Meeting, 24-26 May 2004, at the Instituto de Astronomía, Universidad Nacional Autónoma de México, Mexico City, Mexico
- 80th Meeting, 18-20 April 2005, Monte Mario Osservatorio, INAF, Rome, Italy
- 81th Meeting, 13-23 August 2006, during the IAU XXVI General Assembly, Prague, Czech Republic

The business conducted by the Executive Committee is recorded in the Minutes of these meetings. Summaries of the minutes have appeared in the IAU *Information Bulletin* (IB 94, p. 30; IB 95, p. 27; and IB 97, p. 72-73).

1.3. *Officers' Meetings*

Between the meetings of the Executive Committee, the Officers (President, President-Elect, General Secretary and Assistant General Secretary) met at the IAU Secretariat on 2 - 3 February 2004, on 7 - 8 February 2005, and on 27 - 28 March, 2006.

1.4. *International Year of Astronomy 2009*

At its General Assembly in Sydney in July 2003, the International Astronomical Union voted unanimously in favor of a resolution asking that the year 2009 be declared the *Year of Astronomy* by the United Nations, in recognition of the significance of Galileo's introduction of the astronomical telescope in 1609. The proclamation was subsequently prepared by the 'Year of Astronomy 2009' Working Group, and forwarded to the Executive Board of UNESCO. We were very happy that UNESCO General Conference, in October 2005, recommended that the United Nations General Assembly at its 60th session adopts a resolution declaring 2009 as the *International Year of Astronomy.* In its recommendations to the UN General Assembly, the General Conference of UNESCO recognizes "that the study of the Universe has led to numerous scientific discoveries that have great influence not only on humankind's understanding of the Universe but also on the technological, social and economic development of society" and "that astronomy proves to have great implications in the study of science, philosophy, religion and culture", The Division XII WG on *Communication Astronomy with the Public* agreed to assist the EC in developing a structure of the role and participation in this big global event, in consultation with Commission 41 on *History of Astronomy.* This WG is arranging a plenary session on the IYA in Prague.

1.5. *Definition of a planet*

The discoveries of Trans-Neptunian Object 2003 VB12 (=90377), also referred to as Sedna, and 2003 UB313, which appeared to be comparable in size to Pluto, led to the IAU to establish a definition of a planet. A new Working group of IAU Division III, called *Definition of a Planet* was created with the objective to advise the IAU on this matter. The Working Group, counting 19 members, was chaired by the Division III President, Iwan P. Williams (UK).

The difficulty that brought the whole topic to the forefront of discussion is that the boundary between (major) planet and minor planet had never been defined. In the early days this was not a problem since the gap between Mercury (the smallest planet) and Ceres (the largest minor planet) was enormous. Since Pluto was, at discovery, thought to be larger than the Earth, there still was not a problem. Indeed, even with the current value for the size of Pluto it would not be a problem (the gap from Pluto to Ceres is still large), were it not for the discovery of Trans-Neptunian Objects. Now at least one is larger than Pluto.

The Division III Working Group on *Definition of a Planet* was unable to agree on one definition of a planet, and on 15 November 2005 the WG Chair forwarded to the General Secretary a split recommendation from the WG, containing three alternative definitions which all had varied support among its members. The WG Chair subsequently informed the General Secretary that the WG had agreed that a further discussion would not lead to a consensus. The Organizing Committee of Division III accepted that the WG had completed its mission, and it formally decided that the WG be immediately discontinued. Recognizing the broad interest on this issue beyond the planetary scientist and astronomers, the Executive Committee decided to seek supplementary advice from a broader based, small Advisory Committee, before taking a decision on this matter, which was seen to be of notable concern and interest among the public and in the media.

In March 2006 the Executive Committee appointed an advisory group to the EC, on the planet issue, that should cover history of astronomy, science writing, publishing and education, plus having a solid knowledge and background in planetary sciences. The Committee was chaired by Owen Gingerich (USA). Its six members, including the President of IAU Division III, covered more than one of the named disciplines. The Committee met in Paris, 30 June 1 July 2006, and forwarded its unanimous recommendation on 7 July to the IAU President. The Executive Committee subsequently reviewed the recommendation during a telephone conference on 24 July 2006, and appointed a sub-group with the task to present the Committees recommendation as a Resolution by the EC to be discussed and voted on at the General Assembly in Prague.

2. Membership of the Union

2.1. *National Membership*

During the IAU XXV General Assembly, upon recommendation of the Executive Committee, the General Assembly welcomed Nigeria, Serbia, Romania and Tadjikistan as new National Members of the IAU.

Due to five years of non-payments of dues, the CAAA National Membership was terminated (IAU Statutes §7).

2.2. *Individual Membership*

The number of new Individual Members admitted to the IAU XXV General Assembly was 894, and as of 28 July 2006 the total number of IAU Individual Members is 8 488. In addition, were registered 567 Consultants. In 2006 the IAU National Committees for Astronomy (NCA) were asked to update their membership lists and report to the IAU Secretariat who were no longer active members and who should no longer remain Individual Members. The advice and recommendations thus received from the NCAs led to the current number of Individual Members stated above.

The IAU Executive is saddened to report the death of the following 172 members of the Union, which have been reported to the IAU Secretariat since the IAU XXV General Assembly:

John G. Ables
Tateos A. Agekjan
Ko Aizu
Lawrence Hugh Aller
Gennadij V. Andreev
Horace W. Babcock
John N. Bahcall
James Gilbert Baker
Norman H. Baker
Vassilios Barbanis
Arvind Bhatnagar
Richard G. Bingham
J. G. Bolton
Hermann Bondi
Semion Ya Ag Braude
Nina M. Bronnikova
Anton Bruzek
William Buscombe
Bruno Caccin
Alastair G.W. Cameron
Henri Camichel
John H. Carver
Vittorio Castellani
Joseph W. Chamberlain
Nikolaj S. Chernykh
Yves Chmielewski
Rafael Cid Palacios
G. Colombo
Alan H. Cook
Pierre Cugnon
N. Dallaporta
Leverett Davis Jr
John Alan Dawe
Willem de Graaff
T. de Groot
Juan J. de Orus
Chr. de Vegt
Aleksandr N. Deutsch
Lorant Dezsö
Jerzy Dobrzycki
Geoffrey G. Douglass
Robert A. Duncan
Richard B. Dunn
Nikolai Dzubenko
Hans Elsaesser
Donald J. Faulkner
Walter A. Feibelman
Michel C. Festou
Mikhail S. Frolov
Igor A. Gerasimov
Daniel Gerbal
Robert Glebocki
Nzhet Gökdogan
Thomas Gold
Friedrich Gondolatsch
Shumo Gong
S. I. Gopasyuk
Vitalij G. Gorbatsky
Fumihiko Hagio
Anton Hajduk
R. Glenn Hall
Emilios Harlaftis
Gerald S. Hawkins
Wulff D. Heintz
Helmut Wilhelm Hellwig
Hartmut Holweger
Reiun Hoshi
Charles Latif Hyder
George R. Isaak
Theodor S. Jacobsen
Tadeusz Jarzebowski
Mihkel Jeveer
Henry Emil Kandrup
Boris L. Kashscheev
Sidney Kenderdine
Vera L. Khokhlova
Michael J. Klein
I. G. Kolchinskij

N. S. Komarov
Vladimir A. Kotelnikov
John D. Kraus
L. Kresak
Petr G. Kulikovskij
Barry James LaBonte
Trudpert Lederle
Michael James Ledlow
Vojtech Letfus
Jacques R. Lvy
J. Virginia Lincoln
Alexander M. Lozinskij
Per E. Maltby
Gyorgy Marx
Janet Akyz Mattei
Cornell H. Mayer
Paul J. Melchior
Marie-Odile Mennessier
Klaus Metz
Rolf Mewe
Harry C. Minnet
Ljubisa A. Mitic
Vasilij I. Moroz
Philip Morrison
Mirta B. Mosconi
Andreas B. Muller
C. A. Muller Jr
Sergij Musatenko
Saken O. Obashev
Franco Occhionero
J. Beverley Oke
Mikhail Orlov
J. Oro
Ludwig F. Oster
Lucia Padrielli
John L. Perdrix
Charles L. Perry
Alain Peton
Jack H. Piddington
A. Keith Pierce
Girolamo Pinto
John Polygiannakis
Jason G. Porter
John M. Porter
Neil A. Porter
Kevin H. Prendergast
Helen Dodson Prince
Yurij P. Pskovskij
Tamara B. Pyatunina
Gibson Reaves
James Ring
Ralph Robert Robbins
Brian J. Robinson
Marcello Rodon
Douglas H. Sampson
Hans Schmidt
Egon H. Schroeter
Aleksandr S. Sharov
William M. Sinton
Akira M. Sinzi
George M. Sisson
Humphry M. Smith
Mattheus A.J. Snijders
Gunnar Sorensen
Arnold A. Stepanian
Gerard A. Stevens
Jrgen D. Stock
Ronald Cecil Stone
Aleksandr A. Stotskii
Winardi Sutantyo
Peter A. Sweet
J. T.l. Tavares
Volodymyr Tel'nyuk-Adamchuk
Dirk Ter Haar
Richard Q. Twiss
Anne B. Underhill
Seppo I. Urpo
J. van Nieuwkoop
Paul Verbeek
Franco Verniani
Jean-Pierre Vigier
Yurij I. Vitinskij
Richard L. Walker Jr
Dennis Walsh
Willem Wamsteker
Lai Wan
James A. Westphal
Fred L. Whipple
Raymond E. White
John R. Winckler
Kiyoshi Yabuuti
Boris F. Yudin
Shigeru Yumi
D. Zulevic

3. Divisions, Commissions and Working Groups

3.1. *Working Groups of the Executive Committee*

The EC Working Groups on *Women in Astronomy* and on *The Year of Astronomy 2009* were created at the General Assembly in Sydney. A new Advisory Committee for the IAU Executive Committee, on Impact Threats to the Earth, is being proposed for decision at the IAU XXVI General Assembly in Prague.

3.2. *Working Groups of Divisions and Commissions*

A new Working Group on *Communication Astronomy with the Public* under Division XII was created in 2004.

3.3. *Minor Planet Center*

The present contract between the host of IAU's Minor Planet Center (MPC), the Harvard-Smithsonian Center for Astrophysics (CfA, USA) and the IAU expires by the end of the

IAU XXVI General Assembly in August 2006, unless a new agreement would be in place by then. Extensive discussions with the CfA Director, Dr. Charles Alcock and the IAU Negotiation Team led to a Memorandum of Agreement that will ensure the continued operation of the IAU Minor Planet Center, with financial support from NASA, at Harvard-Smithsonian Center for Astrophysics. This MoA was signed by the CfA Director and by the IAU General Secretary on 5 June 2006.

In July 2006 the IAU General Secretary was informed by the leadership of CfA and the MPC, that Dr. Brian Marsden (USA) would resign as MPC Director by end of the IAU General Assembly in Prague and that Dr. Timothy B. Spahr (USA) would take over, as interim Director.

3.4. *Reviewing of Commissions and Working Groups*

It is an expressed aim of the Union to ensure that it is relevant and serves the best interest of its membership as well as the non-astronomical World. The Commissions, Working Groups and Programme Groups are the bodies upon which the Unions scientific research and educational activities are based. In accordance with the IAU Statutes and Bye-Laws, these bodies shall therefore be reviewed by their parent Divisions every three years. Based on the result of these reviews the Divisions have forwarded their recommendations to the IAU Executive Committee about necessary changes in the current structure in order to meet the demand and challenges for the next triennium and beyond.

This implies that some Commissions, WGs and PGs, may be subject to notable changes, or even be discontinued, and that entirely new bodies will be needed to deal with new fields or activities of concern to IAU. Since all Commissions and the large majority of WGs and PGs have existed for more than six years already, these decisions will be made at the first meeting of the EC at IAU XXVI GA, in consultations with the IAU Division Presidents.

4. Scientific activities

4.1. *Scientific meetings*

The number of IAU Symposia and Colloquia supported in the previous years are, respectively: six and four in 2003, five and three in 2004, six and three in 2005, and there will be a total of nine Symposia in 2006, of which six will be held at the General Assembly. The location and dates of these various events are given from the list of publications below.

Two Regional IAU Meetings were supported in 2005, namely the 9th APRIM in Bali, Indonesia, 26 - 29 July, and the 11th LARIM in Pucón, Chile, 12-16 December.

4.2. *Publications*

Given the large number of international meetings every year of the size of IAU Colloquia, and given the fact that organization of these meetings can usually be handled on a time scale shorter than the IAU would do it, the EC decided at its 80th Meeting in Rome, April 2005, that the IAU support for international meetings will from then on be given to, on the average, nine IAU Symposia every year. The support for two Regional IAU Meetings in years between General Assemblies will remain a high priority.

The following are IAU Proceedings published in 2003 - 2006.

4.2.1. *IAU Transactions, published in 2003 - 2006*

Transactions of the International Astronomical Union
Volume XXVA, Reports on Astronomy 1999 - 2002

Ed. Hans Rickman
(San Francisco: ASP) ISBN 1-58381-137-0

4.2.2. *IAU Symposium Proceedings, published in 2003 - 2006*

Published by the Astronomical Society of the Pacific, USA (print):

IAU S199
The Universe at Low Radio Frequencies
30 November - 4 December 1999, Pune, India
Eds. A. Pramesh Rao, G. Swarup & Gopal-Krishna
(San Francisco: ASP) ISBN 1-58381-121-4, 2003

IAU S201
New Cosmological Data and the Values of the Fundamental Parameters
7-11 August 2000, Manchester, United Kingdom
Eds. A. N. Lasenby & A. Wilkinson
(San Francisco: ASP) ISBN: 1-58381-212-1, February 2006

IAU S202
Planetary Systems in the Universe - Observation, Formation and Evolution
7-11 August 2000, Manchester, United Kingdom
Eds. A. J. Penny, P. Artymowicz, A.-M. Lagrange & S. S. Russell
(San Francisco: ASP) ISBN: 1-58381-176-1, February 2005

IAU S206
Cosmic Masers: from Protostars to Blackholes
5-10 March 2001, Rio de Janeiro, Brazil
Eds. V. Migenes & M. J. Reid
(San Francisco: ASP) ISBN 1-58381-112-5, 2003

IAU S208
Astrophysical Supercomputing Using Particle Simulations
10-13 July 2001, Tokyo, Japan
Eds. J. Makino & P. Hut
(San Francisco: ASP) ISBN: 1-58381-139-7, 2003

IAU S209
Planetary Nebulae: their Evolution and Role in the Universe
19-23 November 2001, Canberra, Australia
Eds. S. Kwok, M. Dopita & R. Sutherland
(San Francisco: ASP) ISBN: 1-58381-148-6, 2003

IAU S210
Modelling of Stellar Atmospheres
17-21 June 2002, Uppsala, Sweden
Eds. N. Piskunov, W. W. Weis & D. F. Gray
(San Francisco: ASP) ISBN: 1-58381-160-5, February 2004

IAU S211
Brown Dwarfs

20-24 May 2002, Honolulu, Hawai'i, USA
Ed. E. Martn
(San Francisco: ASP) ISBN: 1-58381-132-X, 2003

IAU S212
A Massive Star Odyssey: from Main Sequence to Supernova
24-28 June 2002, Lanzarote, Canary Island, Spain
Eds. K. A. van der Hucht, A. Herrero & C. Esteban
(San Francisco: ASP) ISBN: 1-58381-133-8, February 2003

IAU S213
Bioastronomy 2002: Life Among the Stars
8-12 July 2002, Hamilton Island, Great Barrier Reef, Australia
Eds. R. P. Norris & F. H. Stootman
(San Francisco: ASP) ISBN: 1-58381-171-0, July 2004

IAU S214
High Energy Processes and Phenomena in Astrophysics
6-10 August 2002, Suzhou, China
Eds. Xiang Dong Li, V. L. Trimble & Z. R. Wang
(San Francisco: ASP) ISBN: 1-58381-157-5, January 2004

IAU S215
Stellar Rotation
8-12 November 2002, Cancun, Mexico
Eds. A. Maeder & P. R. J. Eenens
(San Francisco: ASP) ISBN: 1-58381-180-X, November 2004

IAU S216
Maps of the Cosmos
14-17 July 2003, Sydney, Australia
Eds. M. Colless & L. Staveley-Smith
(San Francisco: ASP) ISBN: 1-58381-202-4, September 2005

IAU S217
Recycling Intergalactic and Interstellar Matter
14-17 July 2003, Sydney, Australia
Eds. P.-A. Duc, J. Braine & E. Brinks
(San Francisco: ASP) ISBN: 1-58381-166-4, June 2004

IAU S218
Young Neutron Stars and Their Environment
14-17 July 2003, Sydney, Australia
Eds. F. M. Camilo & B. M. Gaensler
(San Francisco: ASP) ISBN: 1-58381-178-8b, December 2004

IAU S219
Stars as Suns: Activity, Evolution and Planets
21-25 July 2003, Sydney, Australia
Eds. A. K. Dupree & A. O. Benz

(San Francisco: ASP) ISBN: 1-58381-163-X, September 2003

IAU S220
Dark Matter in Galaxies
21-25 July 2003, Sydney, Australia
Eds. S. D. Ryder, D. J. Pisano, M. A. Walker & K. C. Freeman
(San Francisco: ASP) ISBN: 1-58381-167-2, July 2004

IAU S221
Star Formation at High Angular Resolution
22-25 July 2003, Sydney, Australia
Eds. M. Burton, R. Jayawardhana & T. Bourke
(San Francisco: ASP) ISBN: 1-58381-161-3, September 2004

Published by Cambridge University Press, UK (print and on-line):
(e-book see: < http://journals.cambridge.org/action/displayJournal?jid=IAU >)

IAU S222
The Interplay among Black Holes, Stars and ISM in Galactic Nuclei
1-5 March 2004, Gramado, Rio Grande do Sul, Brasil
Eds. Th. Storchi-Bergmann, L. C. Ho & H. R. Schmitt
(Cambridge: CUP) ISBN 0-521-84803-2, November 2004

IAU S223
Multi-Wavelength Investigations of Solar Activity
14-19 June 2004, St. Petersburg, Russia
Eds. A. V. Stepanov, E. E. Benevolenskaya & A. G. Kosovichev
(Cambridge: CUP) ISBN 0-521-85195-5, April 2005

IAU S224
The A-Star Puzzle
8-13 July 2004, Poprad, Slovakia
Eds. J. Zverko, W. W. Weiss, J. Ziznovsky & S. J. Adelman
(Cambridge: CUP) ISBN 0-521-85018-5, March 2005

IAU S225
Impact of Gravitational Lensing on Cosmology
19-23 July, 2004, Lausanne, Switzerland
Eds. Y. Mellier & G. Meylan
(Cambridge: CUP) ISBN 0-521-85196-3, June 2005

IAU S226
Coronal and Stellar Mass Ejections
13-17 September 2004, Beijing, China
Eds. K. P. Dere, Jingxiu Wang & Yihua Yan
(Cambridge: CUP) ISBN 0-521-85197-1, June 2005

IAU S227
Massive Star Birth: A Crossroads of Astrophysics
16-20 May 2005, Catania, Sicily, Italy

Eds. R. Cesaroni, M. Felli, E. Churchwell & C. M. Walmsley
(Cambridge: CUP) ISBN 0-521-82198-X, November 2005

IAU S228
From Lithium to Uranium: Elemental Tracers of Early Cosmic Evolution
23-27 May 2005, Paris, France
Eds. V. Hill, P. Francois & F. Primas
(Cambridge: CUP) ISBN 0-521-85199-8, December 2005

IAU S229
Asteroids, Comets, Meteors
7-12 August 2005, Bzios, Rio de Janeiro, Brasil
Eds. D. Lazzaro, S. Ferraz-Mello & J. A. Fernández
(Cambridge: CUP) ISBN 0-521-85200-5, March 2006

IAU S230
Populations of High-Energy Sources in Galaxies
15-19 August 2005, Dublin, Ireland
Eds. E. J. A. Meurs & G. Fabbiano
(Cambridge: CUP) ISBN 0-521-85201-3, May 2006

IAU S231
Astrochemistry: Recent Successes and Current Challenges
29 August - 2 September 2005, Monterey, CA, USA
Eds. D. C. Lis, G. A. Blake & E. Herbst
(Cambridge: CUP) ISBN 0-521-85202-1, March 2006

IAU S232
The Scientific Requirements for Extremely Large Telescopes
14-18 November 2005, Cape Town, South Africa
Eds. P. A. Whitelock, M. Dennefeld & B. Leibundgut
(Cambridge: CUP) ISBN 0-521-85608-6, May 2006

4.2.3. *IAU Colloquium Proceedings, published in 2003 - 2006*

IAU C189
Astrophysical Tides: Effects in the Solar and Exoplanetary Systems
16-20 September, 2002 Nanjing, China
Eds. S. Ferraz-Mello & I. P. Williams
Celestial Mechanics and Dynamical Astronomy 87 (Kluwer), Nos.1-2, 2003

IAU C190
Magnetic Cataclysmic Variables
8-13 December 2002, Cape Town, South Africa
Eds. S. Vrielmann & M. Cropper
ASP-CS 315 (San Francisco: ASP) ISBN 1-58381-170-2, December 2004

IAU C191
The Environments and Evolution of Double and Multiple Stars

3-7 February 2003, Merida, Yucatn, Mxico
Eds. C. Allen & C. Scarfe
(Mexico: UNAM) RevMexAA-SC Vol. 21, ISBN: 970-32-0607-7, August 2004

IAU C192
Supernovae (10 years of SN1993J)
22-26 April 2003, Valencia, Spain
Eds. J. M. Marcaide & K. W. Weiler
Book title: Cosmic Explosions. On the 10th Anniversary of SN1993J
(Heidelberg: Springer) Springer-PP 99, ISBN 3-540-23039-4, October 2004

IAU C193
Variable Stars in the Local Group
6-11 July 2003, Christchurch, New Zealand
Eds. D. W. Kurtz & K. R. Pollard
ASP-CS 310 (San Francisco: ASP) ISBN 1-58381-162-1, May 2004

IAU C194
Compact Binaries in the Galaxy and Beyond
17-22 November 2003, La Paz, B.C. Sur, Mxico
Eds. G. Tovmassian & E. M. Sion
(Mexico: UNAM) RevMexAA-SC Vol. 20, ISBN: 970-32-1185-5, July 2004

Published by Cambridge University Press, UK (print and on-line):
(e-book see: < http://journals.cambridge.org/action/displayJournal?jid=IAU >)

IAU C195
Outskirts of Galaxy Clusters: Intense Life in the Suburbs
12-16 March 2004, Torino, Italia
Ed. A. Diaferio
(Cambridge: CUP) ISBN 0-521-84908-X, October 2004

IAU C196
Transits of Venus: New Views of the Solar System and Galaxy
7-11 June 2004, Preston, U.K.
Ed. D. W. Kurtz
(Cambridge: CUP) ISBN 0-521-84907-1, May 2005

IAU C197
Dynamics of Populations of Planetary Systems
31 August - 4 September 2004, Belgrade, Serbia and Montenegro
Eds. Z. Knezevic & A. Milani
(Cambridge: CUP) ISBN 0-521-85203-X, February 2005

IAU C198
Near-field Cosmology with Dwarf Elliptical Galaxies
14-18 March 2005, Les Diablerets, Switzerland
Eds. H. Jerjen & B. Binggeli
(Cambridge: CUP) ISBN 0-521-85204-8, November 2005

IAU C199
Probing Galaxies through Quasar Absorption Lines
14-18 March 2005, Shanghai, China
Eds. P.R. Williams, Chenggang Shu & B. Mnard
(Cambridge: CUP) ISBN 0-521-85205-6, October 2005

IAU C200
Direct Imaging of Exoplanets: Science and Techniques
3-7 October 2005, Nice, France
Eds. C. Aime & F. Vakili
(Cambridge: CUP) ISBN 0-521-85607-8, May 2006

At the end of 2005, the IAU Colloquium series was terminated, at the benefit of the expansion of the IAU Symposium series.

4.2.4. *Proceedings of Regional IAU Meetings, published in 2003 - 2006*

Proc. 8th Asian-Pacific Regional IAU Meeting, Vol. I
2-5 July 2002, Tokyo, Japan
Eds. S. Ikeuchi, J. Hearnshaw & T. Hanawa
ASP-CS 289 (San Francisco: ASP) ISBN: 1-58381-134-6, 2003

Proc. 9th Asian-Pacific Regional IAU Meeting (APRIM 2005)
26-29 July 2005, Nusa Dua, Bali, Indonesia
Eds. W. Sutantyo, P. W. Premadi, P. Mahasena, T. Hidayat & S. Mineshige
(Bandung: ITB Press) ISBN 979-3507-63-2, March 2006

Proc. 11th Latin-American Regional IAU Meeting (LARIM 2005)
12-16 December 2005, Pucón, Chile
Eds. M. Rubio, L. Infante & S. Torres-Peimbert
Revista Mexicana de Astronoma y Astrofsica - Serie de Conferencias,
(in press, due August 2006)

For a complete list of Proceedings Regional IAU Meetings, please check:
< http://www.iau.org/Regional_Meetings.121.0.html >

4.3. *IAU Publisher*

The contract with the IAU Publisher, the Astronomical Society of the Pacific - Book Publishing Company (ASP), expired in 2003, and the Executive Committee resolved that a new publishing strategy should be worked out in the contract with the a Publisher. It was decided to go for an electronic version of the IAU Proceedings of IAU Symposia and Colloquia, supplemented by a paper. A Call for Tender of the IAU Proceedings Series was issued in the fall of 2003, and a contract was signed with Cambridge University Press (Cambridge, UK), February 2004.

It shall be noted that the IAU is formally a business partner in the new contract with the CUP and thereby will be sharing the net income from the book sales.

An IAU Editorial Board was created to ensure good communication between the Publisher and the Editors of the individual Proceedings, which also would serve to shorten the time between the time of a meeting and the publication of the corresponding Proceedings. The Editorial Board is chaired by the Assistant General Secretary. For the past two years the average lead time for publishing of individual IAU Proceedings range commonly between six and nine months, which signals a very positive development and improvement compared to earlier years.

5. Educational activities

Educational programmes under IAU Commission 46 on *Astronomy Education and Development* represent a high priority activity and responsibility of the IAU. The Executive Committee recognizes the many successful and promising results from its Program Groups.

5.1. *Commission 46 PG International Schools for Young Astronomers*

The Program Group on *International Schools for Young Astronomers* (PG-ISYA) arranged very successful ISYAs in Morocco in 2004 and in Mexico in 2005. We refer to reports of these in IAU *Information Bulletins* no. 96 (pp. 29 - 30) and 97 (pp. 74 - 76).

5.2. *Commission 46 PG World Wide Development of Astronomy*

The Program Group on *World Wide Development of Astronomy* (PG-WWDA) has initiated and conducted wide reaching programs and promotion of interest for astronomy in developing countries. An overview of visits to Mongolia, Kenya, Thailand, Iraq, Cuba, Trinidad and Tobago is given in IAU *Information Bulletin* no. 98, pp. 71 - 74.

5.3. *Commission 46 PGs Teaching Astronomy for Development and Exchange of Astronomers*

The Program Groups on *Teaching Astronomy for Development* and *Exchange of Astronomers* (PG-TAD and PG-EA) initiated a new program at Al Akhawayn University in Ifrane, Morocco in July, 2004.

6. Relations with other scientific organizations

6.1. *ICSU*

The International Council for Science (ICSU) has prepared a strategic plan for 2006 - 2012 with the objective to identify the continued needs for a strong, independent scientific organization like ICSU in a rapidly changing World. Of concern and interest to the IAU are in particular ICSU's establishing of Regional Offices in regions where also the IAU promotes astronomy education, a programme that identifies Natural and Human-Induced Hazards and ICSU's Principle of Universality. The IAU is grouped in a relatively loose cluster structure with 5 Geo-Unions of ICSU. The idea is that grouping of its current 27 member Unions in so-called clusters will stimulate Inter-Union collaboration.

The planning of the International Polar Year 2007 - 2008, which is part of the ICSU strategy for 2006 - 2011, is progressing well.

ICSU is greatly concerned about the continuing problem of obtaining in time visa for scientists from, in particular, Asian and Arabic countries, to attend scientific meetings in some countries. The problem is gradually becoming more wide-spread. ICSU follows the situation closely and it has offered to provide an overview of the increased lead time for visa applications to various countries.

6.2. *CODATA*

Astronomy is in the forefront of taking advantage of new technological opportunities to accelerate the quality and effectiveness of its science. As a result, our science faces great challenges in managing increasing amount of data. Our science is under time pressure to develop and agree on a set of guiding principles for how to manage astronomical data. IAU Commission 5's Working Group on *Astronomical Data*, chaired by Ray Norris, has been actively involved in a consortium under ICSU and CODATA in developing guidelines for safeguarding the future of astronomical data. As astronomers we are also accustomed to enjoy the freedom to establish open access to astronomical data bases in collaboration with colleagues across all boarders on this planet. Many of us have been comfortably unaware of recent legal threats to this privilege and the successful war that has be endured by and won on our behalf by ICSU and CODATA to safeguard this important condition for fostering and promoting scientific co-operations.

6.3. *FAGS*

The ICSU General Assembly, held in Suzhou, China in October 2005, decided to extend ICSU's sponsorship of the Interdisciplinary Services called *Federation of Astronomical and geophysical Data Analysis Services* (FAGS) for an additional period of three years, during which FAGS will continue to fulfill its role of coordination among the Services, while ICSU examines integration of this function within its activities in data and information.

6.4. *UN-COPUOS*

The IAU representative to the UN Committee on Peaceful Use of the Outer Space (UN-COPUOS), IAU Assistant General Secretary Karel A. van der Hucht, attended the annual sessions of the Scientific and Technical Sub-committee of UN-COPUOS. His addresses and overview of relevant IAU activities were well received.

6.5. *OECD Global Science Forum*

The OECD Global Science Forum has arranged two Workshops on *Future Large-Scale Projects and Programmes in Astronomy and Astrophysics* in, respectively, 2003 and 2004, with participation of members from the IAU Executive Committee. The Workshops recognized that ground-based and space-based observatories are vital and complementary for the future astronomy.

6.6. *IAU Representatives to international organizations 2003 - 2006*

	Organization	*IAU Representative*
ICSU	International Council for Science	Oddbjørn Engvold
BIPM	Bureau International des Poids et Mesures	
— CCTF	Consultative Committee for Time and Frequency	Toshio Fukushima
— CCU	Consultative Committee for Units	Nicole Capitaine
CIE	Compagnie Internationale de l'Eclairage	Syuzo Isobe (†)
CODATA	Committee on Data for Science and Technology	Raymond P. Norris
COSPAR	Committee on Space Research	
— Council		Hans Rickman
— SC B	Space Studies of the Earth-Moon System, Planets and Small Bodies of the Solar System	Mikhail Ya. Marov
— SC D	Space Plasmas in the Solar System, including Planetary Magnetospheres	Marek Vandas

— SC E	Research in Astrophysics from Space	
— SC E1	Galactic and Extragalactic Astrophysics	G. Srinivasan
— SC E2	The Sun as a Star	Arnold O. Benz
FAGS	Federation of Astronomical and Geophysical Services	Nicole Capitaine Einar A. Tandberg-Hanssen
IAF	International Astronautical Federation	Henk Olthof
IERS	International Earth Rotation and Reference Systems Service	Jan Vondrak (until 2004) Nicole Capitaine
IGBP	International Geosphere-Biosphere Programme	Richard G. Strom
IHY	International Heliophysical Year	David F. Webb
ISES	International Space Environment Service	Helen E. Coffey
ITU	International Telecommunication Union	
— ITU-R	Radiocommunication Bureau	William J. Klepczynski Tomas E. Gergely Masatoshi Ohishi
IUCAF	Scientific Committee on Frequency Allocations for Radio Astronomy and Space Sciences	Kenneth F. Tapping Darrel T. Emerson Jim Cohen
IUPAP	International Union of Pure and	Hans Rickman
— C4	Commission on Cosmic Rays	Heinrich J. Voelk
— C19	Commission on Astrophysics	Virginia L. Trimble
IVS	International VLBI Service for Geodesy and Astrometry	Patrick T. Wallace
SCAR	Scientific Committee on Antarctic Research	John W.V. Storey
SCOPE	Scientific Committee on Problems of the Environment	Derek McNally
SCOSTEP	Scientific Committee on	Brigitte Schmieder
UN-COPUOS	UN Committee on the Peaceful Uses of Outer Space	Karel A. van der Hucht
URSI	Union Radio-Scientifique Internationale	Luis F. Rodriguez

7. Financial matters

The table in Appendix II comprises the Budget, Income and Expenditure for the years 2003 - 2005, and the Budget for 2006. The proposed Budget for 2007 - 2009 is presented in chapter II, p. 30, (Appendix I) of these Proceedings. The Budget for 2003 - 2006 was approved by the previous General Assembly. Triennial accounts for the years 2003 - 2005 have been certified by the IAU Auditor, Msr. Arnaud de Boisanger.

The Finance Sub-Committee, chaired by Dr. Robert W. Milkey, provided most helpful advice and guidance to the General Secretary on all financial matters of the Union.

7.1. *Comments on Income*

As in previous years, the main source of income is the yearly dues from the Adhering Organizations. An increased minor income is foreseen from sales of IAU Proceedings in accordance with the contract with the new IAU Publisher.

7.2. *Comments on Expenditure*

The expenditure excess of general office expenses is due to the extra cost of creating and implementing a new IAU data base and web server in 2005. The IAU Executive Committee accepted this extra expense at its 81st meeting 18 - 20 April 2005 in Rome.

A major part of the bank expenditure in 2004 and 2005 was balanced in January 2006 by a back-payment from the ABN AMRO bank.

7.3. *Comments on Budget*

The budget for the years 2007 - 2009 was approved by the Executive Committee at its meeting April 18 - 20, 2005, in Rome. The Executive Committee recommended, at the same meeting, an increase of the staff in the IAU Office in Paris, which is reflected in the proposed increase in salaries of the administrative staff.

8. Administrative matters

8.1. *Revision of Statutes, Bye-Laws and Working Rules*

The IAU XXV General Assembly adopted a major revision of the Statutes and Bye-Laws. These changes were followed by a thorough revision of the IAU Working Rules which were adopted by the Executive Committee at its 80th Meeting in Rome 2005. As a result of the revised Statutes and Bye-Laws, the EC resolved to modify the Working Rules accordingly.

A few minor changes in the Statutes and Bye-Laws will be presented for voting at the General Assembly in Prague. These modifications concern the voting on issues of primarily scientific nature at General Assemblies, the nomination of Individual Members and the range number of dues units that are available for the National Members annual contributions to the IAU.

8.2. *Miscellaneous*

At its 80th Meeting in Rome in April 2005, the Executive Committee adopted a set of guidelines to advise countries inviting to host of IAU General Assemblies.

In July 2005 the IAU General Secretary signed a contract with the company Oxford Abstracts to host the server for submission of all abstracts to scientific events of the IAU General Assembly. The submitted abstracts were reviewed for decision separately by the 29 event organizers. At the closing of the abstract submission on 26 June 2006, a total of close to 2,400 accepted abstracts were downloaded from the server and prepared for inclusion in the Abstract Book. This task was previously handled by the General Assembly organizers, and from the experience gained this time the Executive Committee will decide whether it is a solution for future General Assemblies.

The IAU Secretariat has since 1988 enjoyed the privilege to use office space and one storage room in the building of the Institute d'Astrophysique (IAP), 98bis bd Arago, Paris. A new contract between the Centre National de la Recherche Scientifique (CNRS) and the IAU for IAP's continued hosting of the IAU Secretariat was signed in July 2006.

The IAU decided in January 2006 to discontinue its more than 70 years of banking relation with ABN AMRO in Utrecht, the Netherlands, due to a long-lasting dispute concerning transfer of funds to Sydney in 2003. The bank finally accepted its mistake.

8.3. *Update of Data Base and Web Site*

The IAU data base, which initially was located in Uppsala, Sweden, was transferred to a server of the Institute d'Astrophysique, Paris, during the early phase of this triennial.

An upgrading of the IAU web server was due for quite some time and this work started in 2004. The IAU Secretariat became involved in a painstaking detailed planning of the structure and layout of the new data base and server, and the follow-up and checking as the work progressed. The work with the data base and the web site was contracted with the AMEOS company in Strasbourg, and the layout of the actual web pages were done by Ms Alice de Miramon. The local staff of the IAU Secretariat Mme Claire Vidonne and Mme Monique Orine are thanked for their conscientious and invaluable work with the data base and server that has led to the new data base and web site that became available for use of the membership in January 2006.

Among the new features available with the new system is the Individual Member's web page form which enables them to check and, when necessary, notify the IAU Secretariat about changes or corrections of ones personal profile. The updated membership list has led to the issuing an IAU Newsletter at semi-regular intervals, in addition to the IAU *Information Bulletin* that is normally issued twice per year, which serves to keep the membership updated on the work and activities of the Union.

8.4. *Secretariat*

The IAU Secretariat is dealing with a steadily increasing workload on the two employées, Executive Assistant Mme Monique Léger-Orine and Administrative Assistant Mme Claire Vidonne. The dedication and efficiency in their work for the IAU are gratefully appreciated. In order to ensure adequate maintenance and necessary upgrading on the IAU computer systems, one has signed contracts with Msr. Laurent Pezeron, who is paid by the hour. Additional, part time personnel was being hired to assist in the work of the Secretariat when necessary.

The proposed budget for 2007 - 2009 takes note of the need to strengthen the personnel situation of the IAU Secretariat.

APPENDIX II

TRIENNIUM 2003 - 2005

STATEMENT OF INCOME (CHF)

INCOME *BUDGET*	**2003**	**2004**	**2005**	**2003 - '05**	**2006**
unit of contribution	3 260	3 355	3 460		3 580
adjustment for inflation	2.7%	3%	3.2%		3.4%
number of units	260	262	262		262
Adhering Organizations	888 217	776 695	1 169 999	2 834 911	
	847 600	*879 010*	*906 520*	*2 633 130*	*937 960*
royalties IAU proceedings	27 237	13 748	3 831	44 852	
	25 000	*15 000*	*15 000*	*55 000*	*15 000*
ICSU / UNESCO	6 500	- -	- -	6 500	
	- -	- -	- -	- -	- -
ICSU / UNESCO (NEO book project)	- -	114 000	- -	114 000	
	15 000	- -	- -	*15 000*	- -
UNESCO (ISYA Argentina)	1 385	- -	- -	1 385	
	- -	- -	- -	- -	- -
ESO / ESA grants for GA 2003	77 197	- -	- -	77 197	
	- -	- -	- -	- -	- -
PGF Fellowships	- -	90 971	- -	90 971	
	- -	- -	- -	- -	- -
bank interest	7 839	6 843	19 657	34 338	
	25 000	*25 000*	*25 000*	*75 000*	*25 000*
refunds	2 449	66 195	9 869	78 512	
	- -	- -	- -	- -	- -
gain on exchange 1 Jan. of the year	7 557	- -	39 923	47 480	
	- -	- -	- -	- -	- -
gain on internal transfers	3 717	21	- -	3 738	
	- -	- -	- -	- -	- -
total INCOME	**1 022 134**	**1 068 473**	**1 243 279**	**3 333 883**	
	912 600	*919 010*	*946 520*	*2 778 130*	*977 960*

STATEMENT OF EXPENDITURE (CHF)

EXPENDITURE *BUDGET*	**2003**	**2004**	**2005**	**2003 - '05**	**2006**
SCIENTIFIC ACTIVITIES					
General Assemblies					
GA grants	242 412 *220 000*	- - - -	- - - -	242 412 *220 000*	 *235 000*
operations	32 753 *45 000*	8 681 *4 000*	6 743 *4 000*	48 177 *53 000*	 *45 000*
sub-sub-total GAs	275 165 *265 000*	8 681 *4 000*	6 743 *4 000*	290 589 *273 000*	 *280 000*
Meetings					
Symposium / Colloquim grants	198 001 *240 000*	180 000 *280 000*	182 000 *280 000*	560 001 *800 000*	 *280 000*
Regional IAU Meeting grants	- - - -	- - *30 000*	50 000 *30 000*	50 000 *60 000*	 - -
co-sponsored meetings	6 000 - -	5 000 - -	- - - -	11 000 - -	 - -
sub-sub-total Meetings	204 001 *240 000*	185 000 *310 000*	232 000 *310 000*	621 001 *860 000*	 *280 000*
Working Groups					
EC Working Groups	2 135 *5 000*	4 000 *7 000*	- - *7 000*	6 135 *19 000*	 *7 000*
Commission Working Groups	5 820 *5 000*	1 210 *7 000*	9 045 *7 000*	16 075 *19 000*	 *7 000*
Minor Planet Center	6 000 *6 000*	8 000 *8 000*	8 000 *8 000*	22 000 *22 000*	 *8 000*
CB for Astronomical Telegrams	4 000 *4 000*	4 000 *4 000*	4 000 *4 000*	12 000 *12 000*	 *4 000*
Meteor Data Center	- - *1 100*	- - *1 500*	- - *1 500*	- - *4 100*	 *1 500*
NEO study and book	- - - -	114 000 - -	– 3 170 - -	110 830 - -	 - -
sub-sub-total Working Groups	17 955 *21 100*	131 210 *27 500*	17 875 *27 500*	167 040 *76 100*	 *27 500*
sub-total SCIENTIFIC ACTIVITIES	497 121 *526 100*	324 891 *341 500*	256 618 *341 500*	1 078 630 *1 209 100*	 *587 500*

	2003	2004	2005	2003 - '05	2006
EDUCATIONAL ACTIVITIES					
PG-ISYA	--	45 939	44 917	90 856	
	--	*45 000*	*45 000*	*90 000*	--
PG-TAD	40 557	18 636	10 951	70 144	
	40 000	*45 000*	*45 000*	*130 000*	*45 000*
PG-WWDA	10 000	14 532	6 144	30 676	
	--	*10 000*	*10 000*	*20 000*	*10 000*
PG-EA	6 440	3 690	8 605	18 735	
	10 000	*15 000*	*15 000*	*40 000*	*15 000*
PGF Fellowships	37 055	54 430	18 494	109 979	
	--	--	--	--	--
sub-total EDUCATIONAL ACTIVITIES	94 052	137 227	89 111	320 390	
	50 000	*115 000*	*115 000*	*280 000*	*70 000*
DELEGATES TO OTHER UNIONS	9 554	13 470	10 929	33 953	
	10 000	*10 000*	*10 000*	*30 000*	*10 000*
DUES TO OTHER UNIONS					
ICSU	15 283	14 256	16 660	41 199	
	17 000	*20 000*	*20 000*	*57 000*	*20 000*
IERS/FAGS	7 500	7 500	--	15 000	
	7 500	*7 500*	*7 500*	*22 500*	*7 500*
IUCAF	7 500	7 500	7 500	22 500	
	7 500	*7 500*	*7 500*	*22 500*	*7 500*
sub-total DUES TO OTHER UNIONS	30 283	29 256	24 160	83 699	
	32 000	*35 000*	*35 000*	*102 000*	*35 000*
EXECUTIVE COMMITTEE					
Officers' Meetings	14 516	13 334	15 348	43 198	
	6 000	*5 500*	*5 500*	*17 000*	*5 500*
Executive Committee Meetings	62 631	39 050	66 634	168 315	
	75 000	*35 000*	*60 000*	*170 000*	*60 000*
General Secretary	32 054	16 547	29 540	78 141	
	30 000	*30 000*	*30 000*	*90 000*	*30 000*
Assistant General Secretary	489	836	2 803	4 128	
	2 000	*2 000*	*2 000*	*6 000*	*2 000*
President	--	--	3 082	3 082	
	1 000	*2 000*	*2 000*	*5 000*	*2 000*
sub-total EXECUTIVE COMMITTEE	109 690	69 767	117 407	296 864	
	114 000	*74 500*	*99 500*	*288 000*	*99 500*

	2003	2004	2005	2003 - '05	2006
PUBLICATIONS					
Information Bulletin	28 213	10 142	8 436	46 791	
	30 000	*10 000*	*10 000*	*30 000*	*10 000*
Distribution IB	5 571	--	--	5 571	
	10 000	*10 000*	*10 000*	*30 000*	*10 000*
reimbursement ASP GA-XXV preordering	--	61 319	-,-	61 319	
	--	--	--	--	--
sub-total PUBLICATIONS	33 784	71 461	8 436	113 681	
	40 000	*20 000*	*20 000*	*80 000*	*20 000*
ADMINISTRATION					
salaries and charges	175 088	192 772	198 461	566 321	
	175 000	*175 000*	*175 000*	*525 000*	*175 000*
training courses	1 476	--	925	2 401	
	5 000	*5 000*	*5 000*	*15 000*	*5 000*
general office expenses	100 461	82 012	131 237	313 710	
	72 000	*82 000*	*82 000*	*236 000*	*82 000*
auditor	2 401	2 500	--	4 901	
	2 500	*2 500*	*2 500*	*7 500*	*2 500*
bank charges	12 684	25 733	18 128	56 545	
	4 000	*4 000*	*4 000*	*12 000*	*4 000*
loss on internal transfers	--	5 753	1 506	7 259	
	--	--	--	--	--
loss on exchange rate 1st year	--	9 684	--	9 684	
	--	--	--	--	--
sub-total ADMINISTRATION	292 110	318 454	350 257	960 821	
	258 500	*268 500*	*268 500*	*795 500*	*268 500*
total EXPENDITURE	1 066 595	964 527	856 918	2 888 039	
	1 030 600	*864 500*	*889 500*	*2 784 600*	*1 090 500*
net result	**–44 461**	**103 946**	**386 360**	**445 845**	
	–118 000	*54 510*	*57 020*	*–6 470*	*–112 540*
GENERAL FUND	**644 113**	**748 059**	**1 134 419**		

Transactions IAU, Volume XXVIB
Proc. IAU XXVI General Assembly, August 2006
Karel A. van der Hucht, ed.

doi:10.1017/S1743921308023661

CHAPTER V

REPORTS OF DIVISIONS, COMMISSIONS AND WORKING GROUPS

Introduction

This chapter presents the reports of Business Meeting of IAU Divisions, Commissions, Working Groups and Program Groups, held during the IAU XXVI General Assembly in Prague, 14-25 August 2006. The reports have been written by the out-going presidents and chairpersons.

Transactions IAU, Volume XXVIB
Proc. IAU XXVI General Assembly, August 2006
Karel A. van der Hucht, ed.

doi:10.1017/S1743921308023673

DIVISION I FUNDAMENTAL ASTRONOMY

ASTRONOMIE FONDAMENTALE

Division I provides a focus for astronomers studying a wide range of problems related to fundamental physical phenomena such as time, the inertial reference frame, positions and proper motions of celestial objects, and precise dynamical computation of the motions of bodies in stellar or planetary systems in the Universe.

PRESIDENT	**Toshio Fukushima**
VICE-PRESIDENT	**Jan Vondrák**
PAST PRESIDENT	**Nicole Capitaine**
BOARD	**George A. Krasinsky, Andrea Milani, Imants Platais, Veronique Dehant, Demetrios N. Matsakis**

DIVISION I COMMISSIONS

Commission 4	**Ephemerides**
Commission 7	**Celestial Mechanics and Dynamical Astronomy**
Commission 8	**Astrometry**
Commission 19	**Rotation of the Earth**
Commission 31	**Time**

DIVISION I WORKING GROUPS

Division I WG	**General Relativity in Celestial Mechanics, Astrometry and Metrology**
Division I WG	**Re-definition of Universal Time Coordinated**
Division I WG	**Future Developments in Ground-Based Astrometry**
Division I WG	**Precession and the Ecliptic**
Division I WG	**Nomenclature for Fundamental Astronomy**

INTER-DIVISION WORKING GROUPS

Division I-III WG	**Near-Earth Objects (WGNEOs)**
Division I-III WG	**Cartographic Coordinates and Rotational Elements of Planets and Satellites**

PROCEEDINGS BUSINESS MEETINGS on 18 and 23 August 2006

1. Introduction

Division I meetings were organized in two parts during the IAU XXVI General Assembly; one at Sessions 1 and 2 of 18 August, and the other at Sessions 3 and 4 of 23 August 23. This split was designed in order to separate the presentation of proposals on business issues and their amendments and adoption so as to provide the Division members to have enough time of consideration and internal discussion. The timing of the second meeting, which was immediately after the Joint Discussion (JD) 16, was chosen to provide a chance to deepen the understandings of conclusions of the JD.

The first meeting included one session devoted to the reports of the Division and the Inter Division Working Group(WG)s and another for presenting the proposals on the business issues. Thanks to the successful scheduling of the JD, the second meeting was entirely devoted for the discussion of proposals mainly focused on the structural issues of the Division. The SOC of these meetings was composed of the Division I Organizing Committee (OC) for 2003-2006, namely, the president, the vice-president, and the past president of the Division as well as the presidents of five Commissions under the Division.

In the below, we omit the reports of the Working Groups, which are found in *Reports on Astronomy, Transactions IAU Volume XXVIA* (2007, Ed. O. Engvold), and concentrate ourselves on the summary of the presentations and discussions on business issues.

2. First meeting

After the presentation of the activity reports of five Division WGs, the Division officers for 2006-2009 were proposed. They are J. Vondrák as president, D. D. McCarthy as vice-president, who was nominated by the Division Organizing Committee (OC) by e-balloting, and T. Fukushima as past-president.

Next, the enlargement of the Division OC was proposed. The essence was to add the vice-presidents of Commissions under the Division to get them accustomed to the conventions and rules of the IAU and to inherit the experiences from the current presidents to the next ones easily. This was enabled thanks to the amendment of By-Laws to relax the restriction on the number of Division OC members adopted at the first Session of General Assembly.

Third, the summary of the EC Meeting on 13 August was reported by the Division president.

Fourth, a package proposal on the change in the internal structure of the Division was proposed. This consists of (1) the creation of a new Commission under Division I dedicated for resolving the issues on general relativity within the Division, (2) the change of status of WG on Natural Satellites under Commission 20 to a new Inter Division WG related to Divisions I and III, (3) the continuation of the Division WG on FDGBA, and (4) the set-up of two new Division WGs; WGs for the revision of ICRF and for the consideration of astronomical constants.

The first proposal is actually a promotion of the WG on RCMAM to a Commission, the tasks of which has been acknowledged to relate with issues of all the Commissions under the Division and is believed to be so difficult and complicated that a temporal WG will not be able to resolve them. This proposal itself was welcomed, however, the proposed name of the new Commission was questioned. Also, it was requested to clearly specify the scientific goals of the new Commission. The intention of the second proposal is to enlarge the scope of the WG not only within the physical and observational aspects of natural satellites but also their dynamical characters which are well within the scope of Division I. The third proposal was seriously discussed and the change of the scope as well as the name of the WG was requested. As for the new WGs, the detailed rationale were presented by the proposers. Both of the proposals are welcomed, however, further refinement of the scientific goals was required. These homeworks were requested to finish before the second Division Meeting.

3. Second meeting

First, the proposals on the Division officers and on the enlargement of the Division OC were approved.

Also approved was the creation of a new Commission. Its title was, after some debate, fixed as *Relativity in Fundamental Astronomy.* The scientific goals are (1) to clarify the geometric and dynamical concepts of Fundamental Astronomy (FA) within the relativistic framework, (2) to provide adequate mathematical and physical formulations to be used in FA, (3) to deepen the understandings of the above results among astronomers and students, and (4) to promote the researches needed to accomplish these tasks. The Officers of the new Commission were nominated as S. Klioner as Commission president and G. Petit as Commission vice-president.

As a result, the new Division OC members are the following 14 persons; J. Vondrak, D.D. McCarthy, T. Fukushima, J.A. Burns, I.I. Kumkova, A. Brzezinsky, P. Defraigne, S.A. Klioner, G.H. Kaplan, Z. Knezevic, D.W. Evans, Chopo Ma, R. Manchester, and G. Petit.

Next, the change of the status of WG on Natural Satellites was approved. J.-E. Alrot was nominated to chair this new Inter Division WG.

Third, instead of continuing the WG on FDGBA, the Division concluded its abortion and the creation of a new WG named WG on Astrometry by Small Ground-Based Telescopes (WG-ASGBT) to be chaired by W. Thuillot. Its scientific goals are (1) to follow-up the observation of large and fast surveys including the detection of NEOs and the determination of asteroid masses; (2) to set-up a dedicated observation network for *GAIA*; (3) to conduct observation campaign on the mutual events of natural satellites, stellar occultation, and binary asteroids; and (4) to encourage teaching astrometry for the next generation.

Fourth, the two new WGs proposed in the first meeting were both approved and specified in detail. One is titled as the WG on *Second Realization of ICRF* and is chaired by Chopo Ma. Its scientific goal is (1) to oversee the generation of the second realization of the ICRF from VLBI observations of extragalactic radio sources through state-of-the-art astronomical and geophysical models in the analysis; (2) to ensure the selection of defining sources and the mitigation of source position variations and the consistency with the ITRF and the IERS EOP; and (3) to present the second ICRF at the next IAU General Assembly. The other is the WG on *Numerical Standards of Fundamental Astronomy* (NSFA) chaired by B. Luzum. Its scientific goal is to update the IAU Current Best Estimates of Astronomical Quantities while ensuring the conformity with the latest IAU Resolutions on FA, the IERS Conventions, and the concepts/definitions of SI.

Finally, free discussions on the following topics were conducted; the review process of the Division structure, the strategy to propose the future IAU Symposia and Joint Discussions, the arrangement of meetings of Division and Commissions at the next General Assembly, the activation of Division-wide discussions, and the e-communication among Division Members.

Toshio Fukushima
president of the Division

Transactions IAU, Volume XXVIB
Proc. IAU XXVI General Assembly, August 2006
Karel A. van der Hucht, ed.

doi:10.1017/S1743921308023685

DIVISION I / WORKING GROUP NOMENCLATURE FOR FUNDAMENTAL ASTRONOMY

PRESIDENT **Nicole Capitaine**
MEMBERS **Alexandre H. Andrei, Mark R. Calabretta, Véronique Dehant, Toshio Fukushima, Bernard R. Guinot, Catherine Y. Hohenkerk, George H. Kaplan, Sergei A. Klioner, Jean Kovalevsky, Irina I. Kumkova, Chopo Ma, Dennis D. McCarthy, P. Kenneth Seidelmann, Patrick T. Wallace**

PROCEEDINGS BUSINESS MEETING on 16 August 2006

1. Introduction

The IAU Division I Working Group on *Nomenclature for Fundamental Astronomy* (NFA) was established by the IAU XXV General Assembly with the task of providing proposals for new nomenclature associated with the implementation of the IAU XXIV GA resolutions (2000) and to make related educational efforts for bringing the issue to the notice of scientists in the community.

A meeting of this working group was organized during the IAU XXVI GA in order to adopt the final conclusions of the WG and to discuss the status of the WG resolutions proposals to the IAU before they were discussed at the Division I meetings and then submitted to the IAU GA at its second meeting (24 August). This WG meeting was also an opportunity to discuss the final scheme and content of the NFA explanatory document, as well as the plan for publication of the NFA recommendations and of the NFA explanatory document. The future diffusion of the NFA educational documents was also discussed. Finally, a list of scientific issues were identified by the NFA WG to be proposed to Division I as topics for its proposed new WGs. Ten out of fifteen WG members participated in this NFA WG meeting. A few other participants, not belonging to the WG, were invited to attend the meeting, including the upcoming Division I president, Jan Vondrák.

2. The NFA Working Group recommendations on terminology

At the time the WG was established, the series of resolutions on reference systems adopted by the IAU XXIV GA in 2000 (i.e. B1.1 to B1.9) had just been implemented in the IERS Conventions (2003) and SOFA 2003 release (Wallace 2004). Several terminology issues were identified during the Division I meetings at the IAU XXV GA (see IAU Transactions XXVB, 2008), which required further studies and discussions among experts.

During the period 2003-2006, the NFA Working Group has worked on selecting a consistent and well defined terminology for all the quantities based on the IAU XXIV GA (2000) resolutions in order that it will be understood, recognized and adopted by the astronomical community. The terminology and guidelines recommended by the WG have been prepared through WG e-mail discussion and have been described in a number of Newsletters and documents. All the WG documents have been regularly posted on the NFA web site (at: <http://syrte.obspm.fr/iauWGnfa/>), which includes links to other sites relevant to the WG activities. A special page of the web site provides links to educational documents that explain the NFA issues. Two open WG discussions of about 100 participants were organized during the Journées 2004 and 2005 in Paris and Warsaw, respectively, where successive drafts of NFA recommendations were submitted for comment.

There has been detailed WG discussion on the terminology choices related to IAU XXIV GA (2000) resolutions B1.3, B1.5, B1.6, B1.7 and B1.8 with coordination with the International Association of Geodesy for the terrestrial reference system definitions.

The final WG recommendations on terminology that have been adopted are the following:

1. *Using existing terms (e.g., right ascension) in extended ways for the terminology associated with the new paradigm with a clear specification, rather than introducing new names.*

2. *Using "equinox based" and "CIO based" for referring to the classical and new paradigms, respectively.*

Comment: the "Celestial/Terrestrial Intermediate Origin" with the acronym CIO/TIO is proposed here as the updated terminology to replace the IAU XXIV GA (2000) "Celestial/Terrestrial Ephemeris Origin" with the acronym CEO/TEO (see below items 3 and 4 and the proposed resolution).

3. *Using "intermediate" to describe (i) the moving geocentric celestial reference system defined in the IAU 2000 resolutions (i.e. containing the CIP and the CIO), and (ii) the moving terrestrial system containing the CIP and the TIO.*

Comment: the term "intermediate" has been chosen to specify that these systems are intermediary systems between the geocentric celestial system and the terrestrial system, which are realized by using the models, constants and procedures that are conventionally accepted; it conventionally separates the instantaneous celestial orientation of the Earth into components we label polar motion (in the terrestrial system) and precession-nutation (in the celestial system).

4. *Harmonizing the name of the pole and the origin to "intermediate" and therefore changing CEO/TEO to CIO/TIO.*

5. *Using "system" in a broad sense rather than "frame" in this context of the intermediary system/frame.*

6. *Using special designations for particular realizations of the intermediate celestial system.*

Comment: this applies for example to "the IAU 2000A system" to designate the system which is realized by transforming the geocentric celestial system GCRS to the intermediate system using the IAU 2000A precession-nutation and associated frame biases at J2000 (the GCRS being transformed from the BCRS by using the coordinate transformation specified in the IAU 2000 Resolution B1.3).

7. *Keeping the classical terminology for "true equator and equinox" (or "true equinox based") for the classical equatorial system.*

8. *Choosing "equinox right ascension" (or "RA with respect to the equinox") and "intermediate right ascension" (or "CIO right ascension", or "RA with respect to the CIO"), for the azimuthal coordinate along the equator in the classical and new paradigms, respectively.* (Note that right ascensions and declinations with respect to the ICRS are usually designated by α_{ICRS}, δ_{ICRS}).

Comment: this is to be specified only once in the presentation of a paper if there is some risk of misunderstanding. Afterward, "right ascension" alone is sufficient.

9. *Giving the name "equation of the origins" to the distance between the CIO and the equinox along the intermediate equator, the sign of this quantity being such that it represents the CIO right ascension of the equinox, or equivalently, the difference between the Earth Rotation Angle and Greenwich apparent sidereal time.*

10. *Retaining "apparent places" and "mean places" in the equinox based system.*

11. *Not introducing "apparent intermediate places" in the CIO based system, but introducing instead "intermediate places".*

12. *Using "ITRF zero-meridian" to designate the plane passing through the geocenter, ITRF pole and ITRF x-origin and using, if necessary, "TIO meridian" to designate the moving plane passing through the geocenter, the CIP and the TIO.*

13. *Fixing the default orientation of the BCRS so that for all practical applications, unless otherwise stated, the BCRS is assumed to be oriented according to the ICRS axes.*

Comment: Once the BCRS is spatially oriented according to the ICRS, the spatial GCRS coordinates get an "ICRS-induced" orientation.

14. *Re-defining Barycentric Dynamical Time (TDB) as a fixed linear function of TCB*:

$$\mathrm{TDB} = \mathrm{TCB} - \mathrm{L_B} \times (\mathrm{JD_{TCB}} - \mathrm{T_0}) \times 86400 + \mathrm{TDB_0},$$

where $\mathrm{T_0} = 2443144.5003725$,
and $\mathrm{L_B} = 1.550519768 \times 10^{-8}$ and $\mathrm{TDB_0} = -6.55 \times 10^{-5}$ s are defining constants.

Additional points

- *Considering a terminology associated with other types of apparent places, although it may be required for specific use, has not been considered as being essential for common astronomical use and is therefore not part of the NFA WG terminology recommendations.*
- *No WG consensus having been reached for having strict rules for using or not using capitals for names for origins, poles and systems, no recommendation on this issue is proposed by the WG. The policy adopted throughout the NFA document is to capitalize those terms that are defined in IAU or IUGG resolutions.*

3. The NFA resolution proposals to the IAU

The major NFA Recommendations have resulted in two IAU resolution proposals. The first one recommends 1) using the designation "intermediate" to describe the moving celestial and terrestrial reference systems defined in the IAU 2000 resolutions and the various related entities, and 2) fixing the default orientation of the Barycentric and Geocentric celestial reference systems GCRS to that of the ICRS axes. The second one recommends a re-definition of Barycentric Dynamical time (TDB).

Successive versions of the resolution proposals were discussed by the WG during the period 2004-2006 and posted on the NFA WG web page. In March 2006, the final wording of the NFA resolution proposals was adopted by the NFA WG, in coordination with the IAU Working Group on "Relativity in Celestial Mechanics, Astrometry and Metrology", chaired by M. Soffel, and also with G. Petit and M. Standish. The proposals were submitted to Commissions 4, 8, 19, 31 for approval, and then supported by Division 1. In April 2006, the NFA resolutions proposals were officially submitted to the IAU. In July 2006, the resolutions to be presented for voting at the IAU General Assembly, including those proposed by the NFA WG, were posted on the IAU web site and provided in IAU Newsletter 5. They were published in the official GA newspaper of 16 August 2006. The IAU resolutions were submitted to be voted on during the second session of the 26th IAU GA.

The status of the resolution proposals was discussed during the WG meeting in order to decide on the process for further discussion before it was voted by the IAU GA. The comments on the NFA proposals that had been received were discussed and a few changes were proposed to the wording in order to address the concerns that had been expressed. It was agreed that the revised versions should be presented at Division 1 meetings during the GA.

[The two resolutions submitted to the Executive Committee by the Working Group were voted on at the final General Assembly meeting on 2006 August 24. Both resolutions were adopted. They are resolutions 2 and 3 from the XXVIth General Assembly.]

4. The NFA explanatory document

The content of the NFA Explanatory document was presented and discussed. The introduction should present the background on the IAU 2000 resolutions on reference systems and a summary of the terminology work. Part A should review the basis of the IAU resolutions and their implementation. Part B should contain the "NFA IAU 2000 Glossary", which provides a set of detailed definitions that best explain all the terms required for implementing the IAU 2000 resolutions, including a few newly proposed terms. It is completed by additional documents including a chart, the purpose of which is to illustrate the various stages showing the BCRS-to-GCRS-to-ITRS transformation in General relativity (Resolution B1.3) and the parallel CIO and equinox based processes (Resolution B1.8).

The following scheme has been adopted:

- Introduction to the document
- Part A: Basis of the IAU resolutions and their implementation
  Historical review: reasons for the changes
  Detailed explanations on Resolutions B1.3, B1.5, B1.6, B1.7, B1.8, B1.9
  Description of the new paradigm and comparison with the classical one
  Detailed discussion on a few issues especially discussed by the WG
- Part B: Explanation of the proposed terminology
  NFA IAU 2000 Glossary
  Table containing the categorized list of terms
  Chart of the transformation process from ICRS to observed places of stars
  List of abbreviations, acronyms and symbols
- References

Part B of the NFA Explanatory document is already realized, but the other parts of the document have still to be done.

Plans for the future work were discussed. Due to the current rules of the scientific organization of the IAU, this working group, which has met its objectives within the three-year period, should be disbanded at the end of this GA. Hence the WG document will need to be finalized outside the official IAU WG framework.

5. Plan for publication of the NFA working group documents

Preliminary reports of the NFA WG has been published (IAU Transactions A 2006, Capitaine et al. 2005, 2006), and the report of this WG meeting will be part of the Proceedings of the XXVIth IAU GA.

The publication of the final NFA WG document, consisting of the recommendations and the explanatory material, was identified as an important issue and several options were considered during the meeting. The meeting concluded that the final document should be published, both electronically and in a referred journal. However, due to the document size and the fact that it deals with terminology issues, it was recognized that it may not be accepted in any of the journals in which scientific papers in fundamental astronomy are published (e.g. *AJ*, *A&A*, or *Celest. Mech. Dyn. Astr.*). If this occurs, then other options for its publication needs to be considered. One possibility, offered by B. Luzum, was to include the NFA Glossary (part of the final NFA document) in the next release of the IERS Conventions.

6. Plan for diffusion of the NFA educational documents

For educational purposes, a number of presentations made by members of the NFA WG have been realized as PDF files, and are currently on line on the NFA WG website. They include the following:

- The IAU Recommendations on Reference Systems and their applications
- Recent International Recommendations on Reference Systems
- SOFA software support for IAU 2000
- The ICRS, BCRS and GCRS and the ITRS
- Recent progress in astronomical nomenclature in the relativistic framework
- Progress on the implementation of the new nomenclature in "The Astronomical Almanac"
- Latest proposals of the IAU Working Group on Nomenclature for fundamental astronomy
- Développements récents des concepts et des modèles en Astronomie fondamentale
- 3D representation of the Non-Rotating Origin (with movies)

Additionally, an example transformation (by P. Wallace) is currently on line on the NFA WG web site providing an application of the IAU 2000 resolutions concerning Earth orientation and rotation, with the objective of predicting the topocentric apparent direction of a star.

The plan for diffusion of those educational documents has been discussed. Mirror sites may be developed which will include all the files and thus increase accessibility. Other options considered were depositing the files with the CDS or making them available on the IAU website.

7. Concluding remarks

The NFA WG, created at the 2003 IAU General Assembly, has reached its aim with selecting a consistent and well defined terminology for all the quantities based on the IAU 2000 resolutions on reference systems. The conclusion of this work have been reflected in 14 recommendations, which are supported by two explanatory documents and educational documents.

It should be noted that the NFA recommendations on terminology have already been taken into account in the 2006 editions of the Almanac offices that have implemented the IAU 2000 resolutions (Hohenkerk 2005), in the updated IERS Conventions (ongoing), and USNO Circular 179 (Kaplan 2005).

The NFA Recommendations have resulted into two IAU resolution proposals that have been labeled as IAU 2006 Resolutions 2 and 3. [Both resolutions were adopted by the GA.]

The publication of the final NFA WG recommendations and of the NFA explanatory document has to be realized in the most appropriate form. It will be finalized outside the official IAU WG framework.

The scientific issues identified by the NFA WG to be proposed to Division 1 as topics for a new WG or Commission:

- TDB units, astronomical units,
- appropriate nomenclature and description for topocentric reference systems,
- definition of the ecliptic within the GCRS,
- various effects in the dynamics of the rotating Earth and new theories,
- astronomical constants,
- coordination between IERS Conventions and SOFA.

The work accomplished by the NFA WG work has been very fruitful with recommendations that are already followed in recent publications and two resolutions proposed to the IAU. This has been the work of the whole working group and it is my pleasure to thank all the WG members for their efforts and efficient cooperation during this 3-year undertaking.

Nicole Capitaine
chair of the Working Group

References

Capitaine, N., Hohenkerk, C., Andrei, A. H., Calabretta, M., Dehant, V., Fukushima, T., Guinot, B., Kaplan, G., Klioner, S., Kovalevsky, J., Kumkova, I., Ma, C., Mccarthy, D. D., Seidelmann, K., & Wallace, P. 2005, in Proceedings of the "Journées 2004 Systèmes de référence spatio-temporels," N. Capitaine (ed), Observatoire de Paris, p. 161

Capitaine, N., Hohenkerk, C., Andrei, A. H., Calabretta, M., Dehant, V., Fukushima, T., Guinot, B., Kaplan, G., Klioner, S., Kovalevsky, J., Kumkova, I., Ma, C., Mccarthy, D. D., Seidelmann, K., & Wallace, P. 2006, in Proceedings of the "Journées 2005 Systèmes de référence spatio-temporels," A. Breziński et al. (eds), Warsaw, in press

Hohenkerk, C. 2005, in Proceedings of the "Journées 2004 Systèmes de référence spatio-temporels," N. Capitaine (ed), Observatoire de Paris, 168

IAU Transactions (2003) Vol. XXVB, in "Proceedings of the Twenty-Fifth General Assembly", Sydney, ed. O. Engvold, ASP, 2008

IAU Transactions (2003) Vol. XXVIA, "Reports on Astronomy 2003-2005", ed. O. Engvold, CUP, 2007

IERS Conventions (2003), IERS Technical Note 32, D.D. McCarthy & G. Petit (eds), Frankfurt am Main: Verlag des Bundesamts für Kartographie und Geodäsie, 2004

Kaplan, G. H. 2005, *The IAU Resolutions on Astronomical Reference Systems, Time Scales and Earth Rotation Models*, USNO Circular No. 179

Wallace P. T. 2004, SOFA software support for IAU 2000, AAS Meeting 204, #28.02, May 2004

Transactions IAU, Volume XXVIB
Proc. IAU XXVI General Assembly, August 2006
Karel A. van der Hucht, ed.

doi:10.1017/S1743921308023697

COMMISSION 4 EPHEMERIDES

ÉPHÉMÉRIDES

PRESIDENT	**George A. Krasinsky**
VICE-PRESIDENT	**Toshio Fukushima**
PAST PRESIDENT	**Jean Chapront**
ORGANIZING COMMITTEE	**John A. Bangert, Jean Chapront, Catherine Y. Hohenkerk, George H. Kaplan, P. Kenneth Seidelmann, E. Myles Standish, Sean E. Urban, Jan Vondrák**

PROCEEDINGS BUSINESS MEETING on 24 August 2006

1. Introduction

This business meeting was held from 16:00 to 17:30. Toshio Fukushima and George Kaplan were welcomed as the next president and vice-president, respectively. The following, in no particular order, are the summary reports from the various offices. The full versions will be made available on the Commission 4 website at <http://iau-comm4.jpl.nasa.gov/>.

2. Institut de Mécanique Céleste et de Calcul des Éphémérides

2.1. *Recent developments and new ephemerides at IMCCE*

Institut de mécanique céleste et de calcul des éphémérides (IMCCE), formerly "Service des calculs et de mécanique céleste du Bureau des longitudes", is an institute of the Paris Observatory since 1998. It is responsible for computing the official French ephemerides on behalf of the Bureau des longitudes. Our activities include research in the areas of theoretical celestial mechanics, astrometry and planetology, as well as projects in dynamics and applied celestial mechanics with the goal of providing accurate ephemerides.

2.2. *The new dynamical models*

At IMCCE, during the last years, several new dynamical models have been developed. VSOP, the planetary ephemerides (Variations Séculaires des Orbites Planétaires) initially developed by Bretagnon, now improved (VSPO2002b) by Fienga and Simon who have used it to study the perturbations on inner planets by the asteroids. A new numerical planetary ephemerides named INPOP (Intégration Numérique Planétaire de l'Observatoire de Paris) has been produced.

New dynamical models of several planetary satellites systems have been derived. One model NOE (Numerical Orbit and ephemerides) is obtained by a recomposition of the quasi-periodic Fourier series produced from a frequency analysis coupled with digital filtering treatments. This has been used to model the motion of the Galilean satellites (ephemeris L1), and to obtain new ephemerides of the Martian (NOE-4-06) and Uranian (NOE-7-06) satellites. NOE-7-06 has been used to predict the mutual events of the Uranian satellites. A collaboration with the Sternberg State Astronomical Institute of Moscow led to the modelling of, and ephemeris production for 96 outer satellites of Jupiter, Saturn, Uranus and Neptune.

A new dynamical and physical model of the meteoritic streams has been studied.

2.3. *Ephemerides books*

IMCCE provides yearly ephemerides on behalf of Bureau des longitudes. Several books related to various Solar System objects and at different levels of accuracy are published.

The ephemerides of high precision, *Connaissance des temps* has been revitalized. Since 2003, half of the volume contains scientific texts concerning constants, timescale, reference systems and transformations of coordinates. The IAU 2000 precession and IAU 2000A nutation models are also included. Furthermore, the IAU recommendations have been implemented and Earth rotation angle, equation of the origins, Celestial Intermediate Pole coordinates, angle s are included. This volume still gives ephemerides for the Sun, the Moon, the planets, sidereal time and nutations. However, since 2005 the Chebychev coefficients are only used in the accompanying software (CD-ROM). The software also gives the positions of the main planetary satellites, only the differential coordinates, are tabulated, for use in identification. The software also computes topocentric coordinates, rise and set times.

A second yearly book *Guide de données astronomiques - Annuaire du Bureau des longitudes* aims to give ephemerides to medium precision. Data for the Sun, Moon, planets, as well as ephemerides for bright comets and asteroids, stellar occultations by the asteroids and the Moon, phenomena of the Galilean satellites and other phenomena. Since 2003, a scientific booklet is included each year on a topic by a specialist, for example the 2007 issue will contain the new equatorial origin (CIO) by N. Capitaine and B. Guinot.

Three booklets supplement the *Connaissance des temps* and are guides for observers; titled *Suppléments à la Connaissance des temps* giving (1) graphic configurations and dates of the phenomena of the Galilean satellites, (2) the graphic configurations of the first eight satellites of Saturn and (3) positional ephemerides of several faint satellites of Jupiter and Saturn.

For navigation, IMCCE publishes annually a nautical almanac, *Ephémérides nautiques* and ephemerides for air navigation in the *Ephémérides aéronautiques*.

2.4. *Electronic ephemerides*

Ephemerides are available on-line at the web site of IMCCE at <http://www.imcce.fr>. Several improvements to the data and their organization have been done since 2003. The main changes are the introduction of web services with the objectives to provide "self-defined" data and to make these services interoperable. IMCCE has thus been able to join a working group dedicated to the development of software and databases in the Virtual Observatory framework. Their first product SkyBoT (Sky bodies Tracker) has been developed in collaboration with the Centre de données de Strasbourg in France (CDS). This software deals with dynamical models of all the known asteroids and the main natural satellites. It provides weekly updated ephemerides (1949-2009) for identification of solar system objects in star fields and to data mining. Access the software at <http://www.imcce.fr/webservices/skybot/> using query forms or in other user software, or, via the CDS's Star Atlas Aladin at <http://aladin.u-strasbg.fr/aladin.gml> which gives graphic identification from any of the archives, thanks to the interoperability between the CDS and the IMCCE servers.

3. Report of United States Naval Observatory

This report covers activity in the Astronomical Applications (AA) Department since the XXVth General Assembly in Sydney. The AA Department employs 14 scientists in three divisions: The Nautical Almanac Office (NAO), the Software Products Division (SPD), and the Science Support Division (SSD). The SSD was established in January 2004 as the department's research arm. S. Urban was appointed Chief of the NAO in June 2004. M. Efroimsky joined the staff on a permanent basis in March 2005. G. Kaplan, who was Chief of the SSD, retired in October 2005, and was rehired into a part-time position the following month. A. Fredericks joined the SPD staff in December 2005. R. Miller retired from the NAO in January 2006, and was replaced by E. Barron. The same month, J. Hilton moved from the SSD to the NAO.

Hilton served as chair of the Division I Working Group on Precession and the Ecliptic and as a member of the Inter-Division Working Group on Cartographic Coordinates and Rotational Elements. Urban chaired the Commission 8 Working Group on the Densification of the Optical Reference Frame. Kaplan served as a member of the Division I Working Group on Nomenclature for Fundamental Astronomy. J. Bangert served as a member of the Standards of Fundamental Astronomy (SOFA) reviewing board.

Publication of *The Astronomical Almanac* and *The Astronomical Almanac Online*, *The Nautical Almanac*, *The* (U.S.) *Air Almanac*, and *Astronomical Phenomena* continued as a joint activity between Her Majesty's Nautical Almanac Office of the United Kingdom and the NAO. *The Astronomical Almanac* for 2006, released in January 2005, was the first edition to incorporate fully the resolutions on reference frames, Earth rotation models, and time scales adopted by the IAU in 1997 and 2000.

U.S. Naval Observatory Circular 179, *The IAU Resolutions on Astronomical Reference Systems, Time Scales, and Earth Rotation Models: Explanation and Implementation*, was published in October 2005. It is available in print form, and as a PDF file at `<http://aa.usno.navy.mil/publications/docs/Circular_179.html>`.

A major upgrade of the *Multiyear Interactive Computer Almanac*, MICA version 2.0, was completed and released in July 2005. The software is available in two editions for computers running Microsoft Windows and Apple Mac OS operating systems.

Use of the Astronomical Applications Department Web site (`<http://aa.usno.navy.mil/>`) continued to grow during the reporting period, hosting as many as 45000 user sessions per day. The site now contains approximately 1200 pages and numerous interactive calculators.

A new version of the Naval Observatory Vector Astrometry Subroutines (NOVAS) that implements the 1997 and 2000 IAU resolutions is under development, with release anticipated in late 2006 or early 2007.

An active research program is underway within the department. Research topics include gauge functions in celestial mechanics, solid-body tides in orbit theory, asteroid ephemerides and determination of asteroid masses, determination of orbital parameters of one satellite from observations taken from another, and development of instruments to fully automate celestial navigation and surveying.

Other projects underway at USNO, and of interest to Commission 4, include the USNO CCD Astrograph Catalog (UCAC; `<http://ad.usno.navy.mil/ucac/>`) and observations of solar system bodies made with the Flagstaff Astrometric Scanning Transit Telescope (FASTT; `<http://www.nofs.navy.mil/about_NOFS/telescopes/fastt.html>`).

4. Ephemeris astronomy at Institute of Applied Astronomy

4.1. *Almanacs*

Apart from Ephemerides of Minor planets (which are not described here), IAA continues to regularly publish *The Astronomical Yearbook* of Russia, *The Nautical Astronomical Yearbook* and biennially *The Nautical Astronomical Almanac* for ships at long-run sailing.

The Astronomical Yearbook of Russia includes geocentric apparent places of the Sun, Moon, major planets, 779 bright stars and some information on current astronomical phenomena, such as solar and lunar eclipses, planet configurations and so on (686 pages for 2007) and follows the recommendations of IAU whenever possible.

The *Explanatory Supplement to Astronomical Yearbook* (in Russian) has recently been published. It includes a summary of modern trends in ephemeris astronomy and the new theories of the major planets, relativistic time scales, precession-nutation models, relativistic theory of reference systems, and the CIO. A bibliography of over 500 references is included.

In the navigation almanacs, the IAU 2000 resolutions concerning the CIO are deliberately not implemented to ensure continuity.

4.2. *Scientific work*

The work on the publications is done by the Laboratory of the Astronomical Yearbook. Research work to support the published ephemerides is the duty of the Laboratory of Ephemeris Astronomy which is also involved in other research studies. Here the main aim is to attain the highest accuracy, and so the recommendations and standards of IAU or IERS are not always rigorously followed.

In the recent years it was decided that the published ephemerides should be generated after fitting their dynamical models to available high-accuracy observations. The current stage of this approach is given in brief.

The ephemerides, EPM, of the major planets for *The Astronomical Yearbook for the Year 2007* have been fitted (Pitjeva) to the all observations on <http://ssd.jpl.nasa.gov/iau-comm4>, the Commission 4 website. The current version, EPM2006, accounts for the gravitational interaction of the 9 major planets, the Sun, the Moon, 301 biggest asteroids, the perturbations from the asteroid ring of smaller asteroids, and for the perturbations from the solar oblateness. The planetary part has resulted from a least squares adjustment to observational data totaling 437883 observations (1913–2005) including radiometric observations of the planets, spacecraft, and astrometric observations of the outer planets and their satellites.

The lunar ephemerides are obtained by integrating the equations of lunar orbital and rotational motion simultaneously with the equations of planet motion described above. The parameters are estimated from the analysis of LLR observations from 1970–2004 (15599 time delays). About 60 parameters were adjusted, including the lunar Love numbers the tidal lag of the lunar body tides and harmonics of the lunar potential.

As the data are published with 1 mas truncation errors, the published planetary ephemerides coincide with those based on DE405, while for the lunar ephemerides, the differences exceed the truncation error.

The EPM ephemerides are available from <ftp://quasar.ipa.nw.ru/incoming/EPM2004> in the form of Chebyshev's polynomials.

Soon the contents of the *Astronomical Yearbook* will be expanded by including ephemerides of the main satellites of the major planets. This involves, integrated simultaneously, the numerical ephemerides of these satellites and the equations of motion of satellite systems and accounting for their mutual perturbations. At present work has been done for four Galilean satellites, eight satellites of Saturn, and five satellites of Uranus.

For the Earth and major planets the theories of their rotation are to be constructed by numerical integration, presenting the results in the form of Chebyshev's polynomials.

A numerical theory of rotation of the deformable Earth with the fluid core was constructed and fitted to VLBI observed position of the Celestial Pole and Universal Time UT1 (Goddard Space Flight Center series of 1984–2004). The numerical theory provides a better fit than using the adopted IAU 2000 theory of precession-nutation. For the case of Mars, constructing the analogous numerical theory based on the observations of Martian landers is in progress.

4.3. *Software*

The practical preparation of the almanacs, as well as their ephemeris support is carried out by a technology developed by IAA staring in 1982. The approach is based on using a high-level problem-oriented language SLON designed specially for ephemeris and dynamical astronomy. The corresponding programming, version ERA-7 (<ftp://quasar.ipa.nw.ru/incoming/era/> for the 16-bit version) is more advanced both in the functional diversity and in the descriptive power of SLON. The system has been thoroughly tested by a number of practical tasks and has proved its efficiency.

In particular, it provides the user with (a) a unified method of constructing numerical dynamical theories of any body of the Solar system (orbital and rotational motion) and presenting them as Chebyshev polynomials, (b) easy processing of observables thus improving the parameters, and (c) providing a toolkit for updating the constructed numerical ephemerides using Least Squares or by Kalman filtering.

5. HM Nautical Almanac Office, Rutherford Appleton Laboratory

This is the report of HM Nautical Almanac Office (HMNAO) covering the period since the XXVth General Assembly in Sydney. After seven and a half years operating under commercial conditions within CCLRC at the Rutherford Appleton Laboratory (RAL), HMNAO has been transferred to the UK Hydrographic Office (UKHO), an agency of the UK Ministry of Defence (MoD).

HMNAO consists of three staff who are funded by the royalties generated by the sales of the almanacs produced by the office and jointly with the US Naval Observatory and also from the sales of its services. These funds were insufficient to take on and train new staff. As RAL were also unable to provide any support, financial assistance was sought from the Royal Navy/UK MoD. Having satisfied themselves HMNAO's services were needed, the UK MoD requested UKHO to access the viability of taking on the office. Having generated a satisfactory business plan, HMNAO was handed over to UKHO on April 1st 2006. This entire process took the better part of 4 years requiring considerable staff input.

Joint publications with the US Naval Observatory, in particular, *The Nautical Almanac* and *The Astronomical Almanac* have been produced on schedule. A fully navigable pdf version of *The Nautical Almanac* has been produced and extensive changes have been made to *The Astronomical Almanac* to implement the resolutions relating to reference frames, times scales and earth orientation from the previous two IAU's. Improvements have also been made to the calculation and provision of satellite data and phenomena in Section F, to Section K and the provision of lunar eclipse diagrams in Section A.

A new edition of *Navpac and Compact Data 2006-2010* has been produced including some new features. To emphasise its use by the Royal Navy, a joint launch of this product with the Admiralty Manual of Navigation (BR45) was arranged.

Catherine Hohenkerk has served on the IAU Working Group on Nomenclature for Fundamental Astronomy and has given talks at the Journées meetings in Paris (2004) and Warsaw (2005) on the application of this topic to *The Astronomical Almanac*. She has taken an active role on the Software for Fundamental Astronomy (SOFA) board. HMNAO now hosts the SOFA Center web site. She has taken a leading role in the major changes to Section B in the 2006 edition and has also received the USNO Superintendent's Award 2005 for her services to *The Nautical Almanac*. Steve Bell attended the IAU Colloquium 196 on New Views of the Solar System and Galaxy.

Significant effort has gone into the generation of new web site material. Transit information was provided for the 2004 transit of Venus and a new eclipse web site http://www.eclipse.org.uk has been generated giving solar and lunar eclipse information for the period 1500 CE to 2100 CE. A mass participation project for Einstein Year was launched in collaboration with the Institute of Physics to observe the first sighting of new crescent moon involving a web site http://crescentmoonwatch.org. This received significant media coverage. Steve Bell has given several talks around the UK on this subject.

Don Taylor has been involved in work relating to the astronomical application of map projections and the compilation of time zone and daylight saving time rules for web services. He has also continued his worked on solar perturbations for the satellites of Uranus and integration software for cometary ephemerides.

6. Closing remarks

It was agreed that the Almanac Offices and ephemeris producers, the Commission as a whole, should aim to be more pro-active in the next triennium.

George A. Krasinsky
president of the Commission

Transactions IAU, Volume XXVIB
Proc. IAU XXVI General Assembly, August 2006
Karel A. van der Hucht, ed.

doi:10.1017/S1743921308023703

COMMISSION 7 — CELESTIAL MECHANICS AND DYNAMICAL ASTRONOMY

MÉCHANIQUE CÉLESTE ET ASTRONOMIE DYNAMIQUE

PRESIDENT	**Andrea Milani**
VICE-PRESIDENT	**Joseph A. Burns**
PAST PRESIDENT	**John D. Hadjidemetriou**
SECRETARY	**Zoran Knežević**
ORGANIZING COMMITTEE	**Christian Beaugé, Bálint Erdi, Toshio Fukushima, Douglas C. Heggie, Anne Lemaitre, Andrzej J. Maciejewski, Alessandro Morbidelli, Milos Sidlichovsky, David Vokrouhlický, Ji-Lin Zhou**

PROCEEDINGS BUSINESS MEETING on 22 August 2006

1. Opening of the business meeting of Commission 7

1.1. *Address by the outgoing president of the Commission, A. Milani*

In his address to the Commission, the outgoing president A. Milani explained what he considers have been done well in the past triennium, what has been done only in part, and what has not been done at all. Among the things in which the performance was rated good, he mentioned the successful sponsorship and/or co-sponsorship of four meetings (IAU Colloquia 196 and 197, and Symposia 229 and 236) which have been held in the previous period, as well as of the Symposium on exoplanets to be held next year in China. The only failure in this respect was the proposed meeting in India, which failed already at the proposal definition stage. Also, Milani expressed his satisfaction with the triennial report which has been compiled for the occasion, and his gratitude to the collaborating authors.

What was done only in part has to do with the problem of maintenance and regular update of the mailing list of Commission 7 members, and the communication of the OC with the membership. Because of a large number of inaccurate e-mail addresses of individual members, the communication with the members has been mostly commission web-page oriented, which proved to be insufficient and should be improved in the future.

The one thing which was not done at all is the Terms of Reference document, requested by the IAU executives to justify the existence of the commission and the need for it to continue as such in the future.

1.2. *Address by the incoming president of the Commission, J. Burns*

In his brief address, the incoming president J. Burns said that he is aware of the duties and responsibilities of the position he is taking over, and on behalf of the commission congratulated and thanked A. Milani for the excellent job and results he had achieved as president of the Commission.

2. Membership of Commission 7

There were 35 candidates who expressed their interest in becoming members of Commission 7. After a through review of the applications, 31 candidate has been proposed for the membership, while for four of them not enough information had been available to accept them to the membership at this time. The proposal was unanimously accepted.

Commission deeply regrets five of its members who passed away in the previous period.

3. Report on the scientific achievements in 2002–2005

The Report on the scientific achievements in Celestial Mechanics and Dynamical Astronomy in the period 2002–2005 has been prepared in due time and is available from the Commission 7 web page.

4. Report on the Commission activities in the period 2003-2006

4.1. *Past IAU meetings sponsored by the Commission*

Short reports on the Colloquia and Symposia sponsored by the Commission has been given by A. Lemaitre (Coll. 196), Z. Knežević (IAU Coll. No. 197), and by A. Milani (IAU Symposia No. 229 and No. 236). The reports have been acknowledged and accepted.

4.2. *Future IAU meetings sponsored by the Commission*

Detailed information on the future IAU Symposium No. 249 on exoplanets in China, October 2007, has been given by A. Milani and acknowledged by the present commission members.

5. Report on the CMDA journal

The report on the Celestial Mechanics and Dynamical Astronomy journal has been presented by the associated editor B. Erdi, on behalf of the absent Editor-in-Chief, S. Ferraz-Mello. The situation is not fully satisfactory, due to the merger of the previous and current publisher, but it is improving and is expected to be normalized soon enough. The impact factor of the journal is steadily increasing, and only because of the typically somewhat slower response of the readers and of the method of IF computation, the journal is yet to reach IF larger than 1. The report has been unanimously accepted, and is enclosed below.

6. Results of the elections for the new Commission 7 officials

A. Milani has given the details of the procedure of election of new officials of the Commission, and of the new members of the Organizing Committee. The President of the Commission in the period 2006-2009 is J.Burns, and Z. Knežević has been elected vice-president. The new members of the OC are R. Malhotra, S. Peale, and L. Athanassoula.

7. Appointment of the new secretary of the Commission

The secretary will be appointed by the president of the Commission later†.

8. Discussion

8.1. *IAU organization and the role of Commission 7*

WGNEO

The IAU EC has decided to form a new consultative committee to advise the EC on issues connected to the NEO impact risk. This replaces the previous WGNEO, which had expanded to the point that it had become as large as a Commission, with significant intersections of

† The secretary has been nominated: David Vokrouhlický.

competences with Commission 20, Commission 15 and to some extent Commission 7. The former members of the WGNEO are invited to join the relevant permanent Commission, if they do not belong to it yet.

Interdivision activities

The issue of possible change in the collocation of Commission 7 in the IAU structure was raised. A possible option would be to become part of both Division I and Division III: this is allowed by the current rules but has not been experimented yet. After some discussion, the Commission has asked the incoming president to inquire with the Divisions and to consult the Commission membership for a possible proposal in this sense to be discussed at the IAU XXVII GA in 2009.

8.2. *Topics on IAU agenda relevant for Commission 7*

Discovery rules and the MPC

The proponents of a new set of discovery definitions and discovery credit rules (for Solar System objects), including the outgoing President and Secretary of Commission 7, have briefly presented the Celestial Mechanics background of their proposal. It has already been agreed that Division III will be in charge of deciding on a new set of rules, and that this discussion will begin very soon under the supervision of Commission 20. These new rules shall be applied by the MPC, according to the new understanding between IAU and CfA for the operation of the MPC.

Definition of planet

The discussion on the new definition of planet taking place in this GA has been evaluated by the Commission 7 from the point of view of method and contents, especially for the Celestial Mechanics aspects. The current state of the discussion has not been found satisfactory and the Commission 7, which was certainly competent from the scientific point of view, has neither been involved in the preparation nor consulted nor informed at all before the public presentation of a new definition. Thus the Commission has discussed and unanimously approved the following Commission Resolution, to be presented to the Executive and to all the opportunities for discussion inside this GA.

The Commission 7 (Celestial Mechanics and Dynamical Astronomy) of the IAU, at its business meeting at the IAU XXVI General Assembly in Prague, 22 August 2006
CONSIDERING
that the IAU Executive Committee has presented a proposal of resolution on the naming of Solar System bodies, in particular deeply revising the usage of the word "planet";
that the traditional terminology (planet, satellite, minor planet, etc.) is deeply rooted in the history of astronomy as a science and has always had a solid foundation in the basic principles of Dynamical Astronomy;
that Commission 7 has not been consulted in any formal way before making public the EC proposal;
FINDING
that the proposed definitions do not contain any appropriate dynamical argument, in particular lacking whatever consideration of planetary mutual perturbations, dynamical stability and chaos (the historically most important contribution of Astronomy to modern science);
that the proposed definitions do not contain any appropriate cosmogonical argument, in particular the consideration of the sweeping by a planet of its feeding zone;
that the proposed definitions contradict the history of Solar System astronomy, in particular reversing the recognition due to Olbers and Herschel (1804) that Ceres is not to be considered a planet because of the discovery of Pallas with a potentially crossing orbit;
that the proposed definitions contain other inappropriate dynamical arguments, such as unnecessary reference to orbital eccentricity and an unjustified emphasis on the position of the center of mass;
that a thorough consultation including the Dynamical Astronomy community is required,
PROPOSES TO THE IAU EXECUTIVE COMMITTEE
to revise the proposed definition according to the previous findings.

A proposal by M. Bailey of an alternative definition of planet, which would include dynamical arguments but also use the geophysical arguments on the internal structure of the body, was presented and received support from the Commission, with the understanding that the final version would have to be further adjusted to take into account the other points of view.

9. Miscellaneous

There were no additional items proposed for discussion at the business meeting.

Andrea Milani, *president*
Zoran Knežević, *secretary*

Appendix A. Celestial Mechanics and Dynamical Astronomy report 2002-2005

The main fact to report concerns the journal impact factor, published in the Journal of Citation Reports, which has continuously increased during the past 4 years and reached 0.856 in 2005. This factor measures the citations, in a given year, of papers published by the journal in the two previous calendar years. In fact, the citations to CMDA papers peak only in the third year after the publication year, and this profile is not considered in the rule adopted by the Institute for Scientific Information. If we proceed exactly as they do, but considering the papers published by the journal in the three or four previous years, instead of just two, we obtain a factor $\simeq 1.0$, which more properly measures the impact of the journal (see `<http://www.astro.iag.usp.br/~sylvio/celmech.html>`). Since 2004 the journal is also included by JCR in the category Mathematics: Interdisciplinary Applications (in addition to Astronomy and Astrophysics) and ranked there in the 36th position on a set of 76 journals.

Another important fact in the period was the merging of Kluwer and Springer to form a new company. The merging affected the production routines, and successive problems and errors were recorded in the period 2004-2005. These difficulties are almost over and the journal is being published at a pace not far from the nominal schedules. The efforts to increase the diffusion of the journal, initiated by Kluwer, were continued and CMDA reaches now, through electronic and hardcopy subscriptions, some 4,000 institutions. The old issues are now available for free download through the NASA-ADS site. The better diffusion is in direct correspondence with the increase in the impact factor, the growth of one of them directly influencing the growth of the other.

At last, we mention that the on line manuscript submission, review and tracking system is now fully operational allowing much shorter times between the submission of one paper and the final decision concerning it. In the great majority of the cases the decision is taken within some months. These times could be yet shorter if all Associate Editors were collaborating to avoid unreasonable delays to appoint reviewers and decide on received reports. Some extremely long times are still recorded, but they generally correspond to papers demanding time consuming thorough revisions.

Sylvio Ferraz-Mello, *editor-in-chief*

Transactions IAU, Volume XXVIB
Proc. IAU XXVI General Assembly, August 2006
Karel A. van der Hucht, ed.

doi:10.1017/S1743921308023715

COMMISSION 8 — ASTROMETRY

ASTROMÉTRIE

PRESIDENT	**Imants Platais**
VICE-PRESIDENT	**Irina I. Kumkova**
PAST PRESIDENT	**Wen-Jing Jin**
ORGANIZING COMMITTEE	**Edgardo Costa, Christine Ducourant, Dafydd W. Evans, Mario G. Lattanzi, Chunlin Lu, Ralf-Dieter Scholz, Mitsuru Sôma**

DIVISION I / COMMISSION 8 WORKING GROUPS

Division I / Commission 8 / WG	**Astrographic Catalogue and Carte du Ciel Plates (†)**
Division I / Commission 8 / WG	**Densification of the Optical Reference Frame**

PROCEEDINGS BUSINESS MEETING on 21 August 2006

1. Business Meeting (chair I. Platais)

The Business Meeting was opened by the president, Imants Platais. He presented the agenda which was unanimously approved. This session was attended by 40 participants. The meeting approved Dafydd Evans as secretary of minutes.

A minute of silence was observed in memory of the Commission members who had passed away since the Sydney GA: Nina Bronnikova, Jurgen Stock, Ronald Stone and Volodymyr Telnyuk-Adamchuk. Also, Zdenka Kadla who had made significant contributions to astrometry although formally she had not been a member of Commission 8.

1.1. *Commission activities 2003-2006*

Platais briefly described the main work done in this triennium. Most of the information dissemination was conducted via the Commission's WWW homepage at `<http://www.pha.jhu.edu/iau_comm8/comm8.html>`. As of this writing, the Commission's web page now has a new URL at `<http://www.ast.cam.ac.uk/iau_comm8/>`. A total of six electronic newsletters were circulated during the 2003-2006 period. Some Commission members have not reported their e-mail address or have an obsolete one that, unfortunately, resulted in a loss of communications. All members are strongly encouraged to check their personal data in the IAU membership directory and update them. This can also be done via the Commission's WWW homepage.

Following many revisions and modifications of the initial draft version, the Commission now has its own Terms of Reference. This is the main document defining the Objective, Scope, Organization, and Implementation that cover all aspects of the Commission's functions and organization in accordance with the IAU Statutes and Bye-Laws. The Terms of Reference were unanimously approved by the Commission members present at the Business meeting.

The largest task was the compilation of the Triennial Report. Platais thanked the national organizers for their excellent job in collecting the individual reports. The science highlights in this triennium were: (1) completion of large catalogues, such as NPM, USNO-B1.0 and UCAC; (2) re-analysis of the raw Hipparcos data; and (3) discovery of several low-mass "cool neighbours" near the Sun, detected by astrometric means in combination with other techniques.

A large success is the proposal by the Commission and acceptance by the IAU of the astrometric meeting entitled *A Giant Step: from Milli- to Micro-arcsecond Astrometry* (IAU Symposium No. 248) to be held in Shanghai, China PR, 15-19 October 2007. The Commission has also supported JD 16 (*Nomenclature, Precession and New Models in Fundamental Astronomy*), JD 13 (*Exploiting Large Surveys for Galactic Astronomy*) and IAU Symposium No. 240 (*Binary Stars as Critical Tools and Tests in Contemporary Astrophysics*) here at the Prague GA.

1.2. *New Commission members*

The following new Commission members were confirmed by the meeting:

a) IAU members requesting membership of Commission 8

John Bangert (USNO, USA), David Boboltz (USNO, USA), Anthony Brown (Leiden Observatory, the Netherlands), Alan Fey (USNO, USA), Toshio Fukushima (NAOJ, Japan), Valeri Makarov (JPL, USA), Francois Mignard (Observatoire de la Côte d'Azur, France), Myles Standish (JPL, USA), Yoshiyuki Yamada (Kyoto University, Japan), Taihei Yano (NAOJ, Japan).

b) New IAU members requesting membership of Commission 8

Carine Babusiaux (Paris-Meudon Obs., France), Yuri Barkin (Sternberg AI, Russia), Dalia Chakrabarty (University of Nottingham, UK), Maria-Rosa Cioni (University of Edinburgh, UK), Nicholas Cooper (University of London, UK), Vassyl Danylevsky (Kyiv AO, Ukraine), Dana Casetti-Dinescu (Yale University, USA), Marcelo Emilio (Universidade Estadual de Ponta Grossa, Brazil), Zhang Hong (Nanjing University, China PR), Liliya Kazantseva (Kyiv AO, Ukraine), Douglas Mink (Harvard Smithsonian, USA), Jucira Penna (Observatorio Nacional, Brazil), Alexander Rodin (Pushchino Radio AO, Russia), Masahiro Suganuma (NAOJ, Japan), Jonathan Tedds (Leicester University, UK).

c) Commission consultants

The current Commission consultants expressed their wish to discontinue membership. Thus, there are no consultants within Commission 8.

This brings the total membership in the Commission to 241, representing 31 countries.

1.3. *Election results and new Commission Officers*

Platais Platais thanked the outgoing members of the Organizing Committee: Edgardo Costa, Christine Ducourant, Wen-Jing Jin, Mario Lattanzi, and Chunlin Lu for their dedicated service. Elections have been held for the positions of president, vice-president and five vacant Organizing Committee posts. By IAU tradition, the outgoing Vice-President, Irina Kumkova, becomes the new commission president. Out of four nominations for vice-president, only one agreed to remain on the ballot list. Dafydd Wyn Evans was elected unanimously by the Organizing Committee to the position of vice-president for 2006-2009. Eleven commission members were nominated to the new Organizing Committee. The following five were elected (taking into account geographical distribution): Alexandre Andrei, Alain Fresneau, Petre Popescu, Norbert Zacharias, Zi Zhu.

The new Organizing Committee for 2006-2006 was approved by acclamation:
Irina I Kumkova (president, Russia), Dafydd Wyn Evans (vice-president UK). Members: Alexandre H. Andrei (Brazil), Alain Fresneau (France), Imants Platais (USA, ex officio), Petre P. Popescu (Romania), Ralf-Dieter Scholz (Germany), Mitsuru Soma (Japan), Norbert Zacharias (USA), and Zi Zhu (China PR).

2. Working Group reports

In 2003-2006 Commission had two working groups set up to accomplish specific tasks. Their reports and the decisions made about their future are provided below.

2.1. *WG on Densification of the Optical Reference Frame*

Urban (USNO) reported on this Working Group. The main objectives of the WG is the dissemination of information rather than the production of a deliverable. It was pointed out that originally this group was a subgroup under the ICRS Working Group of Division I and that at the Sydney GA it became a Working Group under Commission 8.

Since the Sydney GA, many new catalogues have been measured and released. Also new technology and hardware are becoming available that will be useful for the WG's task, in particular CMOS detectors which have a large dynamic range.

Detailed information on various ground-based and astrometric satellite projects can be found on the WG's WWW homepage at <http://ad.usno.navy.mil/dens_wg/>. Urban asked if the Working Group should continue and requested that, if it did so, it should be under new leadership. Platais gave the group his endorsement and opened the floor for discussion. Among the audience Corbin, Evans, Hilton, Platais, van Altena, Zacharias made several suggestions mainly related to the Working Group's mission in the context of astrometric support to astronomers outside the Commission.

Platais then proposed Zacharias (USNO) as the next chairman of the Working Group and described his work and qualifications. Since there were no objections, Zacharias was approved to this post for 2006-2009.

2.2. *AC/CdC Working Group*

Zacharias (USNO) gave a report on behalf of Beatrice Bucciarelli, the chairwoman of this WG, on the AC/CdC Working Group. The goals of the WG were described and recent results given. Scanning has been carried out for the following CdC zones: Cordoba, San Fernando, Toulouse (photometry), Bordeaux (PM2000) and Sydney. Although no CdC scanning has been carried out at the 1 micron level, it has been done at the 2-3 micron level. More details are provided on the WG's website at <http://www.to.astro.it/AC_CdC/>.

Owing to scheduling issues and prior to this meeting, the Working Group had already become part of the Preservation and Digitization of Photographic Plates (PDPP) Task Force under Commission 5. Its chairwoman Elizabeth Griffin described this Task Force and the history behind it. She also explained why the AC/CdC Working Group would fit very well within the PDPP Task Force. Platais added that Bucciarelli had carried out a poll of AC/CdC Working Group members and they agreed to the move. Griffin also added that ordinary scanners are not good enough for digitizing photographic plates. Harvard have just completed building a new scanner as have Brussels. The way forward is to build and clone new machines, since it is difficult to keep the old machines running. Platais said that many of the existing premier measuring machines were currently out of action and not supported both technically nor financially.

Since no proposals were put forward for new working groups, in 2006-2009 under Commission 8 there will only be the WG on Densification of the Optical Reference Frame. With this, the Business Session was adjourned.

3. Frontiers of Astrometry (chairs J. Kovalevsky & R. Gaume)

In the preceding half a year Platais made all necessary preparations for a very successful science session entitled "Frontiers of Astrometry". The talks were mostly by invitation and covered the main areas of cutting-edge astrometric science and its applications. The abstracts provided here show the scope of the talks but cannot give the details. Therefore, many speakers have agreed to put their presentations on line. The reader is invited to visit the Commission's new website listed above and enjoy these presentations.

3.1. *Abstracts sorted alphabetically by the first author's name*

3.1.1. *A DATABASE OF QSOs AND SURROUNDING STARS FROM THE SDSS, by A. H. Andrei (MCT,UFRJ), J. I. Bueno de Camargo (UFRJ), M. Assafin (UFRJ), D. N. da Silva Neto (UFRJ) & R. Vieira Martins (MCT)*

Sloan Digital Sky Survey (SDSS) Data Release 4 is used to select 62,021 QSOs and the stellar content from a 10-arcmin area surrounding each QSO. Available positional, photometric (and redshift) information for each QSOs and neighbouring stars is stored in the database, along with the error and quality indicators. When compressed the database occupies 4.4 GB only. Simple routines allow us to interrogate the database for the content in a specified area or query individual QSOs and stars. We will demonstrate the database and its use. Further, we will discuss the photometric peculiarities of stellar population around the QSOs. The conclusions we derive may have a significant value for the *Gaia* mission. Our database also enables an investigation of the QSO counterparts and their properties in the USNO B1.0 catalog and, thus, provides insights on the effectiveness of the USNO B1.0 as a research tool for the entire sky. Finally, an estimate of the SDSS astrometric position accuracy is also presented.

3.1.2. *HUBBLE FGS PARALLAXES OF GALACTIC CEPHEIDS AND P-L RELATIONS, by G. F. Benedict (U. of Texas), B. E. McArthur, T. G. Barnes, M. Feast, T. E. Harrison, R. J. Patterson, J. Menzies, and W. Freedman*

We present new absolute trigonometric parallaxes and relative proper motions for 9 Galactic Cepheids: l Car, ζ Gem, β Dor, W Sgr, X Sgr, Y Sgr, FF Aql, T Vul, and RT Aur. We obtained these results with astrometric data from Fine Guidance Sensor 1r (FGS), a white-light interferometer on *Hubble Space Telescope*. We estimate spectral type and luminosity class of the stars comprising each astrometric reference frame from various sources. The derived spectrophotometric parallaxes of reference stars are introduced into our models as observations with an error. We model a volume of space, and our end result is an absolute parallax for each Cepheid. The spectrophotometry also aids in estimating interstellar absorption, required for target absolute magnitudes. Adding our previous absolute magnitude determination for delta Cep, we construct a Period-Luminosity Relation for ten galactic Cepheids. We establish zero-points of the V, I, K, and Wesenheit W(VI) Period-Luminosity relationships with random errors of only 0.03 mag.

3.1.3. *WHERE IS GAIA NOW? by A. Brown (Leiden Observatory)*

Gaia is the European Space Agency mission which will provide a stereoscopic census of our Galaxy through the measurements of high precision astrometry, radial velocities and multi-colour photometry. *Gaia* is scheduled for launch in late 2011 and over the course of its five year mission will measure parallaxes and proper motions for every object in the sky brighter than visual magnitude 20 - amounting to about 1 billion stars, galaxies, QSOs, and solar system objects. It will achieve an astrometric accuracy of 12-25 micro-arcsec at 15th magnitude and 100-300 micro-arcsec at 20th magnitude. Multi-colour photometry will be obtained for all objects by means of low-resolution spectrophotometry between 330 and 1000 nm. In addition, radial velocities with a precision of 1-10 km/s will be measured for all objects down to 17th magnitude, thus providing full six-dimensional phase space information for the brighter sources. *Gaia* thus represents an improvement of several orders of magnitude over *Hipparcos* in terms of numbers of objects, accuracy and limiting magnitude. *Gaia* is fully funded by ESA and the prime contractor, EADS-Astrium, will build both the spacecraft and the scientific payload. The data processing is a task for the scientific community. I will present a brief overview of the current status of the *Gaia* mission with an emphasis on describing the latest EADS-Astrium design of the spacecraft and payload.

3.1.4. *SYNERGY BETWEEN RADIOASTRONOMY AND ASTROMETRY, by E. Fomalont (NRAO)*

High resolution radio interferometry has revolutionized astrometry over the last 30 years. Since the strongest radio sources have compact components smaller than 0.1 mas and are at cosmological distances, they form an excellent set of fixed fiducial points in the sky needed to define a quasi-inertial reference frame. In the 1990's the astronomical community defined the International Celestial Reference System (ICRS), based on a catalog of distant radio sources, and the frame orientation is accurate to 0.02 mas. Other frames (optical, solar-system ephemerides) are now tied to the ICRS. The relative position radio sources within a a few degree region of sky can be determined to an accuracy of 0.01 mas with about five hours of integration time using a large array, such as the VLBA, EVN and VERA. This type of observation have determined the basic astrometric parameters of objects with accuracy much greater than from other astronomical techniques, and rival that of future space interferometry missions. Some examples of putting in work the radio interferometry are: tests of General Relativity; detectable parallax for (radio)stars and pulsars anywhere in the galaxy; dynamics of the galactic center black hole; proper motions of radio-sources in nearby galaxies and accurate distance estimates; orbital motion of binary systems (GPB target); rotation of disks around black holes. Radio interferometry at higher frequencies with ALMA will open new horizons in the galactic and extragalactic research.

3.1.5. *INFRARED ASTROMETRIC SATELLITE JASMINE, by N. Gouda (NAOJ), Y. Kobayashi (NAOJ), Y. Yamada (Kyoto U.), T. Yano (NAOJ), T. Tsujimoto (NAOJ), M. Suganuma (NAOJ), Y. Niwa (Kyoto U.), M. Yamauchi (U. of Tokyo), Y. Kawakatsu (JAXA), H. Matsuhara (JAXA), A. Noda (JAXA), A. Tsuiki (JAXA), M. Utashima (JAXA), A. Ogawa (JAXA), N. Sako (U. of Tokyo) and JASMINE working group.*

We present the Japanese plan for the infrared (z-band: 0.9 mkm) *JASMINE* space astrometry project. *JASMINE* (*Japan Astrometry Satellite Mission for INfrared Exploration*) will measure the parallaxes and apparent motions of stars around the center of the Milky Way with unprecedented 10 micro-arcsec precision for parallaxes and positions and 10 micro-arcsec/yr for proper motions down to z = 14 mag. *JASMINE* will observe about ten million stars belonging to the bulge of our Galaxy, that are hidden by the interstellar dust extinction at optical wavelengths. The anticipated deep and precise mapping of the Milky Way bulge is expected to yield many new exciting scientific results in various fields of astronomy. Presently, *JASMINE* is in the development phase, with a target launch date around 2015. We have adopted a 3-mirror modified Korsch optical system for *JASMINE* with a primary mirror of 1 m. In the focal plane there are dozens of new type CCDs in the z-band to get a wide FOV. The highly-accurate measurements of astrometric parameters require an exceptional stability of the instrument's line-of-sight, including the stability of opto-mechanical parts of the payload. Currently, the overall system (bus) design is ongoing in cooperation the Japan Aerospace Exploration Agency (JAXA).

3.1.6. *DEEP ASTROMETRIC STANDARDS, by I. Platais (JHU), S.G. Djorgovski (Caltech), C. Ducourant (Obs. Bordeaux), A. Fey (USNO), S. Frey (FOMI), Z. Ivezic (UWa), K. Mighell (NOAO), A. Rest (NOAO), R. F. G. Wyse (JHU), N. Zacharias (USNO)*

The advent of next generation imaging telescopes such as LSST and Pan-STARRS - instruments with wide fields and huge Giga-pixel cameras - will soon create a critical need for deep and precise reference frames for astrometric calibrations. The Deep Astrometric Standards (DAS) program aims to establish such a frame by providing astrometry at the 5-10 mas accuracy level in four 10 sq. deg Galactic fields, to a depth of $V = 25$. We use 3-4 m class optical telescopes to set up these standards. The principal source of our reference frame is UCAC2 and VLBI positions of radio-loud QSOs having optical counterparts with $V < 25$. The novelty of the DAS project is a new way of linking our observations to the ICRF. We pre-select the candidate radio-optical link sources from existing radio surveys, then conduct the VLA observations to measure the spectra and spatial compactness and, finally, observe the best 10-15 sources with the VLBI. So far, two out of the four DAS fields are in the advanced stages of construction.

3.1.7. *A QUEST FOR THE NEAREST STARS, by R.-D. Scholz (AIP)*

The stellar census in the Solar neighbourhood is still remarkably incomplete. Even within a very locally set horizon of 10 pc more than 30% of the stars, mainly red and white dwarfs, are missing. In addition, we may be surrounded by large numbers of brown dwarfs from which only few (less than 10%) have been detected so far. In this talk I will give a brief review on various recent activities to foster our knowledge on the nearest stellar and sub-stellar neighbors and will summarise our own efforts: (1) to identify stellar neighbours among known proper motion stars; (2) to extend high proper motion surveys to fainter magnitudes in order to find extremely cool neighbours of different classes (brown dwarfs, cool subdwarfs and cool white dwarfs). The search for faint high proper motion objects based on archival data from SuperCOSMOS Sky Surveys led to the discovery of the nearest known brown dwarf, ϵ Indi B, later resolved as a close pair of T dwarfs and of some of the coolest known subdwarfs, members of the Galactic halo population crossing the Solar neighbourhood at high velocities.

3.1.8. *SIM PLANETQUEST – A SCIENCE AND MISSION UPDATE, by M. Shao (JPL)*

The *Space Interferometry Mission PlanetQuest* (*SIM PlanetQuest*) will be the first space-based long baseline Michelson interferometer designed for precision astrometry. With an accuracy of a few microarcseconds, *SIM* will contribute strongly to many fields including stellar and galactic astrophysics, planetary systems around nearby stars, and the study of quasar nuclei. *SIM* will search for planets with masses as small as a few Earth masses around the nearest stars, using measurements to 1 micro-arcsec precision. It will detect planets around young stars, providing insights into the how planetary systems are born and how they evolve with time. *SIM* will measure positions to 4 micro-arcsec on targets as faint as 19 mag, allowing accurate distances to many types of stars, and will measure stellar masses to 1%, the accuracy needed to challenge physical models. *SIM* will probe the galactic mass distribution, and through studies of tidal tails, the formation and evolution of the galactic halo, using measurements of proper motions. It will use precision astrometry to probe accretion disks and relativistic jets in the variable nuclei of active galaxies. *SIM PlanetQuest* is currently in project Phase B, with a preliminary Design Review in 2007.

3.1.9. *ABRUPT CHANGES IN THE EARTH'S ROTATION SPEED, by M. Soma (NAOJ) and K. Tanikawa (NAOJ)*

From our analyses of total and annular solar eclipses recorded in Asia and Europe we show that the Earth's rotation speed changed abruptly in about AD 500 and AD 900. Specifically the parameter value DeltaT = TT-UT for the Earth's excess rotation changed by more than 3000 s within 160 years around the year 500, whereas it changed by more than 600 s within 40 years around the year 900.

3.1.10. *NEAR-INFRARED PARALLAX PROGRAM AT USNO, by F. Vrba (USNO)*

Beginning in 2000 at the USNO Flagstaff Station we began a program of measuring parallaxes and proper motions at near-infrared wavelengths of brown dwarfs, which are generally too cool and faint to be included in the USNO CCD parallax program. The program began with an initial selection of 40 objects evenly divided between L-type and T-type dwarfs. Preliminary results of the first two years of observations were previously reported by Vrba *et al.* in 2004, with the best parallaxes in the range of 1.5 mas. Since that time, the astrometric accuracies have been greatly improved and the program has been expanded to nearly 80 objects. I will review current astrometric accuracies and the program object list and discuss prospects for the future of the program.

3.1.11. *IAU 2000 RESOLUTIONS FOR THE GENERAL USER, by P. Wallace (SSTD/HMNAO)*

Even before 2000, the *Hipparcos* catalogue had provided a two-orders-of-magnitude increase in the accuracy of the optical frame, and the introduction of the ICRS had, once and for all, broken the link between star catalogues and the orientations of the equator and ecliptic. The IAU 2000 B1 resolutions added various refinements, including a more accurate precession-nutation model and two general-relativity-based reference systems for barycentric and geocentric problems. But to many users the most troubling change was the replacement of the equinox as the zero point for right ascensions and the elimination of sidereal time. The justification for the changed precession-nutation model was clear: a 2-3 orders of magnitude improvement in accuracy. Less obvious was that at these levels of accuracy the traditional equinox based methods had become unwieldy. A symptom of this was the complexity of the GST formula, now requiring both TT and UT and including dozens of correction terms. The "new paradigm" introduced in 2000 cleanly separates Earth rotation from the orientation of the Earth's axis, so that Earth rotation angle, the successor to GST, is simply a two-coefficient linear transformation of UT1. In fact none of this will affect the general user very much; the main consequence is that ordinary astronomers are now shielded from complicated and subtle details and some intimidating nomenclature. The real challenge is getting used to the improvements, and educating new generations of students.

3.1.12. *GROUND-BASED SURVEYS: UCAC AND BEYOND, by N. Zacharias (USNO)*

The all-sky and selected area-based astrometric surveys are reviewed. Recently the Carlsberg/Cambridge and Bordeaux scanning transit circle programs were completed, providing astrometric data for a large fraction of the sky down to about 17th magnitude. The USNO CCD Astrograph Catalog (UCAC) observing program was completed in 2004. Details about the

UCAC3 reductions and products will be presented, including the efforts to measure old photographic plates. The goal of the USNO robotic astrometric telescope (URAT; aperture 0.85 m, FOV 4.5 sq. deg) is to provide positions, proper motions and parallaxes at the 5 mas level in the 14-18 a limiting magnitude of 20. Going even deeper, the Pan-STARRS program will provide a multi-color survey down to 23rd magnitude with a great astrometric potential. Accurate reference stars for calibrating the new generation instruments, including the Large Synoptic Survey Telescope (LSST), will be provided in selected areas by the Deep Astrometric Standards (DAS) program.

Imants Platais
president of the Commission

Transactions IAU, Volume XXVIB
Proc. IAU XXVI General Assembly, August 2006
Karel A. van der Hucht, ed.

doi:10.1017/S1743921308023727

DIVISION I / COMMISSION 8 / WORKING GROUP ASTROGRAPHIC CATALOGUE AND CARTE DU CIEL PLATES

PRESIDENT **Beatrice Bucciarelli**
PAST PRESIDENT **Alain Fresnau**
MEMBERS **Carlos Abad, Robert W. Argyle, James Biggs, Noah Brosch, George V. Coyne, Emmanuael Davoust, Jean-Pierre M. De Cuyper, Edwin L. van Dessel, Christine Ducourant, Ivan H. Bustos Fierro, Michael Geffert, Elena V. Glushkova, Michael J. Irwin, Dayton L. Jones, Kari A. Lumme, J. L. Muiños, Michael Odenkirchen, Rosa B. Orellana, Thierry Pauwels, Theodore J. Rafferty, M. Sanchez, Jörg Sanner, Milcho K. Tsvetkov, Alan E. Vaughan**

PROCEEDINGS BUSINESS MEETINGS on 17 and 21 August 2006

1. Introduction

Various experiments have definitely demonstrated that one-micron accuracy (0.″06) on the definition of stellar images on CdC plates cannot be claimed, as it was speculated back in 1999. More realistically, a 2-3 micron accuracy is achievable, getting worse toward the survey magnitude limit, with an average magnitude error of 0.3. The level of astrometric accuracy corresponds to a 0.″2 - 0.″3 error in position at Epoch 1900, which, once used as first Epoch for proper motion determination in combination with modern epoch observations, can produce errors at the level of 2-5 mas/yr, thereby allowing to detect stellar motions larger than 0.″01/yr, which at a distance of 500 pc from the Sun correspond to $\sim$25-60 km/s tangential velocity. Therefore, the AC/CdC heritage collection can be regarded as a highly valuable first-epoch material, e.g., for the realization of a Tycho-2 extension to fainter magnitudes ($\sim$15 photographic), especially in selected areas where radial velocity data are available, for the exploration of stellar kinematics beyond our solar neighborhood.

In the following I present a possibly non-exhaustive list of efforts that, since the last IAU in Prague, have been undertaken to digitize parts of the AC/CdC collections pertaining to the different zones, with the twofold intent of exploiting the scientific potential of these plates and making the digitized images available to the community.

2. Cordoba zone

The AC/CdC plates of the Cordoba zone have been digitized with a commercial scanner (UMAX Astra1220P) at low resolution for identification and quick-and-raw measurements (Calderón *et al.* 2004); positional accuracy is $\sim$1″, photometric accuracy not yet determined. The digitized plates are available to the community; digitization of selected images at high resolution can be made available upon request.

Moreover, Bustos Fierro & Calderón (2003), have developed a method for the measurement and reduction of AC/CdC plates with the use of a CCD camera, which has been tested on a plate of the Cordoba zone. Their results indicate an astrometric error of 0.″2 - 0.″25 for stars covered by the Tycho-2 catalog, but possibly worse for fainter stars.

3. San Fernando zone

As a result of the collaboration between the institutions CIDA (Venezuela) and ROA (San Fernando, Spain), and after several exploratory tests with other techniques and methods to digitize the plates, ROA has completed the scan with a commercial flatbed scanner (AGFA DuoScan F34) of its collection of AC/CdC plates (declination zone -3° to -9°). Every plate has been digitized in two positions (rotated by 90 degrees unto each other) in order to detect possible systematic trends in the measured coordinates. The two images of every plate have been recorded in FITS format on CD-ROM. The data are recorded on CD-R media located at ROA, under control of both collaborating institutions until the resulting astrometric catalog is completed and made available. The reduction of the Carte du Ciel plates is being accomplished in a two-step process. First, the reduction procedures have been developed based on a subset of the plates, covering eight hours in Right Ascension. The reduction pipeline, from raw pixel data to final stellar coordinates, which includes correction for the distortion introduced by the scanner, is completed. Based on the $\sim$400 plates included in this first stage, the internal precision is 3 microns ($\sim 0.''18$). External comparison with the Tycho-2 Catalog indicates a range of accuracy from $0.''2$ up to $0.''4$, depending on the magnitude of the star as well as its distance from the center of the plate.

4. Toulouse zone

Lamareille *et al.* (2003) investigated the use of a commercial scanner to digitize CdC plates of the Toulouse collection for stellar photometry. The tested scanner is an AGFA SNAPSCAN 1236S with resolution 600×600 dpi. Having calibrated the density-to-intensity transformation with the use of about 100 standard stars per plate, the reported accuracies are of the order of 0.2 to 0.4, depending on the location of the target star on the plate.

5. Bordeaux zone

Ducourant *et al.* (2006) have published the PM2000, a catalog of proper motions for 2,670,974 stars, complete to $V = 15.4$ mag. The proper motions were derived from the reduction of 512 CdC plates of the Bordeaux zone scanned at the APM in Cambridge (Rapaport *et al.* 2006) plus modern-epoch observations with the Bordeaux CCD meridian circle. Reported errors are from 1.5 to 6 mas/yr, depending on magnitude.

6. Sydney zone

Fresneau *et al.* (2003, 2005) have selected, measured (with the APM machine in Cambridge) and analyzed a set of 650 CdC astrographic plates of the former Sydney Observatory along the 4th galactic quadrant. When compared to the GSC 1.2, stars with total annual p.m. larger than $0''.015$ can be considered as 'high' p.m. stars and their distances are derived in order to investigate the differential rotation in the galactic plane up to 500 pc from the Sun. The use of the first Epoch positions provided by these plates in the framework of the Virtual Observatory is currently investigated in order to provide a 'hands-on' experiment when scanning the legacy of the CdC programme in the far southern hemisphere at Macquarie University with a fast scanner.

7. Closing remarks

The Cart du Ciel plate material deserves to be salvaged and digitally recorded to the best accuracy for those astrophysical investigations which can exploit this unique, 100 year-old picture of the sky. In this respect, the scientific interest of such collection is potentially equivalent to that of all the other world-wide astronomical plate archives. For this reason, the goals and objectives of our WG and those of the PDPP task force attached to the IAU Commission 5 on *Astronomical Data* are shared; therefore, to the best interest of both groups, only one WG/Task Force dealing with the preservation and digitization of this heritage from the past should be preserved.

Finally, a sensible issue not yet clearly answered, which would be decisive for a realistic estimation of project resources and timelines, is the level of accuracies achievable from the fits images of AC/CdC plates scanned with commercial scanners. This issue is equally relevant to all astronomical plate collections.

Beatrice Bucciarelli
chair of the Working Group

References

Bustos Fierro, I. H., & Calderón J. H. 2003, *RMAA* 39, 303

Calderón, J. H., Bustos Fierro, I. H., Melia, R., Willimoës, C., & Giuppone, C. 2004, *ApSpSci* 290, 345

Ducourant, C., Le Campion, J. F., Rapaport, M., Camargo, J. I. B., Soubiran, C., Périe, J. P., Teixeira, R., Daigne, G., Triaud, A., Réquième, Y., Fresneau A., & Colin, J. 2006, *A&A* 448, 1235

Fresneau, A., Vaughan, A. E., & Argyle, R. W. 2003, *AJ* 125, 1519

Fresneau, A., Vaughan, A. E., & Argyle, R. W. 2005, *AJ* 130, 2701

Lamareille, F., Thiévin, J., Fournis, B., Grimault, P., Broquet, L., & Davoust, E. 2003, *A&A* 402, 395

Rapaport, M., Ducourant, C., Le Campion, J. F., Fresneau, A., Argyle, R. W., Soubiran, C., Teixeira, R., Camargo, J. I. B., Colin, J., Daigne, G., Périe, J. P., & Réquième, Y. 2006, *A&A* 449, 435

Transactions IAU, Volume XXVIB
Proc. IAU XXVI General Assembly, August 2006
Karel A. van der Hucht, ed.

doi:10.1017/S1743921308023739

COMMISSION 31 — TIME

TEMPS

PRESIDENT	Demetrios Matsaskis
VICE-PRESIDENT	Pascale Defraigne
PAST PRESIDENT	Gérard Petit
ORGANIZING COMMITTEE	Seigfrido Leschiutta
	Gérard Petit
	Mizuhiko Hosokawa
	Zhai Zao-Cheng

PROCEEDINGS BUSINESS MEETING on 21 August 2006

1. Introduction

Most of the Commission's three 90-minute time slots at the General Assembly were devoted to a series of 20 and 50 minute presentations, informally termed "Time and Astronomy". The first part of the meeting was dedicated to time and general relativity. The second part of the session was dedicated to pulsar timing.

2. Appointment of officers for 2006 - 2009

Drs P. Defraigne and R. Manchester were elected as president and vice-president of the Commission for the next term, 2006 - 2009.

3. Scientific session: Time and Astronomy

S. Pireaux presented a basic approach for relativistic equations for future space missions. For spacecraft motions, this approach integrates the relativistic equations of motion numerically with respect to the appropriate metric instead of making relativist corrections to a Newtonian integration. For light, a basic relativistic approach toward laser-links is also relevant to future space missions.

E. Fomalont and S. Kopeikin presented the astrometric test of General Relativity, based on the bending of electromagnetic waves by a moving gravitating body. This was performed by the Very Long Baseline Array in two recent experiments, based on Jupiter and the Sun passing nearly in front of radio sources. They confirmed General Relativity, to a precision three times better than previous VLBI measurements, and the technique promises yet another factor of three improvement in the future.

R. Nelson outlined the fundamental concepts of relativistic time transfer and described the details of the mathematical model. The approximate magnitudes of various relativistic effects for clocks on board the GPS satellites, other representative satellites in Earth orbit, and a clock on the surface of Mars or on the Moon were summarized. A clock on Mars would display period variations with respect to those on earth, with amplitude of 13 ms.

The resolution B3, proposed by the IAU WG on *Nomenclature for Fundamental Astronomy* for the re-definition of Barycentric Dynamical Time, TDB, was then explained by N. Capitaine and discussed by the assembly, with emphasis on the role of TT as a coordinate time, of which TAI or TT(BIPM) are realizations.

G. Petit proposed a comparison between stability of the pulsar time scales and the present stability of TAI and TT; TT(BIPM) has presently accuracy and long-term instability at about

1x10-15 over the recent years. A pulsar's long-term stability may reach a few 10-15, but due to interstellar medium, gravitational effects, geodetic precession, etc., it seems unlikely that a time scale based on pulsars would supersede atomic time scales. However G. Petit emphasized the role of pulsars as the main users of the very long term stability of atomic time scales and in providing possible flywheels to transfer the current accuracy of atomic time to the past, or to the future (if needed).

R. Manchester presented the first results of the Parkes Pulsar Timing Array of which the goal is to detect gravity waves passing over the Earth. As pulsar timing is most sensitive to gravitational waves with frequencies in the nanoHertz region, the most likely astronomical sources gravitational waves that can be detected are binary super-massive black holes in galaxy cores. He noted that the great success of the Parkes pulsar survey should be superseded by an order of magnitude when the Square Kilometer Array comes on line, and the large number of pulsars expected to be discovered may make it possible to create a stable pulsar time scale.

C. Alley presented in detail the Yilmaz theory of Gravity, and its differences with respect to the Einstein theory. This theory adds gravitational binding energy to the stress-energy tensor with the result that black holes do not exist, and the speed of light remains c in a rotating frame. This results in the proper time being independent of latitude, which appears to be more consistent with recent isotropy of one-way speed of light experiments. He is planning a repeat of his one-way speed of light measurement, this time using dark fibers instead of lasers through the lower troposphere.

S. Sheikh and D. Matsakis presented the possibilities of spacecraft position determination using pulsar x-ray signal measurements. The brightest x-ray pulsar by far is in the Crab nebula. If valid timing information from radio observations can be made available, it is possible to use the Crabs X-ray emissions to measure orbits to an accuracy of 500 meters.

I. Stairs proposed some long-term pulsar timing results from Arecibo and Green Bank, as well as some recent scientific results obtained from this database, ranging from neutron-star masses to gravity-wave background limits. She also discussed issues relevant to connecting pulsar data over the several-year observational gaps when Arecibo was redesigned.

Y. Ilyasov presented the millisecond pulsar timing activities at Kalyazin and the resulting database of processed pulsar Times of Arrival (TOA) with refer to the Solar system barycenter for about 10 years period. Although data were single-frequency, he could use second-frequency data from other observatories to model interstellar dispersion.

M. Sazhin discussed the stability of a pulsar time scale based on the long term millisecond pulsar timing, considering the influence of the propagation conditions along the pulsar signal way and the influence of micro-lensing on the TOA, which put natural limits on a pulsar time scale's stability.

4. Laboratory reports

The BIPM time section report was presented by F. Arias, and the report of the time department of NICT was presented by M. Hosokawa. The BIPM reports an overall frequency accuracy of TAI is about 10-15. NICT, formerly CRL (Japan), was recently reorganized. They have great improved their infrastructure and timescale algorithms, so that UTC(NICT) has also improved.

5. Commission web pages

Most of the viewgraphs that were presented in the session can be found on Commission 31s web pages, which as of this submission are in the process of being relocated to the Royal Observatory of Belgium.

Demetrios Matsakis
president of the Commission

Transactions IAU, Volume XXVIB
Proc. IAU XXVI General Assembly, August 2006
Karel A. van der Hucht, ed.

doi:10.1017/S1743921308023740

DIVISION II SUN AND HELIOSPHERE

SOLEIL ET HELIOSPHERE

Division II provides a forum for astronomers studying a wide range of problems related to the structure, radiation and activity of the Sun, and its interaction with the Earth and the rest of the solar system.

PRESIDENT	**David F. Webb**
VICE-PRESIDENT	**Donald B. Melrose**
PAST PRESIDENT	**Arnold O. Benz**
BOARD	**Thomas J. Bogdan, Jean-Louis Bougeret, James A. Klimchuk, Valentin Martinez-Pillet**

DIVISION II COMMISSIONS

Commission 10	**Solar Activity**
Commission 12	**Solar Radiation and Structure**
Commission 49	**Interplanetary Plasma and Heliosphere**

DIVISION II WORKING GROUPS

Division II WG	**Solar Eclipses**
Division II WG	**Solar Interplanetary Nomenclature**
Division II WG	**International Solar Data Access**
Division II WG	**International Collaboration on Space Weather**

PROCEEDINGS BUSINESS MEETING on 21 August 2006

1. Introduction

The scientific fields represented by Division II, solar and heliospheric physics, were well represented at the IAU XXVI General Assembly in Prague, 2006. The Division sponsored or co-sponsored a total of five meetings at the GA, four Joint Discussions and one Special Session. The JDs were: JD01 "Cosmic Particle Acceleration; from Solar System to AGN", JD03 on "Solar Active Regions and 3D Magnetic Structure", JD08 on "Solar and Stellar Activity Cycles", and JD17 on "Highlights of Recent Progress in the Seismology of the Sun and Sun-like Stars". The Special Session 5 was on "Astronomy for the Developing World" and included a subsession on the International Heliophysical Year. In addition, Dr Alan Title presented an Invited Discourse on "The Magnetic Field and its Effects on the Solar Atmosphere as Observed at High Resolution". Thus, there was no need to have a separate scientific meeting of Division II.

On August 21, 2006 an open business meeting of Division II was held during most of the day. About 50 members attended. The meeting included reports from the three Commissions, from the four Working Groups, from the pertinent IAU Representatives, and other matters. This report summarizes the discussions and conclusions. No separate Commission meetings were held. Two of the four Working Groups, Solar Eclipses and International Collaboration on Space Weather, held separate meetings at the GA and these reports appear elsewhere in this volume.

The summary reports of Division II and its Commissions and Working Groups over the period 2003–2006 can be found in *Reports on Astronomy, Transactions IAU Volume XXVIA* (2007, Ed. O. Engvold).

2. Division report

The proposed incoming officers of Division II were presented by the President and confirmed by the attending members. These included for the first time a Division Secretary, Lidia van Driel-Gesztelyi. The Chairs of the four Working Groups and the pertinent IAU Representatives were likewise presented and confirmed by the members. All of these positions were approved by the EC at the General Assembly.

Pertinent results from recent EC meetings were presented and discussed. These included the changes and improvements to the IAU website, improvements in the Commission email address lists, election of new IAU and Commission members, and the revisions in the IAU statutes and by-laws. The importance of updating, maintaining and making effective use of the e-mail lists of Commission members was emphasized. IAU headquarters are trying to assure that the addresses are correct and up to date, but there is room for improvement. We need to request that the e-mail lists of all Commission members be sent to the incoming Presidents. It was suggested that the e-mail lists be routinely tested and email newsletters be sent out on a regular basis. It was proposed that the Division ask the IAU to make aliases for the full Commission membership e-mail lists, which would make contact with the members considerably easier. The new IAU database should allow such alias creation. It was also suggested that better use be made of the web-pages of the Division and of the Commissions. The President noted that the IAU no longer hosts websites of the Divisions or Commissions; the hosting of these must be arranged by each of the Presidents. It is an inefficient system since it means that every three years new sites have to be found and the files transferred.

On elections it was suggested that the Commissions make the elections of new members of the Organizing Committees for the next term, 2009 - 2012, more democratic by widely advertising the nomination process in advance. This process should begin at least half a year before the next GA and could utilize the email/newsletter system. However, since there are currently 200-400 members in each Commission, this system should be tested and streamlined before then.

A sunset clause for Commission OC members was discussed. It was suggested such that no member should have more than three consecutive terms, or that only one third of the OC be renewed after each term, and that inactive members should be removed after one term.

3. Commission reports

Each of the Presidents of the three Commissions summarized the scientific progress in their fields as presented in the Commission reports in the triennial Reports on Astronomy (Transactions IAU Volume XXVIA, 2007, Ed. O. Engvold).

Commission 10 studies various forms of solar activity, including networks, plages, pores, spots, fibrils, surges, jets, filaments/prominences, coronal loops, flares, coronal mass ejections (CMEs), solar cycle, microflares, nanoflares, coronal heating etc., which are all manifestation of the interplay of magnetic fields and solar plasma. Increasingly important is the study of solar activities as sources of various disturbances in the interplanetary space and near-Earth "space weather".

Over the past three years a major component of research on the active Sun has involved data from the *RHESSI* spacecraft and other current and planned solar observations from spacecraft. Significant advances were made in these areas: solar flares with emphasis on new results from RHESSI, two theoretical concepts, magnetic reconnection and magnetic helicity, coronal loops and heating, the magnetic carpet and filaments, and coronal mass ejections and space weather.

Commission 12 covers research on the internal structure and dynamics of the Sun, the "quiet" solar atmosphere, solar radiation and its variability, and the nature of relatively stable magnetic structures like sunspots, faculae and the magnetic network. In large part, the solar magnetic field provides the linkage that connects these diverse themes. The same magnetic field that produces the more subtle variations of solar structure and radiative output over the 11-year activity cycle is also implicated in rapid and often violent phenomena such as flares, coronal mass ejections, prominence eruptions, and episodes of sporadic magnetic reconnection.

The last three years brought significant progress in nearly all the research endeavors touched upon by Commission 12. The underlying causes for this success remain the same: sustained advances in computing capabilities coupled with diverse observations with increasing levels of spatial, temporal and spectral resolution. Significant advances were made in these areas: Irradiance and its Variability, Helioseismology, Surface Magnetism, Structure of the Quiet Sun Chromosphere, Dynamo and Magnetoconvection, Solar Instrumentation, and Steady Coronal Structure.

Commission 49 covers research on the solar wind, shocks and particle acceleration, both transient and steady-state, e.g., corotating, structures within the heliosphere, and the termination shock and boundary of the heliosphere. During the last three years there was considerable progress made in studies of solar energetic particles, compositional and other signatures in the heliosphere, solar wind pickup ions, the termination shock, which was finally crossed by a spacecraft, and the boundary between the heliosphere and interstellar medium, and in solar wind modeling and space weather. Observations from the following spacecraft were extensively used during this period: *Ulysses*, *Cassini*, *Voyager 1* and *2*, *MESSENGER*, *ACE*, *Genesis*, *SOHO*, *Wind*, and *RHESSI*.

4. Working Group reports

The four Working Groups of the Division are *Solar Eclipses*, chaired by Jay Pasachoff, *Solar and Interplanetary Nomenclature*, chaired by Edward Cliver, *International Solar Data Access*, chaired by Robert Bentley, and *International Collaboration on Space Weather*, chaired by David Webb. Each has its own website which is linked under the general Division II site. At the Prague GA each of the Working Groups were extended for at least another three years.

The IAU Working Group on *Solar Eclipses* advises a variety of astronomers and public organizations about the total, annular, and partial solar eclipses visible around the world, and about how to observe them safely. Their Website contains links to articles about eclipses and how to observe them and to maps and guides. Working Group members Fred Espenak (USA) and Jay Anderson (Canada) provide a NASA Technical Publication every few years about the next major eclipse. Michael Gill maintains a Solar Eclipse Mailing List, with daily exchanges of many messages among amateur and professional eclipse observers. The total solar eclipses during the current triennium were in Antarctica (22 November 2003), and across northern Africa, the Mediterranean, mid-Turkey, Russia and Kazakhstan (29 March 2006). IAU Symposium No. 233 was planned in Cairo immediately following this eclipse. Annular eclipses were in Pacific/Panama/northern South America (8 April 2005) and Spain and Africa (3 October 2005). Partial phases are visible in countries hundreds of kilometers to either side, and provide an excellent opportunity for public education, coordinated with IAU Commission 46 on Education and Development.

The Working Group on *Solar and Interplanetary Nomenclature* is chaired by Edward Cliver (USA) and includes Jean-Louis Bougeret (France), Hilary Cane (Tasmania), Sara Martin (USA), Reiner Schwenn (Germany), and Lidia van Driel-Gestelyi (France, UK, Hungary). With the help of the broader community, the WG identifies terms used in solar and heliospheric physics that are thought to be in need of clarification, and then Commissions topical experts to write essays reviewing the origins of terms and their current usage or misusage. The essays are published in EOS, the weekly publication of the American Geophysical Union. The essays published during this triennium were Gradual and Impulsive Solar Energetic Particle Events by Cliver & Cane in 2002, Magnetic Storm – Still an Adequate Name by Daglis in 2003, Terminology of Large-Scale Waves in the Solar Atmosphere by Vrsnak in 2005, and The Last Word: The Definition of Halo Coronal Mass Ejections by St. Cyr *et al.* in 2005.

A new Working Group on *International Solar Data Access* was formed in 2003 to coordinate all solar data access efforts but mainly the international efforts being made on virtual solar observatories. It is chaired by Robert Bentley and has the following members: Frank Hill (USA), Neal Hurlburt (USA), Helen Coffey (USA), Andre Csillaghy (Switzerland), Nadge Meunier (France), Kevin Reardon (Italy), Masumi Shimojo (Japan), Hongqi Zhang (China), Alexander Stepanov (Russia), Luis Sanchez-Duarte (ESA), Dominic Zarro (USA), Aaron Roberts (USA), Adam Szabo (USA), Chris Harvey (France), Chris Perry (UK), David Webb (USA), Franoise Genova and Bob Hanisch (IVOA). The group now covers not only the solar part of Division II but also includes the heliosphere to ensure interoperability among data sets needed to support Space Weather and related studies. Five virtual observatory initiatives from the solar and heliospheric communities are involved in the group. Discussions have focused on data models, descriptions of data resources, and coordinate systems. At least one member of the International Virtual Observatory Alliance (IVOA) participates in the Working Group.

The Working Group for *International Collaboration on Space Weather* has as its main goal to help coordinate the many activities related to space weather at an international level. It is chaired by David Webb. The site currently includes the international activities of the *International Heliospheric Year* (IHY), the *International Living with a Star* (ILWS) program, the CAWSES (Climate and Weather of the Sun-Earth System) Working Group on *Sources of Geomagnetic Activity and Space Weather studies in China*.

The *International Heliospheric Year* is an international program of scientific collaboration being planned for the time period around 2007, the 50th anniversary of the International

Geophysical Year. The physical realm of the IHY encompasses all of the solar system out to the interstellar medium, representing a direct connection between *in-situ* and remote observations. Complete and updated information on the IHY can be found at the main IHY site: `<http://ihy2007.org>`. Nat Gopalswamy is the Chair of the IHY subgroup within the Division II Space Weather WG. David Webb is the IAU Representative for the IHY. Four key activities are being promoted under the IHY program: science activities, the UN Basic Space Sciences (UNBSS) initiative, IGY Gold, and public outreach activities. The science activities are centered around Coordinated Investigation Programs (CIPs), campaigns on focused topics of heliophysical interest in 2007–2008. The IHY disciplines are solar, solar-terrestrial, heliospheric, climate, and atmosphere/ionosphere/magnetospheric sciences. The CIPs will involve as many instruments from space and the ground as possible from around the world. The United Nations IHY effort is being led by Hans Haubold under the UNBSS program. The UNBSS activities include deployment of scientific instruments in developing nations for space science investigations by scientists from developed nations and annual workshops. Within the IAU the IHY was discussed in Special Session 5 at the Prague GA in 2006 on support for astronomy education and research in developing countries. The IGY Gold program recognizes scientists around the world who worked for IGY 1957 programs. Public outreach activities include spreading knowledge of space science and exploration to the public and inspiring the next generation of space scientists.

The CAWSES Working Group on *Sources of Geomagnetic Activity*, also chaired by Nat Gopalswamy, has as its objectives to understand how solar events, such as CMEs and high speed streams, impact geospace by investigating the underlying science and developing prediction models and tools. Other members of this group are B. Jackson (USA), V. Obridko (Russia), A. Prigancova (Slovakia), B. Schmieder (France), K. Shibasaki (Japan), D. Webb (USA), and S. T. Wu (USA).

Finally, the Working Group on *Space Weather Studies in China* is chaired by Jingxiu Wang. As a member of the International Space Environment Service (ISES) Regional Warning Centers, RWC-China provides users with various services, and plays an important role in space environment services for the Chinese Shenzhou series of space missions. Projects initiated to enhance space weather observations include the Geospace Double Star Program (DSP), with coordinated measurements between DSP and Cluster II providing important new results. Solar space projects being developed include the Chinese *Space Solar Telescope* (*SST*), the *Small Explorer for Solar Eruptions* (*SMESE*), and the *Kuafu* Project, a set of three satellites to monitor space weather. The Meridian Chain Project is a Chinese multi-station chain along the 120 E meridian for monitoring space environment variations. Basic research on space weather is supported by the National Natural Science Foundation of China (NSFC), the Chinese Academy of Sciences, and the Ministry of Science and Technology of China. The "Space Weather Research Program" was selected by the NSFC as one priority research area in 2001–2005 and has been extended for another five years. The Chinese Academy of Sciences strongly supports basic research on solar and space sciences aimed at establishing the physical foundation of space weather forecasting.

5. Reports from IAU representatives

The following members of Division II are the current IAU Representatives to these scientific organizations: Oddbjorn Engvold, International Council for Science (ICSU); Marek Vandas, COSPAR Scientific Commission D on Space Plasmas in the Solar System; Arnold Benz, COSPAR Scientific Commission E2 on The Sun as a Star; Helen Coffey, International Space Environment Service (ISES); Nat Gopalswamy, Scientific Committee on Solar-Terrestrial Physics (SCOSTEP); and David Webb, The International Heliophysical Year (IHY).

Helen Coffey from NOAA NDGC presented a report on the ISES. ISES is a permanent service of the Federations of Astronomical and Geophysical Data Analysis Services (FAGS). ISES encourages and facilitates near-realtime international monitoring and prediction of the space environment. It accomplishes this task through: 1) the International URSIgram Service providing standardized rapid free exchange of space weather information and forecasts through its Regional Warning Centers (RWC); 2) preparing the International Geophysical Calendar each year, which gives a list of World Days during which scientists are encouraged to carry out their experiments; and 3) the monthly Spacewarn Bulletins that summarize the status of satellites in earth orbit and in the interplanetary medium.

Coffey and Elena Benevolenskaya discussed long-term databases, such as synoptic solar magnetic field maps, noting that they are very important and should be maintained. An international statement is needed on the importance of maintaining long-term stable data sets, which could be used for example for the calibration of newer data sets. Proposing such an IAU resolution for this GA had been discussed earlier with GS Engvold, but was dropped due to lack of time and uncertainty as to its usefulness. Tom Bogdan pointed out that with limited resources, these programs and data sets should be prioritized by Division II and its Commissions to help guide the bureaucrats in funding decisions.

Nat Gopalswamy gave brief reports on SCOSTEP and on CAWSES, Climate And Weather of the Sun-Earth System. SCOSTEP is an ICSU organization that organizes and conducts international solar-terrestrial programs of finite duration in cooperation with other ICSU bodies. CAWSES is SCOSTEP'S comprehensive international STP program for 2004–2008. Its aim is of significantly enhancing our understanding of the space environment and its impacts on life and society. The main functions of CAWSES are to help coordinate international activities in observations, modeling, and applications crucial to achieving this understanding, to involve scientists in both developed and developing countries, and to provide educational opportunities for students of all levels. It is comprised of four scientific themes represented by Working Groups: WG-1: Solar Influence on Climate, WG-2: Space Weather: Science and Applications, WG-3: Atmospheric Coupling Processes, and WG-4: Space Climatology. Gopalswamy is the new IAU Representative for SCOSTEP, replacing Brigitte Schmieder. The Division thanks Brigitte for her long and excellent service in this position. Gopalswamy also chairs the CAWSES Working Group on Sources of Geomagnetic Activity, a subgroup of WG2.

Dave Webb gave a report on IHY activities within the IAU (see above). He reminded everyone of the meeting of the Division II WG on International Collaboration on Space Weather later during the GA (see the report of this WG elsewhere in this volume).

Marek Vandas gave a report on the solar-heliospheric activities within COSPAR. The science of Division II is represented on two Scientific Committees in COSPAR, D on Space Plasmas in the Solar System and E2 on The Sun as a Star. A. Hady commented that if COSPAR has an IAU Representative then IAGA should also. IAGA, the International Association of Geomagnetism and Aeronomy, is an important international scientific association promoting the study of terrestrial and planetary magnetism and space physics, and is one of the seven Associations of the International Union of Geodesy and Geophysics(IUGG). The next IAGA meeting is with the IUGG General Assembly in Perugia, Italy, 2–13 July 2007. Webb agreed and noted that the Division should take an action through the EC to contact IAGA about this.

6. Other matters

There were a number of other matters that were discussed at the meeting:

6.1. Honoring Z. Svetska and C. de Jager

The Division acknowledged the wonderful and lifelong contributions of Drs Svestka and De Jager to solar physics in general and specifically for starting and maintaining the journal *Solar Physics* over all these years. Svestka was feted at a special luncheon during the GA and in several articles in one of the GA Newsletters. During the Division meeting D. Webb, O. Engvold, L. van Driel-Gesztelyi, J. Klimchuk and F. Farnik gave tributes to these men.

6.2. Status of the journal Solar Physics

Engvold, van Driel-Gesztelyi, and J. Leibacher discussed the status of the journal, *Solar Physics*. These three are either current or past editors of the journal. A year ago the journal was acquired by Springer and it has undergone some problems in the transition. Things are improving, such as that the time from submission to publication of a paper has been reduced to six months on average. Color illustrations are now free for the electronic versions of papers.

6.3. Proposals for new Working Groups

The new Commission 12 President, V. Martinez-Pillet, had considered a new WG on calibration of methods of local helioseismology. However, this proposal was withdrawn because there is already such an initiative in Europe. He asked whether the IAU financially supports WG activities, and Webb answered no but that regional meetings could be organized on WG topics that IAU would support. The new Commission 10 President, J. Klimchuk, commented that the Division should have something like a political action committee to advise and influence the major funding agencies regarding solar-related science. K. Schrijver proposed a WG which would serve as an "advisory committee" under the IAU umbrella informing funding agencies on the importance of space and ground-based facilities, proposals, etc. D. Melrose noted that there are cultural differences between national agencies (e.g., the U.S. *vs* Japan) in their reaction to such influence and that any such approach should be be based on science not politics. J. Leibacher commented that this would have to be an independent international advisory body, different from the advisory committees of the funding agencies. Klimchuk said he would ask Schrijver to draft a "white paper" and lead the WG. A. Hady also suggested that we should have a WG on solar activities in developing countries.

6.4. IAU meeting proposals and publication procedures

O. Engvold, the outgoing GS, noted that there are no longer any IAU Colloquia but just nine IAU Symposia per year, and about 25 scientific meetings at the GA every three years. 25,000 Swiss Francs are provided for each IAU Symposium. For future meetings Webb noted that the Division is coordinating one IAU Symposium for next year, IAU Symposium No. 247 on Waves and Oscillations in the Solar Atmosphere: Heating and Magneto-Seismology in Porlamar, Isla de Margarita, Venezuela, 17 -22 September 2007. The deadlines for the proposed meetings for 2008 are very soon: letters-of-intent are due by 15 September and the full proposals by 1 January (it was later determined that these were due by 1 December!) The proposals should include strong scientific rationale, a draft program, tentative speakers and SOC lists (balanced for gender and geography), and any e-mail correspondence with DPs and CPs. The Division must rank the proposals every year. D. Melrose noted that in the years of the GA, the proposal meetings must cover more than one area. Several meeting proposals for 2008 were discussed. Nat Gopalswamy mentioned a proposed IHY meeting with the place yet unknown and no further details developed yet. S.S. Hasan mentioned one he would lead with Eric Priest that would be held in November 2008 possibly in Bangalore, India on Magnetic coupling between the interior and the atmosphere of the Sun. This is a well-developed proposal.

Problems that editors have had with getting Symposia proceedings published were discussed. Webb reviewed the problem areas that the editors of the two Division-coordinated Symposia in 2004, IAUS 223 and 226, had with the CUP publisher. The editors noted that editors need considerable help in preparing each proceedings. They had to do nearly all the work themselves, including getting the papers refereed, and that the three-month deadline after the meeting for submitting all the papers to the publisher was very short! Also, it was noted that only 40 pages are allowed for any of the Joint Discussions, no matter how many days they were. An action was agreed upon to bring up the problems with the proceedings with the new IAU General Secretary, Karel A. van der Hucht.

It was mentioned that the Division and Commissions devoted much effort to writing the topical reviews for the IAU Reports on Astronomy, but that these are not readily available to anyone, even the authors! Engvold noted that the IAU Transactions (XXVI A) reviewing the previous triennium have still not been published! The Division concurred that these papers should be made freely available on line. Engvold said that the contract with the publisher allows electronic publication of the Reports, and that they will be linked to ADS in fully downloadable form. Van Driel-Gesztelyi said that the reference lists in the Reports are important, so the ADS should be provided not only with the PDF file, but also the electronic reference list for each report. The latest RoA papers should be available soon. It was suggested that we put these reports on the pertinent Division or Commission websites, at least the final draft versions, and an action was taken to do this. Engvold noted that the symposia volumes should also be available on the ADS website.

6.5. OC election procedures

We discussed the current election procedures for both the Commissions and Division. Webb noted that procedures are fairly arbitrary but seem to result in good officers. He suggested that we find a way to make the elections truly democratic by including as many of the Commission members as possible. It is possible now to do this by making full use of the IAU email system, but this will be up to each CP. It was also noted that the Commission OCs should periodically review their membership lists and those of the IAU to make sure that the key scientists they want on the Commission are now or become IAU members and list the Commission(s) of choice. New members can only be approved at the GAs, so this process takes time.

David F. Webb
president of the Division

Transactions IAU, Volume XXVIB
Proc. IAU XXVI General Assembly, August 2006
Karel A. van der Hucht, ed.

doi:10.1017/S1743921308023752

DIVISION II / WORKING GROUP INTERNATIONAL COLLABORATION IN SPACE WEATHER

CHAIR David F. Webb
MEMBERS Nat Gopalswamy, William Liu, David G. Sibeck, Brigitte Schmieder, Jingxiu Wang, Chi Wang

PROCEEDINGS BUSINESS MEETING on 24 August 2006

1. Introduction

The IAU Division II WG on International Collaboration in Space Weather has as its main goal to help coordinate the many activities related to space weather at an international level. The WG currently includes the international activities of the *International Heliospheric Year* (IHY), the *International Living with a Star* (ILWS) program, the CAWSES (Climate and Weather of the Sun-Earth System) Working Group on *Sources of Geomagnetic Activity, and Space Weather Studies in China*. The coordination of IHY activities within the IAU is led by Division II under this working group. The focus of this half-day meeting was on the activities of the IHY program. About 20 people were in attendance. The Chair of the WG, David F. Webb, gave a brief introduction noting that the meeting would have two parts: first, a session on IHY activities emphasizing IHY Regional coordination and, second, a general discussion of the other programs of the WG involving international Space Weather activities.

2. IHY activities

The *International Heliospheric Year* is an international program of scientific collaboration being planned for 2007, the 50th anniversary of the International Geophysical Year. The physical realm of the IHY encompasses all of the solar system out to the interstellar medium. The main IHY site is: <http://ihy2007.org. Nat Gopalswamy is the Chair of the IHY subgroup, and David Webb is the IAU representative for the IHY. Four key activities are being promoted under the IHY program: science activities, the UN Basic Space Sciences (UNBSS) initiative, IGY Gold, and Public outreach activities. The science activities are centered around Coordinated Investigation Programs (CIPs), campaigns on focused topics of heliophysical interest in 2007-2008. The CIPs will involve many instruments from space and from the ground around the world. The United Nations IHY effort is being led by Hans Haubold under the UNBSS program. The UNBSS activities include deployment of scientific instruments in developing nations for space science investigations by scientists from developed nations and annual workshops. The IGY Gold program recognizes scientists around the world who worked for IGY programs in 1957. Public outreach activities include spreading knowledge of space science and exploration to the public and inspiring the next generation of space scientists. The IHY was discussed in Special Session 5 at the Prague GA on 22 August 2006 on support for astronomy education and research in developing countries.

As chair of the WG's IHY subgroup, Nat Gopalswamy gave an overview of the current status of IHY activities. He gave a dedication to the memory of Professor James van Allen, the U.S. space pioneer who died recently. Gopalswamy noted that the IHY consolidates the 50 years of achievements in Space Science following the IGY in 1957 into a knowledge base called Heliophysics. The new word Heliophysics reflects the extension of Geophysics to the current physical scale of in situ investigations. Other daughters of the IGY include Global Geological Campaigns under the International Year of Planet Earth (IYPE), the Third International Polar Year (IPY), and the Electronic Geophysical Year (eGY). IHY has cooperative programs with each of these

*HY's. As the IAU Representative for SCOSTEP, Gopalswamy discussed SCOSTEP activities related to IHY, especially its Climate And Weather of the Sun-Earth System (CAWSES) program which runs from 2004-2008.

The IHY is organized into seven regions worldwide. The coordinators of three of these regions that were in attendance described IHY activities in their region. Jean-Louis Bougeret (with C. Briand), Europe, noted that this region has representatives in about 20 countries. They had their first General Assembly in Paris in January, 2006 and the proceedings have been published. The 2nd European GA is being organized by the Italian Committee (Chair, E. Antonucci) to be held in Torino, Italy, 18-22 June 2007. European IHY activities include developing a DVD about IHY-related scientific topics in three parts: The Heliosphere, The Aurorae, and Space Probes, having an "Open Doors Day" on June 10, 2007, and having science competitions for secondary schools. At the national level activities are being planned in the Czech Republic, Belgium, France, Germany, Ireland, Italy, Bulgaria, Spain, and Turkey/Balkans.

C. Mandrini (with J.-P. Raulin), Latin America, presented the many IHY activities being planned there. The most active countries are Argentina, Brazil, Peru and Mexico. Fourteen institutions are involved in national observatories, national research institutes, laboratories and/or university departments, and 47 principal investigators and their accompanying research groups are involved in instrumentation or observations. IHY Scientific Disciplines include: 1) the Heliosphere and Cosmic Rays: modulation of cosmic rays by the Sun, detection of solar neutrons during large flares, detection and imaging of large-scale solar wind disturbances, and early detection of solar events; 2) Solar Physics: flares, CMEs, active regions, particle acceleration, radiation mechanisms (optical and radio wavelengths); 3) Planetary Ionospheres, Thermospheres, Mesospheres and Climate studies: Ionosphere studies- D-region perturbations, equatorial and mid-latitude quiet and perturbed ionosphere, ionospheric plasma instabilities; and 4) Geomagnetism: Earth magnetic field measurements, South Atlantic magnetic anomaly monitoring, solar induced geomagnetic activity. Regional planning meetings were held in Sao Paulo, Brazil in December 2005, during the SCOSTEP 11th Quadrennial Solar Terrestrial Physics Symposium in Rio de Janerio, Brazil in March 2006, and in Puerto Vallarta, Mexico, November 2006. The region has five CIP programs, and is supporting the SAVnet and SAMA UNBSS programs and is proposing a magnetometer network at the IHY-UNBSS workshop in November 2006. Finally, many outreach and educational activities are being promoted.

A. Stepanov (with V. Obridko), Russia and Eastern Europe, showed many examples of ground-based facilities that are making high quality optical, spectral and radio observations of the Sun and heliosphere. Some of the largest specialized radio antenna systems in the world are in this region including Ratan 600, the Big Pulkova radio telescope and the SSRT. There are also a host of space missions both in operation and being planned. These include Photon, the next in the series of Coronas missions, to be launched late in 2007. An educational website is being organized. The IHY in Russia will be featured in Moscow in October 2007 starting with events celebrating the launch of Sputnik on 4 October 1957. Worldwide this is called the World Space Week 2007 and purports to be the largest public space event in world history, with celebrations in over 50 nations commemorating the 50th anniversary of the Space Age. The Russian IHY group is planning a related IHY meeting in Moscow 8-14 October 2007.

B. Thompson, IHY Director of Operations and Website Coordinator, presented details about the IHY website and the CIPs. She noted that over 45 CIPs have already been proposed. Thompson also leads the IGY Gold (History) program that recognizes scientists around the world who were active in IGY in 1957. There are now more than 50 IHY Gold recipients.

C. Rabello-Soares, IHY Education and Public Outreach Coordinator, presented the many EPO activities being pursued. These start with the IHY/UNBSS instrument program which provides wonderful and varied opportunities for EPO activities in developing countries. The IHY outreach program is dedicated to bringing heliophysics education to all the people of the world. Its goals are to assist in increasing the visibility and accessibility of existing programs, providing greater international exposure for existing programs, promoting new international partnerships, determining the need for multi-lingual adaptations of educational resources and facilitating their translation, and leaving a world-wide legacy of connectivity between those engaged in Education and Outreach efforts related to IHY science. Each IHY National Coordinator for Education and Outreach will act as the liaison between those working on education and public outreach in their country and the IHY secretariat, as well as to the national coordinators of the other nations. Currently 18 National EPO Coordinators have been appointed. We are planning to create a resource CD of selected exemplary education and outreach materials from around the world. It will have a companion booklet, and both will be distributed by the UN to all Member States by the end of 2007. An EPO planning meeting, "Globalizing Space Science Education and Outreach", was held during the COSPAR Assembly in Beijing, July 2006. One of the goals is to make IHY multilingual resources available worldwide. Solar eclipses are an excellent way to promote IHY EPO activities. The March 2006 eclipse in Africa and Europe provided a testbed for such activities and was very successful. Viewing stations with EPO materials for viewing eclipses were set up along the path of totality in these countries: Brazil, Ghana, Nigeria, Egypt, Turkey, Georgia and others. An IHY Schools Program is also part of the IHY EPO effort (see below).

David Webb discussed the IAU's involvement in IHY. These activities were mentioned above. Attempts to get IAU support for funding for the IHY Schools and for the IHY/UNBSS workshops was noted. We need to increase our collaborative efforts with other affiliated groups such as CAWSES, SCOSTEP, European COST 724, NASA, and NSF. Webb discussed with M. Gerbaldi, chair of the IAU's ISYA (International School for Young Astronomers) program, the possibility of linking the ISYA and IHY programs, at least in 2007-08, for travel of students and young astronomers to the IHY schools. We also discussed with Peter Willmore the idea of collaborative educational efforts with COSPAR. Since the IAU is a sponsor of the IHY/UNBSS workshops, it is possible to use IAU travel funds to support students to come to those meetings.

The IHY Schools Program is organized by an IHY Schools Committee (ISC) chaired by David Webb. It has been established to provide undergraduate and graduate students with expert training and understanding of heliophysical universal processes. Four main IHY schools are being planned in 2007 for North America (Boulder, CO, USA), Europe/Africa (Trieste, Italy), Latin America (Brazil), and somewhere in the Asian Pacific region. A website for the Schools program is at: `<http://www.ihy2007.org/outreach.shtml>`. Funding for the schools is an issue both for travel support for students attending the schools and for the support of the schools themselves. The ISC is in the process of developing a uniform curriculum for the schools. Other IHY-related schools are being organized in various countries and will be coordinated as needed by the ISC. The time line of the schools was discussed; they can occur anytime in 2007 but also in 2008 when IHY data sets and results will be better established. The duration and dates of each school will depend on circumstances in the regions/countries where they are held. The dates of the schools should be separated enough in time from each other and also from major international meetings to permit the same key scientists and IHY personnel to attend.

Possible upcoming IHY-related meetings were discussed. The IHY keeps a very good list of past and future meetings on its website at: `<http://www.ihy2007.org/events/events.shtml>`. Major future IHY meetings include The Second UN/NASA Workshop on the International Heliophysical Year and Basic Space Sciences in Bangalore, India, 27 November - 1 December 2006, the IHY Opening Ceremony and Kickoff Meeting at the United Nations STSC Meeting in Vienna, 19-20 February 2007, the "Open Doors" Day at IHY Observatories and Museums on 10 June 2007, regional meetings in Italy, 24-29 June 2007, and Moscow, 8-14 October 2006, and the third IHY/UN/NASA Workshop in Japan, in November 2006. We will propose an IAU Symposium on IHY science in Greece in 2008. In addition, Division II is sponsoring IAU Symposium No. 247 on "Waves and Oscillations in Solar Atmosphere: Heating and Magneto-seismology" to be held at Porlamar Isla de Margarita, Venezuela, 17-22 September 2007.

On 25 August 2006 at the 82nd EC meeting at the Prague GA, David Webb made a presentation to the EC about IHY activities. These activities are summarized in the final section of the IAU Information Bulletin No. 99 to be published in January 2007.

3. Discussion of other programs of the Working Group

Besides the IHY program, the Space Weather WG also includes the *International Living with a Star* program, the CAWSES WG on *Sources of Geomagnetic Activity, and Space Weather studies in China*. The CAWSES WG is chaired by Nat Gopalswamy who discussed it at our meeting. This WG has as its objectives to understand how solar events, such as CMEs and high speed streams, impact geospace by investigating the underlying science and developing prediction models and tools. CAWSES has directed three observing campaigns and many workshops have been held. IHY planners are working to collaborate with and not overlap with the CAWSES activities. Toward this end a joint IHY-CAWSES Observation Database is being developed.

The next major meeting will be the International CAWSES Symposium in Kyoto, Japan, 23-27 October 2007.

J. Wang (with W. Gan) of the Chinese Academy of Sciences described Space Weather studies in China in what is being called a new epoch in Chinese space science. From 2000-2005 the Chinese government initiated a new, well-funded program of space weather research. This has just been extended through 2010. The program has resulted in a large amount of coordinated observations of the Sun and many new revelations from data analysis. Major goals of the next 5-year phase include exploration of the cause-effect chain relationship in space weather initiation and evolution, developing a new scientific framework of space weather forecast and magnetic synoptic meteorology, identifying and attacking the common physical problems in different plasma domains, e.g., 3D magnetic reconnection, shocks and particle acceleration, and creating new concepts for future Chinese space missions. Identified future missions are: *SMESE* – aimed at observing flares and and Coronal Mass Ejections (CME) for the next solar maximum through wavelength coverage from the gamma rays to the far-infrared; *KuaFu* – aimed at continuous surveillance of the Sun and heliosphere and the geomagnetic response to the Suns explosions, and supporting fundamental research in solar and space-plasma physics underlying space weather; and *SST* – to achieve a breakthrough advance in solar physics through coordinated observations of transient and steady MHD processes with broad wave range, high spatial resolution and continuous temporal coverage. *SMESE* (Small Mission on Exploration of Solar Eruption) is a joint France-China mission. *KuaFu* is a Sun-Earth System Explorer project and is concentrated on exploring the complex global behavior of the Sun-Earth system. It will have three spacecraft, one at the L1 point and two in Earth orbit. *SST*, the Solar Space Telescope, will be characterized by a diffraction-limited optical telescope of 1-meter aperture and a two-dimensional polarization spectrometer including a multi-channel birefringent filter.

We ended the session with a discussion about incorporating into the WG information and contacts of the emerging space weather programs and projects that are being developed in other countries and regions. Examples include the fairly extensive programs in Europe, India and South America. As a start David Webb mentioned that we should begin to add this material to the current WG website. He then led a discussion about what should be on the website and what future projects the WG should be involved in.

David F. Webb
chair of the Working Group

Transactions IAU, Volume XXVIB
Proc. IAU XXVI General Assembly, August 2006
Karel A. van der Hucht, ed.

doi:10.1017/S1743921308023764

DIVISION III PLANETARY SYSTEM SCIENCES

SCIENCES DES SYSTÈMES PLANETAIRES

Division III gathers astronomers engaged in the study of a comprehensive range of phenomena in the solar system and its bodies, from the major planets via comets to meteorites and interplanetary dust.

PRESIDENT	**Iwan P. Williams**
VICE-PRESIDENT	**Edward L. G. Bowell**
PAST PRESIDENT	**Mikhail Ya. Marov**
SECRETARY	**Guy J. Consolmagno**
BOARD	**Michael F. A'Hearn, Alan P. Boss, Dale P. Cruikshank, Anny-Chantal Levasseur-Regord, David Morrison, Christopher G. Tinney**

DIVISION III COMMISSIONS

Commission 15	**Physical Study of Comets and Minor Planets**
Commission 16	**Physical Study of Planets and Satellites**
Commission 20	**Positions and Motions of Minor Planets, Comets and Satellites**
Commission 21	**Light of the Night Sky**
Commission 22	**Meteors, Meteorites and Interplanetary Dust**
Commission 51	**Bioastronomy**

DIVISION III WORKING GROUPS

Division III Service	**Minor Planet Center**
Division III WG	**Committee on Small Bodies Nomenclature**
Division III WG	**Planetary System Nomenclature**

DIVISION III INTER-DIVISION WORKING GROUPS

Division III-I WG	**Cartographic Coordinates and Rotational Elements**
Division III-I WG	**Natural Satellites** (formerly Div.III- C20 WG)
Division III-I WG	**Near Earth Objects**

DIVISION III COMMISSION WORKING GROUPS

Div.III/Comm.15 WG	**Physical Study of Comets**
Div.III/Comm.15 WG	**Physical Study of Minor Planets**
Div.III/Comm.20 WG	**Motions of Comets**
Div.III/Comm.20 WG	**Distant Objects**
Div.III/Comm.22 WG	**Professional-Amateur Cooperation in Meteors**
Div.III/Comm.22 WG	**Task Group, Meteor Shower Nomenclature**

PROCEEDINGS BUSINESS MEETINGS on 18 and 22 August 2006

1. Introduction

This report outlines the discussions and conclusions reached at Division III business meetings during the IAU XXVI General Assembly. Five meetings were held in total, with one of them being devoted entirely to a discussion of the definition of a planet. We do not give an account of this meeting here since reports of it have been widely circulated.

2. Business Meeting, Friday 18 August 2006, 11:00 hr

Division President, Iwan Williams, opened the meeting by welcoming all members to what was going to be an exciting series of meetings.

The minutes of the last meeting of the Division, held during the General Assembly in Sydney, that had been circulated were accepted with correction that the Minor Planets Management Committee did not report to Division III in the period 2003-2006.

The President informed the meeting that a memorandum of agreement between the Smithsonian Astrophysical Institution and the IAU regarding the governance and running of the Minor Planet Center (MPC) had now been signed. According to this, a Minor Planet Committee is formed which reports to and is elected by Division III. Its role is to advise and instruct the Minor Planet Center on its day-to-day operations, and as a way for the MPC to communicate with Division III.

2.1. *Discussion on the future structure within the Division*

Under the present IAU rules, all Commissions are appointed for six years and Working Groups for three years. They then cease to exist unless a case is made for their continuation. The Division Board and Commission Organizing Committees had been discussing future organization over the last triennium and Williams gave a summary of the conclusions. All the recommendations of Division III have been accepted by the Executive Committee (EC). These were:

- Commission 16 continues unaltered for another six years.
- Commission 51 changes its name to *Bioastronomy* and continues for another six years.
- Commission 15 and 20 are renewed for three years each and are encouraged to discuss with each other whether they should then join as a single commission. The default position is that they will combine after three years unless they make a case to the contrary.
- Commission 21 and 22 are renewed for three years and strongly encouraged to plan on merging as one commission by 2009.
- The Working Group on Extrasolar Planets (WG-ESP) becomes a Commission on Extra-solar Planets.
- The Working Group on Cartographic Coordinates and Rotational Elements (WGCCRE) was elevated to a commission by the EC at the recommendation of Division I. The reasons for this were not entirely clear, and there would be further discussion on this issue (it is reported later that this will indeed continue as a Working Group of both Divisions).
- Much of the current remit of Working Group on Near Earth Asteroids (WGNEA) has been taken over by a new working group of the EC so that this Division Working group is discontinued.
- A discussion on whether a replacement structure on NEOs should be formed under Commissions 15 or 20 followed. Further discussion was left for the Commissions in question.
- The Working Group on Planetary System Nomenclature (WG-PSN) will continue.
- The Committee on Small Body Nomenclature (CSBN) will continue.
- There was a request from the Working Group on Natural Planetary Satellites (WG-NPS), currently a WG of Commission 20, to become a joint WG of Division I and Division III (it is also reported later that this was accepted by the EC).

Williams pointed out that streamlining of the Division was desirable. If the Board of the Division includes all current presidents of the Commissions and Working Groups and immediate past presidents, this already more than fills up the canonical 12 places.

2.2. *Reports from Working Groups*

Tichá gave the report of the CSBN: - In the past triennium, 2511 minor planets, 674 comets, and 4 satellites of minor planets have been named. Special note was taken of the minor planets named for the crew of the lost Columbia shuttle, minor planets numbered from 51823 to 51829; the interesting solar system bodies Sedna and Apophis (the latter an NEO which will make a close approach to Earth in 2029) and Romulus and Remus, moons of minor planet Sylvia, making the first known triple minor planet.

- The 5th edition of The Dictionary of Minor Planet Names edited by Lutz Schmael, along with an addendum, has been published. Spahr, Marsden, Fernández, and Williams discussed the issue of attribution for the suggestion of names, particularly for Centaurs. Centaurs are named for centaurs in classical literature; there are very limited number of names (as with Trojans) and any on that list not yet used are appropriate. The attribution of the naming of a particular Centaur with a name from this list by an astrologer reflects the fact that anyone is allowed to propose from this list of names.

Archinal gave a report from the Working Group on Cartography Coordinates and Rotational Elements:
- This Working Group is a joint working group of an organization external to IAU (Geodesy). Changes are under consideration by the working group for the Sun, Moon, Mars, and Saturn, and significant advances have occurred in the coordinates of imaged minor planets and the newly discovered or imaged satellites. A final report should be completed and published by the end of the calendar year.

Morrison gave a brief report from Working Group on Near Earth Asteroids:
- As its role as a Division Working Group is completed and much of the time was spent in discussions regarding the new structure.

Boss gave the report from the Working Group on Extrasolar Planets:
- The remit of the current working group was narrowly defined, namely to define and maintain a list of extrasolar planets. This has been done for the past six years, using the relatively strenuous criterion that the discovery must be published in a refereed journal. Approximately 170 such bodies have been confirmed so far, the majority (approximately 150) from radial velocity measurements on their stars but also including others discovered by pulsar timings, transits, and microlensing. Relatively small mass objects are beginning to be found, including a 5.5 Earth mass (with large uncertainty) being the smallest suggested. In addition, there are two possible planetary mass objects around brown dwarfs; one case is a possibly directly detected object, the other an object that has an uncertain mass (1 to 42 Jupiter masses) and thus the system could actually be a binary brown dwarf.
- Now that there is to be a commission on the study of extrasolar planets, it is expected that a subset of this commission, either a committee or subcommittee, will be needed to maintain this list. In addition, the commission will need to work out how to coordinate its work with other commissions (such as Commission 51) and other Divisions of the IAU.
- The nomenclature for such objects has been debated 'to a standstill' with the competing suggestions either to use new unique names, or continue to use the binary star terminology with lower case letters given in order the planets are discovered. The latter becomes unwieldy when the star itself is not concisely named (which happens especially with planets found during transit searches). There is no consensus. Pet names can be used to talk to the press but these do not have an official IAU designation.
- In response to questions, Boss noted that there is at least one possible binary planet (a 7 Jupiter mass object in orbit around a 14 Jupiter Mass object, which within errors might be planetary), and that in the future a consistent definition will be needed to classify such. It is also to be noted that several dozen stars are known to have more than one planet, with one having at least four planets.

2.3. *Election of next Division III president*

Following the general rule that the vice-president becomes the next president, Ted Bowell was confirmed by acclamation as the next president of Division III.

2.4. *Other business*

Harris noted that the upcoming definition of a planet might have implications for the organization of committees responsible for planet and minor planet nomenclature, and both the WG-PSN and CSBN agreed to discuss different possibilities (including creating an umbrella committee, or having the CSBN be subordinated to the WG-PSN) at their respective Working Group and Committee meetings.

3. Business Meeting, Friday 18 August 2006, 14:00 hr

This session of the Division had specially been reserved for the a discussion of the proposed definition of a planet by members of the Division.

There was a turnout of several hundred members which included the President, President-Elect, designated future President-Elect of the IAU as well as several past or present General Secretaries. Williams remarked in his opening comments that this must clearly be a record for any Division business meeting indicating the wide-spread interest in the topic. There was a lively discussion that has been widely report, and no attempt at reproducing this is given here. However, it should be noted that an alternative resolution to that from the EC and published earlier, emerged from this meeting (and essentially carried at the Final general Assembly)

4. Business Meeting, Friday 18 August 2006, 16:00 hr

The president, Williams, open the session by reminding who those gathered that the main purpose of this session was to discuss priority concerning discoveries of small solar system objects. Their numbers are expected to increase dramatically within the next three years as the next generation of automated surveys come on line.

4.1. *A proposal for a definition proposed by Milani*

Milani explained that his proposal was an attempt to be fair to those who not only discover these objects but provide sufficient follow-up observations to produce a reliable orbit, while at the same time allowing individuals outside the automated programs to be given credit.

A wide ranging discussion following this proposal pointing out the desirability for simplicity, ability to deal with special cases, the encouragement of amateurs, and the need to guard against those who might either prematurely announce discoveries in the hopes that they were correct as opposed to those who keep discoveries secret until full orbits can be established, thus preventing others from doing useful science on those objects.

It was decided that the incoming Division president will discuss the matter with WG-PSN and CSBN and come to a conclusion which will then go to the Division Board for approval.

4.2. *Planets and Dwarf Planets*

Noll began a discussion of how the newly designated planets or dwarf planets are to be named. After significant discussion of the possibilities, Williams decreed that this should properly be held to the next meeting of the Division, on Tuesday, so that this discussion could be announced to the wider membership- given them the opportunity to contribute.

5. Business Meeting, Tuesday 22 August 2006, 14:00 hr

Williams opened the meeting by pointing out that this session should be devoted essential to business, and not be diverted to discuss planets other than the specific discussion initiated by Noll.

5.1. *Elections for 2006-2009*

Following an election amongst the outgoing Board members held by e-mail, the Division confirmed by acclamation the nomination of Karen Meech, the current president of Commission 51, as the incoming vice-president of the Division, to become president in three years time.

By the IAU rules, Commission presidents are appointed by Division. Following the recommendations of the Commissions, the following new presidents were appointed: Commission 15: Huebner; Commission 16: Courtin; Commission 20: Fernández; Commission 21: Witt; Commission 22: Spurný; Commission 51: Boss.

Major was a likely president for the newly established commission on extrasolar planets, but this will need to be discussed by the Commission itself.

The following chairs of Committees and Working Groups were appointed:
WG on Coordinates and Rotation: Archinal.

(Williams stated that he has written to the General Secretary, and will present to the EC that Division III believes the WG-CCRE was elevated to a commission in error and so it was dealt with here on that assumption that it is a Working Group).
CBSN: Tichá.
WG-PSN: Schultz.

5.2. *The Division board*

One issue for consideration was that the IAU rules say that such Boards normally will have 8-12 members. However, Division III has seen the need to exceed 12 members, given the large number of commissions and regular officers.

Of the outgoing commission presidents, only Gustafson and Valsecchi expressed a willingness to serve, the others are already on board by virtue of holding other officers, or have declined to serve. It was proposed by A'Hearn that two people who have not held other major offices within the Division in the last three years should be elected. This was accepted, which generates a board of 15 members.

Nominations were solicited with an eye for both geographical and scientific balance. Watanabe, Marov, Schulz, and Levasseur-Regourd were proposed. Rather than choosing only two, Bowell asked that all four be accepted, as all four are needed for balance. The Division accepted this.

Thus the new Division III Board is: Edward L. G. Bowell (USA, president), Karen J. Meech (USA, vice-president), Iwan P. Williams (past president, UK), Guy J. Consolmagno (secretary, Vatican City State), Board: Walter F. Huebner (USA, C15 president), Régis Courtin (France, C16 president), Julio Fernández (Uruguay, C20 president), Adolf N. Witt (USA, C21 president), Pavel Spurný (Czech Republic, C22 president), Alan P. Boss (USA, C51 president), Michel Mayor (Switzerland, C53 president), Bo A.S. Gustafson (USA, outgoing C21 president), Giovanni B. Valsecchi (Italy, outgoing C20 president), Anny-Chantal Levasseur-Regourd (France, additional), Mikhail Ya. Marov (Russia, additional), Rita M. Schulz (Germany, additional), and Jun-Ichi Watanabe (Japan, additional),

5.3. *Re-organizing Working Groups*

A suggestion had been made that there should be one Working Group to deal with naming. This could have sub-groups dealing with specific object types.

A straw vote between discussing a change or continuing under the present system led to a vote of 19-16 in favor of no change.

Considering the role that both naming groups will have in the naming of the yet-to-be-determined planetary objects, it was agreed that the two committees involved would discuss the matter among themselves. Their meetings, to follow on Wednesday, are open to all members, although voting will be only by the members of the working group or committee.

6. Business Meeting, Thursday 24 August 2006, 11:00 hr

The main purpose of this session was to wind up administrative items, including reports from WGs and Commissions that had not previously reported or that had new items to report.

6.1. *Update on Working Group and Commission status*

Williams reported that on August 23 the EC had accepted that the WG on Cartography and Rotational Elements should stay as a WG, and that the WG on Satellites should become a joint WG of Divisions I and III.

6.2. *Outstanding reports*

WG-PSN given by Aksnes

The WG held a workshop in Norway in September 2005 to revise its guidelines, as reported in IAU Transactions. With the exception of this face-to-face meeting, most of its business was normally done by e-mails.

There were 228 names of surface features approved during triennium, and 24 new satellite names. Formal approval of lists by the Division followed, by acclamation.

6.3. *The matter of Provisional Names*

It was proposed that we should cease to use 'provisional' names since this does not serve a useful purpose but creates extra work for those maintaining the name database. This was agreed to (bringing WGPSN in line with CSBN practice).

6.4. *Minor changes to the Terms of Reference for WG-PSN*

Minor but necessary changes from 2000 version were approved, including the change from a 'president' to a 'chairperson'. It was noted that outside 'consultants' are still needed for the WG to function. Though this class of membership is no longer recognized by the IAU, it was noted that since consultant no longer had an official IAU meaning the term could continue to be used in the ToR as a common descriptive term for these experts. It was also found that task groups of the WG in several cases needed up to seven members, rather than the six originally allowed. In section 4, the words 'major and minor planets' becomes 'planets', and 'comets' becomes 'small solar system objects' in anticipation of likely changes in the definition of a planet by the General Assembly.

Concerning the issue of 'provisional' names, it was proposed that section 4b would be revised to read: "Names will be made available quarterly in an official web site for public review. Any objections to these names based on significant substantive problems must be forwarded in writing or email to the Div III president within three months of the placing of these names on the website. Valid objections do not include personal preferences of the discoverers or other individuals." The results of such appeals will be applied as before by the still-standing terms of section 4c.

It was noted that, given the uncertain status of the definition of a planet, references to dwarf planets, etc. would be premature. The WG-PSN along with the CSBN recommend in the particular case of 2003 UB313, the discoverer will be solicited for an appropriate name, which will be given jointly to the CSBN and WG-PSN for approval; the Division III board and then the EC will have the final say on any name. Bowell noted that this removes a level of bureaucracy and has more people look at it. There will be pressure to name it clearly; the hope is that a name can be approved within one month.

The issue of approaching the discoverer, rather than soliciting names from the public, was accepted unanimously.

6.5. *CSBN matters*

New membership of the committee includes Bowell (ex-officio, Division III president), Schulz (ex-officio, WG-PSN chair) and the addition of new members Syuichi Nakano and Keith Noll. In addition it was agreed to publish both names and citations of names of satellites of minor planets in the MPCs. The committee continues to search for the best way of limiting the naming process, and to concentrate on a smaller number of meaningful names/important bodies, rather than attempting to approve 100,000 new minor planet names. It is also searching for a way to provide a free website containing names and citations; this needs resolving copyright questions with the IAU and Springer, who publish the Dictionary of Minor Planet Names.

6.6. *Commission on Extrasolar Planets*

Williams reported that he had chaired a brief meeting at 12.00 hr to which 25 people showed up who were interested in joining the new Commission. It was agreed that all interested people should e-mails Boss.

6.7. *Other brief statements*

Commission 16: The question was raised if the physical study of Dwarf Planets come under Commission 15 or 16? Either 15 loses Ceres, or 16 loses Pluto. Williams notes that, in any event, the Division eventually will have to deal with the new situation, but that will depend on what passes at the General Assembly.

Commission 21: Progress is being made toward eventual combination with Commission 22; this will be discussed further at a scientific meeting on meteors to be held next year in Spain.

Commission 22 had adopted a formal process for naming meteor showers, and defined a task group to define how individual showers are defined, to come up official names for the next triennium. The Division agreed to this proposal without dissent.

6.8. *Any other business*

The upcoming vote on the definition of a planet at the General Assembly was discussed, and Williams agreed to present the concerns of the Division to the Executive Committee.

Finally, Bowell proposed two votes of thanks. One was to Williams for his stewardship during these tumultuous three years, who replied that "this tumult is why I attend IAUs, I think it's a lot of fun. We've been described in bad terms because we're contentious but that's why we enjoy what we do."

The second person to thank was Brian Marsden who after 28 years at the MPC is to become director emeritus. He noted that Marsden had steered that organization amazingly; from a time when there were only 2,000 numbered asteroids when to 134,000 now, an enormous task. He also noted that the catalog or orbits has gone from 3,000 to 300,000, and we may see a further increase by a factor of 100 in the next ten years.

Iwan P. Williams
president of the Division

Transactions IAU, Volume XXVIB
Proc. IAU XXVI General Assembly, August 2006
Karel A. van der Hucht, ed.

doi:10.1017/S1743921308023776

DIVISION III / WG COMMITTEE SMALL BODIES NOMENCLATURE

LA NOMENCLATURE DES PETITS CORPS CÉLESTES

CHAIR **Jana Tichá**
SECRETARY **Brian G. Marsden (MPC)**
MEMBERS **Daniel Green (CBAT,** for comet names only**),**
Rita M. Schulz (WG-PSN),
Michael F. A'Hearn, Julio A. Fernandez,
Pamela Kilmartin, Daniela Lazzaro,
Syuichi Nakano, Keith S. Noll,
Lutz D. Schmadel (editor of DMPN),
Viktor A. Shor, Gareth V. Williams,
Donald K. Yeomans, Jin Zhu

PROCEEDINGS OF THE BUSINESS MEETING

1. Membership

After a brief discussion it was agreed that this membership is well distributed in terms of geography, gender, and fields of planetary work.

2. Results of previous triennium

The Chair briefly commented on the CSBN work during the triennium. Regarding minor planets, 2511 names were approved and published in the Minor Planet Circulars. 674 comets received names, the great majority of them being SOHO comets. In accordance with its new task, the CSBN also approved the names of five satellites of minor planets. No appeals concerning CSBN decisions were reported to Division III during this triennium. The fifth edition and the Addendum 2003-2005 of the *Dictionary of Minor Planet Names* edited by CSBN member L. D. Schmadel have been published by Springer.

3. Naming of "dwarf planets"

The naming of certain special objects, notably of the "dwarf planet" 2003 UB313 and its satellite, was discussed. The process will consist of the following steps: to ask discoverer M. Brown for reasonable name proposal(s), to discuss the outcome in the CSBN and to transmit the CSBN conclusions to the WG-PSN for further consideration; the preferred name(s) will then be examined by the Division III Organizing Committee, and the choice will be passed to the IAU EC for final adoption; this whole process should be carried out as rapidly as possible because of the intense public interest, with the final name(s) approved within a month if all will go well.

4. Satellites of minor planets

The CSBN agreed to publish both names and citations for satellites of minor planets in the Minor Planet Circulars and the DMPN.

5. Limiting of names

At the previous meeting in Sydney in 2003 the CSBN recognized the need to limit the number of namings of minor planets. It is better to concentrate on having a small number of meaningful names having broad international appeal than to name all the tens of thousands of newly numbered main-belters. During 2003-2005 individual discoverers and teams were requested to propose no more that two names each two months. The majority of them cooperated. Exceptions were allowed in a few cases, but on the whole the restriction seemed to work. At the Prague meeting the discussion on this topic continued. The most important matter seems to be to name unusual bodies of extraordinary scientific and public interest, i.e., Earth approachers, mission targets, transneptunian objects for which there have been detailed physical studies, binary or multiple systems, and so on. The CSBN members agreed to continue to examine ways of limiting the naming process by concentrating on the more significant bodies.

6. Discussion on public access to names and citations

This session of the Division had specially been reserved for the a discussion of the proposed definition of a planet by members of the Division.

The final matter discussed was a website containing minor planet names and citations. We are living in an "internet world" now. Members of the public ask about the meanings of specific names. There are the printed editions of the DMPN (now an official IAU reference work) as well as the batches of Minor Planet Circulars. On the other hand, one can find various websites containing citations that appear at least partly without appropriate references and acknowledgment of copyright. The CSBN members agreed to search for a satisfactory way to establish a perhaps freely available website containing names and citations, bearing in mind in particular the copyright arrangement between the IAU and Springer.

Jana Tichá
chair of the CSBN

Transactions IAU, Volume XXVIB
Proc. IAU XXVI General Assembly, August 2006
Karel A. van der Hucht, ed.

doi:10.1017/S1743921308023788

DIVISION III / WG PLANETARY SYSTEM NOMENCLTURE

LA NOMENCLATURE DU SYSTÈME PLANÉTAIRE

CHAIR **Kaare Aksnes**

MEMBERS **J. Ellen Blue, Juergen Blunck, George A. Burba, Guy J. Consolmagno, Mikhail Ya. Marov, Brian G. Marsden, Tobias C. Owen, Mark S. Robinson, Rita M. Schulz, Bradford A. Smith, Iwan P. Williams**

PROCEEDINGS BUSINESS MEETING on 23 August 2006

1. Introduction

The meeting was attended by six from the WG (K. Aksnes, J. Blunck, G. Consolmagno, B. Marsden, R. Schulz, V. Shevchenko) and two from the Task Groups (D. Morrison, J. Watanabe). Also the incoming WG members E. Bowell and R. Courtin, as well as some guests, attended.

The first session was devoted to a review of the nomenclature for which approval was sought from the IAU through Division III. A total of 252 names have in the triennium been approved by the WG: 228 surface features have been named on Venus (33), Moon (8), Mars (83), Itokawa (3), Io (16), Europa (11), Ganymede (3), Callisto (2), Titan (45) and Phoebe (24), and 24 satellites of Jupiter (10), Saturn (5), Uranus (6), Neptune (1) and Pluto (2) have received names. These 252 names and their attributes are listed at the end of this report. The names and some new name categories and descriptor terms are also listed on the website <http://planetarynames.wr.usgs.gov/>.

In the second session, new terms of reference (see below) for the WG were discussed and agreed upon, in consultation with the president of DivisionIII. The most important change is that names approved by the WG-PSN no longer need to be labeled 'provisional'. New names will immediately upon approval by the WG be displayed on the mentioned website. Any objections to the names must be forwarded in writing or by e-mail to the DivisionIII which will rule on the objections.

Finally, changes in the membership of the WG were agreed on. Rita Schulz becomes the new WG-PSN chairperson, switching place with Kaare Aksnes on the WG. The new Division III president, Edward Bowell, the new Commission 16 president, Regis Courtin, and the new chairperson of the Outer Solar System Task Group, Rosaly Lopes, replace, respectively, Iwan Williams, Guy Consolmagno and Tobias Owen on the WG. The new membership is listed below.

2. Terms of Reference

In consultation with Division III, the following new terms of reference have been agreed on for the WG:

1. The IAU Working Group for Planetary System Nomenclature (WG-PSN) is appointed by IAU Division III under similar procedures as Divisional Working Groups (IAU Working Rules 38). It reports to the General Assembly through the president of Division III. For necessary meetings, the Working Group is eligible for travel support from the IAU.

2. The membership of the Working Group includes:
a. A chairperson of the WG-PSN and a minimum of eight other members of IAU Division III, including all the chairpersons of the Task Groups (see *g*), representing a diversity of geographical

and cultural backgrounds.
b. The president of Division III. The president may delegate the divisional representation to another Division member.
c. The president of IAU Commission 16 on *Physical Study of Planets and Satellites* or his/her representative.
d. The chairperson of the IAU Committee on *Small Body Nomenclature*, or his/her representative.
e. Additional members up to a maximum of 15, including a maximum of three consultants who possess special expertise but are not IAU members.
f. A member may be in more than one of the above categories.
g. The Working Group may appoint Task Groups of up to eight members, including a chairperson, to assist in the nomenclature work for the various celestial bodies, as the need arises.

3. Appointment and terms of service:
The members, including the chairperson, will be appointed for three years at each General Assembly by Division III, upon the advice of the outgoing Working Group. The Working Group may appoint or dismiss Task Group members at any time. Continuity in the work is important, so Working Group and Task Group members are encouraged to serve through several triennia.

4. Names and guidelines:
a. The Working Group will develop, maintain, and publish guidelines for naming natural satellites of planets and surface features on all the solar sytem bodies, based on the established guidelines. Significant changes in the guidelines should be submitted by the Working Group to Division III for discussion and approval before being put into effect. Minor changes may be approved by the Division III president.
b. The Working Group will periodically approve lists of new nomenclature, with accompanying explanatory notes. The names will be made available immediately on approval in an official website for public review. Any objections to these names based on significant, substantive problems must be forwarded in writing or e-mail to the Division III president within three months of the placing of these names on the website. Valid objections do not include personal references of the discoverers or the individuals. Division III will rule on objections.
c. Three months before each General Assembly, the Working Group will submit to the IAU General Secretary, through Division III, a list of all names approved in the immediately preceding three calendar years. The names will be published or referenced in the Proceedings of the General Assembly.

3. WG members 2006-2009

R. M. Schulz (chair, the Netherlands). Members: K. Aksnes (Norway), J. Blue (USA), J. Blunck (Germany), E. Bowell (USA), G. A. Burba (Russia), R. Courtin (France), R. M. Lopes (USA), M. Ya. Marov (Russia), B. G. Marsden (USA), M. Robinson (USA), V. V. Shevchenko (Russia), and B. A. Smith (USA).

4. New nomenclature

FEATURE NAME	LAT	LON	DIAM	DESCRIPTION
VENUS				
CHASMA				
Hanwi Chasma	10.5N	247.0E	1800.0	Oglala (Sioux) moon and sky goddess.
CORONAE				
Achall Corona	31.2S	259.6E	265.0	Celtic earth and nature goddess.
Asintmah Corona	25.9N	208.0E	150.0	Athabaskan (W. Canada Subarctic) Earth and nature goddess; the first woman on Earth.
Benzozia Corona	27.5N	204.5E	185.0	Basque mother goddess.
Chanum Coronae	29.2S	245.5E	330.0	Kachin (Tibetan people of Burma/Myanmar) creator goddess.
Embla Coronae	28.9N	205.4E	132.0	Scandinavian Earth goddess, creator of life.
Kulimina Corona	27.8S	261.9E	170.0	Arawakan (Brazil, Venezuela) creator goddess who created women.
Madalait Corona	37.6N	206.4E	150.0	Australian creator goddess; "Creator of life."
Nimba Corona	32.8N	204.5E	88.0	Guinea (West Africa) Earth and mother goddess.

Name	Lat.	Long.	Diam.	Origin
Qakma Corona	35.5N	207.1E	130.0	Bella Coola/Nuxalk (SW Canada) creator of life, the first woman.
Sitapi Coronae	36.5S	246.8E	270.0	Indonesian earth, nature, and creator goddess.
CRATERS				
Clementina	35.9N	208.6E	4.0	Portuguese form of Clementine, French first name.
Denise	14.4S	94.7E	2.0	Greek first name.
Gail	16.1S	97.5E	10.0	Hebrew first name.
Lisa	29.0N	182.0E	4.5	Short form of Elizabeth, Hebrew first name.
LINEA				
Agrona Linea	40.0N	280.0E	2300.0	Welsh goddess of slaughter, destroyer of life.
MONTES				
Ninisinna Mons	25.7N	197.5E	110.0	Mesopotamian goddess of health and healing.
Shala Mons	39.4N	208.0E	90.0	Canaanite (Phoenicia) storm goddess.
Toma Mons	12.9S	232.0E	80.0	Tibetan goddess of intelligence and creativity.
Waka Mons	26.3N	207.7E	60.0	Polynesian lizard goddess.
Xtoh Mons	39.7N	194.2E	110.0	Quiche (Guatemala) goddess of weather and rain.
PATERAE				
Darcle Patera	37.4S	263.8E	15.0	Hariclea; Romanian soprano singer (1860-1939).
Destinnov Patera	31.5S	250.2E	15.0	Ema (pseudonym of Emilia Kittlova); Bohemian/Czech singer, also known as Emmy Destinn (1878-1930).
Dutrieu Patera	33.8N	198.5E	80.0	Helene; Belgian/French pioneer aviatrix (1877-1961).
Garland Patera	32.7N	206.8E	45.0	Judy; American singer and actress (1922-1969).
Lindgren Patera	28.1N	241.4E	110.0	Astrid; Swedish author (1907-2002).
Nikolaeva Patera	33.9N	267.5E	100.0	Olga V.; Russian planetologist/geochemist (1941-2000).
Witte Patera	25.8S	247.6E	35.0	Wilhelmine; German astronomer (1777-1854).
THOLI				
Apakura Tholus	40.3N	205.8E	10.0	Maori (New Zealand) goddess of justice.
Azimua Tholi	34.0S	249.3E	40.0	Sumerian underworld goddess.
Monoshi Tholus	37.7S	252.0E	15.0	Bengal goddess of snakes.
Otohime Tholus	32.0S	268.2E	20.0	Japanese goddess of the arts and beauty.
Wohpe Tholus	41.4N	288.1E	40.0	Lakota goddess of order, beauty, and happiness.
VALLIS				
Ganga Valles	4.8N	53.0E	200.0	Hindu goddess of the sacred river Ganges.
MOON				
CRATERS				
Chawla	42.8S	147.5W	15.0	Kalpana; American astronaut, Space Shuttle Columbia Mission Specialist (1961-2003).
D. Brown	42.0S	147.2W	15.0	David McDowell; American astronaut, Space Shuttle Columbia Mission Specialist (1956-2003).
Husband	40.8S	147.9W	29.0	Rick Douglas; American astronaut, Space Shuttle Columbia Commander (1957-2003).
L. Clark	43.7S	147.7W	16.0	Laurel Blair Salton; American astronaut, Space Shuttle Columbia Mission Specialist (1961-2003).
M. Anderson	41.6S	149.0W	17.0	Michael Phillip; American astronaut, Space Shuttle Columbia Payload Commander (1959-2003).
McCool	41.7S	146.3W	21.0	William Cameron; American astronaut, Space Shuttle Columbia Pilot (1961-2003).
Ramon	41.6S	148.1W	17.0	Ilan; Israeli astronaut, Space Shuttle Columbia Payload Specialist (1954-2003).
Ryder	44.5S	143.2E	17.0	Graham; United Kingdom-born, American geologist (1949-2002).
MARS				
CATENA				
Ophir Catenae	9.6S	59.1W	577.0	Classical albedo feature name.
CAVI				
Amenthes Cavi	16.0N	245.0W	1340.0	Classical albedo feature name.
Boreum Cavus	84.5N	20.5W	50.0	Classical albedo feature name.
Ganges Cavus	10.2S	51.5W	42.3	Classical albedo feature name.
Olympia Cavi	85.0N	178.0W	860.0	Classical albedo feature name.
Ophir Cavus	10.0S	55.1W	36.8	Classical albedo feature name.
Tenuis Cavus	84.7N	0.0W	35.0	Classical albedo feature name.

Name	Lat.	Long.	Diam.	Origin
CHAOS				
Baetis Chaos	0.2S	60.5W	55.0	Classical albedo feature name.
Ganges Chaos	9.7S	46.2W	120.0	Classical albedo feature name.
Iamuna Chaos	0.2S	40.6W	18.0	Classical albedo feature name.
Oxia Chaos	0.2N	39.9W	26.5	Classical albedo feature name.
Xanthe Chaos	11.8N	42.2W	34.0	Classical albedo feature name.
CHASMATA				
Promethei Chasma	82.5S	217.5W	300.0	Classical albedo feature name.
Ultimum Chasma	81.1S	209.9W	350.0	Classical albedo feature name.
COLLES				
Chryse Colles	7.9N	42.1W	60.0	Classical albedo feature name.
CRATERS				
Bacolor	33.0N	241.4W	20.8	Town in the Philippines.
Beloha	39.5S	303.4W	33.5	Town in Madagascar.
Boola	81.2N	105.8W	17.0	Town in Guinea.
Bronkhorst	10.7S	55.2W	17.9	Town in the Netherlands.
Castril	14.7S	184.8W	2.2	Town in Spain.
Crotone	82.3N	70.0W	6.4	Town in Italy.
Culter	8.8S	54.0W	4.6	Village near Aberdeen, Scotland, also called Peterculter.
Davies	46.0N	0.0W	49.2	Merton Edward; American engineer, planetary geodesist (1917-2001).
Deseado	80.6S	289.7W	27.0	Town in Argentina.
Dilly	13.2N	202.9W	1.3	Town in Mali.
Dokka	77.1N	146.0W	52.5	Town in Norway.
Eberswalde	24.0S	33.3W	65.3	Town in Germany.
Elim	80.1S	263.4W	43.0	Town in South Africa.
Gamboa	40.7N	44.4W	33.0	Town in Panama.
Gratteri	17.7S	160.2W	7.3	Town on the island of Sicily, Italy.
Johnstown	9.8S	51.1W	3.2	Town in Pennsylvania, USA.
Jojutla	81.5N	169.6W	19.0	Town in Mexico.
Jumla	21.3S	273.6W	45.0	Town in Nepal.
Karzok	18.3N	131.9W	15.6	Village in Kashmir.
Katoomba	79.0S	232.5W	53.0	Town in Australia.
Mojave	7.5N	33.1W	58.5	Town in California, USA.
Morella	9.7S	51.4W	78.9	Town in Spain.
Okotoks	21.2S	275.7W	22.6	Town in Alberta, Canada.
Pangboche	17.2N	133.6W	10.4	Village in Nepal.
Persbo	8.5N	203.2W	19.5	Town in Sweden.
Puyo	83.9N	222.4W	10.4	Town in Ecuador.
Saheki	21.7S	286.9W	85.0	Tsuneo; Japanese amateur astronomer (1916-1996).
Somerset	9.7S	51.3W	3.3	Town in Pennsylvania, USA.
Toconao	20.9S	74.8W	17.7	Town in Chile.
Tombaugh	3.5N	198.2W	60.3	Clyde William; American astronomer (1906-1997).
Tomini	16.3N	234.2W	7.4	Town in Indonesia.
Tooting	23.1N	152.4W	27.5	Town in England.
Udzha	81.9N	282.8W	45.0	Village in northern Russia.
Wallula	9.9S	54.4W	12.5	Town in Washington, USA.
Winslow	3.8S	300.8W	1.0	Town in Arizona.
Zumba	28.6S	133.2W	3.3	Town in Ecuador.
FOSSA				
Idaeus Fossae	37.0N	51.9W	235.0	Classical albedo feature name.
LABYRINTHUS				
Tyrrhenus Labyrinthus	16.0S	258.9W	93.0	Classical albedo feature name.
LINGULAE				
Australe Lingula	83.8S	289.4W	430.0	Classical albedo feature name.
Hyperborea Lingula	80.0N	53.5W	80.0	Classical albedo feature name.
Promethei Lingula	83.0S	240.0W	560.0	Classical albedo feature name.
Ultima Lingula	76.3S	215.0W	560.0	Classical albedo feature name.
MENSAE				
Abalos Mensa	81.1N	75.6W	130.0	Classical albedo feature name.
Australe Mensa	86.8S	5.0W	200.0	Classical albedo feature name.
Ganges Mensa	7.2S	48.8W	140.0	Classical albedo feature name.
Olympia Mensae	78.4N	236.0W	380.0	Classical albedo feature name.
Tenuis Mensa	81.1N	93.8W	130.0	Classical albedo feature name.
MONS				
Xanthe Montes	18.4N	54.5W	500.0	Classical albedo feature name.

Name	Lat.	Long.	Diam.	Origin
PALUS				
Cerberus Palus	5.7N	212.1W	480.0	Classical albedo feature name.
PLANA				
Aeolis Planum	0.8S	215.0W	820.0	Classical albedo feature name.
Amenthes Planum	3.2N	254.3W	960.0	Classical albedo feature name.
Zephyria Planum	1.0S	206.9W	550.0	Classical albedo feature name.
SCOPULI				
Abalos Scopuli	80.7N	75.7W	110.0	Classical albedo feature name.
Australe Scopuli	83.5S	115.0W	530.0	Classical albedo feature name.
Boreales Scopuli	88.2N	105.0W	850.0	Classical albedo feature name.
Gemini Scopuli	81.0N	335.0W	1100.0	Classical albedo feature name.
Ultimi Scopuli	77.5S	180.0W	1100.0	Classical albedo feature name.
SULCUS				
Australe Sulci	85.0S	225.0W	400.0	Classical albedo feature name.
THOLUS				
Cerberus Tholi	4.8N	196.1W	600.0	Classical albedo feature name.
VALLES				
Columbia Valles	9.5S	42.9W	94.0	River in Washington, USA.
Coogoon Valles	17.3N	21.8W	300.0	River in Australia.
Daga Vallis	12.1S	42.3W	70.0	River in Burma.
Grjot Valles	15.6N	194.6W	370.0	River in Iceland.
Lethe Vallis	4.0N	206.5W	225.0	River in Katmai National Monument, Alaska, USA.
Rahway Valles	9.4N	186.2W	500.0	River in New Jersey, USA.
Silinka Vallis	9.0N	28.2W	140.0	River in Russia.
Vichada Valles	19.6S	271.9W	430.0	River in Colombia.
Walla Walla Vallis	9.9S	54.5W	24.0	River in Washington, USA.

ITOKAWA

Name	Lat.	Long.	Diam.	Origin
REGIONES				
MUSES-C Regio	70.0S	60.0	0.3	MUSES-C, the name of the Hayabusa spacecraft prior to lauch.
Sagamihara Regio	80.0N	15.0	0.23	Town in Japan where the Institute of Space and Astronautical Science is located.
Uchinoura Regio	40.0N	90.0	0.07	Town in Japan, launch site of the Hayabusa spacecraft.

IO

Name	Lat.	Long.	Diam.	Origin
ERUPTIVE CENTER				
Thor	39.2N	133.1W	239.4	Norse god of thunder.
MENSAE				
Prometheus Mensa	1.9S	151.9W	184.0	Greek fire god.
Tvashtar Mensae	61.6N	119.9W	326.4	Indian sun god and smith who forged the thunderbolt of the thunder god Indra.
MONTES				
Gish Bar Mons	18.6N	87.7W	110.0	Babylonian sun god.
Monan Mons	15.2N	104.5W	297.0	Brazilian god who destroyed the world with fire and flood.
Pillan Mons	8.8S	246.7W	163.0	Araucanian thunder, fire, and volcano god.
PATERAE				
Ah Peku Patera	10.3N	107.0W	84.0	Mayan thunder god.
Chors Patera	68.5N	249.9W	65.0	Slavic sun god.
Estan Patera	21.6N	87.7W	95.0	Hittite sun god.
Llew Patera	12.2N	242.3W	78.0	Celtic sun god.
Rarog Patera	41.7S	304.4W	104.3	Czech fire deity.
Reshef Patera	27.7N	158.1W	62.0	Phoenician god of lightning, sun, and thunder.
Thomagata Patera	25.7N	165.9W	59.0	Chibcha storm god, a terrifying fire spirit who flew through the air changing men into animals.
Vivasvant Patera	75.1N	294.0W	83.2	Hindu god of the morning sun.
REGIO				
Bulicame Regio	34.8N	190.8W	498.0	Hot sulphur spring, the water of which sinful women were permitted to use in "The Inferno."
VALLIS				

Tawhaki Vallis	0.5N	72.8W	190.0	Maori lightning god.

EUROPA

CRATERS

Amaethon	13.8N	177.4W	1.7	Celtic god of agriculture.
Bress	37.6N	98.6W	10.0	Beautiful son of Elatha in Celtic mythology.
Dagda	37.3N	168.7W	9.8	One of the chief deities of the Tuatha de Danann in Irish mythology.
Eochaid	50.4S	233.3W	10.6	King of the Fir Bolgs in Celtic mythology.
Gwern	9.1N	344.5W	22.2	Son of Branwen in Celtic mythology.
Luchtar	40.2S	257.5W	19.9	Celtic god of carpentry.
Lug	27.9N	44.3W	11.0	Irish omnicompetent god.
Midir	3.6N	338.7W	37.4	Gaelic fate and underworld deity.
Ogma	87.4N	287.8W	5.0	Celtic god of eloquence and literature, a son of Dagda.
Tuag	59.9N	172.3W	15.2	Irish dawn goddess.

LINEA

Yelland Linea	16.7S	196.0W	186.0	Stone row in England.

GANYMEDE

CRATERS

Damkina	30.0S	5.0W	180.0	Babylonian sky and health deity, queen of the gods, and mother of Marduk.
Menhit	36.5S	140.5W	140.0	Egyptian lion and war goddess.
Saltu	14.2S	352.7W	40.0	Babylonian goddess of discord and hostility.

CALLISTO

CRATER

Debegey	10.2N	166.2W	125.0	Yukagir (NE Siberia) mythological hero, the first man.

FACULA

Kol Facula	4.5N	282.7W	390.0	Icelandic frost or storm giant.

TITAN

ALBEDO FEATURE

Aaru	10.0N	340.0W	0.0	Egyptian abode of the blessed dead.
Adiri	10.0S	210.0W	0.0	Melanesian afterworld where life is easier than on Earth.
Aztlan	10.0S	20.0W	0.0	Mythical land from which the Aztecs believed they migrated.
Belet	5.0S	255.0W	0.0	Malay afterworld reached by a flower-lined bridge.
Ching-tu	30.0S	205.0W	0.0	Chinese Buddhist paradise where those who attain salvation will live in unalloyed happiness.
Dilmun	15.0N	175.0W	0.0	Sumerian garden of paradise, primeval land of bliss.
Fensal	5.0N	30.0W	0.0	In Norse mythology, magnificent mansion of Frigga, to which she invited all married couples who had led virtuous lives on Earth to enjoy each other's company forever.
Mezzoramia	70.0S	0.0W	0.0	Oasis of happiness in the African desert, from an Italian legend.
Quivira	0.0N	15.0W	0.0	Legendary city in the American Southwest; site of a fabulous treasure sought by Coronado and other explorers.
Senkyo	5.0S	320.0W	0.0	Japanese ideal realm of aloofness and serenity, freedom from wordly cares and death.
Shangri-la	10.0S	165.0W	0.0	Tibetan mythical land of eternal youth.
Tsegihi	40.0S	10.0W	0.0	Navajo sacred place.
Xanadu	15.0S	100.0W	3400.0	An imaginary country in Coleridge's "Kubla Khan."

ARCUS

Hotei Arcus	28.0S	79.0W	600.0	One of the seven gods of happiness in Japanese Buddhism. He is the god of contentment, good fortune, cheerfulness, and he is always smiling.

CRATERS

Menrva	20.1N	87.2W	392.0	Etruscan goddess of wisdom.
Sinlap	11.3N	16.0W	80.0	Kachin (N. Burma) wise spirit who dwells in the sky and gives wisdom to his worshippers.

FACULAE

Antilia Faculae	11.0S	187.0W	260.0	Archipelago corresponding to the mythical island of

				Antilia, once thought to lie midway between Europe and the Americas.
Bazaruto Facula	11.6N	16.1W	215.0	Mozambique island.
Coats Facula	11.1S	29.2W	80.0	Canadian island.
Crete Facula	9.4N	150.1W	680.0	Greek island.
Elba Facula	10.8S	1.2W	250.0	Italian island.
Kerguelen Facula	5.4S	151.0W	135.0	French subantarctic island.
Mindanao Facula	6.6S	174.2W	210.0	Philippine island.
Nicobar Faculae	2.0N	159.0W	575.0	Indian archipelago.
Oahu Facula	5.0N	166.7W	465.0	Hawaiian island.
Santorini Facula	2.4N	145.6W	140.0	Greek island also known as Thira.
Shikoku Facula	10.4S	164.1W	285.0	Japanese island.
Sotra Facula	12.5S	39.8W	235.0	Norwegian island.
Texel Facula	11.5S	182.6W	190.0	Dutch island.
Tortola Facula	8.8N	143.1W	65.0	Island in the British Virgin Islands.
Vis Facula	7.0N	138.4W	215.0	Croatian island.
LACUS				
Ontario Lacus	72.0S	183.0W	235.0	Lake on the border between Canada and the United States.
LARGE RINGED FEATURE				
Guabonito	10.9S	150.8W	55.0	Taino Indian (Antilles) sea goddess who taught the use of amulets.
Nath	30.5S	7.7W	95.0	Irish goddess of wisdom.
Veles	2.0N	137.3W	45.0	Slavic god of housekeeping wisdom.
MACULAE				
Eir Macula	24.0S	114.7W	145.0	Norse goddess of healing and peace.
Elpis Macula	31.2N	27.0W	500.0	Greek goddess of happiness and hope.
Ganesa Macula	50.0N	87.3W	160.0	Hindu god of good fortune and wisdom.
Omacatl Macula	17.6N	37.2W	225.0	Aztec god of good cheer and lord of banquets.
REGIO				
Tui Regio	20.0S	130.0W	0.0	Chinese goddess of happiness, joy, and water.
VIRGAE				
Bacab Virgae	19.0S	151.0W	485.0	Mayan rain god.
Hobal Virga	35.0S	166.0W	1075.0	Arabian rain god.
Kalseru Virga	36.0S	137.0W	630.0	NW Australian rainbow serpent, bringer of rain.
Perkunas Virgae	27.0S	162.0W	980.0	Lithuanian god of rain, thunder, and lightning.
Shiwanni Virgae	25.0S	32.0W	1400.0	Zuni rain god.

PHOEBE

CRATERS				
Acastus	9.6N	148.5W	34.0	Argonaut, son of the Thessalian king Pelias, took part in the Calydonian boar hunt.
Admetus	11.4N	39.1W	58.0	Argonaut, founder and king of Pherae in Thessaly.
Amphion	27.0S	1.8W	18.0	Argonaut, son of Hyperasius and Hypso.
Butes	49.6S	292.5W	29.0	Argonaut, son of Teleon, bee-master.
Calais	38.7S	225.4W	31.0	Argonaut, son of Boreas, the north wind.
Canthus	69.6S	342.2W	44.0	Argonaut, son of Kanethos or Cerion, the only member of the expedition to die in combat.
Clytius	46.0N	193.1W	52.0	Argonaut, son of Eurytus, skilled archer who was killed by Apollo for challenging the god to a shooting match.
Erginus	31.6N	337.1W	38.0	Argonaut, son of Neptune, helmsman of the Argo after the death of Tiphys.
Euphemus	31.3S	331.1W	23.0	Argonaut, son of Neptune and Europa.
Eurydamas	61.5S	281.6W	19.0	Argonaut, son of Ctimenus.
Eurytion	30.4S	8.0W	14.0	Argonaut, son of Kenethos or Cerion.
Eurytus	39.7S	177.2W	89.0	Argonaut, son of Mercury and Antianira.
Hylas	7.9N	354.5W	30.0	Argonaut, son of Theiodamas/Theodamas, king of the Dryopes.
Idmon	67.1S	197.8W	61.0	Argonaut, son of Apollo and the nymph Cyrene, or of Abas, a prophet.
Iphitus	27.2S	293.3W	22.0	Argonaut, son of Eurytus, Jason's host during his consultation with the Oracle at Delphi.
Jason	16.2N	317.7W	101.0	The leading argonaut, son of the Thessalian king Aeson, delivered the Fleece.
Mopsus	6.6N	109.1W	37.0	Argonaut, prophesying son of Apollo.
Nauplius	31.5N	241.5W	24.0	Argonaut, son of Neptune and Amymone, or of Klytoneos.
Oileus	77.1S	96.9W	56.0	Argonaut, king of the Locrians, renowned for his courage in battle.

Peleus	20.2N	192.2W	44.0	Argonaut, son of Aeacus, father of Achilles.
Phlias	1.6N	359.1W	14.0	Argonaut, son of Dionysus.
Talaus	52.3S	325.2W	15.0	Argonaut, son of Teleon, or of Bias and Pero.
Telamon	48.1S	92.6W	28.0	Argonaut, son of Aeacus, took part in the Calydonian boar hunt.
Zetes	20.0S	223.0W	29.0	Argonaut, son of Boreas, the north wind.

CHANGES TO APPROVED NAMES

VENUS

Breksta Dorsa changed to Breksta Linea

SATELLITES OF JUPITER

Hegemone	= Jupiter XXXIX	= S/2003 J 8
Mneme	= Jupiter XL	= S/2003 J 21
Aoede	= Jupiter XLI	= S/2003 J 7
Thelxinoe	= Jupiter XLII	= S/20023J 22
Arche	= Jupiter XLIII	= S/2002 J 1
Kallichore	= Jupiter XLIV	= S/2003 J 11
Helike	= Jupiter XLV	= S/2003 J 6
Carpo	= Jupiter XLVI	= S/2003 J 20
Eukelade	= Jupiter XLVII	= S/2003 J 1
Cyllene	= Jupiter XLVIII	= S/2003 J 13

SATELLITES OF SATURN

Narvi	= Saturn XXXI	= S/2003 S 1
Methone	= Saturn XXXII	= S/2004 S 1
Pallene	= Saturn XXXIII	= S/2004 S 2
Polydeuces	= Saturn XXXIV	= S/2004 S 5
Daphnis	= Saturn XXXV	= S/2005 S 1

SATELLITES OF URANUS

Francisco	= Uranus XXII	= S/2001 U 3
Margaret	= Uranus XXIII	= S/2003 U 3
Ferdinand	= Uranus XXIV	= S/2001 U 2
Perdita	= Uranus XXV	= S/1986 U 10
Mab	= Uranus XXVI	= S/2003 U 1
Cupid	= Uranus XXVII	= S/2003 U 2

SATELLITES OF NEPTUNE

Psamathe	= Neptune X	= S/2003 N 1

SATELLITES OF PLUTO

Nix	= Pluto II	= S/2005 P 2
Hydra	= Pluto III	= S/2005 P 1

Kaare Aksnes
chair of the Working Group

Transactions IAU, Volume XXVIB
Proc. IAU XXVI General Assembly, August 2006
Karel A. van der Hucht, ed.

doi:10.1017/S174392130802379X

COMMISSION 15 — PHYSICAL STUDY OF COMETS AND MINOR PLANETS

ÉTUDE PHYSIQUE DES COMÈTES ET DES PETITS PLANÈTES

PRESIDENT	**Walter F. Huebner**
VICE-PRESIDENT	**Alberto Cellino**
PAST PRESIDENT	**Edward F. Tedesco**
SECRETARY	**Daniel C. Boice**
ORGANIZING COMMITTEE	**Dominique Bockelée-Morvan, Yuehua Ma, Harold J. Reitsema, Rita M. Schulz, Petrus M. M. Jenniskens, Dmitrij Lupishko, Gonzalo Tancredi**

DIVISION III / COMMISSION 15 WORKING GROUPS

Division III / Commission 15 WG	**Physical Stydy of Comets**
Division III / Commission 15 WG	**Physical Study of Minor Planets**

PROCEEDINGS BUSINESS MEETING on 22 August 2006

1. Introduction

This report of the business meeting at the 2006 IAU GA is based on notes provided by Edward Tedesco, past president, and on the minutes taken by Petrus Jenniskens, secretary of Commission 15 in the triennium 2003 to 2006, with additional notes from the current secretary of Commission 15, Daniel Boice. For political reasons the incoming President, Walter Huebner, was unable to attend the General Assembly.

Ed Tedesco gave the President's report. In the necrology section he acknowledged the loss of three outstanding members of Commission 15: Vasilij I. Moroz, Fred L. Whipple, and George Wetherill. In describing the activities of the Commission he referred to the Commission 15 website where detailed reports can be found. Activities included the proceedings of Joint Discussions 14 and 19 at the IAU XXIV General Assembly (GA) in Sydney, as well as support of meetings unrelated to the GA: IAU Symposium No. 229 on *Asteroids, Comets, Meteors* (ACM), Rio de Janeiro, Brazil, 7-12 August 2005, the meeting on *Dust in Planetary Systems*, Kauai, Hawaii, 26-30 September 2005, and IAU Symposium No. 231 *Astrochemistry throughout the Universe*, Monterey, California, 29 August - 2 September 2005. The president's report was followed by the reports from the Working Groups (WGs). Dominique Bockelée-Morvan presented the Comet WG report and Alberto Cellino gave the Minor Planets WG report.

2. Task Group reports

The following suggestions and general statements were offered:
1. A "task group" should have specified deliverables for the triennium. Tasks should be clearly defined.
2. There could be a task group on physical properties of comets and one on physical properties of asteroids.

3. Recalibration of albedo-polarization law is a possible task group (proposed by Ted Bowell, Ed Tedesco, and Alberto Cellino).
4. Magnitude definition for asteroids and comets and about polarization are possible task groups, as proposed by Gonzalo Tancredi.
5. Rita Schulz suggested a task group on comets with a report due by October 2008.

2.1. *Comet Task Group*

1. Magnitudes for comet brightnesses need to be publicly available. Dan Green, will look into this matter in the next triennium.
2. Magnitudes for comets need to be better defined. No progress was reported from the Commission 20 task group. Michael A'Hearn will correct $Af\rho$ values of dust for phase angle.
3. Anny-Chantal Levasseur-Regourd commented that apertures should be translated into distance from the comet nucleus.
4. A depository of information on individual comets, similar to EARN for NEOs should be created.

2.2. *Minor Planet Task Group*

1. There are problems with the consistency of the magnitude scale. Ted Bowell commented that future surveys will retire this problem in the next two to three years. However, Ed Tedesco commented that three different surveys will use different bands. Groups should look into making them agree.
2. Michael A'Hearn recommended that information on asteroid spin rates and pole orientations should be made available on one website.
3. Anny-Chantal Levasseur-Regourd suggested a task group on physical properties for TNOs with a report due October 2008.
4. Ted Bowell pointed out that the IAU is good at having people adopt a given rule or calibration that then becomes a baseline. That calibration does not need to be better than we know now.

3. Commission 15 membership and other business

1. Commission 15 has about 202 confirmed members, including 41 new members joining during GA XXVI. This membership count falls short by about 150 members of the official IAU e-mail list for Commission 15 members. The IAU Commission 15 membership list is out of date. The IAU has an electronic interactive system that can help people update their biography and address information.
2. Many active researchers, who currently are not members of the IAU, should be encouraged to become members.
3. Commission websites should be accessible through a central IAU page. "Useful links" should be established. It is left to the Commission president to decide which links are useful.
4. E-mails should be sent out when candidacy for offices of Commission 15 are proposed.

4. Commission 15 elections

1. In addition to electing the new president and vice-president, an election was held for the chair of the Minor Planets Working Group (MP-WG) and for members of the Organizing Committee (OC). Ricardo Gil-Hutton was elected chair of the MP-WG. Usually, all nominations for the OC were made during the business meeting, but the by-laws adopted in a recent General Assembly require now that they should be made prior to the meeting. However, difficulties with contacting members via e-mail made some notifications challenging. Thus, at Mike A'Hearn's suggestion, and with the concurrence of the Commission president, the decision was made to allow additional nominations during the Commission 15 Business meeting. Julio Fernandez withdrew his nomination because of other commitments.
2. At the suggestion of the IAU Executive Committee, the total number of members serving on the OC was reduced so that only three new members were added to bring the size of the OC down to that required by the current by-laws.
3. New members were elected to the Organizing Committee. They are: Petrus M. M. Jenniskens, Dmitrij Lupishko, and Gonzalo Tancredi. All members of the Organizing Committee are listed at the beginning of this report.

4. Thanks were given to the outgoing members of the OC: Alan Harris, Nikolai Kiselev, Lucy Ann McFadden, Tadeusz Michalowski, Karri Muinonen, and Mark Sykes. Also all members who stood for election to the OC, including Diane Wooden and Dan Green were thanked.
5. Daniel Boice was appointed by the incoming president to be the new secretary.

5. Proposed merging of Commissions 15 and 20 in 2009

1. It was pointed out that some limited overlap exists between the two Commissions. However, the merged commission would have about 400 members. This may be unwieldy. Alternatively, small joint interest groups and task groups with members of both commissions could be formed. Such joint groups might be more effective than a very large Commission.
2. Regarding the writing of reports, discussion centered on whether report writing is rewarding. It was concluded that it is not rewarding if reports are written by one person. However, if reports are written by the Commission membership, many gain insight in work that has been done and that needs to be done.
3. Regarding possible interest and task groups, discussion centered on members to contribute suggestions.
4. Ed Tedesco proposed that Commission 15 discussions include the Commission 20 Organizing Committee and *vice versa*. This may aid in the trial merging of the two Commissions at the next General Assembly.
5. Tedesco also proposed that the president of the combined trial Commission (15 and 20) be decided by a coin toss between those two Commission's current vice-presidents and that the one not-becoming president will serve as vice-president.

6. Closing remarks

I want to thank the past president, Edward Tedesco, and the past secretary, Petrus Jenniskens, for their untiring commitment and service to Commission 15.

Walter F. Huebner
president of the Commission

Transactions IAU, Volume XXVIB
Proc. IAU XXVI General Assembly, August 2006
Karel A. van der Hucht, ed.

doi:10.1017/S1743921308023806

DIVISION III / COMMISSION 15 / WORKING GROUP PHYSICAL STUDY OF MINOR PLANETS

CHAIR **Walter F. Huebner**
VICE-CHAIR **Alberto Cellino**
PAST CHAIR **Edward F. Tedesco**
MEMBERS **Dominique Bockelee-Morvan, Yuehua Ma, Petrus M. M. Jenniskens, Dmitrij F. Lupishko Harold J. Reitsema, Gonzalo Tancredi, Rita M. Shulz**

PROCEEDINGS BUSINESS MEETING on 24 August 2006

The meeting of the Physical Properties of the Minor Planets Working Group of IAU Commission 15 took place on 24 August, and was devoted to purely scientific matters, since other topics (organization of the Minor Planet WG, need of a new web page, election of the new chairman) had been already discussed during the business meeting of Commission 15, on 22 August. A brief summary of the talks given during the meeting is given in what follows.

1. Introduction

The former chairman of the Minor Planet Working Group, A. Cellino, made a very short introduction, to introduce the main subjects of the discussion. The meeting was divided into two major sections, whose titles were "Asteroid Radiometry and Polarimetry" and "The Renaissance of Polarimetry", respectively. As the titles clearly show, the two parts were intimately correlated, and they were chosen because they deal with some delicate issues of the current state of the art in asteroid science, which deserve careful attention and, very likely, some possible actions by IAU Commission 15 in the next triennium.

2. Asteroid radiometry and polarimetry

E. Tedesco gave two consecutive talks, devoted to two complementary subjects. The first talk had the title "Issues with the Radiometric Method for Small Asteroids" and was a summary of the state of the art in the field of thermal radiometry observations of asteroids, with particular emphasis on what has been learned by the most recent observing campaigns devoted to small objects, both in the Main Belt and among the near-Earth population. In particular, the presentation emphasized the contrast between the fairly good performances of thermal radiometry as far as the determination of sizes for objects larger than 20 km is concerned, and the much more uncertain results concerning the determination of sizes and, even more, albedos, for objects smaller than the above value. In particular, it was emphasized that the large error bars in the determination of the albedos for small asteroids observed by means of thermal radiometry (up to 60% for objects having only one single thermal radiometry measurement) come from two main sources: the uncertainty in the thermal models to be used, and the large uncertainties in the knowledge of the absolute magnitudes of the objects.

The second talk just focused on the problem of the large errors in the absolute magnitudes (the H values, following the H,G system recommended by the IAU) as listed by the official sources, like the Minor Planet Center. These errors turn out to be increasingly important for objects of increasing faintness. Different investigations carried out in recent years by different authors have led to conclude that the listed values of H are affected by both random and systematic errors, which are as large as 0.5 magnitudes for objects with an estimated H between 14 and 15. These large uncertainties on the absolute magnitudes have very negative effect on the performances of thermal radiometry in the determination of the albedos, and of polarimetry in the determination of the sizes (see below).

3. The renaissance of polarimetry

A. Cellino gave a talk summarizing the current state of the art in the field of asteroid polarimetry. Recent years have seen a substantial blossoming of polarimetric studies, taking profit of the availability of new polarimeters, large telescopes and theoretical and laboratory advances. Some examples are given by the extensive observing campaigns carried out at the CASLEO observatory, that led, among other interesting results, to the discovery of objects exhibiting previously unknown properties like (234) Barbara, and the determination of the average albedo of (25143) Itokawa before the *rendez-vous* with the *Hayabusa* space probe. After a presentation about "A polarimetry-based asteroid taxonomy", prepared by A. C. Levasseur-Regourd and given by A. Cellino (due to the fact that the author could not be present at the meeting) E. F. Tedesco gave a talk focusing about the problem of the calibration of the so-called polarimetric slope *versus* albedo relation, which is a hot topic because in the current phase of renaissance of polarimetry different authors are still using different calibrations, leading to unnecessary confusion.

4. Conclusions

Due to the fact that both the problems of the systematic errors in the H absolute magnitude and the calibration of the polarimetric slope - albedo relation are important and deserve careful analysis and quick actions, it was decided, as also suggested by the new President of IAU Division III, E. Bowell, to set up urgently two Task Groups to deal with these two issues. A. Cellino took the commitment to prepare two short documents to be submitted to the President of Commission 15, in order to start the actions needed to establish these two Task Groups.

Alberto Cellino
former chair of the Working Group

Transactions IAU, Volume XXVIB
Proc. IAU XXVI General Assembly, August 2006
Karel A. van der Hucht, ed.

doi:10.1017/S1743921308023818

COMMISSION 16 — PHYSICAL STUDY OF PLANETS AND SATELLITES

ÉTUDE PHYSIQUE DES PLANÈTES ET DES SATELLITES

PRESIDENT — **Guy J. Consolmagno**
VICE-PRESIDENT — **Regis Courtin**
PAST PRESIDENT — **Dale P. Cruikshank**
ORGANIZING COMMITTEE — **Carlo Blanco, Dale P. Cruikshank, Leonid V. Ksanfomality, Melissa A. McGrath, David Morrison, David Morrison, Keith S. Noll, Tobias C. Owen, Maarten C. Roos-Serote, John R. Spencer, Victor G. Tejfel**

PROCEEDINGS BUSINESS MEETING on 23 August 2006

1. Introduction

Commission 16 held its business meeting during the General Assembly in Prague, on Wednesday August 23, 2006, with nine members present. The meeting was called to order at 14:00 hr by president Guy Consolmagno. A moment of silence was observed in memory of those Commission (or Division) members deceased since the last General Assembly. They are Joseph W. Chamberlain, Michel Festou, Thomas Gold, Cornell H. Mayer, Vasilij I. Moroz, William M. Sinton, Willem Wamsteker, James A. Westphal, and Fred L. Whipple.

2. Report from the president

In introducing his report, Guy Consolmagno explained that the president of Commission 16 sits *ex officio* on several committees and working groups (e.g., WG-SPN, the Working Group on Planetary System Nomenclature), essentially representing the users, i.e., astronomers or other scientists who make use of the nomenclature and/or coordinates produced by these committees. The WG-SPN has been quite active in naming surface features on Venus, the Moon, Mars, Itokawa, Io, Europa, Ganymede, Callisto, Titan, and Phoebe. Commission 16, through its president and several Organizing Committee members, also contributed to the efforts aiming at a definition of a planet, which will hopefully come to fruition at the end of the present General Assembly. Guy Consolmagno expressed his personal view that, although the task is difficult, a definition is sorely needed, and the IAU should rapidly come to terms with this issue.

Among the various objectives put forward at the Commission meeting in Sydney for the 2003-06 triennium, most were met with reasonable success:

• convening a scientific meeting in Prague: Guy Consolmagno thanks the Organizing Committee for supporting and helping to organize the Joint Discussion 10 "Progress in Planetary Exploration Missions" held during the present General Assembly, and which was very well attended;

• defining a continued role for the Commission: on the basis of the activity report and set of objectives presented for the next triennium, Division III has approved the continuation of Commission 16;

• updating the membership list: the roster has been updated, but there are still numerous e-mail addresses missing, and correcting for errors is a frustrating process;

• meeting regularly with OC members: this has been only partly successful since about half of the members are usually present at the same time at major planetary science meetings (particularly, the AAS/DPS meeting);

• setting up a regular news bulletin or "blog": this has been held back mostly because of the difficulty of completing the e-mail address list;

• managing the Commission web site: this has been successful with some help provided by Maarten Roos-Serote, who unfortunately has announced his intention to resign from the position of Secretary. Many thanks are expressed for his work.

3. New membership

The list of 29 new IAU members who expressed their wish to join Commission 16 was presented. The large diversity of the countries of origin (Australia, France, Germany, Greece, Italy, Portugal, Russian Federation, South Korea, Switzerland, the Netherlands, Ukraine, United Kingdom, USA) is a gratifying reflection on the constant development and vitality of planetary sciences throughout the world. The list of new Commission members was unanimously approved. This brings the membership to a total of 271. It is interesting to note that 52% of the members are also members of the AAS/DPS.

4. New officers

The next president and vice-president of the Commission will be Régis Courtin (Observatoire de Paris, France), and Melissa A. McGrath (NASA-Marshall Space Flight Center, USA). No candidate could be definitely proposed for the position of Secretary during the Commission meeting, but a search conducted by Régis Courtin among the membership subsequently led to the designation of Luisa M. Lara (Instituto de Astrofisica de Andalucìa-CSIC, Granada, Spain).

Several members of the Organizing Committee have asked or agreed to rotate out: Dale Cruikshank, Keith Noll, Tobias Owen, Maarten Roos-Serote. New members were proposed by the assembly to replace the outcoming ones. However, the IAU bye-laws now limit the number of OC members to eight (excluding the Commission president). With the addition of Luisa M. Lara as Secretary, and Guy Consomalgno as immediate past President, the new OC membership matches this requirement ; therefore and regrettably, it was not deemed possible to include the suggested individuals. Thus, the composition of the Organizing Committee for the triennium 2006-09 is the following: Carlo Blanco (Italy), Guy Consolmagno (Vatican City State), Leonid Ksanfomality (Russia), Luisa Lara (Spain), Melissa McGrath (USA), David Morrison (USA), John Spencer (USA), and Victor Tejfel (Kazakhstan).

5. Additional matters

Although it is not directly under the purview of Commission 16, Guy Consolmagno mentioned the proposal – by the Working Group on Cartographic Coordinates and Rotational Elements – of a new coordinate system for Mars and of a new rotation period for Saturn. On the latter topic, Dennis Matson (*Cassini* Mission Scientist) remarked that no firm conclusion has been drawn so far from the relevant Cassini investigations.

A discussion was held on the content and format of the scientific session to be proposed and organized by the Commission at the next General Assembly. In retrospect, it was found that in the case of Joint Discussion 10, restricting the topic to space missions posed some problems, since the invited speakers (the Mission Scientists) were not necessarily the most knowledgeable experts in the discipline. Also, the organization of the Poster Sessions left much to be desired: the exhibit room was far from the meeting room and there were no brief presentations of the posters. The next General Assembly (Rio de Janeiro) will pose problems of its own because of probable travel limitations. David Morrison suggested finding "hot topics" that would attract a large audience. Dennis Matson remarked that in the case of Cassini, it is more appropriate to regroup instrument talks by discipline (atmosphere, fields and particles, surfaces, etc.). Finally, other suggestions were to hold a Joint Discussion with Commission 51 (Bioastronomy) or to organize a Symposium.

Because of the likelihood of the adoption of the Scientific Resolution defining a new class of solar system objects, namely "dwarf planets", the question arises as to which Commission will "own" this new category of objects. Guy Consolmagno proposed that Commission 16 be the one Commission in charge. The assembly agreed unanimously; the proposal was passed on to the new Division III president, Ted Bowell.

Leonid Ksanfomality raised the issue of scientific communication with the new Commission "Extrasolar planets". The assembly agreed that it would be very useful to set up formal links for the benefit of both Commissions. The new Organizing Committee will be taking up this matter.

The meeting was adjourned at 15:30 hr.

Guy J. Consolmagno
president of the Commission

Transactions IAU, Volume XXVIB
Proc. IAU XXVI General Assembly, August 2006
Karel A. van der Hucht, ed.

doi:10.1017/S174392130802382X

COMMISSION 20 — POSITION AND MOTION OF MINOR PLANETS, COMETS AND SATELLITES

POSITION ET MOUVEMENT DES PETITES PLANÈTES, DES COMÈTES ET DES SATELLITES

PRESIDENT	**Giovanni B. Valsecchi**
VICE-PRESIDENT	**Julio Fernandez**
PAST PRESIDENT	**Edward L. G. Bowell**
ORGANIZING COMMITTEE	**Jean-Eudes Arlot, Edward L. G. Bowell, Yulia A. Chernetenko, Steven R. Chesley, Daniela Lazzaro, Anne Lemaitre, Brian G. Marsden, Karri Muinonen, Hans Rickman, David J. Tholen, Makoto Yoshikawa**

DIVISION III / COMMISSION 20 WORKING GROUPS

Division III / Commission 20 / WG	**Service: Minor Planet Center**
Division III / Commission 20 / WG	**Motions of Comets**
Division III / Commission 20 / WG	**Distant Objects**

PROCEEDINGS BUSINESS MEETING on 22 August 2006

1. Membership of the Commission

A total of 16 among the new IAU members have asked to join Commission 20; they are: Jerome Berthier, Nicholas J. Cooper, Marco Delbò, Romina P. Di Sisto, Michael W. Evans, Tetsuharu Fuse, Ludmila Hudkova, Yurij N. Krugly, Elena N. Polyakhova, Zhanna Pozhalova, Alessandro Rossi, Qi Rui, Jonathan D. Shanklin, Slawomira E. Szutowicz, Gino Tuccari and Hong-Suh Yim. Moreover, two requests to join the Commission have been received by astronomers that are already IAU members: Peter De Cat and Ricardo A. Gil-Hutton.

The Commission stands briefly in silence, in memory of the members deceased since the IAU XXV General Assembly, Nikolaj S. Chernykh and Fred L. Whipple.

2. Past IAU meetings sponsored by the Commission

In the past triennium, three major meetings of interest to the Commission have taken place:

- IAU Colloquium No. 197 "Dynamics of Populations of Planetary Systems", 31 August - 4 September 2004, Belgrade, Serbia and Montenegro;
- IAU Symposium No. 229 "Asteroids, Comets, Meteors 2005", 7-12 August 2005, Búzios, Rio de Janeiro, Brazil;
- IAU Symposium No. 236 "Near Earth Objects, our Celestial Neighbors: Opportunity and Risk", 14-18 August 2006, Prague, Czech Republic (at this General Assembly).

3. Proposal for a future IAU Symposium of interest for Commission 20

Julio Fernández describes a proposal for a Symposium to be proposed for the next IAU General Assembly; the provisional title is "Icy bodies and the early solar system".

4. New officers and OC members

The president presents the list of possible new officers of the Commission; before any decision on the new officers, the Commission discusses the opportunity that members of the Organizing Committee step down, under normal circumstances, after two triennia. This is decided by a nearly unanimous vote (one abstention). The following members of the current OC agree to step down: Jean-Eudes Arlot, Edward L. G. Bowell, Anne Lemaitre, Brian G. Marsden, Hans Rickman.

The Commission then appoints the new members of the OC: Alan Gilmore, Petr Pravec, Tim Spahr, Jana Ticha, Jin Zhu, votes on the new Secretary, that is Steve Chesley, and appoints the new president and vice-presidents, that are Julio Fernández and Makoto Yoshikawa, respectively.

5. The present and future of the Commission

The Commission then discusses a number of relevant issues: among these, particular attention is devoted to the problem of assigning discovery credits for minor bodies, especially in view of the revolution that is likely to happen as a consequence of the new surveys. It is decided that the new president of Commission 20 will appoint a task group with the purpose of deciding in a reasonable time frame (before the start of operations of the new surveys) if new rules are needed, and what they should be.

Finally, the problem of the possible merging with Commission 15 in 2009 is discussed. A decision will have to be taken within the next triennium; to accumulate experience on what such a merging could bring, it is decided that the OC of Commission 20 will keep informed of all of its decisions the OC of Commission 15 (the reciprocal will also happen, in agreement with the Officers of Commission 15), so that both OCs will gather sufficient information on the business of both Commissions, to be able to make a wise decision in due time.

Giovanni B. Valsecchi
president of the Commission

Transactions IAU, Volume XXVIB
Proc. IAU XXVI General Assembly, August 2006
Karel A. van der Hucht, ed.

doi:10.1017/S1743921308023831

COMMISSION 21 — LIGHT OF THE NIGHT SKY

LUMIÈRE DU CIEL NUCTURNE

PRESIDENT	**Bo A. S. Gustafson**
VICE-PRESIDENT	**Adolf N. Witt**
PAST PRESIDENT	**Philippe Lamy**
ORGANIZING COMMITTEE	**Eli Dwek, Philippe Lamy, Richard C. Henry, Ingrid Mann**

PROCEEDINGS BUSINESS MEETING on 22 August 2006

1. New members

Applications have been received from Dr. Peter Wheatley (proposed by UK), Prof. Harald Schuh (proposed by Austria), and Dr. Busaba Kramer (proposed by Thailand). All applications were endorsed, with the caveat that Dr. Kramer's application needs to be endorsed by at least one of the other commissions since she has not a publication record in our field.

Members of the IAU who's names have been forwarded to the IAU Secretariat as new Commission 21 members by the outgoing president and therefore need no vote to join our Commission are: Dr. Munetaka Ueno (Japan), Dr. Akiko Nakamura (Japan), and Dr. Junichi Watanabe (Japan).

All new members were enthusiastically welcomed.

2. New officials

The EC proposed as officials of the Commission for the next triennium: president: Prof. Adolf Witt (USA); and vice-president: Dr. Jayant Murthy (India). The proposal was voted upon and unanimously endorsed.

3. Proposal for a future IAU Symposium of interest for Commission 20

Julio Fernández describes a proposal for a Symposium to be proposed for the next IAU General Assembly; the provisional title is "Icy bodies and the early solar system".

4. Organizing Committee

Discussions took into account the recommendation for a better geographical distribution made at the preceding business meeting (2003 GA in Sydney) and also the need for members with experience and knowledge about the commission's long term goals as a reorganization is contemplated. This led to the following proposal for composition of the OC for the coming triennium:
Dr. Ingrid Mann, Kobe University, Japan
Dr. Eli Dwek, NASA/GSFC, USA
Dr. Junichi Watanabe, National Astronomical Observatory of Japan
Prof. Kalevi Mattila, University of Helsinki Observatory, Finland
Prof. Jack Baggaley, University of Canterbury, Christchurch, New Zealand
Prof. Anny-Chantal Levasseur-Regourd, University of Paris VI, France
Prof. Bo Gustafson, University of Florida, USA.

The proposal was voted upon and endorsed.

5. Working Groups

Working groups established or continued at the 2003 Business Meeting in Sydney had not sent in reports although at least one organized a meeting. It was proposed that all past triennium working groups be terminated. It was further proposed to establish a working group to review the need for an update of the comprehensive review of the "Light of the Night Sky": "1997 Reference of Diffuse Night Sky Brightness" to be chaired by A.-Ch. Levasseur-Regourd. The proposal was voted upon an endorsed.

Bo A. S. Gustafson
president of the Commission

Transactions IAU, Volume XXVIB
Proc. IAU XXVI General Assembly, August 2006
Karel A. van der Hucht, ed.

doi:10.1017/S1743921308023843

COMMISSION 22 — METEORS, METEORITES AND INTERPLANETARY DUST

MÉTÉORES, MÉTÉORITES ET POUSSIÈRE INTERPLANÉTAIRE

PRESIDENT	**Pavel Spurný**
VICE-PRESIDENT	**Jun-ichi Watanabe**
PAST PRESIDENT	**Ingrid Mann**
ORGANIZING COMMITTEE	**William J. Baggaley, Jiří Borovička, Peter G. Brown, Guy J. Consolmagno, Peter Jenniskens, Asta K. Pellinen-Wannberg, Vladimír Porubčan, Iwan P. Williams, Hajime Yano**

DIVISION III / COMMISSION 22 WORKING GROUPS

Division III / Commission 22 / WG	**Professional-Amateur Cooperation in Meteors**
Division III / Commission 20 / TF	**Meteor Shower Nomenclature**

PROCEEDINGS BUSINESS MEETING on 22 August 2006

1. Attendance

Present were the members Abe, Babadzhanov, Baggaley, Borovička, Bowell, Ceplecha, Consolmagno, Jenniskens, Kolomiets, Koten, Marsden, Porubčan, Spurný, Valsecchi, Watanabe, and Williams.

After a short introduction by I. Williams, the current Division president (DP), who is also OC member of Commission 22, the business meeting was managed by P. Spurný, the current Commission 22 vice-president (VP).

In the beginning the meeting stood in silence in remembrance of deceased Commission 22 members: A. Hajduk, B. L. Kashscheev, K. N. Kramer, G. W. Wetherill, F. L. Whipple.

2. Commission triennium report

The vice-president informed that unfortunately the standard Commission triennium report for the IAU Transactions A was not prepared this time. Thanks to the Division president, at least a very limited version, edited by DP, VP and J. Borovička, was prepared. It was decided that such a situation is not acceptable in the future and that at the end of the next triennium the Commission report in the standard form has to be compiled.

3. Report of the Commission 22 Pro-am Working Group

P. Jenniskens presented his report describing the heavy commitment of members in many campaigns and noted the important role of amateurs in meteor astronomy and their strong activity. The WG was contributing to commission work, was very active and should continue at least as a newly based Interest Group.

4. Election of new members

The IAU executive was represented by the current and forthcoming Division III presidents. The following new members to IAU and to C22 had been proposed: Margaret D. Campbell-Brown (Canada), Yuri Gorbanec (Ukraine), Pavel Koten (Czech Republic), Katharina Lodders (USA), Ian C. Lyon (UK), Andrei B. Makalkin (Russian Federation), Michele Maris (Italy), Juergen Rendtel (Germany), Jérémie Vaubaillon (France), Masayuki Yamamoto (Japan), and Josep M. Trigo-Rodríguez (Spain). Their acceptance was proposed and accepted with acclamation by the meeting.

5. Election of executive

Because of the absence of the Commission 22 president, it was proposed by the present OC members as well as by the Division president that Pavel Spurný, the current Commission 22 VP be the next president. Pavel's election was carried with acclamation.

The new president proposed as a very suitable candidate for the next vice-president, Jun-ichi Watanabe from Japan, who is well known for his long-lasting work in meteor astronomy. The proposal was carried with acclamation. The president also proposed J. Borovička to be the new secretary of the Commission 22.

6. Commission 22 Organizing Committee.

Noting the IAU Executive guidelines on numbers and their recommendation to rotate members within the constrictions of expertise, discussion converged to the election of the following members of the OC: Jun-ichi Watanabe (Japan, vice-president), Jiří Borovička (Czech Republic, secretary), William J. Baggaley (New Zealand), Peter G. Brown (Canada), Guy J. Consolmagno (Vatican City State), Petrus M. M. Jenniskens (USA), Asta K. Pellinen-Wannberg (Sweden), Vladimír Porubčan (Slovakia), Iwan P. Williams (UK), and Hajime Yano (Japan).

7. Establishing and election of new Task Group for Meteor Shower Nomenclature

The Commission has established the Task Group for Meteor Shower Nomenclature with the objective to formulate criteria for the definition of meteor showers and to prepare a list of established meteor showers that can receive official names during the next IAU General Assembly. The objective of this action is to uniquely identify all existing meteor showers and thus facilitate the establishment of associations between meteor showers and parent bodies among the many Near-Earth Objects that are being discovered.

The Task Group for Meteor Shower Nomenclature will work from a working list of ~230 showers compiled from past publications. This working list will be posted at the website of the IAU Meteor Orbit Data Center: `<http://www.astro.sk/ ne/IAUMDC/Ph2003/>`. The working list can be extended with newly identified meteor showers. A subset of these showers will be selected for inclusion in the list of established meteor showers. The list of established meteor showers shall also be posted at the IAU Meteor Orbit Data Center website after the approval by the Commission. Proposed members of the task group: Peter Jenniskens (USA, head of the TG MSN), Pavel Spurný (Czech Republic, president Commission 22), Jürgen Rendtel (Germany, IMO representative), Shinsuke Abe (Japan), Bob Hawkes (Canada), Jack Baggaley (New Zealand), Vladimír Porubčan (Slovakia, IAU Meteor Orbit Datacenter representative), and Tadeusz Jopek (Poland). The proposal was carried with acclamation.

8. Other business

The proposed merging of the Commission 22 with Commission 21 was discussed and it was decided that the next opportunity for discussion between representatives of both commissions will be during the Meteoroids 2007 conference in Barcelona, 11-15 June 2007.

Pavel Spurný
president of the Commission

Transactions IAU, Volume XXVIB
Proc. IAU XXVI General Assembly, August 2006
Karel A. van der Hucht, ed.

doi:10.1017/S1743921308023855

COMMISSION 51 — BIOASTRONOMY

BIOASTRONOMY

PRESIDENT	**Karen J. Meech**
VICE-PRESIDENT	**Alan P. Boss**
PAST PRESIDENT	**C. Stuart Bowyer**
ORGANIZING COMMITTEE	**Fabrizio Capaccioni, Pascale Ehrenfreund, Carlos Eiroa, William B. Hubbard, David W. Latham, Eduardo L. Martín, Michel Mayor, David Morrison, Raymond P. Norris, Jill C. Tarter**

PROCEEDINGS BUSINESS MEETING on 18 August 2006

1. Introduction

Commission 51 met on 18 August 2006. President Karen Meech chaired the meeting and there were 16 members present, including vice-president Alan Boss, and OC member David Morrison.

2. Bioastronomy 2007

One of the premier activities of our Commission is the organization of a triennial Bioastronomy meeting. The series of international Bioastronomy meetings has been ongoing since 1984. Bioastronomy, or Astrobiology, is a rapidly growing field, and there are an increasing number of international meetings which cover this subject area. One of the major activities during the past three years has been to work with members of the Federation of Astrobiology Organizations to try to reduce the number of meetings by looking for opportunities to hold joint congresses.

Table 1. Commission 51 Bioastronomy meetings

year	IAU sponsorship	location
1984	IAU Symposium No. 112	Boston, MA, USA
1987	IAU Colloquium No. 99	Balaton, Hungary
1990		Val Cenis, France
1993		Santa Cruz, CA, USA
1996	IAU Colloquium No. 151	Capri, Italy
1999		Kohala Coast, HI, USA
2002	IAU Symposium No. 213	Hamilton Island, Australia
2004		Reykjavik, Iceland
2007		San Juan, Puerto Rico

To this end representatives from IAU Commission 51, ISSOL (International Society for the Study of the Origin of Life), AbSciCon (Astrobiology Science Conference), and the NAI (NASA Astrobiology Institute) met in September 2004 to formulate a plan for meeting coordination. The outcome of the 9/04 meeting was an agreement between ISSOL and IAU Commission 51 to hold a simultaneous meetings in 2008 (one year later than the nominal Commission 51 meeting schedule) with some joint sessions at a site to be mutually selected. At the 2005 summer business meeting in China, ISSOL accepted an offer from hosts in Florence Italy for the 2008 meeting and they invited participation by Commission 51. In spite of many discussions, by March 2006 it was realized that the arrangements would not work, and Commission 51 had to search for an independent venue. Proposals to host the meeting were entertained from: Mexico, Tenerife, NRAO Virginia, Yellowstone, Flagstaff and San Juan Puerto Rico. The selection of the site in March 2006 was based on cost and the availability of the Arecibo facilities as well as a unique environment for astrobiology field excursions.

President Meech described the plans for the next Bioastronomy meeting, Bioastronomy 2007 on *Molecules, Microbes and Extraterrestrial Life*, to be held in San Juan, Puerto Rico from 12-16 July, 2007. The meeting website, `<www.ifa.hawaii.edu/UHNAI/bioast07.htm>` is available for registration, hotel booking and abstract submission. The meeting site, the Condado Plaza Hotel, is conveniently located 10 minutes from the international airport within walking distance of old San Juan. The conference facilities are excellent, with 9100 square feet of space for both oral sessions and poster papers. A block of rooms at an excellent group conference rate is being held until June 2007.

Meech is the chair of the Local Organizing Committee and William Irvine is the chair of the Scientific Organizing Committee.

Table 2. Local and Scientific Organizing Committee members

LOC member	institution	SOC member	institution
Jose Alonso	Univ. Puerto Rico	Peter Backus	SETI Institute
Daniel Altschuler	Arecibo Observatory	John Baross	Univ. Washington
Hector Ayala del-Rio	Univ. Puerto Rico	Alan Boss	Carnegie Inst.
Frank Drake	SETI Institute	Nader Haghighipour	Univ. Hawaii
Edna DeVore	SETI Institute	William Irvine	Univ. MA
Sixto Gonzales	Arecibo Observatory	Karen Meech	Univ. Hawaii
Pamela Harman	SETI Institute	Michael Mumma	NASA Goddard
Karen Meech	Univ. Hawaii	Carol Oliver	Austr. Ctr. Astrobio.
Carlos Rodriguez	Univ. Puerto Rico	John Rummel	NASA HQ
Michele Sonoda	Univ. Hawaii	Janet Siefert	Rice Univ.
Melita Thorpe	MWT Associates, Inc.	Thorsteinn Thorsteinsson	Orkustofnun
		Dan Werthimer	UC Berkeley

Irvine reported on the content of the invited speaker's program for the meeting and on the status of the acceptances to speak. The program consists of both invited and contributed talks. The session topics included: (*i*) synthesis of molecules and organics; (*ii*) comets and meteorites; (*iii*) outer solar system; (*iv*) Mars; (*v*) development of cells and complex life; (*vi*) life in extreme environments; and (*vii*) extrasolar planets and habitability.

In addition to the science program the meeting will have a full Education and Public Outreach component, lead by Daniel Altschuler. This will consist of a teacher workshop on Wednesday 18 July, the meeting which will offer lessons and materials related to the meeting, as well as the opportunity to interact with meeting scientists. Bioastronomy education papers will also be integrated into the main program. Tim Slater will be offering a course on "Teaching Introductory Level Astronomy and Astrobiology in the College Classroom" to meeting participants with a special workshop held July 13-15. In addition, AbGradCon will be holding a meeting for Astrobiology Graduate students on 14-15 July, just prior to the meeting. The early registration deadline for the meeting is set at 15 January 2007, the abstract and travel support request deadline is 15 February 2007, and the program will be available 1 April 2007.

The meeting sponsors include: The Institute for Astronomy, University of Hawaii, the NASA Astrobiology Institute, NAIC/Arecibo Observatory, Puerto Rico Space Grant Consortium, and the Australian Centre for Astrobiology.

3. Bioastronomy 2011

The idea of holding a joint meeting with ISSOL in 2011 was also discussed. Venues for such a joint meeting need to be solicited soon, in order for a joint venue to be decided upon in time for the 2008 ISSOL meeting in Florence, Italy. It was decided that the Commission should take the lead in negotiating with ISSOL regarding this possibility, rather than relying on the efforts of third parties, such as the Federation of Astrobiology Organizations, who have offered to help in the past. There was a brief discussion of possible sites for the 2011 meeting, including Lake Baikal, however, given the recent locations of the meetings, a site in the central US or Canada might be appropriate for even geographic distribution.

4. New members of Commission 51

Twenty new members were accepted into the Commission at the business meeting.

Table 3. New Commission 51 members

name	country	research field
Ian A. Bond	New Zealand	Astrophysics
Rolando P. Cardenas	Cuba	Cosmology
MilanM. Cirkovic	Serbia	Astrobiology, galaxy evolution
MariaR. Cunningham	Australia	Star formation
William Dent	UK	Star formation
HelenJ. Fraser	UK	Astrochemistry
Ray Jayawardhana	Canada	Star formation
Charles H. Lineweaver	Australia	Exoplanets
Ian C. Lyon	UK	Interstellar grains
Giuseppe Marzo	Italy	Astrobiology
MarlaH. Moore	USA	Ice irradiation
Dominique Naef	Chile	Extrasolar planets
Ralf Neuhaeusen	Germany	Star formation
Don Pollacco	UK	Extrasolar planets
Nuno MiguelC. Santos	Portugal	Extrasolar planets
PeterJ. Sarre	UK	ISM molecules
MarkR. Sims	UK	Space Missions
In-OkSong	Korea	Astrochemistry
Andrew J. Walsh	Australia	Star formation, Astrochemistry
Wilfred M. Walsh	Australia	Radio Astronomy

5. Election of officers and new Organizing Committee members

The Commission elected William Irvine to be the next vice-president and Alan Boss begins his term as Commission president at the end of the 2006 IAU XXVI General Assembly. Karen Meech was asked to continue to serve the Commission as past-president, given her institutional knowledge of the Commission for the last six years. One or two new Organizing Committee members will be sought, in order to widen the geographical representation. The Commission 51 membership list will be added to the web pages. Commission 51 needs to broaden its membership among IAU members, and a means to do this needs to be identified.

6. Closing remarks

There was consensus among the membership that our Commission is making progress in addressing the issues of consolidation of science meetings, and that we need to continue to be proactive. The members were keen to more actively participate in the Commission activities, and suggested that there be more frequent communication with the membership either via email or a newsletter. David Morrison publically thanked president Meech for all her efforts on behalf of the Commission for the past six years.

Karen J. Meech
president of the Commission

Transactions IAU, Volume XXVIB
Proc. IAU XXVI General Assembly, August 2006
Karel A. van der Hucht, ed.

doi:10.1017/S1743921308023867

DIVISION IV STARS

ÉTOILES

Division IV organizes astronomers studying the characterization, interior and atmospheric structure of stars of all masses, ages and chemical compositions.

PRESIDENT	**Dainis Dravins**
VICE-PRESIDENT	**Monique Spite**
PAST PRESIDENT	**Beatriz Barbuy**
BOARD	**Christopher Corbally, Wojciech Dziembowski, William I. Hartkopf, Christopher Sneden**

DIVISION IV COMMISSIONS

Commission 26	**Binary and Multiple Stars**
Commission 29	**Stellar Spectra**
Commission 35	**Stellar Constitution**
Commission 36	**Theory of Stellar Atmospheres**
Commission 45	**Stellar Classification**

DIVISION IV WORKING GROUPS

Division IV WG	**Abundances in Red Giants**
Division IV WG	**Massive Stars**

INTER-DIVISION WORKING GROUPS

Division IV-V WG	**Active B-type Stars**
Division IV-V WG	**Ap and Related Stars**

PROCEEDINGS SCIENCE AND BUSINESS MEETING on 21 August 2006

1. Division IV

The meeting of Division IV (Stars) during the IAU XXVI General Assembly in Prague was organized as mainly a science session, preceded by a short business part. In addition, its Organizing Committee had an informal meeting some day earlier. A comprehensive report on the activities of all its Commissions and Working Groups during the preceding triennium is in *Reports on Astronomy, Transactions IAU Volume XXVIA* (2007, ed. O. Engvold). In an evaluation of the activities of those bodies, it was concluded that they all were active and needed, and should continue in basically the same format also during the next triennium. One item for possible consideration, however, is the question of whether – in the longer run – perhaps some kind of merger of Commissions 29 and 36 should be considered.

2. Election of new officials

In the months preceding the IAU General Assembly, elections for a new Division president and vice-president were organized. Following discussions within the Organizing Committee, a format was chosen to assure a broader participation than was common in the past (elections

Figure 1. The Division IV Organizing Committee of 2003-2006 assembled during the IAU XXVI General Assembly in Prague. Left to right: Christopher Corbally (Vatican & Tucson; president Comm. 45; Div. IV vice-president-elect); Monique Spite (Paris-Meudon; president Comm. 36; Div. IV president-elect); Wojciech Dziembowski (Warsaw; president Comm. 35); Christopher Sneden (Austin; president Comm. 29); Beatriz Barbuy (São Paolo; former Div. IV president), William Hartkopf (Washington, DC; president Comm. 26); Dainis Dravins (Lund; Div. IV president).

held during the General Assembly, not always with very many participants), while still keeping the procedure manageable and having some assurance that the electorate be familiar with the duties expected for those serving on IAU bodies.

The electorate was defined as all members of all Organizing Committees of all Commissions within the Division, supplemented with the chairpersons of the Division-level Working Groups. Each person was to vote only once, even if holding multiple appointments. This electorate consisted of 58 persons, representing a broad range in geographic locations and thematic specialties. The votes were sent as e-mails to the Division, with 'backup' copies to the IAU Assistant General Secretary.

It was seen natural to elect the Division president from among the current Commission presidents, and each was contacted with an invitation to be a candidate. In those cases where the current presidents declined, the [immediate] past presidents were invited to be candidates. This procedure resulted in four excellent candidates (each of which received numerous votes), among whom Monique Spite was elected as the new Division president for 2006-2009, and Christopher Corbally as the new vice-president.

3. Stellar astrophysics in the 2010s

It was decided to devote most of the Division IV meeting to 'visionary science', with topics not covered at other General Assembly meetings. The title *Stellar Astrophysics in the 2010's: Extrapolating scientific challenges some decade into the future* was chosen, with the intention not to present recent results, but rather try to extrapolate (or even speculate) some decade into the future of what the scientific challenges then might be. The desire not to overlap with topics covered elsewhere meant that, e.g., problems such as the hydrodynamic modeling of entire stars were not included, since related problems were to be treated during Symposium 239 on *Convection in Astrophysics*.

During the well-attended meeting (organized and chaired by Dainis Dravins, Lund Observatory), the following talks were given:

Jason Aufdenberg (Embry-Riddle Aeronautical University, Florida)
Interferometric stellar imaging: thoughts on probing stellar photospheres at high-resolution in the 2010's.

As steps toward the full imaging of stellar surfaces, interferometric diameters and limb darkening measures may be obtained for stars across much of the HR diagram. Solar limb darkening was measured in the 1870's, and interpreted by Karl Schwarzschild a century ago: the center-to-limb profile for the Sun was then derived with a radiative (i.e., not adiabatic) temperature structure. Limb darkening is thus a key to understanding stellar atmospheric structure, and there is a need for multi-wavelength observations extending toward the blue and violet. Models are most sensitive at short wavelengths in segregating, e.g., among models with, or without convective overshoot. The disks of rapidly rotating stars become oblate, and some have already been mapped through interferometry: Altair (A7 V) has an equator-to-polar axis ratio = 1.14; Regulus (B7 V) has 1.32, and Archenar (B3 Vpe) has 1.56! Full imaging requires a good sampling of the (u,v) plane, becoming feasible with aperture masking on extremely large telescopes, which will offer hundreds of baselines. Supergiants such as Betelgeuse and Antares will be imaged already quite soon by interferometric instruments on the Large Binocular Telescope. Prospects include the imaging of odd-shaped and never-before-seen stars, perhaps differentially rotating rapid rotators shaped like butterflies? For further information on optical stellar interferometry, see links from `<http://olbin.jpl.nasa.gov/intro/>` .

Andreas Quirrenbach (Landessternwarte Königstuhl, Heidelberg)
Interferometric spectroscopy across spatially resolved stellar disks
Ordinary spectroscopy of stars observed as point sources blurs all surface features together, making those awkward or impossible to disentangle. However, combining the spatial resolution of stellar disks with high-resolution spectroscopy promises to be particularly powerful. Already now, giant stars are seen to be noticeably larger in molecular bands such as TiO, than in the continuum. Interferometry may also be used to assure uniqueness in other stellar surface image reconstructions, in particular to break the north-south degeneracy often seen in reconstructions from Doppler tomography. In Mira-type stars, shocks may be traced, and surface structures identified. The orientation of stellar rotation axes on the sky can be obtained from interferometer observations in different positions. By combining such measurements with exoplanet orbital inclinations, new information on their possible alignments can be obtained. Further, the possible axis alignments of stars in binaries or clusters may be studied. A spectral resolving power R $\approx$ 60,000 is required, feasible to realize in a proposed project utilizing the UVES spectrometer at ESO's Very Large Telescope (VLT). A fiber link would bring two phase-stable beams from the VLT interferometer to UVES, where the light would be spectrally dispersed just as from any other source. For a barely resolved star of magnitude 7, signal-to-noise ratios of 30 should be reachable for each spectral element, using interferometer telescope apertures of 75 cm. Observational issues include the handling of phase delay changes during integration, caused by the terrestrial atmosphere. Further information on various VLT interferometer projects can be found at `<http://www.eso.org/projects/vlti/>` .

Rafael Bachiller (Observatorio Astronómico Nacional, Madrid)
Stellar evolution: prospects with ALMA
Currently under construction at 5,000 m altitude on the Andean altiplano in northern Chile, the Atacama Large Millimeter Array (ALMA) is a radio interferometer to comprise up to 64 twelve-meter diameter antennas with about 8,000 m^2 of aperture (including the ALMA Compact Array system). With its good coverage of the mm and the sub-mm ranges, ALMA will greatly expand the field of stellar astronomy and, in particular, will revolutionize our knowledge of the two ends of stellar evolution. On one hand, ALMA will be the first telescope capable of resolving the key processes taking place during the collapse of star-forming clouds, the structure of protostars, the origin of proto-stellar outflows, and the physics and chemistry of proto-planetary disk systems. On the other hand, ALMA will allow the direct observation of

dust formation processes in AGB stars and will provide crucial details on the structure and composition of the dust and molecular material in the circumstellar envelopes of AGB stars, post-AGB objects, and planetary nebulae, leading to an unprecedented and detailed view of the late stellar evolution. Further information on the highly international ALMA project can be found at <http://www.eso.org/projects/alma/>, <http://www.cv.nrao.edu/naasc/>, <http://www.nro.nao.ac.jp/alma/>, and further links from these.

Karl Menten (Max-Planck-Institut für Radioastronomie, Bonn)
Submillimeter analyses of stellar photospheres

ALMA will have about 100 times better continuum sensitivity than existing radio arrays, and will open up radio studies of also 'normal' stars. Detection of a nearby star such as α Cen A (which, at some 20 milliarcsecond [mas] diameter remains spatially unresolved) with a flux of some 27 mJy at 345 GHz will be possible already within seconds of integration. Adaptive calibration using stellar photospheres of nearby stars will permit very deep integrations, offering the possibility of radio detection of extrasolar planets. A possible 'Jupiter' around α Cen would be 4 arcsec away, have a flux of 6 μJy at 345 GHz, and be potentially detectable. Giant and supergiant stars are expected to have quite different sizes (and shapes?) for their 'optical photosphere', 'radio photosphere', 'molecular photosphere', and the outer shells of SiO and H_2O masers. Maser emission sources could further be used as 'guide stars' for calibrating a possibly very weak photospheric continuum emission. The inner envelopes of at least supergiant stars will be imaged with ALMA, revealing the structure of the radio photosphere; of atmospheric chemistry; of dust formation and depletion, and showing the beginning stellar-wind outflow. Other future radio facilities of significance for stellar studies include the Expanded Very Large Array (maximum spatial resolution to be improved to some 5 mas; <http://www.aoc.nrao.edu/evla/>, and of course the future Square Kilometer Array (<http://www.skatelescope.org/>).

Lennart Lindegren (Lund Observatory)
Browsing for rare stellar types in the Gaia catalog

One lesson learned from the previous space astrometry mission *Hipparcos* is that astrometric data, however comprehensive, must be complemented by photometric and spectroscopic information to become widely usable. The *Gaia* space astrometry mission is to measure more than a billion stars, and is to collect also such complementary data. Photometry will be made in the wide G-band (330-1050 nm) at full spatial resolution, and in blue and red portions using a dispersing prism. Expected precision per observation for a 15 mag star is 0.002 mag in G, and 0.02 mag per spectral resolution element in the prismatic modes. Radial-velocity measurements will be made for stars brighter than 17 mag, using the infrared Ca II triplet for accuracies of 1-10 km s^{-1}. During the five-year mission, each star will typically be observed during $\approx$ 100 epochs.

The *Gaia* database will offer globally calibrated data for many millions of objects. For example, the number of stars with relative uncertainty in distance better than 1% is expected to exceed 10 million. Probably, stars in very rare evolutionary phases (e.g., post-AGB) will be found, as will such of especially high or low luminosity; of high or low metal content. Perhaps one will identify radial orbits in the Galaxy for the first stars? Or black holes in non-interacting binaries (not visible in X-rays) that need astrometry for identification? Nearby stars with (accidentally) small proper motions might emerge; as could very wide binaries or loose stellar groups. However, efficient data-mining algorithms will be required since data will still be inhomogeneous in terms of accuracy (dependent on the apparent magnitude), and since spurious outliers in the data can not be avoided (e.g., such caused by detector noise): even a very low frequency of such will produce many among the more than a billion objects studied. With a currently planned launch date in 2011, nominal observations would start in 2012 and, following some early data releases, the final catalog would become available in 2020. Further information on the Gaia mission can be found at <http://www.esa.int/science/gaia> .

Ken Carpenter (NASA Goddard Space Flight Center) & SI Vision Mission Study Team
Stellar Imager (SI) space mission: stellar magnetic activity

The *Stellar Imager* (SI) is a planned ultraviolet and optical, space-based interferometer, designed to enable spectral imaging of stellar surfaces and stellar interiors (via asteroseismology) and of the Universe in general, with a 0.1 milliarcsecond spatial resolution. *SI* was identified as a 'Flagship and Landmark Discovery Mission' in the 2005 Sun Solar System Connection (SSSC)

Roadmap and as a candidate for a 'Pathways to Life Observatory' in the Exploration of the Universe Division (EUD) Roadmap (May, 2005). The ultra-sharp images from the *Stellar Imager* will revolutionize our view of many dynamic astrophysical processes: The 0.1 mas resolution of this deep-space telescope will transform point sources into extended ones, and snapshots into evolving views. *SI*'s science focuses on the role of magnetism in the Universe, particularly on magnetic activity on the surfaces of stars like the Sun. *SI*'s prime goal is to enable long-term forecasting of solar activity and the space weather that it drives in support of the 'Living With a Star' program in the 'Exploration Era'. *SI* will also revolutionize our understanding of the formation of planetary systems, of the habitability and climatology of distant planets, and of many magneto-hydrodynamically controlled processes in the Universe. A mission architecture has been defined that can meet the science goals, involving a group of 1-m class telescopes flying in formation, and forming an optical Fizeau interferometer with a light-combining hub at its center. Additional information on the *SI* mission concept and related technology development can be found at <http://hires.gsfc.nasa.gov/si/> .

Webster Cash (presented by Eric Schindhelm) (CASA, University of Colorado, Boulder)
MAXIM - coronal structure with X-ray interferometry
MAXIM (Micro-Arcsecond X-ray Imaging Mission) is a proposed space mission with an imaging X-ray interferometer producing images with 0.1 microarcesond resolution. Under such resolutions, many stellar coronae become highly extended sources, with a potential for detailed imaging. The short wavelengths of X-rays imply that the required baselines for a given spatial resolution can be 100 to 1,000 times smaller than those in the optical and infrared bands. Diffraction-limited X-ray optics with a size of meters can yield an angular resolution of milliarcseconds, which would be sufficient to resolve coronal structures on nearby stars, while the X-ray interferometry envisioned for *MAXIM* may improve resolution by further orders of magnitude. Formation-flying spacecraft are envisioned, positioned to 25 μm precision. One main scientific aim of the mission is to image black holes and their X-ray-emitting immediate vicinities (only few black holes have their Einstein radii greater than some microarcsec). The full mission could be preceded by a 'MAXIM Pathfinder' spacecraft, providing 100 microarcsecond resolution. Alternative mission designs include normal-incidence optics, while technical issues include questions such as how to point a microarcsecond telescope. Additional information on the MAXIM mission concept can be found at <http://maxim.gsfc.nasa.gov/> .

4. Closing remarks

Following the Division IV meeting, it was generally agreed that stellar astrophysics has an exciting future for the 2010s and beyond. The advent of extremely large optical telescopes and interferometers, will make many stars accessible as extended surface objects rather than the mere point sources of the past. Radio interferometers are becoming sufficiently large and sensitive to enable the study of also 'ordinary' stars, not only the most active ones detectable in the past. Large-scale surveys of stars in both our own, and neighboring Galaxies will enable the identification of also rare types of stars, or stars in short-lived evolutionary stages, while space missions further ahead promise to enable a quite detailed imaging of active-region structures across stellar surfaces, and their evolution over both short timescales, and over stellar activity cycles. We indeed look forward to experiencing such a stellar future!

Dainis Dravins
president of the Division

Transactions IAU, Volume XXVIB
Proc. IAU XXVI General Assembly, August 2006
Karel A. van der Hucht, ed.

doi:10.1017/S1743921308023879

DIVISION IV / WG ABUNDANCES IN RED GIANTS

CHAIR **John C. Lattanzio**
MEMBERS **Pavel A. Denissenkov, Roberto Gallino, Josef Hron, Uffae G. Jorgensen, Claudine Kahane, Sun Kwok, Jacobus Th. van Loon, Verne V. Smith, Christopher Tout, Robert F. Wing, Ernst Zinner**

PROCEEDINGS BUSINESS MEETING on 22 August 2006

1. Introduction

Unfortunately the Business Meeting clashed with interesting sessions on stellar convection theory that were very relevant to most members of this Working Group. Hence the attendance was very small, and some preliminary discussions were later followed up by email among the Organising Committee members.

2. Recent activities

It has been a busy year for the WG. On top of organising JD11 *Pre-Solar Grains as Astrophysical tools* at the GA, we were closely involved in the planning for a scientific meeting hosted by the Vienna Observatory. This meeting took place from 7 to 11 August the topic "Why Galaxies Care About AGB Stars" and was well attended with over 100 participants. We are currently discussing the possibility of extending this into a series of meetings.

3. Plans for 2007

There are a number of important meetings planned for 2007 and the WG is represented on the SOC for two of these. These include "First Stars III" to be held in the USA in July 2007 as well as the meeting "Nuclear Astrophysics - Beyond the first 50 Years". This meeting will be hosted at Clatech in July 2007 and will mark the 50th anniversary of two truly seminal papers defining the birth of the field of nucleosynthesis: Burbidge, E. M., Burbidge, G. R., Fowler, W. A., & Hoyle, F. 1957, *Rev. Mod. Phys.* 29, 547; and Cameron, A. G. W., 1957, *PASP*, 69, 201.

4. Plans for 2008 and 2009

The WG is involved in proposals for meeting in 2008 and we are discussing how best to serve our needs by the time of the next IAU-GA in 2009.

5. Closing remarks

The field of nucleosynthesis is very active at present, and we expect a lively future.

John C. Lattanzio
incoming chair of the Working Group

Transactions IAU, Volume XXVIB
Proc. IAU XXVI General Assembly, August 2006
Karel A. van der Hucht, ed.

doi:10.1017/S1743921308023880

DIVISION IV-V / WG ACTIVE B-TYPE STARS

CHAIR **Stanley P. Owocki**
PAST CHAIR **Stanislav Štefl**
MEMBERS **Conny Aerts, Beatriz Barbuy, Juan Fabregat, Douglas R. Gies, Edward F. Guinan, Hubertus F. Henrichs, Geraldine J. Peters, John M. Porter († 2005), Thomas Rivinius, Stanislav Štefl**

PROCEEDINGS BUSINESS MEETING on 18 August 2006

1. Introduction

The meeting of the Working Group on Active B-type Stars consisted of a business meeting followed by a scientific meeting containing invited and contributed talks. The titles of the talks and their presenters are listed below. We plan to publish a series of articles containing summaries of these talks in Issue No. 39 of the *Be Star Newsletter*.

Outgoing chair of the Working Group, Stan Owocki, welcomed the attendees. He presented a brief update on the status of the proceedings from the meeting *Active OB-Stars: Laboratories for Stellar and Circumstellar Physics* held in Sapporo, Japan, 29 August to 2 September 2005. The editors of the proceedings are S. Štefl, S. Owocki and A. Okazaki. A proposal to continue IAU recognition of the Working Group during the 2006-09 triennium was submitted by Stan Owocki by the Spring deadline.

2. Business Meeting

2.1. *Be Star Newsletter*

The *Be Star Newsletter*, which is published in hard copy at Georgia State University for the Working Group on Active B-type Stars, continues to be the main source of information on new discoveries, ideas, manuscripts, and meetings on active B-type stars. G. Peters, D. Gies, and D. McDavid continue, respectively, as Editor-in-Chief, Technical Editor, and Webmaster. Abstracts and announcements are usually posted on our website (`<http://www.astro.virginia.edu/~dam3ma/benews/>`) within 48 hrs of being received.

Articles submitted for publication have been refereed since the year 2000. Although this practice has resulted in fewer full articles, the quality of the published material has been vastly improved. In 2005 we introduced a new section called Community Comments, in which Working Group members can voice opinions or ideas on which the community can submit rebuttal. In March 2005 we published Issue No. 37 that contains the proceedings from the scientific session held during the business meeting of the Working Group on Active B-type Stars at the IAU XXV General Assembly in Sydney, Australia, 2003.

2.2. *OC election results*

In August 2006 an election was held by e-mail ballots that were sent to all current IAU members of the Working Group on Active B-type Stars. The Organizing Committee (OC) for the 2006-09 triennium is:

Term expiring in 2009: Juan Fabregat, Douglas Gies, Huib Henrichs, and David McDavid.
Term expiring in 2012: Karen Bjorkman, Coralie Neiner, Geraldine Peters, and Philippe Stee.
Non-voting: president IAU Division IV (*Stars*), president IAU Division V (*Variable Stars*), and Stan Owocki(outgoing WG chair).

2.3. *Action items*

2.3.1. *Bye-laws for the Working Group*

The bye-laws would include a mission statement and rules governing the election of the OC. Suggestions were made that non-IAU members should be allowed to vote in elections if they have participated in at least one meeting on active B-type stars within a six year period, and that a non-Ph.D. researcher might be recognized as voting member if his/her National Committees agrees.

2.3.2. *Other items*

There was a discussion on how to stimulate more participation from Working Group members. In 2006 there was a slight increase in the voter turnout of 25% of the eligible voters compared with 15% in 2003. The number of articles submitted to the Be Star Newsletter has declined, but submissions of abstracts and news items have increased.

2.4. *Rules governing the election of the OC in 2006*

The top priority of the OC during the 2006-09 triennium will be to establish a set of bye-laws including a set of formal rules governing the election of the Organizing Committee. The following rules were followed for the 2006 election:

1. The Working Group (WG) is considered to consist of all subscribers of the Be Star Newsletter, Electronic Edition, who are IAU members.

2. Every member of the WG can nominate four persons as candidates for the new Organizing Committee (OC) and send them to the Election Officer (chosen from the editors of the Be Star Newsletter.) The Election Officer selects the 10 candidates with the highest number of votes. If several individuals have the same number of nominations for the last spot on the ballot, they are all accepted and the number of candidates can be higher than ten. According to the general IAU rules, only IAU members, or new members pending approval at the current General Assembly, can be accepted as candidates for OC membership, and balanced regional representation should be taken into account in the nomination of the candidates.

3. The Election Officer must verify that all nominees to be listed as candidates on the ballot are willing to serve if elected.

4. The four new OC members are elected from the OC candidates by the members of the WG, who may vote for up to four different persons.

5. The four persons with the largest numbers of valid votes are elected for a six-year term. The four new and the four continuing OC members determine among themselves the new Chairperson.

3. Scientific session

Session 1 (D. McDavid, chair)

09:20	*Observations of the B[e] Star MWC 349 with mid-infrared interferometry* A. Quirrenbach
09:35	*The Kepler mission's guest opportunity program: opportunity for optical monitoring of B-type stars?* M. Smith
10:00	*Rotational velocities of Be-type stars* I. Howarth
10:15	*Analysis of the high temperature region in B-type stars* A. Torres

Session 2 (G. Peters, chair)

11:00	*Statistical Properties of a Sample of Periodically Variable B-Type Supergiants: Evidence of a Pulsation-Mass Loss Connection?* K. Lefever, J. Puls , C. Aerts*
11:15	*The Comings and Goings of Be Stars* V. McSwain

11:30 *HD 61273: a new binary system with a hot component showing an Hα emission line*
D. Briot*, F. Royer

11:45 *Spectropolarimetry of Be stars with FORS1 at the VLT revisited*
R. Yudin

12:00 *Circumstellar disks from rotating stars with and without magnetic fields*
S. Owocki*, A. ud-Doula, R. Townsend

* Presenter

4. Closing remarks

I would like to thank those who participated in the Working Group business meeting through their attendance and the presentation of a set of very interesting talks. Hearty thanks are due to outgoing OC chair Stan Owocki, who skillfully led our Working Group for the past three years and was one of the primary organizers of the meeting on active B-type stars in Sapporo, Japan in 2005.

We are looking forward to seeing you again in 2009 at the meeting of the Working Group on Active B-type Stars at the IAU XXVII General Assembly in Rio de Janeiro.

Geraldine J. Peters
incoming co-chair of the Working Group

Transactions IAU, Volume XXVIB
Proc. IAU XXVI General Assembly, August 2006
Karel A. van der Hucht, ed.

doi:10.1017/S1743921308023892

COMMISSION 26 — BINARY AND MULTIPLE STARS

ÉTOILES DOUBLES ET MULTIPLES

PRESIDENT — **William I. Hartkopf**
VICE-PRESIDENT — **Christine Allen**
PAST PRESIDENT — **Colin D. Scarfe**
ORGANIZING COMMITTEE — **John Davis, Francis C. Fekel, Patricia Lampens, Josefina F. Ling, Edouard Oblak, Terry D. Oswalt**

DIVISION IV / COMMISSION 26 WORKING GROUP

Division IV / Commission 26 / WG — **Binary and Multiple System Nomenclature**

PROCEEDINGS BUSINESS MEETING on 17 August 2006

1. Introduction

The Business Meeting of Commission 26 took place on 17 August 2006, during the IAU XXVI General Assembly in Prague. It was chaired by its outgoing president, William Hartkopf, and was attended by the following members and guests: H. Abt, C. Allen, W. van Altena, R. Argyle, G. van Belle, D. Bisikalo, A. Bradley, T. ten Brummelaar, D. Cline, T. Corbin, R. Costero, J. Davis, H. Dickel, J. Docobo, J. Echevarria, F. Fekel, E. Griffin, P. Harmanec, W. Hartkopf, P. Lampens, O. Malkov, B. Mason, H. McAlister, T. Oswalt, D. Pourbaix, A. Poveda, D. Raghavan, W. Sanders, C. Scarfe, W. Tango, A. Tokovinin, S. Urban, R. Wilson, and H. Zinnecker.

2. Membership and elections

Hartkopf began the meeting with a report on events during the past triennium. He announced the deaths, during that period, of members Geoffery G. Douglass, Wulff D. Heintz, and Richard L. Walker. The meeting observed a few moments of silence in their memory.

The Commission then welcomed the following 26 new members:

Javier A. Ahumada (Argentina),
Charles D. Bailyn (USA),
Zorica D. Cvetković (Serbia),
Robert Dukes, Jr. (USA),
Vladimir Elkin (UK),
Lars M. Freyhammer (United Kingdom),
Simon P. Goodwin (UK),
Todd C. Hillwig (USA),
Liliya Kazantseva (Ukraine),
Ludmilla Kisseleva-Eggleton (USA),
Spyridon Kitsionas (Germany),
Wilhelm Kley (Germany),
Jeremy Lim (China Taipei),
Yuri Lyubchik (Ukraine),
Vladislava Marsakova (Ukraine),
David A. McDavid (USA),
Maciej P. Mikolajewski (Poland),
Ignacio Negueruela (Spain),
Ralf Neuhäuser (Germany),
Dieter E.A. Nürnberger (Chile),
Victor V. Orlov (Russia),
Monika G. Petr-Gotzens (Germany),
Donald Pollacco (UK),
Lewis C. Roberts, Jr. (USA),
Allyn J. Smith (USA), and
Alexander P. Zheleznyak (Ukraine).

This brings the total membership of Commission 26 to 151, from 35 countries.

The president then reported on the recent election, in which just over half the membership voted. For the period 2006-2009, the president will be Christine Allen, and the vice-president

José Docobo. The Organizing Committee will be made up of new members Yuri Balega, Brian Mason, Dimitri Pourbaix, and Colin Scarfe, along with continuing members John Davis, Edouard Oblak, and Terry Oswalt. Four new OC members were added rather than the usual three, due to a tie in the vote. Outgoing OC members Francis Fekel, Patricia Lampens, and Josefina Ling were thanked for their service over the past six years.

Hartkopf gave a short summary of developments at the Commission's website `<http://ad.usno.navy.mil/wds/dsl.html>`, which contains meeting announcements, bibliographies of double star papers, commission membership lists, and links to catalogs of interest to C26 members. Suggestions for further improvements were invited.

Docobo gave a report on Commission 26 circulars, which saw their 50th anniversary during the past triennium and which have been published since 1993 by Docobo and Josefina Ling of the Observatorio Astronómico 'Ramón María Aller'. These circulars, emailed to all C26 members and also available on the Commission's website, include orbits (some 160 during the past 3 years), measurements, notes, bibliographies, obituaries, and miscellaneous announcements and commentaries.

3. Past and future meetings

Andrei Tokovinin gave a brief account of the ESO workshop *Multiple Stars across the H-R Diagram*, held 12-15 July 2005 in Garching. Topics discussed during this meeting included

- ⋆ Observing multiple stars (using interferometry, speckle, AO, X-ray, spectroscopy)
- ⋆ Dynamics and evolution of multiple stars
- ⋆ Formation, young systems, statistics
- ⋆ Multiple stars with special components

Proceedings of this meeting are now available on line at `<http://www.eso.org/gen-fac/meetings/ms2005/proc_papers.html>`.

Announcements were then made of upcoming meetings, including IAU Symposium No. 240, *Binary Stars as Critical Tools and Tests in Contemporary Astrophysics*, to be held later during the Prague GA, and Symposium No. 248, *A Giant Step: from Milli- to Micro-arcsecond Astrometry*, to be held 15-19 October 2007 in Shanghai.

4. Catalogs

Reports on various catalogues and journals of interest to Commission 26 followed:

4.1. *USNO CD-ROM 2006.5*

Hartkopf described this update to the 2001.0 CD, which includes current versions (as of 30 Jun 2006) of the four USNO double-star catalogs on that CD *(Washington Double Star Catalog, 6th Orbit Catalog, 4th Interferometric Catalog*, and *2nd Photometric Catalog)*, as well as a *Catalog of Linear Elements* and a brief history of double star activities at the USNO. Copies of the USNO CD were made available to all meeting participants.

4.2. *Ninth Spectroscopic Orbit Catalog*

Pourbaix briefly described the latest version of the catalog (see Pourbaix et al. 2004), which consists of 3,191 orbits of 2,684 systems. He estimates that ~75% of systems with published orbits are now included in the catalog. SB9 may be accessed through (`<http://sb9.astro.ulb.ac.be>`) either on a star-by-star basis (via various identifiers) or as a tarball containing all orbits and notes. Manpower to work on SB9 is limited, so help in data entry is welcome.

4.3. *Multiple Stars Catalog*

Tokovinin announced the latest version of the MSC (Tokovinin 1997), which now has WMC designations implemented, and is available at `<http://www.ctio.noao.edu/~atokovin/msc/>`.

4.4. *Catalogue of Orbits and Ephemerides of Visual Double Stars*

Docobo described this catalog which is available at `<http://www.usc.es/astro>`. As of 7 August 2006 the catalog comprised 1,844 orbits of 1,497 binaries, and includes ephemerides and orbit grades.

4.5. *Catalogue of Eclipsing Binaries*

Oleg Malkov described this new catalog, whose goal is to not only catalog all known eclipsing variables, but to develop a procedure for their classification (detached, near-contact, contact; spectral type and luminosity class; description of eclipse depths and durations, etc.). A description of the catalog and classification work has been recently published (see Malkov et al. 2006).

5. Observing reports

Robert Argyle followed with a discussion of amateur projects in double-star measurement, including the efforts of the French Astronomical Society, the Webb Society, the new electronic *Journal of Double Star Observations*, the Spanish/South American LIADA group, and Garraf Observatory in Spain. Amateurs are responsible for over 1/3 of the measures added to the WDS database during the past six years, and many of these measures are of quite high quality. Argyle highlighted some of the major amateur observers, their equipment, and examples of their observations. He ended his talk with a plea encouraging observers to work in the long-neglected southern hemisphere.

Reports were given on various observing facilities and programs, including the speckle programs of Calar Alto (Docobo), the USNO (Brian Mason), PISCO (Argyle), and RYTSI/WIYN (William van Altena). Theo ten Brummelaar gave an update on the CHARA Array, which is now operating with six 1-meter telescopes over a 350-m baseline.

The business portion of the meeting then concluded with an outline of plans for the next triennium, by incoming president Allen.

6. Science session

Following a break, the science portion of the Commission meeting was comprised of six interesting talks:

⋆ Theo ten Brummelaar: *The use of an outlying star as a calibrator for close pairs in interferometry*

⋆ Rafael Costero: *The Double–Lined Spectroscopic Binary θ^1 Ori E: an Intermediate-Mass, Pre-Main-Sequence System* (R. Costero, A. Poveda, & L. Echevarría)

⋆ Jose Docobo: *Coplanarity in the triple system Gliese 22* (J. A. Docobo, V.S. Tamazian, G. Weigelt, D. Schertl, Y. Y. Balega, M. Andrade, & P. P. Campo)

⋆ Elizabeth Griffin: *The Binary Star Gamma Persei: Bright, but Ill-understood...*

⋆ Patricia Lampens: *Spectral disentangling and combined orbit analysis of the Hyades binary θ^2 Tau* (P. Lampens, Y. Frémat, P. De Cat, & H. Hensberge)

⋆ Brian Mason: *Double Stars and the Terrestrial Planet Finder Mission*

After a few words of thanks by the outgoing president, the meeting adjourned to the steps of the Prague Congress Centre for a group photo .

William I. Hartkopf
president of the Commission

References

Malkov, O., Oblak, E., Avvakumova, E., & Torra, J. 2006, *A&A*, 446, 785
Pourbaix, D., et al., 2004, *A&A*, 424, 727
Tokovinin, A. 1997, *A&AS*, 124, 75

Transactions IAU, Volume XXVIB
Proc. IAU XXVI General Assembly, August 2006
Karel A. van der Hucht, ed.

doi:10.1017/S1743921308023909

COMMISSION 35 — STELLAR CONSTITUTION

CONSTITUTION DES ÉTOILES

PRESIDENT	**Wojciech A. Dziembowski**
VICE-PRESIDENT	**Francesca D'Antona**
PAST PRESIDENT	**Don A. VandenBerg**
ORGANIZING COMMITTEE	**Corinne Charbonnel, Joergen Christensen-Dalsgaard, Joyce A. Guzik, Norbert Langer, Richard B. Larson, James W. Liebert, Georges Meynet, Ewald Müller, Hideyuki Saio**

PROCEEDINGS BUSINESS MEETING on 16 August 2006

1. Introduction

The session was brief and quite informal as there were only six participants. The agenda included my report on organizational activities of the Commission during the 2003-2006 term and Virginia Trimble's presentation *Presence of binary stars in the current astronomical literature.* I summarize below the most important part of my report.

2. Membership

The Commission has now over 350 members. Our excellent website (<http://iau-c35.stsci.edu>) contains up-to-date information about the Commission activity and a number of useful links to astronomical resources of interest for the members. The site has been designed and maintained by Claus Leitherer from Space Telescope Science Institute, who deserves for that our special thanks.

3. Organizing Committee

The election of all four new members of the Organizing Committee was conducted before the General Assembly through a web poll. All the Commission members with valid e-mail addresses have been first asked to nominate candidates and then to vote for eight candidates, who received more than two nomination and agreed to stand for the election. Nearly one hundred members took part in the poll. Soon after, the present and newly elected OC members chose the next vice-president.

The resulting composition of the Organizing Committee for the 2006-2009 looks as follows:
Francesca D'Antona (president, Italy), Corinne Charbonnel (vice-president, France) Gilles Fontaine (Canada), Richard B. Larson (U.S.A), John Lattanzio (Australia), James W. Liebert (U.S.A.), Ewald Müller (Germany), Achim Weiss (Germany), and Lev R. Yungelson (Russian Federation).

Of 925 new IAU members proposed by their National IAU Committees, 32 expressed their wish of joining Commission 35. Here are their names with proposing countries in the brackets: Torben Arentoft (Denmark), Sydney A. Barnes (USA), Sarbani Basu (USA), Michael Bazot (Denmark), Ignazio Bombaci (Italy), Brian C. Chaboyer (USA), Gilles Chabrier (France), Stephane Charpinet (France), Jadwiga Daszyńska-Daszkiewicz (Poland), Jean Dupuis (Canada),

Leo A. Girardi (Italy), Raphael Hirshi(Swiitzerland), Jes K. Joergensen (USA), Li Qingkang KANG (China), Matthias Liebendoerfer (Switzerland), Francois Lignieres (France), Marco Limongi (Italy), Keiichi Maeda (Japan), David A. McDavid (USA), Gijs Nelemans (the Netherlands), Joana M. Oliveira (UK), Simon J. O'Toole (Australia), Klaus M. Pontoppidan (USA), Maarten Reyniers (Belgium), Leonardo G. Sigallotti (Venezuela), Takuma Suda (Japan), Victor P. Utrobin (Russian Federation), Paolo Ventura (Italy), Jorick S. Vink (UK), Ken Yi (Republic of Korea), Takashi Yoshida (Japan), and Maxim V. Yuskin (Russian Federation).

They all have been welcomed to Commission 35.

Wojciech A. Dziembowski
president of the Commission

Transactions IAU, Volume XXVIB
Proc. IAU XXVI General Assembly, August 2006
Karel A. van der Hucht, ed.

doi:10.1017/S1743921308023910

COMMISSION 36 — THEORY OF STELLAR ATMOSPHERES

THÉORIE DES ATMOSPHÈRES STELLAIRES

PRESIDENT — **Monique Spite**
VICE-PRESIDENT — **John D. Landstreet**
ORGANIZING COMMITTEE — **Martin Asplund, Thomas R. Ayres, Suchitra C. Balachandran, Dainis Dravins, Peter H. Hauschildt, Dan Kiselman, K. N. Nagendra, Christopher Sneden, Grazina Tautvaišienė, Klaus Werner**

PROCEEDINGS BUSINESS MEETING on 16 August 2006

1. Introduction

The business meeting of Commission 36 was held during the General Assembly in Prague on 16 August. It was attended by about 15 members. The issues presented included a review of the work made by members of Commission 36, and the election of the new Organising Committee. We note that a comprehensive report on the activities of the commission during the last triennium has been published in Reports on Astronomy, Transactions IAU Volume XXVIA. The scientific activity of the members of the commission has been very intense, and has led to the publication of a large number of papers.

2. Activities

The Commission 36 Organising Committee evaluated and recommended several proposals for IAU meetings during the last triennium. It is pleased that qualified programs of stellar physics could be offered at the 2006 Prague General Assembly.

3. Membership

There are now more than 300 members of the Commission 36: 142 from Western Europe, 113 from North America, 33 from Asia Pacific, 20 from Eastern Europe and 9 from South America. Europe provides an important fraction of members, and this number will even increase in the future, since among the 33 new IAU members who have chosen Commission 36 as their primary Commission, 25 are European and only one is from North America.

4. Elections

In order to enhance the participation of members in commission matters and to have in the OC only members really interested in participating to the work of the Commission, we sent an e-mail to all the members of the Commission asking them if they would be interested in being members of the future OC. They were 15 candidates; from these candidates, 5 were elected by the OC of Commission 36. The newly elected OC members are: Svetlana V. Berdyugina

(Finland/Switzerland), Hans-G. Ludwig (Germany/France), Lyudmila I. Mashonkina (Russian Federation), Joachim Puls (Germany), and Sofia Randich (Italy).

Then the 12 OC members and the five new members have elected the new vice-president, Martin Asplund. The previous elected vice-president, John Landstreet, becomes president of the OC. The composition of the OC for the next triennium is: John D. Landstreet (president, Canada), Martin Asplund (vice-president, Australia). Members: Suchitra C. Balachandran (USA), Svetlana V. Berdyugina (Switzerland), Peter H. Hauschildt (Germany), Hans G. Ludwig (France), Lyudmila I. Mashonkina (Russian Federation), K.N. Nagendra (India), Joachim Puls (Germany), Sofia Randich (Italia), Monique Spite (past president, France), and Grazina Tautvaišiené, (Lithuania).

5. Possibility of merging Commissions 29 and 36?

During the meeting in Prague the possibility of merging Commissions 29 and 36 was discussed. It had been noted previously that there is an important overlap in interests of these commissions: Commission 36 representing mainly the theoretical side of the "stellar atmospheres" and Commission 29 the observational side.

A merger of these commissions could improve the discussion and organisation of symposia. However there were diverse views among the members of the Commission and in particular, there was concern that the point of view of the theoreticians could be diluted. Moreover since only 20% of the members of the commission 36 also belong to Commission 29, the resulting Commission would be "huge" (more than 500 members). Finally it was decided to continue to explore this possibility, and to have further discussion at the next General Assembly in Rio de Janeiro. Such discussion should reasonably be preceded by a concrete proposal, and will depend on the policies of the IAU, particularly as they concern large Commissions.

6. Conclusion

The outgoing president thanked the outgoing OC and welcomed John Landstreet as the new president of the Commission 36 for the next triennium.

Monique Spite
president of the Commission

Transactions IAU, Volume XXVIB
Proc. IAU XXVI General Assembly, August 2006
Karel A. van der Hucht, ed.

doi:10.1017/S1743921308023922

COMMISSION 45 STELLAR CLASSIFICATION

CLASSIFICATION STELLAIRE

PRESIDENT	**Christopher J. Corbally**
VICE-PRESIDENT	**Sunetra Giridhar**
PAST PRESIDENT	**Thomas H. Lloyd Evans**
ORGANIZING COMMITTEE	**Coryn A.L. Bailer-Jones, Roberta M. Humphreys, Joseph D. Kirkpatrick, Xavier Luri, Dante Minniti, Laura E. Pasinetti, Vytautas P. Straižys, Werner W. Weiss**

DIVISION IV / COMMISSION 45 WORKING GROUP

Division IV / Commission 45 / WG Standard Stars

PROCEEDINGS BUSINESS MEETING on 16 August 2006

1. Business

The Business Meeting of Commission 45 was held on 16 August 2006. It was attended by the president and vice-president of the Commission as well as by twenty other members of the Commission. Attendance was limited, as usual, by the unavoidable occurrence of parallel sessions.

The memories of our colleague Dr. Charles Perry were recalled at the meeting. Subsequently in September 2006, and with deep regret, we learnt of the death of Dr. Laura Pasinetti, one of the retiring members of the Organizing Committee.

New members of the IAU who are known to be joining Commission 45 are Carlos Allende Prieto, Cassio L. Barbosa, Johan Holmberg, Yuri V. Pakhomov, and Allyn John Smith. These were warmly welcomed.

The President reported that a proposal, initiated by the Commission, for an IAU Joint Discussion during the IAU XXVI General Assembly on *Exploiting Large Surveys for Galactic Astronomy* had been approved by the Executive Committee. Vigorous support from four participating Divisions and five additional Commissions, leading to a Scientific Organizing Committee of ten persons, had resulted in a stimulating programme. The main goals of this Joint Discussion were to review the major surveys in astronomy whose targets are stars, to appreciate the scientific goals and the byproducts of these surveys with respect to our understanding of Galactic astronomy, to consider how well the techniques employed achieve these goals, and to exchange ideas on how best to exploit the opportunities of future surveys. A report of this meeting is given in *Highlights of Astronomy, Volume 14* (CUP, 2007, ed. K.A. van der Hucht), and full proceedings are published in the Memorie della Societá Astronomica Italiana, volume 77, n.4.

The president also drew the attention of the meeting to the Commission 45 contribution to *Transactions IAU XXVIA, Reports on Astronomy 2002-2005* (CUP, 2007, ed. O. Engvold), where activity in stellar classification over the last several years is cited, and he thanked all the contributors to this report. Two major developments have been the unification of the T-dwarf standards and an ongoing project, the new general catalogue of stellar spectral classification by Brian Skiff. See `<http://cdsweb.u-strasbg.fr/viz-bin/VizieR?-source=III/233B>`.

The third major activity of Commission 45, the Working Group on *Standard Stars*, and the production of its newsletter is reported in the following pages of these Transactions.

2. New Organizing Committee and president

The Organising Committee (OC) for the next triennium was arranged by e-mail prior to the General Assembly. Nominations from the Commission Membership at large were just sufficient to fill vacancies arising from the usual process of rotation, so it was not necessary to hold an election. Retiring members of the OC were sincerely thanked, and the incoming members and officers were acclaimed by those present.

The constitution of the new OC 2006 - 2009: Sunetra Giridhar (president, India), Richard O. Gray (vice-president, USA). Members: Coryn A.L. Bailer-Jones (Germany), Christopher J. Corbally, PP (Vatican City State), Laurent Eyer (Switzerland), Michael J. Irwin (UK), Joseph D. Kirkpatrick (USA), Steven R. Majewski (USA), Dante Minniti (Chile), and Birgitta Nordström (Denmark).

3. Science

The scientific programme started with a provocative talk by H. Levato (Complejo Astronmico el Leoncito) on *Spectral classification in the southern hemisphere: old fashioned technique? Or the best one for astrophysical insight?* This led to considerable discussion on the role of Commission 45 for promoting astronomy in the coming decade or so. Areas of influence included further extensions of classification to peculiar stars, to those of non-solar metallicity, and into spectral domains other than the classical blue-green region. These extensions were particularly important to implement for the automated classification of stellar spectra.

The larger surveys were devising new sets of photometric filters or, like the *Gaia* survey, new systems of spectrophotometric bins. There was discussion on the role of Commission 45 to stimulate the best development of these new systems and to promote communication among the various survey teams.

The incoming president undertook that the OC would, in consultation with the whole membership, produce a new statement of purpose for Commission 45, and that it would outline steps to implement this for the general benefit of astronomy.

4. Closing remarks

The president thanked those present for their participation in this meeting and for their support of Commission 45's work in the last three years, and he wished them every success in their work and projects during the coming triennium.

Christopher J. Corbally
president of the Commission

Transactions IAU, Volume XXVIB
Proc. IAU XXVI General Assembly, August 2006
Karel A. van der Hucht, ed.

doi:10.1017/S1743921308023934

DIVISION IV / COMMISSION 45 / WORKING GROUP STANDARD STARS

CHAIR **Christopher J. Corbally**
NEWSLETTER EDITOR **Richard O. Gray**

PROCEEDINGS BUSINESS MEETING on 16 August 2006

1. Business

The meeting of the Working Group on standard stars (WG-SS) opened under the chairmanship of Corbally. Gray, as editor, gave a report on the Standard Stars Newsletter (SSN). Through a round of thanks given to Gray, it was recognized by the two dozen present that this publication is the major vehicle between WG-SS meetings for achieving its tasks, namely promoting and communicating work on standard stars. Copies of the newsletter can be downloaded from <http://stellar.phys.appstate.edu/ssn>.

2. Discussion

The meeting broadly considered the status of work on standard stars and then was largely devoted to a discussion on evolving the WG-SS and the SSN with the times. It was felt that, with the development of new photometric and spectrophotometric systems by the large surveys, there was even more need for communication among the various survey teams and from these teams to the individual researchers, who would use the survey databases. Practical proposals to further such communication included:

(*a*) promoting a web-based bibliography of papers on standard stars, such as the current **S**tandard **O**bjects **F**or **A**stronomy site (<http://sofa.astro.utoledo.edu/SOFA/>);
(*b*) expanding the WG-SS membership, especially to those working on the new, large survey systems;
(*c*) making and promoting a web-based discussion forum for standard star issues;
(*d*) changing to the ApN's strategy of posting contributions immediately to a website, and "crystallizing" these into a newsletter issue after six months;
(*e*) communicating the changed structure and possibilities of interaction to the whole IAU membership, perhaps via the IAU Information Bulletin.

In the light of these discussions those present considered that an updated WG-SS would continue to fulfill a unique service to astronomy. To help achieve this, Corbally was prevailed upon to continue as chairperson and Gray as newsletter editor.

Christopher J. Corbally
chair of the Working Group

Transactions IAU, Volume XXVIB
Proc. IAU XXVI General Assembly, August 2006
Karel A. van der Hucht, ed.

doi:10.1017/S1743921308023946

DIVISION V VARIABLE STARS

ÉTOILES VARIABLES

Division V provides a joint forum for the study of stellar variability in all its manifestations, whether due to pulsation, surface inhomogeneities, evolutionary changes, or to eclipses and other phenomena specifically related to double and multiple stars.

PRESIDENT	**Jørgen Christensen-Dalsgaard**
VICE-PRESIDENT	**Alvaro Giménez**
PAST PRESIDENT	**Edward F. Guinan**
BOARD	**Conny Aerts, Luis A. Balona, Jorge Sahade**

DIVISION V COMMISSIONS

Commission 27	**Variable Stars**
Commission 42	**Close Binary Stars**

DIVISION V WORKING GROUP

Division V WG	**Spectroscopic Data Archiving**

INTER-DIVISION WORKING GROUPS

Division V-IV WG	**Active B-type Stars**
Division V-IV WG	**Ap and Related Stars**

PROCEEDINGS BUSINESS MEETING on 16 August 2006

1. Introduction

Division V organized a brief Business Meeting during the IAU XXVI General Assembly, in addition to Business Meetings, reported separately, of the two constituent Commissions. Furthermore, the Division held a longer Science Meeting which covered activities of both Commissions; the list of talks in this meeting, giving a good indication of the scope of the activities within the Division, is included below.

2. Overview of activities and events in the past triennium

With great regret the deaths of four prominent scientists within the research fields covered by the Division were noted: Professor Marcello Rodonó, former president of Division V, in October 2005; Professor Norman Baker, former president of Commission 27, also in October 2005; Professor George Isaak, in June 2005; and Professor John Bahcall, in August 2005. They all made seminal contributions to astrophysics and will be greatly missed, also, by many of the members of the Division, on a personal level.

The scientific activities in Division V predominantly take place through the two Commissions who both have long traditions and a continued high level of activity and a healthy development of the membership. Thus it is proposed to maintain an unchanged Commission structure of the Division. On the other hand, the Working Group on Variable and Binary Stars in Galaxies, established at the previous General Assembly, has not developed as hoped, although the area is certainly relevant and active. It has been decided to continue these activities on an informal level, while investigating the potential for formalizing them in a Working Group at the next General Assembly. The Working Group on Spectroscopic Data Archiving is discussed below.

The Division was involved in a number of conferences in the past triennium, including substantial activity at the IAU XXVI General Assembly: IAU Symposium No. 239 on *Convection in Astrophysics* and IAU Symposium No. 240 on *Binary Stars as Critical Tools and Tests in Contemporary Astrophysics*, as well as three Joint Discussions at the GA.

The Division and Commission web sites play an important role in the activities of these bodies. We are very grateful to Andras Holl of the Konkoly Observatory for setting up and maintaining these sites.

3. Election of new officers

The elections for officers for the coming triennium resulted in the following Organizing Committee:

- Alvaro Gimenéz, president
- Steven Kawaler, vice-president, president of Commission 27
- Conny Aerts, Division secretary
- Slavek Rucinski, president of Commission 42
- Michel Breger
- Edward Guinan
- Jørgen Christensen-Dalsgaard, ex officio, past president of Division V.

The president thanked the departing members of the Organizing Committee, Jorge Sahada and Luis Balona for the long contributions to the activities of the Division. In addition he warmly thanked Conny Aerts for unofficially serving as Division secretary during the past triennium, a status now officially recognized. He welcomed the new Organizing Committee, particularly the incoming president, with his best wishes for the activities in the coming period.

4. Discussion items

The discussions during the meeting centered on three key issues, raised in abstracts contributed before the meeting.

(a) Spectroscopic Data Archiving Working Group, by Elizabeth Griffin

Abstract: The general improvement in willingness and capability, in more recent times, to make data archiving – at least, saving the bits – an accepted telescope routine has encouraged the WG to move its attention to the worst bottleneck in the present progress, which it has identified as pipeline processing. Accordingly, the WG is now looking for fresh members with specific expertise and experience in that field to join the WG. The objective of the modified WG will be the automatic creation of accessible archives of spectra in scientifically-meaningful units. Members of the WG will be considering and recommending, both generally and specifically, how to establish pipeline processing of spectra in a way that will be efficient, robotic and as "correct" as is possible, benefiting from all extant knowledge about the design and operation of the spectrograph in question.

This was recognized as a natural development of the activities within the Working Group, and Elizabeth Griffin was encouraged to pursue the proposed reorganization. It is evident that these activities are also very closely related to the establishment of 'Virtual Observatories', an issue of great prominence at this General Assembly.

(b) Prospects and problems in cataloging variable stars, by Nikolai Samus

Abstract: I am planning to speak about the current situation with variable star catalogs. The Moscow GCVS team is continuing its effort but encounters a number of problems. To make the catalog a really GENERAL one, we have to merge it with many new catalogs presented by automatic surveys. We have made the important preliminary step by providing accurate coordinates for virtually all old GCVS stars. Then, the automatic classifications provided in the survey catalogs are too superficial from the GCVS point of view, and the GCVS classification system seems too complex. I will suggest a simpler classification system based on the GCVS experience, having in mind that experts in particular variability types will be able to supplement it in their specialized catalogs. I need the opinion of the Division for planning the future of the GCVS.

This is an issue common to both Commissions of the Division and it was discussed in both the corresponding Business Meetings. As mentioned in the Commission 27 report, an informal working group will be established to discuss this further, with the aim of setting up an IAU Working Group under Division V at the next General Assembly.

(c) Information Bulletin on Variable Stars, by Katalin Olah

Abstract: The editorial work of IBVS during the past three years will be briefly summarized, with emphasis on the two special, collected papers dealing with discoveries and observations of variable stars, appearing as the last two papers in every 100 issues. The experience in editing such contributions will be discussed. Finally, the appearance of IBVS in electronic databases will be presented.

The highly valuable work of Katalin Olah and the group at the Konkoly Observatory in maintaining this important publication was greatly appreciated by the meeting.

5. Division V science meeting

The science meeting was held on 17 August, 9:00-15:30 hr. The following talks were presented:

- Welcome (Jørgen Christensen-Dalsgaard, Denmark)
- Characteristics of post-outburst novae spectra: the Quadratic Zeeman Effect? (Robert Williams, USA)
- RS Ophiuchi and its 6th(?) outburst (Stuart Eyres, UK)
- Evidence for precession of a white dwarf in a close binary system (Gaghik Tovmassian, Mexico)
- A complete study of the pulsating eclipsing binary (Haili Hu, the Netherlands)
- Productivity and impact data on binary and variable star astronomy (Virginia Trimble, USA)
- UU Cnc and BM Cas: two long-period eclipsing binaries involving supergiant primaries: intrinsic variability as a circumstantial evidence on on-going common envelope activity (Izold Pustylnik, Estonia)
- Evidences of the donor star heating in precessional and orbital variabilities of SS 433 (T. Irsmambetova, Russia)
- A study of two peculiar interacting binaries from NURO observations: VV CVn and V965 Cygni (Ron Samec, USA)
- Creating a statistically significant sample of stars showing the flip-flop phenomenon (Heidi Korhonen, Germany)
- Light curves and high-resolution spectroscopy of Blazhko stars (Katrien Kolenberg, Austria)
- Long-term photometric monitoring with the MERCATOR telescope (Peter De Cat, Belgium)
- Amplitude variability and pulsation (Michel Breger, Austria)
- Spatial asymmetry and temporal aperiodicity in bright Mira variables (Lee Anne Willson, USA)
- Summary of the meeting (Jørgen Christensen-Dalsgaard, Denmark)

6. Closing remarks

The incoming president expressed his appreciation of the work done by the present Organizing Committee during the last triennium to foster the development of the field and the cooperation between scientists working in different areas. Concerning the activities for the next three-year period, he identified two areas as requiring special attention by the Division: the emerging activities in the study of pulsating stars in binary systems and the cataloging and classification of variable stars coming from large surveys.

Jørgen Christensen-Dalsgaard
president of the Division

Transactions IAU, Volume XXVIB
Proc. IAU XXVI General Assembly, August 2006
Karel A. van der Hucht, ed.

doi:10.1017/S1743921308023958

COMMISSION 27 — VARIABLE STARS

ÉTOILES VARIABLES

PRESIDENT	**Conny Aerts**
VICE-PRESIDENT	**Steven D. Kawaler**
PAST PRESIDENT	**Jørgen Christensen-Dalsgaard**
ORGANIZING COMMITTEE	**Timothy R. Bedding, Carla Cacciari, Peter L. Cottrell, Margarida Cunha, Gerald Handler, Peter Martinez, Dimitar D. Sasselov, Seetha Somasundaram, Douglas L. Welch**

PROCEEDINGS BUSINESS MEETING on 16 August 2006

1. Overview of activities and achievements

The meeting started at 16h00. The president welcomed the 24 participants to the business meeting of Commission 27. After the approval of the agenda, she gave an overview of the activities of Commission 27 of the past three years.

Large emphasis was put on contact with the community, through the creation of an up-to-date operational email list (311 entries) and a refurbishment of the Commission's web site. The president gave special thanks (by means of Belgian chocolates) to Dr. Andras Holl from Konkoly Observatory (Hungary) for his continuous efforts as Commission 27 web master. On average a monthly message was sent through the email list, concerning the distribution of news, meeting announcements, newsletters, deadlines for proposals, etc. The feedback from the community was rather large (some 1000 messages were received by the president over the three years).

Strong attention was also paid to keep close contact with the IAU secretariat. The president collaborated with the secretariat on the debugging of the new IAU website.

A total of 18 IAU meeting proposals was evaluated by the OC. The OC did this evaluation by means of a ranking system, and not on a case-by-base basis, to ensure the most objective procedure.

A specific effort was made to make sure that the triennial report on Research in Variable Stars and Commission 27 activities was as representative as possible. In order to achieve this, a draft version written by the president was screened by the OC, and, subsequently, distributed to all the Commission 27 members with a call for feedback and comments. About 60 reactions were received according to which the draft was revised and submitted for publication in the transactions.

2. Election of new officers

Elections were held to compose the new Organising Committee of Commission 27 for the coming triennium. The results are as follows:

- New president: Steve Kawaler (USA).
- New vice-president: Gerald Handler (Austria).
- Retiring OC members: Jørgen Christensen-Dalsgaard (Denmark), Carla Cacciari (Italy), Peter Cottrell (New Zealand), Dimitar Sasselov (USA), and Doug Welch (Canada).
- Continuing OC members: Conny Aerts (ex officio), Tim Bedding (Australia), Margarida Cunha (Portugal), Peter Martinez (SA), and Seetha Somasundaram (India).

• New OC members: Marcio Catalan (Chile), Laurent Eyer (Switzerland), Simon Jeffery (Northern Ireland), Karen Pollard (New Zealand), and Katalin Olah (Hungary).

The president expressed her gratitude to Jørgen Christensen-Dalsgaard, Carla Cacciari, Peter Cottrell, Doug Welch and Dimitar Sasselov, who served on the OC of Commission 27 for at least three terms of three years each with a lot of motivation and dedication. She thanked them for the fruitful collaboration. She welcomed Marcio Catelan, Laurent Eyer, Simon Jeffery, Karen Pollard and Katalin Olah as new OC members and wished them good luck with their OC work under the leadership of the new president.

3. New members

A total of 62 among the 925 approved new IAU members chose Commission 27 as their prime commission. All of them received a welcome mail from the outgoing president straight after the GA. Their contact addresses have been added to the email list, which was passed on to the new president.

4. Contributed talks

Two talks were submitted during the preparation of the Business meeting. They were considered of high relevance to the commission and thus time was devoted to them.

4.1. *IAU Archives of Unpublished Observations of Variable Stars by Michel Breger*

Abstract. Several of the IAU Archives of Unpublished Observations of Variable Stars are essentially unavailable at present. We propose a solution to this problem. In the past, data files were sent as paper copies to three places (Royal Astronomical Society in UK, Odessa Astronomical Observatory in Ukraine, and CDS in France). Recent communications suggested that the retrieval problem is still not solved and that modern access to the valuable and irreplaceable data needs to be established. A cursory inspection of a hundred files shows about half to be under 10 pages, but a few as large as 57 pages. We propose the following solution:
1. Peter Hingley (Librarian RAS) has kindly agreed to make paper copies of the files to be sent to a centre where they are digitized.
2. Arne Henden (AAVSO) has kindly agreed to digitize the data.
3. Michel Breger updates the cover sheets (e.g., listing publications and the occasional revisions) and descriptions.
4. The data are made publicly available through the AAVSO, Commission 27 and Vienna web sites with the AAVSO site being long-term.
We would like to receive the approval of Commission 27 for this procedure before starting the work.

The proposal by Michel Breger to digitize the remaining Archives of Unpublished Observations of Variable Stars for which only a paper copy exists at present, was approved and the president thanked Peter Hingley, Arne Henden, Andras Holl and Michel Breger for their efforts.

4.2. *Prospects and problems of the General Catalogue of Variable Stars by Nikolai Samus*

Abstract. I am planning to speak about the current situation with variable star catalogs. The Moscow GCVS team is continuing its effort but encounters a number of problems. To make the catalog a really GENERAL one, we have to merge it with many new catalogs presented by automatic surveys. We have made the important preliminary step by providing accurate coordinates for virtually all old GCVS stars. Then, the automatic classifications provided in the survey catalogs are too superficial from the GCVS point of view, and the GCVS classification system seems too complex. I will suggest a simpler classification system bases on the GCVS experience, having in mind that experts in particular variability types will be able to supplement it in their specialized catalogs. I need the opinion of the commission/division for planning the future of the GCVS.

The plan was made to create an informal Working Group on variable star classification and future variable star catalog issues. This working group should be composed of members from both Commission 27 and Commission C42. Depending on the achievements during the coming three years, and the needs thereafter, an official IAU Working Group Status will be applied for during the next GA in Rio de Janeiro. A call for participation will be made in the fall of 2006. The president thanked Nikolai Samus for taking up this initiative and his lively discussion.

5. Plans for future meetings

It was decided that Division V (Commission 27 *Variable Stars* and Commission 42 *Close Binaries*) should join their efforts and start preparing a proposal for a Symposium on Variable Stars at the IAU XXVII GA in Rio de Janeiro, 2009. It was felt that this should be done in close contact with the Working Groups that act under the auspices of Division V, as these had already expressed their interest to join this initiative during the GA.

6. Closing remarks

The president thanked the Commission 27 community for the numerous constructive reactions and encouragements she received from members. She ended the meeting by wishing the Commission 27 members a good continuation of their work under the leadership of the new OC. She pointed out that the minutes of the meeting, as well as the slides of the talks, will be made available through the Commission 27 website.

The meeting ended at 17h15 with a telephone conversation with the new president, who was unfortunately unable to attend the GA, giving him a summary of the meeting.

Conny Aerts
president of the Commission

Transactions IAU, Volume XXVIB
Proc. IAU XXVI General Assembly, August 2006
Karel A. van der Hucht, ed.

doi:10.1017/S174392130802396X

COMMISSION 42 CLOSE BINARIES

ÉTOILES DOUBLES SERRÉES

PRESIDENT	**Alvaro Giménez**
VICE-PRESIDENT	**Slavek M. Rucinski**
PAST PRESIDENT	**Paula Szkody**
ORGANIZING COMMITTEE	**Douglas R. Gies, Young-Woon Kang, Jeffrey L. Linsky, Mario Livio, Nidia Morrell, Ronald W. Hilditch, Birgitta Nordström, Ignasi Ribas, Edward Ribas, Sonja Vrielmann, Colin D. Scarfe**

DIVISION V / COMMISSION 42 WORKING GROUP

Division V / Commission 42 / WG	**Accretion Physics in Interacting Binaries**

PROCEEDINGS BUSINESS MEETING on 16 August 2006

1. Introduction

The president of the Commission welcomed the participants in the business meeting and provided an overview of the activities carried out during the past triennium 2002-2005. A good number of meetings have been held during this period on close binaries, about two per year, including both classical and interacting systems. One specific Symposium at the General Assembly in Prague, devoted to binary stars as astrophysical tools, showed the vitality of the field and the trend of cooperation between scientists studying close binaries and those specialized in visual double stars. The study of very low-mass binaries, including those containing planet-sized components also received much attention as well as the analysis of massive objects in nearby galaxies.

Concerning instrumental developments, they can be divided in space and ground based facilities. From space, the launch of *Integral* and *Spitzer* have provided important new insight in evolved and as well newly formed close binaries. The continuing operations of *Chandra* and *XMM-Newton* allowed the study of physical processes in interacting binaries and the launch of *COROT*, later in the year, is expected to open a new field of very-high precision photometry for the analysis of light curves. From the ground, it is interesting to note the increasing number of pulsating stars discovered in binary systems. Their analysis opens new methods to look into the stellar interiors. Newly discovered close binaries in different environments, thanks to large data bases created through continuing surveys, are requiring additional efforts and special techniques for their exploitation.

A summary of these activities is published as part of the *IAU Transactions XXVIA, Reports on Astronomy* (Ed. O. Engvold, 2007, CUP) for the triennium.

2. Commission activities

Very useful bibliographic notes on Close Binaries (BCB) are being produced under the coordination of C. Scarfe and posted in the commission web pages. In fact, the web pages of

Commission 42 have been modified to fit the look and accessibility of the Division V and Commission 27 pages (all hosted at Konkoly Observatory). A. Hall has been key to the success of this transfer and is maintaining them in a very efficient way.

In this process, a full list of e-mail addresses for members of the Commission has been produced and an interactive method for updates was set up. Concerning membership, 35 new members of the IAU have requested their acceptance in Commission 42, keeping it among one of the large ones within our Union.

On the sad side, we have to note some departures. Among them, the unfortunate early pass away of one of our past Commission president Marcello Rodono in 2005. Within weeks another active IAU member, as Commission and Division president, with close connection to our Commission, also left us, Willem Wamsteker.

3. New organizing committee

A new list of members of the organizing committee of the commission was presented: Slavek Rucinski (president, Canada), Ignasi Ribas (vice-president, Spain). Members: Ronald W. Hilditch (UK), Birgitta Nordstrom (Denmark), Edward M. Sion (USA), Sonja Vrielmann (Germany), Petr Harmanec (Czech Republic), Janusz Kaluzny (Poland), Panayiotus Niarchos (Greece), Katalin Olah (Hungary), Guillermo Torres (USA), and Mercedes T. Richards (USA). The six last members are new in the organizing committee while the rest serve for one more term. Ex-officio members included the continuation of Colin D. Scarfe (Canada) as editor of the BCB and A. Giménez (the Netherlands) as ex-president of the Commission.

The new commission Organizing Committee was elected by the previous members of the committee following the tradition of the commission procedures. The proposed list for the OC serving in the next triennium was endorsed by the participants in the business session and the outgoing members where thanked for their work during the last term. Nevertheless, it was discussed during the meeting that, given the availability of reliable membership lists and checked e-mail addresses, for the next time the OC should propose a system involving the vote of all members. A clear recommendation in this sense was made and accepted by the new OC.

Since the approval of the New Organizing Committee for Commission 42, the new president Slavek Rucinski took over as chairman of the session.

4. Presentations to the Business Meeting

Ignasi Ribas made a short presentation about the future of the WG on *Variable Stars in Galaxies*. Despite the initial interest and support, the activities of the working group could not be consolidated and it was proposed to discontinue it until a large enough number of commission members reconsider the situation.

The situation about catalogues of close binaries was discussed with presentations by N. Samus and O. Malkov. It is clear that a practical proposal has to be made with a small number of cases to be included in catalogues of variable stars. The current classification by types of light curves is far from satisfactory and new ideas are welcomed. It was recommended to discuss, within Division V, due to the similar problems encountered for variable stars by Commission 27, this matter with the idea to propose a solution by next IAU General Assembly.

The current status of the Information Bulletin on Variable Stars (IBVS) was presented by K. Olah and A. Hall were again common experience and problems were identified with Commission 27. The publication of the Bulletin was acknowledged and supported by our Commission.

5. Closing remarks

Justification for the continuation of the Commission and its current structure was made and conveyed to Division V though there was some discussion about the separation of interests between close binaries with degenerate or non-degenerate components. It was finally emphasized the need to increase the cooperation with Commission 27, under the umbrella of Division V, in the coming years through common activities and scientific meetings.

Alvaro Giménez
president of the Commission

Transactions IAU, Volume XXVIB
Proc. IAU XXVI General Assembly, August 2006
Karel A. van der Hucht, ed.

doi:10.1017/S1743921308023971

DIVISION VI INTERSTELLAR MATTER

MATIÈRE INTERSTELLAIRE

Division VI gathers astronomers studying the diffuse matter in space between stars, ranging from primordial intergalactic clouds, via dust and neutral and ionized gas in galaxies, to the densest molecular clouds and the processes by which stars are formed.

PRESIDENT	**John Dyson**
VICE-PRESIDENT	**Thomas J. Millar**
PAST PRESIDENT	**Bo Reipurth**
BOARD	**You-Hua Chu, Gary J. Ferland, José Franco, Chon Trung Hua,Susana Lizano, Antonella Nata, Bo Reipurth, Yoshiaki Sofue, Grazyna Stasińska, José M. Vilchez**

DIVISION IV COMMISSIONS

Commission 34	**Interstellar Matter**

DIVISION IV WORKING GROUPS

Division VI WG	**Star Formation**
Division VI WG	**Astrochemistry**
Division VI WG	**Planetary Nebulae**

PROCEEDINGS BUSINESS MEETING on 23 August 2006

An informal discussion touching on areas such as Division/Commission membership and scope of possible future meetings took place.

John Dyson
president of the Division

Transactions IAU, Volume XXVIB
Proc. IAU XXVI General Assembly, August 2006
Karel A. van der Hucht, ed.

doi:10.1017/S1743921308023983

DIVISION VII GALACTIC SYSTEM

SYSTÈME GALACTIQUE

Division VII provides a forum for astronomers studying our home galaxy, the Milky Way, which offers a unique laboratory for exploring the detailed structure of the stellar and gaseous components of galaxies and the process by which they form and evolve.

PRESIDENT	**Patricia A. Whitelock**
VICE-PRESIDENT	**Ata Sarajedini**
PAST PRESIDENT	**Georges Meylan**
BOARD	**Ortwin Gerhard, Miroslav Giersz, Georges Meylan, Birgitta Nordström, David N. Spergel, Rainer Spurzem**

DIVISION VII COMMISSIONS

Commission 33	**Structure and Kinematics of the Galactic System**
Commission 37	**Star Clusters and Associations**

DIVISION VII WORKING GROUP

Division VII WG	**Galactic Centre**

PROCEEDINGS BUSINESS & SCIENCE MEETINGS on 21 August 2006

1. Introduction

Most of the day was devoted to the scientific session with a brief business meeting for Division VII and Commission 33 just before lunch. Commission 37, the other 'half' of the Division, had already held their business session on 17 August.

2. Business Meeting

The president described the various activities of the Division which primarily involve sponsoring IAU Symposia and Colloquia. The suggestion for members of the new Board (see below) were approved and most of the discussion centred around the proposed Working Group (see below).

3. Working Group on the Galactic Centre

At the meeting it was decided that Division VII would form an IAU Working Group on the *Galactic Centre*. The first chair would be Joseph Lazio who would select an organizing committee taking into account those at the meeting who expressed an interest, the usual considerations of equity and representation, but primarily ensuring that all the major parties working in the area were represented.

There had been some debate on the possibility of making this an inter-Divisional Working Group on *Galactic Centres*, in order to encourage the exchange of ideas between those working on related Galactic and on extra-galactic problems. In the end this was not the path chosen, but noting the considerable interest from Commission 28 (Galaxies) it was decided that participation from this group would be encouraged. The possibility of widening the scope of the Working Group would be re-examined after three years, at the time of the next GA.

4. Science session – new results on the Galactic Centre

This topic was chosen because it is of broad interest to members of the Division and because it was a subject in which there had been considerable progress over the preceding three years. Furthermore, the Division VII Board had considered in detail the proposal that the Division should create an IAU Working Group to work on the *Galactic Centre*. The IAU General Assembly in Prague offered an ideal opportunity to guage the level of interest in such a group.

A special organizing committee was set up for the occasion, comprising: Reinhard Genzel (Germany), Ortwin Gerhard (Germany), Cornelia Lang (USA) Joseph Lazio (USA) and Patricia Whitelock (South Africa, chair).

The science session comprised the following invited and contributed presentations:

- Super-massive black hole at the Centre of our Galaxy — Mark Morris (USA)
- Radio imaging of Sgr A* — Avery Broderick (USA)
- Recent near-IR/X-ray and the first time-resolved AO near-IR polarization measurements of Sgr A* — Andreas Eckart (Germany)
- Sgr B2 region in $H^{13}CO^+$ and SiO lines — Masato Tsuobi (Japan)
- Galactic Centre region as seen by HESS in very high-energy γ-rays — Loic Rolland (France)
- Theory of accretion and flares — Sera Markoff (the Netherlands)
- Origin of the young stars in the Galactic Centre — Thibaut Paumard (Germany)
- Origin of the young stars in the central parsec — Simon Portegies Zwart (the Netherlands)
- *Spitzer* mid-infrared survey of the inner $2^\circ \times 1^\circ.5$ of the Galaxy — Howard Smith (USA)
- Star clusters and star formation in the Galactic Centre — Angela Cotera (USA)
- Centres of disk dominated galaxies — Marcella Carolla (Switzerland)

This proved a very interesting meeting with the lecture room full to capacity for much of the day and lively discussion following each of the presentations.

5. New Division Board

The established practice of Division VII is for the Board to comprise the presidents, vice-presidents and past presidents of Commissions 33 and 37. This process will be continued, but the chair of the Working Group will also be nominated for membership of the Board.

The past practice of the Division has been for the presidency to alternate between Commissions 33 and 37 from one triennium to the next. Within these terms the new president would have been from Commission 37, but on this occasion the Commission 37 president had not previously been on the Divisional Board and decided she did not wish to take up the Divisional presidency as well as that of the Commission. It was therefore agreed that the new president of Commission 33 would also take up the leadership of the Division.

The new Division VIII Board is, therefore, as follows:

- Ortwin Gerhard (Germany, president, president Commission 33)
- Despina Hatzidimitriou (Greece, vice-president, president Commission 37)
- Charles J. Lada (USA, vice-president Commission 37)
- Joseph Lazio (USA, chair WG Galactic Centre)
- Ata Sarajedini (USA, past president Commission 37)
- Patricia A. Whitelock (South Africa, past president Commission 33)
- Rosemary Wyse (USA, vice-president Commission 33)

Patricia A. Whitelock
president of the Division

Transactions IAU, Volume XXVIB
Proc. IAU XXVI General Assembly, August 2006
Karel A. van der Hucht, ed.

doi:10.1017/S1743921308023995

COMMISSION 33 — STRUCTURE AND DYNAMICS OF THE GALACTIC SYSTEM

STRUCTURE ET DYNAMIQUE DU SYSTÈME GALACTIQUE

PRESIDENT — **Patricia A. Whitelock**
VICE-PRESIDENT — **Ortwin Gerhard**
PAST PRESIDENT — **David Spergel**
ORGANIZING COMMITTEE — **Yurij N. Efremov, Wyn Evans, Chris Flynn, Jonathan E. Grindlay, Birgitta Nordström, Michael A. C. Perryman, Rosemary F. Wyse, Chi Yuan**

PROCEEDINGS BUSINESS MEETING on 21 August 2006

1. Introduction

Commission 33 held a brief business session on 21 August within the science session organized jointly with its parent Division VII.

2. New Commission Organizing Committee

The meeting approved the following new Organizing Committee for the Commission: Ortwin Gerhard (president, Germany), Rosemary F. Wyse (vice-president,USA). Members: Yurij N. Efremov (Russia), Wyn Evans (UK), Chris Flynn (Finland), Jonathan E. Grindlay (USA), Joseph Lazio (USA), Birgitta Nordström (Denmark), Patricia A. Whitelock (South Africa), and Chi Yuan (Taiwan).

3. Science session: new results on the Galactic Centre

Details of the presentations given at this very successful meeting are given in the report from Division VII.

Patricia A. Whitelock
president of the Commission

Transactions IAU, Volume XXVIB
Proc. IAU XXVI General Assembly, August 2006
Karel A. van der Hucht, ed.

doi:10.1017/S1743921308024009

COMMISSION 37 STAR CLUSTERS AND ASSOCIATIONS

AMAS STELLAIRES ET ASSOCIATIONS

PRESIDENT	**Ata Sarajedini**
VICE-PRESIDENT	**Rainer Spurzem**
PAST PRESIDENT	**Georges Meylan**
MEMBERS	**Russell D. Cannon, Vittorio Castellani, Gary S. Da Costa, Kyle McC. Cudworth Licai Deng, Miroslav Giersz, Despina Hatzidimitriou, Charles J. Lada**

PROCEEDINGS BUSINESS MEETING on 17 August 2006

Summary

The Commission business meeting was held on 17 August 2006. Approximately 20 people attended, including the Commission vice-president, Rainer Spurzem. The sole member of the Commission Organizing Committee that was present was Gary Da Costa, who, along with Patricia Whitelock, the outgoing president of Division VII and Commission 33, served as chairs of the meeting.

The nominations of Despina Hatzidimitriou as Commission president and Charles Lada as vice-president were approved by the members of the Commission Organizing Committee prior to the business meeting. Those present at the business meeting agreed with this recommendation.

The membership of the Organizing Committee (OC) of Commission 37 was also revised. Rainer Spurzem and Mirek Giersz requested to leave the OC; they were replaced by Young-Wook Lee and Monica Tosi. Vittorio Castellani passed away earlier in the year. Thus, the OC is now composed of Russell Cannon, Kyle Cudworth, Gary Da Costa, Licai Deng, Young-Wook Lee, Ata Sarajedini (past president), and Monica Tosi.

Ata Sarajedini
president of the Commission

Transactions IAU, Volume XXVIB
Proc. IAU XXVI General Assembly, August 2006
Karel A. van der Hucht, ed.

doi:10.1017/S1743921308024010

DIVISION VIII GALAXIES AND THE UNIVERSE

LES GALAXIES ET L'UNIVERSE

Division VIII gathers astronomers engaged in the study of the visible and invisible matter in the Universe at large, from Local Group galaxies via distant galaxies and galaxy clusters to the large-scale structure of the Universe and the cosmic background radiation.

PRESIDENT	**Francesco Bertola**
VICE-PRESIDENT	**Sadanori Okamura**
PAST PRESIDENT	**Virginia L. Trimble**
BOARD	**Mark Birkinshaw, Françoise Combes, Simon J. Lilly, John A. Peacock, Elaine M. Sadler, Rachel L. Webster**

DIVISION VIII COMMISSIONS

Commission 28	**Galaxies**
Commission 47	**Cosmology**

DIVISION VIII WORKING GROUP

Division VIII WG	**Supernovae**

PROCEEDINGS BUSINESS MEETING on 18 August 2006

1. Report of Business Meeting

1.1. *Attendance*

F. Bertola, E. dalla Bonta, F. Combes, E. M. Corsini, F. Firneis, J. Funes, R. de Grijs, R. W. Hunstead, A. Koekemoer, P. C. van der Kruit, V. J. Martínez, R. H. Miller, S. Okamura, J.-C. Pecker, E. Sadler, S. Simkin, R. L. Webster, H. van Woerden, and P. T. de Zeeuw. This was the joint meeting of Division VIII and its two Commissions, 28 and 47.

1.2. *Division VIII officers, triennium 2006-2009*

The Division VIII Board includes the president, vice-president and past president, and the presidents and vice presidents of the two Commissions. These appointments were approved at the meeting. Dr. Mark Birkinshaw has been asked to act as webmaster. The new Board of Division VIII is as follows: Sadanori Okamura (president), Elaine Sadler (vice-president), Francesco Bertola (past president), Françoise Combes (president Commission 28), Roger Davies (vice-president Commission 28), Thanu Padmanabhan (vice-president Commission 47), Rachel Webster (president Commission 47), and Mark Birkinshaw (Division webmaster).

2. Scientific meetings

The main activities of the Division have been concerned with the review of proposals for IAU Symposia and Colloquia. There are always more proposals from these active research fields than can be accommodated. During the last triennium, 12 out of the 26 IAU Symposia and Colloquia were dealing with subjects of interest to Division VIII:

2004

• IAU Symposium No. 222, *The Interplay Among Black Holes, Stars and ISM in Galactic Nuclei*, Gramado, Rio Grande do Sul, Brazil, 1-4 March 2004

• IAU Colloquium No. 195, *Outskirts of Galaxy Clusters: Intense Life in the Suburbs*, Torino, Italy, 12-16 March 2004

• IAU Symposium No. 225, *Impact of Gravitational Lensing on Cosmology*, Lausanne, Switzerland, 19-23 July 2004

2005

• IAU Colloquium No. 199, *Probing Galaxies through Quasar Absorption Lines*, Shanghai, China PR, 14-18 March 2005

• IAU Colloquium No. 198, *Near-Field Cosmology with Dwarf Elliptical Galaxies*, Les Diablerets, Switzerland, 14-18 March 2005

IAU Symposium No. 228, *From Lithium to Uranium: Elemental Tracers of Early Cosmic Evolution*, Paris, France, 23-27 May 2005

• IAU Symposium No. 230, *Populations of High-Energy Sources in Galaxies*, Dublin, Ireland, 15-19 August 2005

• IAU Symposium No. 231, *Astrochemistry throughout the Universe: Recent Successes and Current Challenges*, Monterey, CA, USA, 29 August - 2 September 2005

2006

• IAU Symposium No. 234, *Planetary Nebulae in our Galaxy and Beyond*, Waikoloa Beach, HI, USA, 3-7 April 2006

• IAU Symposium No. 235, *Galaxy Evolution across the Hubble Time* Prague, Czech Republic, 14-17 August 2006

• IAU Symposium No. 238, *Black Holes: from Stars to Galaxies - across the Range of Masses*, Prague, Czech Republic, 21-25 August 2006

• IAU Symposium No. 241, *Stellar Populations as Building Blocks of Galaxies*, La Palma, Canary Islands, Spain, 10-14 December 2006

There are many conferences on topics covered by Division VIII that are not directly supported or sponsored by the IAU. These also ensure effective discussion within the research community.

One of the main tasks of the Division presidents is to judge all proposals for Symposia. The Union gives only access to the presidents for the proposals, and this will probably continue in this way. An effort has been made within Division VIII to share this information with the OC of the Division.

3. Discussion

There is some discussion about the increasing interaction between cosmology and physics. Developments in cosmology continue to attract the attention of physicists, especially as new experimental techniques are developed, such as gravitational wave detection and γ-ray telescopes. Many of these people are not IAU members, and therefore the role of the Division should be to foster the interaction.

At the end of the meeting the president thanked all Board members for the pleasant and productive collaboration over the past triennium, and welcomed his successor and the new members of the Board. In turn, the newly elected president of the Division VIII, Sadanori Okamura, addressed some words to the participants. He stressed that the Division and both Commissions can run very well, he also thanked all the past Board members and the past Division president, Francesco Bertola, for their productive activity and expressed the wish in a fruitful work of the new members of the Board.

Francesco Bertola
president of the Division

Transactions IAU, Volume XXVIB
Proc. IAU XXVI General Assembly, August 2006
Karel A. van der Hucht, ed.

doi:10.1017/S1743921308024022

DIVISION VIII / WG SUPERNOVAE

CO-CHAIRS **Wolfgang Hillebrandt, Brian P. Schmidt**

PROCEEDINGS BUSINESS MEETING on 21 August 2006

1. Report on past WG-SN activities (W. Hillebrandt, B. Schmidt)

Objectives of the 'new' WG-SN when it was re-established in 2002:

- Coordination and planning of observational activities.
- Supernova data archiving and availability.
- Interface to VO-activities.
- Electronic newsletter.

As far as coordination and planning of observational activities are concerned, the main need for WG-SN activities seem to lie in the field of suitable follow-ups to the many on-going and planned search programs (in particular for SNe Ia's; LOTOSS, NGSS, SNfactory, SNLS, ESSENCE, GOODS, PanStarrs, ...). With the new robotic 2m-class telescopes (e.g., the Las Cumbres Observatory Global Telescope Network and other robotic telescopes) photometric follow-ups do not seem a major problem, but for most of the searches spectroscopic follow-up requires 4m-class telescopes at least which will become rare in the future. Possible ways out were discussed.

Contacts between the SN community and VO-activities (AVO, NVO, Astro-Grid, GAVO, ...) were established in the past years. In longer terms the goal is to set up a central data archive to allow multiple data use and to establish a platform to link model "data".

2. Joint CBAT/WG-SN supernova webpage (D. Green)

On behalf of the IAU's Central Bureau for Astronomical Telegrams (CBAT), Dan Green reported on activities supported by the Harvard-Smithsonian Center for Astrophysics to set up a webpage with information concerning supernovae. This webpage could include among other things new (and not yet confirmed) supernova discoveries. The SNWG will join these activities.

3. New and/or planned systematic programs (B. Schmidt)

We discussed the various systematic SN search and follow up studies likely to be undertaken by the community over the coming five years. These included the existing amateur, KAIT, SN Factory, SDSS-II, Essence, SNLS studies, as well as new studies associated with Pan-Starrs and SkyMapper. It was noted that the number of SN discovered was going to continue to rise, and that it would be useful to have a more automated way of dealing with SN discoveries, SN observations, and SN IAU certification.

4. Supernova nomenclature (O. Bartunov)

Since in the near-future the number of newly discovered supernovae will increase to several hundreds per year and more. Many of these supernovae will remain unclassified and for many of them a few images will be the only information. Therefore, several questions arise:

- Will we need a new nomenclature for supernovae, given the fact that we may get more than can be dealt with in the present system?
- Do we need a way to distinguish (spectroscopically) confirmed supernovae from the others?

The WG-SN will work out a suggestion and present them to the IAU. One possibility that was discussed is to name all supernova candidates according to the year and month of discovery plus their coordinates, and leave the present nomenclature for spectroscopically confirmed ones. But there may be better solutions.

Wolfgang Hillebrandt & Brian Schmidt
co-chairs of the Working Group

Transactions IAU, Volume XXVIB
Proc. IAU XXVI General Assembly, August 2006
Karel A. van der Hucht, ed.

doi:10.1017/S1743921308024034

COMMISSION 28 GALAXIES

GALAXIES

PRESIDENT	**Elaine M. Sadler**
VICE-PRESIDENT	**Françoise Combes**
PAST PRESIDENT	**Sadanori Okamura**
ORGANIZING COMMITTEE	**James J. Binney, Anthony P. Fairall, Timothy M. Heckman, Simon J. Lilly, Valentina Karachentseva, Renée C. Kraan-Korteweg, Gillian R. Knapp, Bruno Leibundgut, Jayant V. Narlikar**

PROCEEDINGS BUSINESS MEETING on 18 August 2006

1. Introduction

The members of Commission 28 on *Galaxies* were very busy during this General Assembly, with the Commission involved in two Symposia (IAU Symposium No. 235 *Galaxy Evolution across the Hubble Time*, IAU Symposium No. 238 *Black Holes: from Stars to Galaxies*), and two Joint Discussions (JD07 *The Universe at $z > 6$*, JD15 *New Cosmology Results from the Spitzer Space Telescope*). Therefore, the Business Meeting was combined with the Division VIII Business Meeting, which included a short information session on the new Commission 28 Organizing Committee. The triennial report of the Commission for 2003-2005 was also distributed, and is available on the Commission 28 web site.

2. New Commission Organizing Committee

The outgoing president of Commission 28, introduced the new Organizing Committee for 2006-2009: Françoise Combes (president, France), Roger L. Davies (vice-president, UK). Members: Avishai Dekel (Israel), Marijn Franx (the Netherlands), John S. Gallagher (USA), Valentina Karachentseva (Ukraine), Gillian R. Knapp (USA), Rene C. Kraan-Korteweg (South Africa), Bruno Leibundgut (Germany), Naomasa Nakai (Japan), Jayant V. Narlikar (India), Monica Rubio (Chile), and Elaine M. Sadler (Australia).

Françoise Combes thanked the outgoing president for her smooth and positive action during the past three years.

All information concerning Commission 28 (Newsletters, reports, etc.) will be posted on the new website: `<http://aramis.obspm.fr/IAU28/index.php>`

Elaine M. Sadler
president of the Commission

Transactions IAU, Volume XXVIB
Proc. IAU XXVI General Assembly, August 2006
Karel A. van der Hucht, ed.

doi:10.1017/S1743921308024046

DIVISION IX — OPTICAL AND INFRARED TECHNIQUES

TECHNIQUES OPTIQUES ET INFRAROUGES

Division IX provides a forum for astronomers engaged in the innovation, development and calibration of optical instrumentation and observational procedures including data processing.

PRESIDENT	**Christiaan L. Sterken**
VICE-PRESIDENT	**John B. Hearnshaw**
PAST PRESIDENT	**Arlo U. Landolt**
BOARD	**Martin Cullum, Michel Dennefeld, Rolf-Peter Kudritzki, Arlo U. Landolt, Peter R. Lawson, Peter Martinez, Birgitta Nordström, Su Ding-qiang, Andrei A. Tokovinin, Stéphane Udry**

DIVISION IX COMMISSIONS

Commission 9	**Instrumentation and Techniques (†)**
Commission 25	**Stellar Photometry and Polarimetry**
Commission 30	**Radial Velocities**

DIVISION IX WORKING GROUPS

Division IX WG	**Optical and Infrared Interferometry**

INTER-DIVISION WORKING GROUP

Division IX-X WG	**Encouraging the International Development of Antarctic Astronomy**

PROCEEDINGS BUSINESS MEETING on 18 August 2006

1. Introduction

Division IX Board members expressed their ideas and opinions on the role and structure of the Division, and discussed measures to improve its functionality. This report includes the summary of the discussions on future organization of the Division within the revised by-laws of the IAU. The reports of Division IX Working Groups and Commissions over the period 2003–2006 can be found in *Reports on Astronomy, Transactions IAU Volume XXVIA* (2007, Ed. O. Engvold).

The Division IX structure reflects a historical evolution: it encompasses two rather small Commissions (25 and 30), and one very large Commission (9). Division IX differs from most other Divisions in the sense that it does not have very well-focused scientific topics of its own. This also explains why the Division rarely introduces proposals for Symposia.

The Board members are all aware of the fact that many of the highly technical areas – like adaptive optics and detectors – have other active forums outside the IAU, and that most of the key developments are carried out by non-IAU members. Division IX, though, can hardly be expected to become a driver for innovations as we see happen in SPIE and other organisations, even if funds would be made available.

Observing techniques, especially standardisation and calibration, are transversal to all Divisions. The Division must take a leading role in the science and calibration work that is not done by SPIE and other organisations, and support this activity by providing a dynamical information service which is accessible to everyone at no cost. In fact, the situation is not unlike the situation in the commercial world: private industries and consortia develop products, but governmental and international organisations regulate standardisation through setting norms, rules and requirements.

Standardisation is not to be left to technicians and engineers alone: standardisation is closely linked to expertise and is driven by those scientists who have the knowledge and the means to find out how to do it. Globalisation, in the sense of all-sky surveys, large-scale data facilities and observations from space, creates a large demand for standardisation, which becomes increasingly complicated as it surpasses the level of individual approaches. But standardisation has long been one of the tasks of our Commissions (especially C25 and C30), and as the IAU acts as a rule-making body in many fields, Division IX should support this role and strengthen its activities in all matters of calibration and standardisation in astronomical techniques.

2. Commission 9

Commission 9 covers a huge spectrum of techniques and assembles people with very diverse interests – its mandate is, in fact, as wide as the mandate of the Division, and one can even say that the two other Commissions deal with techniques that could be a sub-activity of Commission 9. On the other hand, Commission 9 has never been the center of international discussions on instruments. A profound re-structuring of Commission 9 is more than desirable, but is also extremely difficult and should, by all means, be much more than a simple rearrangement of the administrative structure.

As a first step, a shortlist of disciplines to be covered by Working Groups was compiled. Following this, the right people will be found to take charge of the development of these WGs, and proposals will be formulated to create a WG on *Adaptive Optics*, a WG on *Instrumentation for Site Testing*, a WG on *Extremely Large Telescopes*, and a WG on *Small Telescopes* (i.e., 2-meter class). It is expected that some of these Working Groups will evolve to Commission status by the time of the IAU XXVII General Assembly.

In view of the wide re-structuring of Commission 9, the decision was taken to remove the historical redundancy in the mandates of this Commission and its parent Division by dissolving the Commission, and by moving its WG *Sky Surveys* and WG *Detectors* (consisting largely as a link to an external forum) to become WGs of the Division.

3. Creation of a new Commission

The Divisional WG it Optical & Infrared Interferometry was created in 2000. At this moment, the goals of the WG have been met, and the WG has matured to become a new Commission.

The Commission on *Optical/Infrared Interferometry* will coordinate international collaborations on scientific and technical matters relating to long-baseline optical and infrared interferometry. As a Commission within Division IX its focus will be to establish scientific and technical standards that will facilitate the future growth of the field. The Commission will take an active role in promoting the science of interferometry through collaborations with individual Commissions within the IAU, most particularly with the Commissions on *Astrometry* (C 8), *Double and Multiple Stars* (C 26), *Variable Stars* (C 27), and *Theory of Stellar Atmospheres* (C 36).

The work of the Commission will take place primarily within the Commission's Working Groups. Although they have not yet been formally proposed, these working groups would likely include a WG on *Interferometry Data Standards*, a WG on *Imaging Algorithms*, a WG on *Calibrator Stars* and a WG on *Future Large Arrays*.

The Commission will also continue the work begun through the Working Group on *Optical/IR Interferometry*, which will include further development of the Optical Interferometry Data Exchange format and its supporting software. The Commission's website will be hosted at the Optical Long Baseline Interferometry News (OLBIN).

4. Scientific and technical meetings

Commissions 9 and 25 organised two scientific meetings: IAU Symposium No. 232 *Scientific Requirements for Extremely Large Telescopes*, and a workshop *The Future Of Photometric, Spectrophotometric and Polarimetric Standardization*. The latter proposal fell below the cut-off line for IAU support, mainly because of the fear that such technical matter would not attract a large audience, at the same time, the meeting overpassed the scope of a Joint Discussion. Both scientific meetings organised from inside Division IX show that there is a strong interest in the technical fields covered by the Division, and that the issue of standardisation and calibration is considered to be important by designers and users of instruments, whatever the location, size and extent of the observing facility. The existing and new Commissions and WGs are expected to enhance their support service to the community through stimulation by enhancing their websites, and through the organisation of technical meetings.

The concept of IAU-sponsored meetings is continuously revised and dynamically adapted to the changing needs of the IAU global community. It is perhaps time to reconsider the requirement on the size of the audience – at least for meetings of a technical nature – so that proposing WGs, Commissions and Divisions can count on IAU support to convene their experts on a *regular* basis. Such technical meetings, properly documented by proceedings in the IAU Publications, will promote and safeguard the science based on observational astronomy. Not to forget the tremendous benefit for students and young astronomers: we all know that education and training in calibrated experiment today falls dramatically short of what is needed to make optimal use of the gigantic tools that will be used by them in a decade from now.

5. Closing remarks

The change in Division structure will necessarily reflect in the composition of the new Division Board. In order to keep its extent manageable, and to allow access to younger colleagues, a new class of Board members called *Advisers* (mainly past presidents of the Division and its Commissions) is created. Those advisers will participate in appropriate discussions but will not drive the Division's business.

Christiaan L. Sterken
president of the Division

Transactions IAU, Volume XXVIB
Proc. IAU XXVI General Assembly, August 2006
Karel A. van der Hucht, ed.

doi:10.1017/S1743921308024058

INTER-DIVISION IX-X / WORKING GROUP ENCOURAGING THE INTERNATIONAL DEVELOPMENT OF ANTARCTIC ASTRONOMY

CHAIRPERSON **Michael G. Burton**
MEMBERS **Maurizio Busso, Eric G. Fossat, James P. Lloyd, Mark J. McCaughrean, Christian Spiering, Shoji Tori**

PROCEEDINGS BUSINESS MEETING on 23 August 2006

1. Introduction

The business session of the Working Group followed the completion of Special Session 7, *Astronomy in Antarctica*. The proceedings of this meeting are published in *Highlights in Astronomy*, Volume 14 (Ed. K. A. van der Hucht, 2007, CUP). The session involved 18 papers spread over 5 sessions, together with a further 18 poster papers. A dinner was also held in Prague following the first day of the Special Session.

2. Proposal to establish an Antarctic Astronomy research programme in SCAR

Currently there are five Scientific Research Programs within SCAR, the Scientific Committee for Antarctic Research. A motion has been put forward to SCAR to make Astronomy the sixth. This would place Astronomy on an equal footing with the other science disciplines in Antarctica for the first time. A preliminary proposal needs to be prepared for the meeting of the SCAR Executive in May 2007. Assuming this is accepted, a full proposal will then to go to the 2008 SCAR meeting.

The following background briefing was provided by Jeremy Mould (NOAO) and John Storey (UNSW):

Astrophysical observations require minimum interference from the Earth's atmosphere, low thermal background, low absorption, and high angular resolution. The moderate "launch costs" for Antarctic plateau observatories make them an attractive alternative to space.

Astronomy from the Antarctic came of age in the last decade with a cosmological result of major significance. Balloon-borne millimetre observations of the cosmic microwave background from the first BOOMERANG flight led directly to the discovery of the zero-curvature Universe. Sub-millimetre astronomy has also prospered in the Antarctic: the South Pole Telescope is expected to deliver a large-area survey of the hot gas in clusters of galaxies with a uniquely uniform redshift distribution; this will probe the nature of 'dark energy', the biggest constituent of a 'flat' Universe.

Now is the time for SCAR to initiate, as its sixth program, Astronomy & Astrophysics from Antarctica, aimed at understanding the overarching ecological processes in the Universe, the birth of stars and of planetary systems around other stars, the return of heavy element enriched materials to the interstellar medium, and the formation of molecular clouds.

SCAR will add value by fostering international collaboration in order to permit goals to be achieved that are beyond those of single national programs. SCAR's approach to broad scientific programs is to define themes within the program. Some themes for AAA (Antarctic Astronomy & Astrophysics) will be exoplanet biosignatures, high angular resolution, time domain astrophysics, microwave cosmological background radiation studies, and the physics of molecular clouds.

In the next two years the AAA Planning Group will consult with the community, clarify the objectives of the research in these and other proposed astrophysical themes, and create a roadmap that will allow groups to make progress toward achieving these goals.

The notion of precursor projects is a useful one in road-mapping. National goals will differ for facilities of common interest. It is important to pursue the scientific goal of a facility all the way to the science. A multi-wavelength approach will be necessary to meet SCAR's expectations for a full AAA program. Multidisciplinary links outside astronomy also need to be addressed in the roadmap.

SCAR can enhance the scientific value of Antarctic astronomy by moving to establish the AAA Scientific Research Programme at this time. The benefits of coordination and international collaboration will be keenly felt. A strong AAA program will also strengthen the accomplishments of SCAR, which exists to promote frontier science driven coordination and collaboration.

The AAA Scientific Research Programme planning group will be made up of members of the existing AAA Action Group plus other contributors, and will produce a preliminary proposal for the 2007 SCAR Executive with a view to submission of a full, formal proposal to the 2008 SCAR meeting in St Petersburg. Once the AAA Scientific Research Programme is established, the AAA Expert Group will be dissolved.

Following a discussion, a call for volunteers to participate in writing the proposal was made. One aspect of this proposal that was discussed, in particular, was its role in providing a roadmap for future astronomical developments in Antarctica.

3. Proposal for the IAU to join SCAR as an ICSU member

Following discussions held at the recent SCAR meeting in Hobart, Australia (July 2006), the meeting considered whether IAU should be approached to join SCAR as an ICSU member. The meeting resolved to ask the IAU Executive to make such a request. Note: the ensuing discussions have indeed resulted in the IAU Assistant General Secretary, Ian Corbett, writing to SCAR, on behalf of the Executive, with such a membership request.

4. Limiting artificial backgrounds in Antarctica

Albrecht Karle spoke about the need to consider limiting artificial electromagnetic backgrounds in Antarctica so as not to interfere with pristine sites. The issue has arisen at the South Pole because of the possible interference of the Super Darn radar with the ICECUBE neutrino telescope. Chris Martin described discussions within the South Pole Users Committee (SPUC) regarding the issue. As developments continue at other sites over the plateau it is important to ensure that pristine sites for astronomy do not get compromised by inadvertent artificial backgrounds. It was noted that liaison with Commission 50, Protection of Existing and Potential Observing Sites, would be useful in this regard.

5. Continuation of the Working Group

The meeting discussed the continuation of the Working Group, and agreed that it provides a useful role and should continue. Michael Burton was re-elected as chair.

Michael G. Burton
chair of the Working Group

Transactions IAU, Volume XXVIB
Proc. IAU XXVI General Assembly, August 2006
Karel A. van der Hucht, ed.

doi:10.1017/S174392130802406X

DIVISION IX / COMMISSION 9 / WORKING GROUP DETECTORS

CHAIR **Timothy M. C. Abbott**
MEMBERS **Dennis Crabtree, Craig D. Mackay, Christopher G. Tinney**

PROCEEDINGS BUSINESS MEETING on 18 August 2006

1. Optical detectors

The same general claims hold true as in the last report and the field is as busy as ever. Community focus is on the deployment of large mosaics (up to 10^8 pixels) to telescopes, the pursuit of programs to build next generation mosaics (of order 10^9 pixels), and the investigation of novel variations in CCD design.

The standard mosaic tile format is still 2048 pixels wide by 4096 or 4608 pixels tall, with pixels either 15 or 13.5 μm square. Production is dominated by e2v technologies, although Fairchild Imaging show promise with catalog devices up to 4k×4k pixels. The MIT Lincoln Labs consortium is winding down its best effort program after experiencing mixed results. Other small-scale best-effort programs continue, prominent among them being Lawrence-Berkley Laboratory's production of devices for DECam and SNAP.

Conventional CCD performance has stabilized with readout rates of 10^6 pixels/sec or faster with readout noise less than 10 electrons per pixel; typical readout noise floors at slower speeds have been pushed below 2 electrons per pixel, although typically they are around 3 electrons per pixel, and full well capacities are around 10^5 electrons. Backside thinning technologies result in typical responses of over 80% QE spanning wavebands 200-300 nm wide.

The demands of large mosaic production translate into tight specifications for detector manufacturers and particularly for consistent device properties in large batches. Flat-field uniformity of a few percent and surface flatness of a few tens of microns are both sought-after characteristics. Charge transfer inefficiency of better than 10^{-5} is routine and dark current is neglible below -100°C for most applications.

The first wave of large mosaics was comprised of some twelve cameras containing 8192×8192 pixels now in place at various observatories around the world. A second wave is already well under way with Megaprime at CFHT and Megacam at MMT in operation for some time and ESO's Omegacam due to begin operations in 2007. A third wave is building with cameras planned and under development, including the Dark Energy Survey's DECam (0.5 Gpixel), WYNN/ODI (1 Gpixel), LSST (3 Gpixel), Pan-STARRS (4 × 1.4 Gpixel), NASA's *SNAP* (1 Gpixel) and ESA's *GAIA* (1 Gpixel); the latter two are space-based systems. HETDEX/VIRUS at McDonald Observatory deserves mention for its plans to incorporate 145 spectrographs.

The limited space and massively parallel architecture inherent in these systems mean that controller development is tending towards custom systems for each camera. Nevertheless, NOAO has established the collaborative, open source controller program, Monsoon. Turnkey, off-the-shelf CCD systems are closing the gap between amateur and professional demands and such systems are increasingly in evidence at observatories for the less challanging tasks.

Best-effort, wafer-run detector production programs concentrate on the development of novel device architectures or technologies. In particular, worthy of mention are the orthogonal transfer CCD from MIT, Lincoln Labs, USA, to be used in WYNN QUOTA and ODI and PanSTARRS; high-resisitivity, fully depleted CCDs from Lawrence-Berkley National Laboratory, USA to be used in DECam and SNAP; and the L3CCD photon-counting readout architecture from e2v technologues, UK. The development of high speed sensors for active optics systems is also a busy sub-genre.

CMOS detectors are receiving some attention, with various hybrid designs showing promise, perhaps bridging the optical/near-IR divide, but are not yet serious contenders at the observatory besides their use as multiplexers for infrared detectors.

2. Infrared detectors

Teledyne Imaging Systems (previously Rockwell Scientific Company) and Raytheon Vision Systems remain the primary players in the manufacture of infrared detectors. Teledyne continues the development of its HAWAII series of detectors, now working on devices as large as 6k×6k with 10 μm pixels and 4k×4k with 15 μm pixels. Several observatories are already operating mosaics of 4 HAWAII-2 arrays with more in development. The UK's VISTA telescope will sport a camera containing 16 of Raytheon's VIRGO arrays. Ever larger focal planes are being developed for space-based systems, in particular *SNAP* which will fly as many as 36 2k×2k IR detectors.

Raytheon are also moving mid-IR detectors into the megapixel range with the 1k×1k Aquarius Si:As IBC array.

Historically, infrared detector development has driven hybrid technologies in the pursuit of higher senstivities over broader wavebands and of Readout Integrated Circuits for faster and more flexible readout schemes and lower power consumption. This trend continues and there are signs of transfer of these technologies to CCD and other visible detector development.

3. Community interaction

No less than three periodic conferences offer venues for community interaction, of which the 3-yearly workshops "Scientific Detectors for Astronomy" continue to form the core. Most recently, they were held on Hawaii (2002) and on Sicily (2005). The proceedings from the 2005 conference include three overview papers providing comprehensive coverage of the state of existing and planned instruments and detectors at ground based observatories (Simons et al. 2005), CCD technology (Burke, Jorden and Vu, 2005) and CMOS detector technology (Hoffman, Loose and Suntharalingham, 2005). SPIE conferences every two years are additional opportunities for community discussion; most recently, they have been at Glasgow, UK in 2004, and Orlando, Florida, USA in 2006. The International Conference on Scientific Optical Imaging has been held on Cozumel, Mexico in 2003 and 2006. The proceedings of all of these conferences provide excellent road maps for following ongoing developments in astronomical detector technology.

The CCD-world mailing list continues to serve the community's day-to-day needs for exchanging news and information on all aspects of detector development. Now hosted at CTIO, its web pages and archives may be found at `<http://www.ctio.noao.edu/CCD-world/>`.

The Working Group will continue as a Division IX Working Group.

Timothy M. C. Abbott
chair of the Working Group

References

Simons, D. A., Amico, P., Baade, D., Barden, S., Campbell, R., Finger, G., Gilmore, K., Gredel, R., Hickson, P., Howell, S., Hubin, N., Kaufer, A., Kohley, R., MacQueen, P., Markelov, S., Merrill, M., Miyazaki, S., Nakaya, H., O'Donoghue, D., Oliva, T., Richichi, A., Salmon, D., Schmidt, R., Su, H., Tulloch, S., Vargas, M. M. Wagner, R. M., Wiecha, Ol, Ye, B. 2005, in: J. E. Beleteic, J. W. Beletic, P. Amico (eds.), *Scientific Detectors for Astronomy, 2005*, Proc. Workshop, Taormina, Sicily, 19-25 June 2005, (Springer) p. 13

Burke, B., Jorden, P., Vu, P. 2005, in: J. E. Beleteic, J. W. Beletic, P. Amico (eds.), *Scientific Detectors for Astronomy, 2005*, Proc. Workshop, Taormina, Sicily, 19-25 June 2005, (Springer) p. 225

Hoffman, A., Loose, M., Suntharalingam, V. 2005, in: J. E. Beleteic, J. W. Beletic, P. Amico (eds.), *Scientific Detectors for Astronomy, 2005*, Proc. Workshop, Taormina, Sicily, 19-25 June 2005, (Springer) p. 225

Transactions IAU, Volume XXVIB
Proc. IAU XXVI General Assembly, August 2006
Karel A. van der Hucht, ed.

doi:10.1017/S1743921308024071

COMMISSION 25 — STELLLAR PHOTOMETRY AND POLARIMETRY

PHOTOMETRIE ET POLARIMETRIE STELLAIRE

PRESIDENT — **Arlo U. Landolt**
VICE-PRESIDENT — **Peter Martinez**
ORGANIZING COMMITTEE — **Pierre Bastien, Sergei N. Fabrika, Ronald L. Gilliland, Frank Grundahl, Carme Jordi, Ulisse Munari**

PROCEEDINGS BUSINESS MEETING on 18 August 2006

1. Introduction

Commission 25 is one of three Commissions under the umbrella of Division IX on *Optical and Infrared Techniques*. It is a technique oriented Commission.

2. Business Meeting

Commission 25's Business Meeting convened on Friday, August 18th during Session 2, with 18 individuals in attendance. Landolt reported that the Commission membership was 211, plus an additional 10 new members who would be admitted at the current General Assembly. He briefly reviewed the activity of the Commission during the past three years.

Landolt reported that the new president and vice-president of the commission were Peter Martinez and Eugene F. Milone. The membership of the Commission had been involved in nominating, and then voting, upon an Organizing Committee (OC) for the triennium 2006-2009. The new OC members are: Carme Jordi, Alexei Mironov, Edward G. Schmidt, Qian Sheng-Bang, and Christiaan J. Sterken.

C. Sterken, outgoing president of Division IX, reported on Division affairs. He described the dissolution of Commission 9, and its reconstitution as a series of Working Groups. He noted that a new Commission within Division IX was created during the IAU Executive Committee meeting (EC 81). The new Commission's name will be "Optical & Infrared Interferometry". Sterken informed the attendees that Commission 25 needed to better articulate its scientific mandate.

P. Martinez, incoming Commission 25 president, stated that he felt the Commission was an important one. He invited members to ensure that their and colleagues' e-mail addresses and other personal information in the IAU data base were up-to-date, for Landolt had noted the large number of bounces during attempts to contact the membership by e-mail. Martinez was interested in the possibility of organizing a symposium, across the electromagnetic spectrum, on standardization topics, for the IAU XXVII General Assembly in Rio de Janeiro.

K. Sekiguchi proposed the creation of a Commission 25 endorsed standard star database. The final product would be a one stop location for all standard star data access. He stated that there was a flood of new standard star information available from large survey projects. It would be useful to have a "One Stop" data depository or a gateway to the various recommended data servers via internet access. Such a database should be quality controlled by a group of experts, so that observers could easily choose suitable standard stars that they need to use for their observations. Also, it was suggested that the database should comply with the "Virtual

Observatory (VO)" standard, in order to take full advantage of the suite of data manipulating tools developed by the International VO Alliance group. It also was suggested that the standard star web page not only have the standard data information, but also should provide educational material on how to properly use those standards. Bessell noted that those involved with Virtual Observatory efforts already were considering such problems. Breger agreed that a cookbook on methods of doing and using photometry would be useful. He was in favor of such information, some of which will appear in Sterken's workshop proceedings "The Future of Photometric, Spectrophotometric and Polarimetric Standardization", the result of a meeting held in Blankenberge, Belgium in May 2006. Lub stated concerns regarding the transformability of data between different photometric systems.

Sekiguchi argued that to construct such a database may need considerable efforts. It was noted that an expert(s) would be required to compile and verify such standard star information. The new Commission president Peter Martinez and Kaz Sekiguchi will discuss this matter further to form a Working Group to pursue this idea.

3. Commission 25's Science Session contained six reports

C. Sterken reported on the content of a workshop "The Future of Photometric, Spectrophotometric and Polarimetric Standardization" which took place in Blankenberge, Belgium, from 8 to 11 May 2006. The SOC consisted of members of C25 (Chris Sterken, Arlo Landolt, Carme Jordi, John Landstreet, Gene Milone, Ulisse Munari, Ralph Bohlin, Alex Mironov, Alan Tokunaga, and Pierre Bastien).

The following topics were explored:

- The current status of optical, infrared, spectrophotometric and polarimetric standard stars.
- Problems in defining standard systems: from observational viewpoint, from theoretical viewpoint and from practical viewpoint.
- Standardization needs as driven by small and large telescopes.
- The line between standardization and correct reduction algorithms.

There were 65 participants (1/3 from N America, many from Europe, S. America, Australia, New Zealand, Japan), and more than 50 papers were presented. Proceedings (ed. C. Sterken) will be published by ASP *Conference Series*) early 2007.

A. G. D. Philip and V. Straizys, reported by Philip, described progress in their program of classifying faint stars through use of the Strömvil photometric system. Since 1996 a group of astronomers has been working on setting up and then using the Strömvil photometric system, a combination of the four Strömgren and three Vilnius system filters. The system was announced in Straizys *et al.* (*Baltic Astron.* 5, 83, 1996). The major ability of the Strömvil system is that, from photoelectric measures alone, one can determine the reddening, temperature, gravity and metallicity of stars. With all the new surveys that have been made and ones yet to be made, such a system will be of great use to identify the nature of the new faint stars that will be identified and classify them by stellar type. And since the reddening can be calculated for each region, the intrinsic properties of these stars can be determined.

The main observational programs underway in the Strömvil system at present are:

- Setting up the primary standards. Kazlauskas *et al.* (*Baltic Astron.* 14, 465, 2005) have published a list of 780 photoelectric standards in the northern hemisphere.
- At the Vatican Advanced Technology Telescope on Mt. Graham, Boyle and Philip have been making CCD Strömvil measures of open and globular clusters. Observations are taken in each run of the rich open cluster M 67. These measures are matched to the high-accuracy CCD photometry of Laugalys *et al.* (*Baltic Astron.* 13, 1, 2004) by constraining the corrections to each flatfield to provide the needed one percent photometry in new program fields with only a few standards for zero-point calibration.
- At Casleo, in Argentina, Philip and Pintado have been observing clusters with the 2.15m telescope.
- On the data reduction side Janusz and Boyle have written the CommandLog which automates the process of data reduction for members of our group. This will ensure that all observations will be reduced in exactly the same way.

F. Vrba described near-IR H-band photometry of 18 early-L to early-T dwarf objects. These data come from observations in the U.S. Naval Observatory near-IR astrometry program which has now obtained observations of more than 40 objects for longer than 5 years. He noted that while brown dwarfs of this spectral range are thought to not possess the mechanisms of variability found in late-M stars, they could be variable over timescales of months or years due to global atmospheric changes, such as dust formation and destruction. Thus the USNO database is ideal to test long term variability of these objects. He found no evidence for light variability in the IR on timescales of days to years. In fact, his data indicated a high level of brightness consistency, statistically more constant than surrounding field stars. This is in contrast to at least some studies which have found variability of as high as 50% for a subsample of these same objects.

A. Landolt reported on the status of *UBVRI* broad-band standard star sequences that were nearing completion around the sky. Observations and data reductions have been completed for updated sequences around the sky at the celestial equator, and for a new set of sequences centered at −45 degrees declination. Manuscripts are in preparation. Sequences at +45 degrees declination, begun at Kitt Peak National Observatory telescopes, are being completed at Lowell Observatory. The sequences at all three declinations are photoelectrically based, and cover a range of color indices and the approximate magnitude interval $10 < V < 15$. Data for CCD-based *UBVRI* photometric sequences are being obtained around the celestial equator. These sequences approach 20th magnitude.

E. Craine and D. L. Crawford, in a short report read by Landolt, told about *The Global Network of Astronomical Telescopes* (GNAT). GNAT is in the process of implementing a distributed network of scan mode telescopes. This system is being used to photometrically monitor stars in fixed declination bands, typically within about 10 degrees of the celestial equator. Continuous imaging is obtained with time delay integration of the 1K $\times$ 1K CCD cameras, and stars between $19 < m < 11$ are recorded. Monitoring has been underway for some six years, typically 2.5 to 3 years in each declination band. Observations are being made on nearly every clear photometric night of about 2.5 million stars during the course of a year. Craine and Crawford report that they have discovered about 50,000 new variable stars, but equally interesting, have extensive observations of equatorial band stars which do not appear to vary within the sensitivity of the survey (a function of stellar brightness). Craine and Crawford note that this is a ripe area for collaborators to assist in the creation of large catalogues of relatively faint comparison stars and secondary standard stars. Additional information may be found at < `www.eGNAT.org` >.

B. Smalley gave a status report on the ASTRA Spectrophotometer Project. The ASTRA (*Automated Spectrophotometric Telescope Research Associates*) Cassegrain Spectrophotometer and its automated 0.5-m f/16 telescope are being integrated at the Fairborn Observatory near Nogales, Arizona. The spectrograph uses both a grating and a cross-dispersing prism to produce spectra from both the first and the second orders simultaneously. The square 30 arc second sky fields for each order do not overlap. The resolution is 7Å in second and 14Å in first order. The wavelength range is of approximately $\lambda\lambda$3300-9000. Vega will be the primary flux standard and the anticipated internal star-to-star precision will be better than 1%. The first test observation of the Solar Spectrum was recently taken and scientific observations are expected to begin in 2007. Once fully operational there will be a call for collaborative projects.

Arlo U. Landolt
president of the Commission

Transactions IAU, Volume XXVIB
Proc. IAU XXVI General Assembly, August 2006
Karel A. van der Hucht, ed.

doi:10.1017/S1743921308024083

DIVISION IX / COMMISSION 25 / WORKING GROUP INFRARED ASTRONOMY

CHAIR Eugene F. Milone
VICE-CHAIR Andrew T. Young
MEMBERS Roger A. Bell, Michael S. Bessell, Martin Cohen, Robert F. Garrison, Ian S. Glass, John A. Graham, Lynne A. Hillenbrand, Robert L. Kurucz, Matthew Mountain, George H. Rieke, Stephen J. Schiller, Douglas A. Simons, Michael F. Skrutskie, C. Russell Stagg, Christiaan L. Sterken, Roger I. Thompson, Alan T. Tokunaga, Kevin Volk, *et al.*

PROCEEDINGS BUSINESS MEETING on 17 August 2006

1. Introduction

The WG-IR was created following a Joint Commission Meeting at the IAU General Assembly in Baltimore in 1988, a meeting that provided both diagnosis and prescription for the perceived ailments of infrared photometry at the time. The results were summarized in Milone (1989). The challenges involve how to explain the failure to systematically achieve the milli-magnitude precision expected of infrared photometry and an apparent 3% limit on system transformability. The proposed solution was to redefine the broadband Johnson system, the passbands of which had proven so unsatisfactory that over time effectively different systems proliferated although bearing the same *JHKLMNQ* designations; the new system needed to be better positioned and centered in the atmospheric windows of the Earth's atmosphere, and the variable water vapour content of the atmosphere needed to be measured in real time to better correct for atmospheric extinction.

The WG-IR was formalized by Ian McLean, then president of Commission 25, at the Buenos Aires IAU General Assembly in 1991, and Milone formally appointed to the chair. A subcommittee had been formed almost immediately in 1988 to look at ways to implement the recommendations put forward in Milone (1989). It established the procedure and criteria for judging the performance of existing infrared passbands and began experimenting with passband shapes, widths, and placements within the spectral windows of the Earth's atmosphere. The method and coding were initiated and largely carried out by Andy Young, with Milone running the simulations and Stagg assisting with profiles.

By 1993, preliminary recommendations were presented at the photometry meeting in Dublin (Young, Milone, & Stagg 1993). The full details of the criteria and results of the numerical simulations were presented by Young, Milone, & Stagg (1994). Subsequent work, described in WG-IR and/or Commission 25 reports, included the use of a new MODTRAN version (3.7) to check and extend previous work. This part of the program proved so successful in minimizing the effects of water vapour on the source flux transmitted through the passband that the second stage, real-time monitoring of IR extinction, was not pursued, although this procedure remains desirable for unoptimized passbands designed for specific Astrophysical purposes.

2. Developments within the past triennium

The WG-IR has the had the policy of being open to input from its members at all times following the initial consultations with all segments of the infrared community.

During the 2003-2006 triennium, the WG concentrated on gathering and presenting evidence of the usefulness of the WG-IR infrared passband set. For the near infrared portion of the WG-IR set (namely iz, iJ, iH, iK), field trials were conducted over the years 1999-2003 with the 1.8-m telescope at the Rothney Astrophysical Observatory of the University of Calgary. The results of those trials and the details of further work done to that date were presented in Milone & Young (2005). This paper contained, for the first time, evidence that not only were the WG-IR passbands more useful to secure precise transformations than all previous passbands, but that they were also superior in at least one measure of the signal to noise ratio. This evidence was further refined in Milone & Young (2007). As a consequence, the original purpose of the WG-IR largely has been achieved but opposition to the new passband system is still strong, and passbands that somewhat compromise the WG-IR recommendations have been advanced in order to provide more throughput, at the cost of precision and standardization. As a consequence, non-optimized passbands are still in use at the highest altitude infrared sites. The situation is described (and decried!) in Milone & Young (2007).

On the other hand, the *need* for improved IR passbands has been accepted by the community, by far and large. The work of Salas, Cruz-González, & Tapia (2006) in making use of Padé approximants to simulate atmospheric extinction as they define a new set of unoptimized intermediate IR passbands illustrates this well. In the near infrared, the work of Simons & Tokunaga (2002) and Tokunaga, Simons, & Vacca (2002) typify the new attitude; the work by this group was summarized recently by Tokunaga & Vacca (2007).

Perhaps the most important aspect of the WG-IR's work is the promise that the near-IR WG-IR passband set holds for highly precise photometry at intermediate and even low elevation sites.

3. Closing remarks

The future activities require fabrication and field testing of the remainder of the WG-IR passband set, namely the passbands iL, iL', iM, iN, in, and iQ. It is also desirable to extend the list of near infrared standard stars presented in Milone & Young (2005) to a fuller all-season set, and to extend them to fainter stars, as, for example, Landolt (1992) (and in earlier papers cited therein) has done for the visual Johnson-Cousins passbands.

Eugene F. Milone
chair of the Working Group

References

Landolt, A. 1992, *AJ* 104, 340

Milone 1989, in: E. F. Milone (ed.), *Infrared Extinction and Standardization*, Proc., Two Sessions of IAU Commissions 25 and 9, Baltimore, MD, USA, 4 August 1988, *Lecture Notes in Physics*, Vol. 341 (Heidelberg: Springer), p. 1

Milone, E. F., & Young, A. T. 2005, *PASP*, 117, 485

Milone, E. F., & Young, A. T. 2007, in: C. Sterken (ed.), *The Future of Photometric, Spectrophotometric, and Polarimetric Standardization*, Proc. Intern. Workhop, Blankenberge, Belgium, 8-11 May 2006, *ASP-CS*, 364, 387

Salas, L., Cruz-González, I., & Tapia, M. 2006, *RMAA*, 42, 273

Simons, D. A., & Tokunaga, A. T. 2002, *PASP*, 114, 169

Tokunaga, A. T., & Vacca, W. D. 2007, in: C. Sterken (ed.), *The Future of Photometric, Spectrophotometric, and Polarimetric Standardization*, Proc. Intern. Workhop, Blankenberge, Belgium, 8-11 May 2006, *ASP-CS*, 364, 409

Tokunaga, A. T., Simons, D. A., & Vacca, W. D. 2002, *PASP*, 114, 180

Young, A. T., Milone, E. F., & Stagg, C. R. 1993, in: C. J. Butler & I. Elliott (eds.), *Stellar Photometry – Current Techniques and Future Developments*, Proc. IAU Coll. No. 136, Dublin, Ireland, 4-7 August 1992 (Cambridge: University Press), p. 235

Young, A. T., Milone, E. F., & Stagg, C. R. 1994, *A&AS*, 105, 259

Transactions IAU, Volume XXVIB
Proc. IAU XXVI General Assembly, August 2006
Karel A. van der Hucht, ed.

doi:10.1017/S1743921308024095

COMMISSION 30 — RADIAL VELOCITIES

VITESSES RADIALES

PRESIDENT	**Birgitta Nordström**
VICE-PRESIDENT	**Stéphane Udry**
PAST PRESIDENT	**Andrei A. Tokovinin**
ORGANIZING COMMITTEE	**Dainis Dravins, Francis C. Fekel, Elena V. Glushkova, Hugo Levato, Dimitri Pourbaix, Myron A. Smith, Laszlo Szabados, Guillermo Torres**

DIVISION IX / COMMISSION 30 WORKING GROUPS

Division IX / Commission 30 / WG	**Radial-Velocity Standard Stars**
Division IX / Commission 30 / WG	**Stellar Radial-Velocity Bibliography**
Division IX / Commission 30 / WG	**Catalog of orbital Elements of Spectroscopic Binary Systems**

PROCEEDINGS BUSINESS MEETING on 18 August 2006

1. Report by the president

The president welcomed all the participants of the Business Meeting and remarked that several of the major ongoing and planned Radial Velocity projects were well represented.

1.1. *Membership*

A number of very large radial velocity surveys are ongoing and planned. Several of the key persons in these projects are new members in Commission 30.

The following 16 colleagues had applied for membership of Commission 30 and were unanimously welcomed as members: Richard A. Arnold (New Zealand), Francoise Crifo (France), Vladimir Elkin (UK), Johan Holmberg (Germany), David A. Katz (France), Geoffrey W. Marcy (USA), Douglas J. Mink (USA), Dominique Naef (Chile), Nicola R. Napolitano (Italy), Mikhail E. Sachkov (Russian Federation), Nuno Miguel C. Santos (Portugal), Anja C. Schroder (UK), Zaggia Simone (Italy), Matthias Steinmetz (Germany), Catherine Turon (France), and Tomaz Zwitter, Slovenia.

1.2. *Elections*

The Nominating Committee had solicited nominations to the Commission membership for the five vacant posts in the Organizing Committee (OC). The working rules concerning elections of members of the OC had been followed. No election was necessary as the nomination procedure resulted in five candidates which was equal to the number of vacancies. The new members of the Organizing Committee are: Ken Freeman (Australia), Goeff Marcy (USA), Catherine Turon (France), Robert Mathieu (USA), and Tomaz Zwitter (Slovenia).

Outgoing members of the OC are: Dainis Dravins, Hugo Levato, Birgitta Nordström, Myron Smith, and Laszlo Szabados.

Thus the new officers and organizing committee for 2006 - 2009 are:

Stephane Udry (president), Guillermo Torres (vice-president), Birgitta Nordström (past president), Francis C. Fekel, Kenneth C. Freeman, Elena V. Glushkova, Goeffrey W. Marcy, Robert D. Mathieu, Dimitri Pourbaix, Catherine Turon, and Tomaz Zwitter.

2. The triennial report 2002-2005

The very lively activity in the field of radial velocities is shown in the extensive triennial report to which several of the OC members made great contributions. See (Nordström 2006), and <http://www.ctio.noao.edu/science/iauc30/iauc30.html>.

3. Reports from the Working Groups

The chairpersons of the three Working Groups of the commission gave reports on the work done during the past triennium as well as their plans for the future:
- Radial velocity standard stars new needs for exoplanets, *Gaia*, *SIM*, etc. by Stephan Udry;
- Stellar radial velocity bibliography by Hugo Levato;
- Catalogue of orbital elements of spectroscopic binary systems by Dimitri Pourbaix.

4. Reports from new radial velocity surveys planned and ongoing work

- 'New results in exoplanet work' by Stéphane Udry;
- 'The Solar neighbourhood' by Johan Holmberg;
- 'The RAVE project' by Tomaz Zwitter.

5. End of meeting

The president thanked the outgoing members of the Organizing Committee for their contributions, constructive ideas and very active participation in discussions on Commission 30 matters. She also thanked the speakers and the participants at the meeting.

Birgitta Nordström
president of the Commission

References

Nordström 2006, in: *Commission 30, Radial Velocities*, IAU Transactions XXVIA, Reports on Astronomy 2002 - 2005, O. Engvold (ed.), 2007 (Cambridge: CUP).

Transactions IAU, Volume XXVIB
Proc. IAU XXVI General Assembly, August 2006
Karel A. van der Hucht, ed.

doi:10.1017/S1743921308024101

DIVISION IX / COMMISSION 30 / WORKING GROUP 9TH CATALOGUE OF SPECTROSCOPIC BINARY SYSTEMS

CHAIR **Dimitri Pourbaix**
MEMBERS **Alan H. Batten, Frank C. Fekel, William H. Hartkopf, Hugo Levato, Nidia I. Morrell, Andrei A. Tokovinin, Guillermo Torres, Stéphane Udry**

PROCEEDINGS BUSINESS MEETING on 21 August 2006

1. Progress report

The current content of the database was presented, emphasising the substantial progress accomplished since the IAU XXIV General Assembly in Manchester, 2000. More than 1 200 stellar systems have been added to the 8th Catalogue over the past six years, for a total of 540 papers compiled. A first paper was published to make the community aware of this facility (Pourbaix *et al.* 2004).

The usefulness of the work carried on by the Working Group was assessed through the statistics of the access to the database. More than twenty thousands plots have been retrieved since January 2001 and 350 copies of the tar ball with the database content have been downloaded since August 2003. The 22 citations to the 2004 paper were also a good indication of our success.

2. Future

A cross matching SB9 with the bibliographic references by H. Levato and co-workers reveals that the completeness of SB9 is about 75% of the published orbits. Strategies were therefore discussed in order to maximise the completeness minimising the work (i.e., rank the contribution of each missing papers in term of benefit for SB9 and enter the paper according to that rank).

The case of extrasolar planets was also discussed. In one hand, those in favour of adding the spectroscopic orbits of extrasolar planets without indication of the planetary nature of the companion. The drawback of this approach is that planetary and stellar companions would be mixed together when plotting, for instance, a (e-$logP$) diagram. In the other hand, those in favour of clearly identifying the planetary systems. The drawback is here that the planetary nature of the companion cannot be assessed in most cases. It was finally decided to add the extrasolar planets with a special tag (at the epoch of the discovery).

3. Closing remarks

D. Pourbaix thanked all the contributors, inside and outside the work group, for their help over the past three years.

Dimitri Pourbaix
chair of the Working Group

References

Pourbaix, D., Tokovinin, A. A., Batten, A. H., Fekel, F. C., Hartkopf, W. I., Levato, H., Morrell, N. I., Torres, G., & Udry, S. 2004, *A&A*, 424, 727

Transactions IAU, Volume XXVIB
Proc. IAU XXVI General Assembly, August 2006
Karel A. van der Hucht, ed.

doi:10.1017/S1743921308024113

DIVISION X RADIO ASTRONOMY

RADIOASTRONOMIE

Division X provides a common theme for astronomers using radio techniques to study a vast range of phenomena in the Universe, from exploring the Earth's ionosphere or making radar measurements in the Solar System, via mapping the distribution of gas and molecules in our own Galaxy and in other galaxies, to study the vast explosive processes in radio galaxies and QSOs and the faint afterglow of the Big Bang itself.

PRESIDENT	**Luis F. Rodriguez**
VICE-PRESIDENT	**Ren-Dong Nan**
PAST PRESIDENT	**Lucia Padrielli** († 22 December 2003)
BOARD	**Philip J. Diamond, Gloria M. Dubner, Michael Garrett, W. Miller Goss, Anne Green, Masato Ishiguro, A. Pramesh Rao, Russell A. Taylor, Jose M. Torrelles, Jean L. Turner**

DIVISION X COMMISSIONS

Commission 40	**Radio Astronomy**

DIVISION X WORKING GROUPS

Division X WG	**Interference Mitigation**
Division X WG	**Astrophysically Important Spectral Lines**
Division X WG	**Global VLBI**

INTER-DIVISION WORKING GROUPS

Division X-IX WG	**Encouraging the International Development of Antarctic Astronomy**
Division X-XII WG	**Historic Radio Astronomy**

PROCEEDINGS BUSINESS MEETING on 18 August 2006

1. Introduction

The Division X business meeting took place in two 1.5 hour sessions during the afternoon of 18 August 2006. The meeting started with the president presenting the agenda of the meeting to an audience of about 70 participants. The president noted that at the moment of the meeting the Division counts with 889 members and that there were 85 candidates for membership, that would take the number of members to 974 at the end of the General Assembly.

2. The Organizing Committee

The duties of the Organizing Committee were listed for the audience and some time was spent explaining the handling of the requests for support for IAU Symposia. The president noted that some 20 requests were received every year and that the Executive Committee approves about

9 of them, based upon the recommendations of the Division presidents. It was also reminded that the support is of 25,000 Swiss Francs for every IAU Symposium (to be used for grants for participants) and that the previous category of IAU Colloquia no longer exists.

Among the IAU Symposia for 2007, Division X is the main sponsor of IAU Symposium No. 242 on *Astrophysical Masers and their Environments*, in Alice Springs, central Australia, during 12-17 March 2007. This will be the third in a series of international symposia on this topic and it was suggested that the Division should try to keep these symposia as a tradition to take place every several years, like the successful series of Planetary Nebula symposia sponsored by the IAU.

The newly appointed members of the Organizing Committee were presented, with Ren-Dong Nan being the next president of the Division and Russ Taylor the next vice-president. It was also noted that the five older members of the Organizing Committee were replaced by other members of the Division.

3. In search of more involvement from the membership

Until the Prague General Assembly, the election of new members of the Organizing Committee has been made by the Committee itself. The Executive Committee of the IAU wants a larger involvement of the general membership on these decisions and in the next election previous to the Rio General Assembly in 2009, the Organizing Committee will take advantage of email to ask the membership for candidates to the positions.

4. Commissions

While some IAU Divisions like Division X (Radio Astronomy) and Division XI (Space & High Energy Astrophysics) have only one Commission that in practice is equal to the Division itself, others have many commissions, with the case of Division IV (Stars) having as much as five Commissions (Double & Multiple Stars, Stellar Spectra, Stellar Constitution, Theory of Stellar Atmospheres, and Stellar Classification). The audience was informed that the discussion inside the Organizing Committee favored maintaining the present structure, in the sense that Commission 40 (Radio Astronomy) should not disappear. There were comments from several members of the audience supporting this position.

5. Working Groups

The duties of the three Working Groups of the Division were summarized and a discussion opened on the convenience of transforming, disappearing, of merging the present working groups. There was considerable disagreement on this topic and it was concluded that we will keep the present structure, asking the Organizing Committee to look into these possibilities in more detail and come with a more clear point of view for the next General Assembly.

6. Final remarks

The president expressed that in the following few days there was going to be an important voting on the *'Endorsement of the Washington Charter for Communicating Astronomy with the Public* and that the participation of the members of the Division was expected. He also noted that it appears quite likely that 2009 will be declared the *International Year* of Astronomy by the United Nations and that all IAU members should make an effort to participate with outreach to the public.

7. Presentations of the Working Groups

After the previous points were presented, we had two short presentation by Masatoshi Ohashi (Astrophysically Important Spectral Lines) and Tasso Tzioumis (Interference Mitigation), on the work made by these groups over the last three years. Ohashi emphasized the work made in preparation for the important 2007 World Radiocommunication Conference, while Tzioumis discussed the new challenges faced by interference mitigation technologies.

The meeting concluded with seven short presentations on recent developments in radio observatories:
'The SubMillimeter Array' (James M. Moran),
'RATAN-600 at the CROSS ROAD' (Yuri N. Parijskij),
'The Large Millimeter Telescope' (Alfonso Serrano),
'NAIC/Arecibo' (Murray Lewis),
'FAST - Five hundred meter Aperture Spherical radio Telescope' (Ren-Dong Nan),
'EVLA: The Expanded Very Large Array' (Bryan Butler), and
'JIVE' (Mike Garrett).

Luis F. Rodriguez
president of the Division

Transactions IAU, Volume XXVIB
Proc. IAU XXVI General Assembly, August 2006
Karel A. van der Hucht, ed.

doi:10.1017/S1743921308024125

DIVISION XI — SPACE & HIGH-ENERGY ASTROPHYSICS

ASTROPHYSIQUE SPATIALE & DES HAUTES ENERGIES

Division XI connects astronomers using space techniques or particle detectors for an extremely large range of investigations, from *in-situ* studies of bodies in the solar system to orbiting observatories studying the Universe in wavelenghts ranging from radio waves to γ-rays, to underground detectors for cosmic neutrino radiation.

PRESIDENT	**Haruyuki Okuda**
VICE-PRESIDENT	**Guenther Hasinger**
PAST PRESIDENT	**Ganesan Srinivasan**
BOARD	**Monique D. Arnaud, Sidney A. Bludman, João Braga, Noah Brosch, Leonid I. Gurvitz, Hashima Hasan, George Helou, Ian D. Howarth, Hajime Inoue, Luigi Piro, Ganesan Srinivasan**

DIVISION XI COMMISSION

Commission 44	**Space & High-Energy Astrophysics**

DIVISION XI WORKING GROUPS

Division XI WG	**Particle Astrophysics**
Division XI WG	**Astronomy from the Moon**

PROCEEDINGS BUSINESS AND SCIENCE MEETING on 17 August 2006

1. Introduction

A business meeting of Division XI / Commission 44 was held during the IAU XXVI General Assembly, Prague, Chezc Republic, on August 17, 2006. Chair of the meeting was Prof. Haruyuki Okuda, president of the Division and the Commission 44. Vice-president Prof. Guenther Hasinger was also present.

2. Report of the Business Meeting

At the beginning, a brief report on the scientific activities of the Division in the past period (2003-2006) was given by Prof. Okuda, in which he concluded that the past period was very successful and fruitful in every relevant fields of the Division; UV, X-ray/γ-ray, optical, IR/sub-millimeter as well as radio regions, having provided a variety of interesting results by the many ongoing missions as well as by newly launched missions, such as *ASTRO-E/Suzaku*, *SIRTF/Spitzer*, and *ASTRO-F/Akari*. The ground-based Cherenkov facilities such as HESS and MAGIC have started exciting observations of extremely high-energy γ-ray radiation. There are also many challenging missions to be launched in the near future, such as *SOLAR-B* (Sun),*TAUVEX* (UV), *GLAST* (γ-ray), *Herschel/Planck* (IR/sub-mm), and *LISA Pathfinder* (gravitational waves).

After an introductory talk by the president, some discussion took place on the organization matter of the Division. The Division is composed of a wide complement of disciplines from X-ray/γ-ray to IR/sub-mm and radio regions. The Division has started from rather technique oriented subjects, such as Space and High-Energy Astrophysics, and thus covers quite a wide range of astronomy, particularly due to the rapid progress of observational technique in IR/sub-mm range. On the other hand, the membership is highly biased towards the high energy community. This situation has introduced some difficulty in arranging scientific activities in the Division, such as proposing and hosting scientific meetings. The problem had been discussed among the Board members and EC members prior to the General Assembly. There had been a variety of opinions and suggestions of possible amendments of the Division, re-definition and restructuring of the Division, e.g., re-unification to High-Energy Astrophysics exclusively, or re-inforcement of the low energy part by adding a new Commission for IR/sub-mm and radio astronomy. However, it was concluded that the problem is too difficult to be solved in a short time and the issue should be watched continuously in the coming period. This basic attitude has been approved in the business meeting.

3. Working Groups

Two Working Groups, WG *Astronomy from the Moon* and WG *Particle Astrophysics* have been run in the Division.

The former WG has been created for international communication and coordination of the Lunar missions proposed by many countries. Prof. Sallie Baliunas gave a short report on the current status of the Moon Exploration programs. Recently, it has met a new situation by the US campaign of the Moon and the Mars Exploration. It has been decided in the Science Session that the Working Group will be re-organized into a Inter-Division WG including Division IX and X.

The Science Meeting of the WG *Particle Astrophysics* was held, but unfortunately poorly attended and many invited talks were cancelled, only brief reports were given on the current activities of the ground based observations of neutrino and Tev-energy γ-rays. Given this situation, Prof. Reinhard Schlickeiser, the chair of the WG, has proposed a future re-organization of the WG.

4. Board members 2006 - 2009

Taking balance among the fields and reflecting the discussion of the amendment of the Division organization, Prof. Chiristine Jones was proposed for vice-president and the following board members were proposed by the new and former presidents and agreed in the meeting. All members have been approved in the General Assembly.
President: Guenter Hasinger (BRD). Vice-president: Christine Jones (USA). Board: Joao Braga (Brasil), Noah Brosch (Israel), Thijs de Graauw (the Netherlands), Leonid I. Gurvitz (the Netherlands), George Helou (USA), Ian D. Howarth (UK), Hideyo Kunieda (Japan), Thierry Montmerle (France), Haruyuki Okuda (Japan), Marco Salvati (Italy), and Kulinder P. Singh (India).

Haruyuki Okuda
president of the Division

Transactions IAU, Volume XXVIB
Proc. IAU XXVI General Assembly, August 2006
Karel A. van der Hucht, ed.

doi:10.1017/S1743921308024137

DIVISION XI / WG PARTICLE ASTROPHYSICS

CHAIR **Reinhard Schlickeiser**
MEMBERS **Roger D. Blandford, Alain Brillet, Masa-Katsu Fujimoto, Piero Madau, Angela V. Olinto, Marco Salvati, Bernard F. Schutz, Peter F. Smith, Michel Spiro, Arnold A. Stepanyan, Yoji Totsuka, Sylvaine Turck-Chieze, Heinrich J. Voelk**

PROCEEDINGS BUSINESS MEETING on 25 August 2006

1. Introduction

The business meeting of the Division XI Working Group on *Particle Astrophysics* took place in the morning of 25 August 2006, and was attended by 14 participants.

From the five planned talks three (by Blandford, Ellison, Waxman) were withdrawn on rather short notice; in two cases (Ellison, Waxman) the invited speakers refused to come after finding out about the huge registration fee of this IAU General Assembly.

One additional talk by Dr. Spiering was withdrawn on very short notice because of illness of the speaker. Two graduate students from the ICECUBE experiments (M. Duvoort and T. Castermans), who were present in Prague, substituted Dr. Spiering.

So the only invited talk, that took place as planned, was the talk by Dr. Paula Chadwick about new results from TeV γ-ray astronomy.

2. Conclusions

As a consequence of this disastrous meeting, I informed Dr. Hasinger, the new president of Division XI, that I will resign as Working Group chairman. It seems to me after this experience that most particle astrophysicist do not recognize the IAU (and their meetings) as an important agency for the advancement of their field.

In the preparation of this meeting and during the meeting I had no support from any of the members of the Working Group despite several e-mail contacts. My working time is to valuable to waste it on a hopeless issue.

Also from my biased point of view, the IAU agency is too big and inflexible to provide the necessary support for chairpersons of small Working Groups. I was never properly informed about deadlines; when asking for registration fee waivers for the invited speakers the only response I got was that this is now much too late.

Reinhard Schlickeiser
chair of the Working Group

Transactions IAU, Volume XXVIB
Proc. IAU XXVI General Assembly, August 2006
Karel A. van der Hucht, ed.

doi:10.1017/S1743921308024149

DIVISION XI / WG ASTRONOMY FROM THE MOON

CHAIR **Sallie L. Baliunas**
VICE-CHAIR **Yoji Kondo**
PAST CHAIR **Norio Kaifu**
MEMBERS **Oddbjorn Engvold, Norio Kaifu, Haruyuki Okuda, Yervant Terzian, Willem Wamsteker** († 24 November 2005)

PROCEEDINGS BUSINESS MEETING on 23 August 2006

1. Introduction

The Business Meeting opened with a recall of the memory of a member of the Organizing Committee, Willem Wamsteker. N. Kaifu, past president of the Working Group, was thanked for his outstanding service.

2. Inter-divisional status of Working Group proposed

Discussion ensued of broadening the Working Group to an inter-divisional status, should the Working Group status be renewed by Division XI (*Space and High-Energy Astrophysics*).

It was proposed that the Working Group ask to be extended to Divisions IX (*Optical and Infrared Techniques*), Division X (*Radio Astronomy*), as well as the mother Division XI.

G. Hasinger subsequently related that the IAU EC had agreed to extend the Working Group to inter-divisional status to span Divisions IX, X and XI.

3. Working Group tasks

H. Okuda remarked on the welcome, new opportunities emerging across the world for lunar research. Such projects necessarily will be large in scope and infrastructure, requiring international cooperation. The Working Group could provide a forum for the cooperative, international exchange of information.

H. Okuda also asked the Business Meeting's participants to discuss and generate proposals for the 2009 IAU XXVII General Assembly concerning astronomy from the moon.

It was agreed that the Working Group would continue to discuss, largely electronically, several matters:
(*i*) a WG proposal for a Joint Discussion at the 2009 IAU GA, mindful of the fact that the proposal should be oriented toward science results, and that the due date for proposals is 15 December 2007;
(*ii*) summarizing the objectives of the reorganized WG;
(*iii*) consideration of information on the impact of electromagnetic pollution in the lunar environment on scientific research plans; and
(*iv*) advancing the exchange of information among those interesting in WG activities.

The website <http://www.cfa.harvard.edu/moon> was created for the distribution of WG material.

4. Science presentations

Following the Business Meeting was an excellent series of science presentations, as listed below:

SMART-1 and *Aurora*, B. Foing *et al.* (read by L. Gurvits)

SELENE, by M. Kato

SELENE II, by M. Kato

Chang'e 1, by M. Huang

LOFAR on the moon - a lunar low-frequency radio telescope and more, by H. Falcke

Radio astronomy view at the Moon and from the Moon, by L. Gurvits

Investigation in Japan on astronomy from the Moon, by N. Kawano

Long Wavelength Astrophysics from the Moon, by J. Lazio

Optical and UV Interferometry from the Moon, by P. Chen, Y. Kondo, E. Guinan, F. Bruhweiler and S. Baliunas

The Moon – 2012+, A. Gusev

5. New Organizing Committee

Sallie Baliunas (chair, USA) and Yoji Kondo (vice-chair,USA). Members: Oddbjorn Engvold (Norway), Karel A. van der Hucht (Netherlands), Norio Kaifu (Japan), Haruyuki Okuda (Japan), and Yervant Terzian (USA).

6. Closing remarks

The meeting was adjourned at the close of science presentations and discussion. I am grateful for the enthusiasm of participants in delivering what were excellent talks.

Sallie L. Baliunas
chair of the Working Group

Transactions IAU, Volume XXVIB
Proc. IAU XXVI General Assembly, August 2006
Karel A. van der Hucht, ed.

doi:10.1017/S1743921308024150

DIVISION XII UNION-WIDE ACTIVITIES

ACTIVITÉS D'INTÉRÊT GÉNÉRAL DE L'IAU

Division XII consists of Commissions that formerly were organized under the Executive Committee, that concern astronomers across a wide range of scientific sub-disciplines and provide interactions with scientists in a wider community, including governmental organizations, outside the IAU.

PRESIDENT	**Virginia L. Trimble**
VICE-PRESIDENT	**Johannes Andersen**
BOARD	**Kaare Aksnes, Françoise Genova, Alexander A. Gurshtein, Sveneric Johansson, Jay M. Pasachoff, Malcolm G. Smith**

DIVISION XII COMMISSIONS

Commission 5	**Documentation and Astronomical Data**
Commission 6	**Astronomical Telegrams**
Commission 14	**Atomic and Molecular Data**
Commission 41	**History of Astronomy**
Commission 46	**Astronomy Education and Development**
Commission 50	**Protection of Existing and Potential Observatory Sites**

INTER-DIVISION WORKING GROUP

Division X-XII WG	**Historic Radio Astronomy**

PROCEEDINGS BUSINESS MEETING on 23 August 2006

A very brief Division XII meeting was held during the IAU XXVI General Assembly, which provided opportunities for the incoming presidents and vice-presidents of the individual Commissions and of the Division to get to know each other, to set up informal inter-Commission meetings during the General Assembly and to meet with – and thank – the outgoing Division president Virginia Trimble and vice-president Johannes Andersen. It was agreed that Commission 50 should seek ways to work more closely with the WG on *Radio Interference Mitigation* of Division X (*Radio Astronomy*).

Informal meetings, held later during the General Assembly, started to develop ways in which the various Commissions of this somewhat catchall Division can work more closely together. One possibility is to look for ways that each Commission and Working Group can unite, prepare for and contribute to the International Year of Astronomy in 2009. In this context it is clear that the new Commission 55 on *Communicating Astronomy with the Public* under the leadership of its first president, Ian Robson, is a particularly welcome addition to the Division.

The business and science covered in the individual Commissions is set out elsewhere in this publication, as are obituaries for two presidents of Commission 50, Jim Cohen and Hugo Schwarz, who have both died since the meeting in Prague.

Malcolm G. Smith
incoming president of the Division

Transactions IAU, Volume XXVIB
Proc. IAU XXVI General Assembly, August 2006
Karel A. van der Hucht, ed.

doi:10.1017/S1743921308024162

COMMISSION 5 — DOCUMENTATION AND ASTRONOMICAL DATA

DOCUMENTATION ET DONNÉES ASTRONOMIQUES

PRESIDENT — Françoise Genova
VICE-PRESIDENT — Raymond P. Norris
PAST PRESIDENT — Olga B. Dluzhnevskaya
ORGANIZING COMMITTEE — Michael S. Bessel, H. Jenker, Oleg Yu. Malkov, Fionn Murtagh, Koichi Nakajima, François Ochsenbein, William D. Pence, Marion Schmitz, Roland Wielen, Yong Heng Zhao

COMMISSION 5 WORKING GROUPS

Div. XII / Commission 5 WG	Astronomical Data
Div. XII / Commission 5 WG	Designations
Div. XII / Commission 5 WG	Libraries
Div. XII / Commission 5 WG	FITS
Div. XII / Commission 5 WG	Virtual Observatories, Data Centers and Networks
Div. XII / Commission 5 TF	Preservation and Digitization of Photographic Plates

PROCEEDINGS BUSINESS MEETING on 23 August 2006

1. Introduction

Commission 5 has been very active during the IAU XXVI General Assembly in Prague: the Commission, its Working Groups and its Task Force held business meetings. In addition, Commission 5 sponsored two Special Sessions: Special Session 3 on *The Virtual Observatory in Action: New Science, New Technology, and Next Generation Facilities* which was held for three days 17–22 August, and Special Session 6 on *Astronomical Data Management*, which was held on 22 August. Commission 5 also participated in the organisation of Joint Discussion 16 on *Nomenclature, Precession and New Models in Fundamental Astronomy*, which was held 22-23 August. The General Assembly and Commission 5 web sites provides links to detailed information about all these meetings.

This paper summarizes Commission 5 Business Meeting discussions and provides reports from the Working Groups and Task Force.

2. Structure of Commission 5

During the last triennium, Commission 5, which was previously directly attached to the IAU Executive, has been part of Division XII *Union-Wide activities*. This has been very positive, by providing a forum for discussion with other Commissions and an official communication path to the Executive for proposing meetings, new individual members, etc.

The following structure is proposed for Commission 5 Working Groups and Task Force:

- WG *Astronomical Data*, chair R. P. Norris
- WG *Designations*, chair M. Schmitz
- WG *Libraries*, co-chairs U. Grothkopf and F. Murtagh
- WG *FITS*, chair W. D. Pence, Vice-chair F. Ochsenbein
- WG *Virtual Observatory*, chair R. J. Hanisch
- TF *Preservation and Digitization of Photographic Plates*, chair E. Griffin

The new OC proposed to Division XII includes all WG chairs. It is composed of: president R. P. Norris, vice-president M. Ohishi, past-president F. Genova; and members U. Grothkopf, R. J. Hanisch, O. Malkov, W. D. Pence, M. Schmitz, and X. Zu.

Commission 5 expresses its gratitude to retiring officers, M. S. Bessel, O. Dluzhnevskaia, H. Jenkner, F. Murtagh, K. Nakajima, F. Ochsenbein, R. Wielen, and Y. H. Zhao, and to R. Garstang who has asked that his name be dropped from the list of Commission 5 members, for their action for Commission 5.

Four consultants who play an important role for Commission 5 and IAU have been proposed for IAU membership through Division XII: Suzanne Borde (France), Monica Gomez (Spain), Uta Grothkopf (Germany), Doug Tody (USA).

Commission 5 welcomes incoming members: Jerome Berthier (France), Ian Bond (New Zealand), Hsiang-Kuang Chang (Taiwan), Jacqueline Chapman (Australia), Lucio Chiappetti (Italy), John Cuniffe (Ireland), Sylvia Dalla (UK), Adam Dobrzycki (Germany), Daniel Durand (Canada), Davide Elia (Italy), Ducan Fyfe (UK), Vladimir Garaimov (Russia) , Andrew Hopkins (Australia), Ludmila Hudkova (Ukraine), Robert G. Mann (UK), Douglas Mink (USA), Rhys Morris (UK), Tara Murphy (Australia), Yuri Pakhomov (Russia), Kevin Reardon (Italy), Vladimir Samodurov (Russia), David Schade (Canada), Anja Schroder (UK), Jonathan Tedds (UK), and Hon-Jin Yang (Korea), who have expressed their willingness to join through the procedure managed by the IAU Executive, and D. Green who has directly contacted the Commission.

This constitute a significant increase in the number of members (124 on the present list at the IAU Secretariat), and is twice the number of new members (12) who joined Commission 5 at the Sydney GA. Taking advantage of the increased flexibility of IAU procedures, the new OC will check the list of Commission members and contact IAU members who play an important role in the field, to propose them to join Commission 5.

A huge and reasonably successful effort has allowed to check and update the email addresses of Commission 5 members after Sydney GA. However, about ten addresses in the present IAU database are not valid, and the effort to contact these members will be pursued. In particular, checks can be made with the membership directories of the national astronomy associations.

3. Working Groups and Task Force reports

3.1. *WG Astronomical Data*

The Working Group on *Astronomical Data* (WG-AD) held three sessions, the first of which was to enact business and the last two were for discussions on data science and management. The agenda and presentations (where available) for these latter two sessions are on <http://www.atnf.csiro.au/people/rnorris/WG-AD/>.

The first agenda item of the first session was to review the role and operation of the WG-AD. The WG-AD has two distinct but interwoven roles. One is to act as a conduit between the IAU and CODATA, the ICSU Committee on Data in Science and Technology. The WG-AD chair is also the IAU delegate to CODATA, and uses the WG-AD to discuss issues arising from CODATA activities, or issues in which some input should be made to CODATA. The other role is to act as a forum for discussing issues in data science. In this respect the WG-AD has been very active in the last year, conducting a series of electronic discussions on <http://www.ivoa.net/twiki/bin/view/Astrodata/> and <http://tech.groups.yahoo.com/group/astro-data/>, leading up to Special Session 6 at the IAU GA on *Data Management*. There was a consensus that the WG-AD was acting effectively in both these roles, that those participating had found it a useful and fruitful exercise, and that the WG-AD should continue in those roles.

The second agenda item was to review the process of these electronic discussions. It was agreed that these discussions had been productive and stimulating, and should continue as a forum in the future. Several suggestions were made as to how the participation in the discussions could be enhanced, and the WG-AD chair agreed to draft a protocol to guide participation in these discussions. A concern was voiced that the outcomes of these discussions need to be promulgated to the wider community and it was agreed that we must search for better ways for the 'data enthusiasts' to engage with the wider astronomical community. This will introduced as a topic for a future e-discussion.

Membership of the WG-AD has steadily increased over the last three years, with a process for admitting new members which involves polling existing members, with the result that new members can quickly engage with the activities of the WG-AD. Several participants at the meeting expressed a wish to join the WG-AD, and their names will be proposed to the membership through this process.

A report was given on behalf of the Task Force for *Preservation and Digitization of Photographic Plates* (TF-PDPP), and members invited to attend the subsequent meeting of the TF-PDPP.

3.2. *WG Designations*

The WG *Designations* currently has 20 members, including journal editors. At the 2003 Sydney GA, Marion Schmitz took over the chair succeeding Hélène Dickel. The WG *Designations* clarifies existing astronomical nomenclature and help astronomers avoid potential problems when designating their sources.

The most important function of WG *Designations* during the period 2003–2006 was overseeing the IAU *Registry for Acronym*, which is sponsored by IAU and operated by the Centre de Données de Strasbourg (CDS). An on-line form for pre-registering newly discovered astronomical sources of radiation is available at `<http://vizier.u-strasbg.fr/viz-bin/DicForm>`. The Clearing House, a subgroup of the WG, screens the submissions for accuracy and conformity to the IAU Recommendations for Nomenclature, see `<http://vizier.u-strasbg.fr/Dic/iau-spec.htx>`. From its beginning in 1997 through August 2006, there have been 132 submissions and 111 acceptances. Attempts to register asterism, common star names, and suspected variable stars were rejected. The three past years saw 61 acronyms submitted with 50 of them being accepted.

Assistance was provided for inquiries about nomenclature for planet-candidates discovered through micro-lensing events; a short history of naming stars (posted on the Commission 5 Web site); and designations of Cosmic Microwave Background structures.

E-mail discussions included the changing of position-based names when new measurements or reprocessing indicate a different position. Strong support from WG members was to retain the policy of NOT changing the designation once one has been established. Another e-mail discussion concerned nomenclature for follow-up observations of an area of the sky by different telescopes.

Hélène Dickel and Ray Norris submitted several short newspaper articles for the IAU newspaper *Nuncio Sidereo III* during Prague General Assembly.

3.3. *WG FITS*

The Working Group *FITS* was substantially reorganized in 2003 with the appointment of a new chairman and vice-chairman, and the addition of 12 new members (for a total of 22 members). The major actions of the WG over the past 3 years include:

- Established written rules and procedures for conducting business;
- Appointed a 4th Regional FITS committee to provide advice to the WG, representing Australia and New Zealand. The other three regional committees represent North America, Europe, and Japan;
- Approved a proposal to create two new MIME-types for use when transmitting FITS files over the Internet;
- Officially approved two widely used conventions related to FITS binary tables that had previously only been described in an unofficial appendix;
- Approved the third paper on world coordinate systems, dealing with spectral coordinates;

• Approved the addition of 64-bit integers to the list of numerical data types that can be used in FITS files;

• Released a new version of the FITS Standard document that incorporates the recently approved changes;

• Established a new 'Registry of FITS Convention' for documenting the many existing FITS keyword conventions that have been developed by various projects around the world. It is hoped that this registry will become the central and authoritative place to document existing usage in FITS data files.

During the next triennium, planned activities include:

• Continue documenting existing FITS usage in the Registry of FITS Conventions;

• Convene a technical panel to update and clarify the FITS Standard document;

• Produce a new version of the FITS Users Guide to document the many changes since the last version was released in 1997;

• Promote the development of new FITS conventions as needed by the FITS community (e.g., for representing colour images, or for describing TIME coordinates in FITS data files).

3.4. *WG Libraries*

IAU Commission 5 WG on *Libraries* held a Business Meeting jointly with the Working Group on *Publishing* on 18 August.

Françoise Genova (CDS) described the 'Use and Validation of the IAU Astronomy Thesaurus in Ontologies', a collaboration of French IT laboratories in the frame of the *Massive Data in Astronomy* project. Uta Grothkopf (ESO) and Brenda Corbin (USNO) presented a summary of the LISA V (*Library and Information Services in Astronomy V*) conference, which had been held in Cambridge, MA, in June 2006. Brenda Corbin also gave an 'Overview of Observatory Archives' with a special focus on the setup and maintenance of the archives of Lowell Observatory, the National Radio Astronomy Observatory and Yerkes Observatory. Guenther Eichhorn (ADS) presented an update on the ADS Abstract Service and reported on the close 'Cooperation between the ADS and Libraries', both in the early days of the ADS as well as nowadays. 'The New Structure of the IAU Proceedings Series' was explained by Karel A. van der Hucht (IAU AGS); after the re-organization of IAU Symposia and Colloquia (with the IAU Colloquia having been merged into the IAU Symposium series), the WG *Publishing* is now pleased to report much shorter production times for proceedings volumes. Finally, Uta Grothkopf reported on the 'Use of Bibliometrics by Observatories', based on a questionnaire distributed among major observatories. PDF versions of all presentations can be found on the pages of the WG *Libraries* at <`http://www.eso.org/libraries/iau06/`>.

3.5. *WG Virtual Observatories, Data Centers and Networks*

The VO community has been very active during the last triennium, and is now at a stage where the objectives and organization of the WG Virtual Observatory have been fully clarified: its primary role is to provide an interface between *International Virtual Observatory Alliance* (IVOA) activities, in particular IVOA standards and recommendations, and other IAU standards, policies, and recommendations. It raises VO-related topics (e.g., symposia, GA sessions, ...) that should be handled by the IAU (Commission 5, Division XII and Executive level). It is responsible for approving the standards proposed by IVOA, after checking that there has been a process of consultancy according to the IVOA procedures, and that the proposed standards are consistent with other IAU approved standards (e.g., FITS, coordinate standards, etc.). The IAU WG VO brings to the attention of the IVOA Executive any topics it considers to be important for the IVO. It can be consulted by the IVOA Executive on any topic relevant to the international development of the VO.

The WG-VO Business Meeting was held on 24 August. It was attended by about 25 participants, some from VO projects, and others who were looking for information about the Astronomical Virtual Observatory. The meeting was organized as a walk through the web pages of IVOA and of several national VO projects and answers to the participant questions, with input from the project representatives and IVOA participants which were attending the meeting. A list of relevant Web links was sent to the participants after the meeting.

Special Session 3 organized during the General Assembly has been an important milestone for the international Virtual Observatory, with presentation and discussion of the present status of the VO in its different aspects - implementation by data centres, scientific usage by the community, technical challenges. The meeting proceedings have been be published in *Highlights of Astronomy, Volume 14* (ed. K.A. van der Hucht, 2007, Cambridge: CUP).

3.6. *Task Force on Preservation and Digitization of Photographic Plates*

The Task Force on Preservation and Digitization of Photographic Plates met on 17 August. In its 3-hour meeting the PDPP Task Force first received reports on (*a*) digitizing plates from the USNO plate archive of Natural Satellite observations; (*b*) calibration tests of the USNO StarScan instrument and commercial scanners; (*c*) digitizing the Bordeaux Carte du Ciel plates and generating astrometric catalogues; and (*d*) the new Harvard rapid digitizer, including a video of the instrument in operation. The meeting was informed about the latest version of the TF's Newsletter, SCAN-IT #4, now available at the PDPP website. It then accepted a proposed amalgamation with the 'Astrographic Catalogue & Carte du Ciel' WG of Commission 8, such that the PDPP absorbs new members who are not already members of both groups.

The rest of the meeting was given over to a proposal from the chair for a cooperative effort to digitize plates in Europe on a major scale. The TF has been encouraging efforts to digitize plates wherever feasible, and while there have been a few noble individual programmes, continuity of funding has remained a serious obstacle. The TF is, therefore, proposing a comprehensive collaboration involving as many European observatories as possible, offering storage and scanning facilities at a central site and seeking funding from the EU. Such a scheme was planned in 2000 in Brussels, but a prototype rapid scanner had first to be constructed to demonstrate proof-of-concept. The meeting agreed that the timing for a major collaboration is now appropriate, and encouraging signals for the proposal were indicated by potential participating observatories.

The PDPP Web site is available at `<http://www.lizardhollow.net/PDPP.htm>`.

4. Commission 5 charter

The charter of Commission 5 was discussed. Commission 5 main role is to host Working Groups essential to define, promote and apply an IAU policy in different domains pertaining to astronomical data and documentation: Astronomical Data (general policy issues); Designations; FITS, the international authority for the FITS data format; Libraries; Virtual Observatories. It represents these topics in IAU.

Commission 5 will continue to fulfill its duties during the next triennium. It is expected that with the evolution of the Virtual Observatory toward implementation the WG VO role will increase since it is responsible for approval of the Virtual Observatory standards.

One important role of Commission 5 is to foster communication channels toward the astronomical community, which includes proposing and organizing specific meetings at General Assemblies, such as SPS3 and SPS6 at this GA. But this is not enough since there is always the risk to 'preach only to the converted'. Commission 5 has thus to increase lobbying to encourage exposure of all aspects covered by Commission 5 in 'normal' scientific symposia.

Another remark on general policies is that Commission 5 is ready to shelter the Electronic Publishing WG, which is presently a WG of the IAU Executive, as already proposed in Sydney, if the IAU Executive wishes so. One can note, as explained above, that in Prague this WG has hold its session in common with WG Libraries.

Françoise Genova
president of the Commission

Transactions IAU, Volume XXVIB
Proc. IAU XXVI General Assembly, August 2006
Karel A. van der Hucht, ed.

doi:10.1017/S1743921308024174

DIVISION XII / COMMISSION 5 / WORKING GROUP DESIGNATIONS

CHAIR **Marion Schmitz**

MEMBERS **Heinz J. Andernach, Suzanne Borde, Kirk D. Borne, Anne P. Cowley, Helene R. Dickel, Pascal Dubois, John S. Gallagher, Françoise Genova, Paul W. Hodge, Richard W. Hunstead, Marie-Claire Lortet, Donald A. Lubowich, Oleg Yu. Malkov, Tetsuya Nagata, François Ochsenbein, Sean E. Urban, Ethan T. Vishniac, Wayne H. Warren, Norbert Zacharias**

PROCEEDINGS BUSINESS MEETING on 23 and 24 August 2006

Summary of three years activity

At the 2003 Sydney IAU meeting, Marion Schmitz (Caltech, USA) took over the chair of the Commission 5 Working Group Designations, succeeding Helene Dickel. The Working Group Designations of IAU Commission 5 clarifies existing astronomical nomenclature and helps astronomers avoid potential problems when designating their sources. The most important function of WG Designations during the period 2003-2005 was overseeing the IAU REGISTRY FOR ACRONYMS (for newly discovered astronomical sources of radiation: see the website `<http://cdsweb.u-strasbg.fr/cgi-bin/DicForm>`) which is sponsored by the WG and operated by the Centre de Données de Strasbourg (CDS). The Clearing House, a subgroup of the WG, screens the submissions for accuracy and conformity to the IAU Recommendations for Nomenclature (`<http://cdsweb.u-strasbg.fr/iau-spec.html>`). From its beginning in 1997 through August 2006, there have been 132 submissions and 111 acceptances. Attempts to register asterisms, common star names, and suspected variable stars were rejected. The past three years saw 61 acronyms submitted with 50 of them being accepted. (GIRL - yes; WOMEN - no).

Assistance was provided for inquiries about nomenclature for planet-candidates discovered through microlensing events; a short history of naming stars (which has been posted on the Commission 5 web site `<http://cdsweb.u-strasbg.fr/IAU/starnames.html>`; and designations of Cosmic Microwave Background structures.

E-mail discussions included the changing of position-based names when new measurements or reprocessing indicates a different position. Jay Gallagher and Ethan Visniac were added to the e-mail distribution list for such discussions. Strong support from WG members was to retain the policy of NOT changing a designation once one had been established.

E-mail discussions concerning follow-up observations of an area of sky by different telescopes. Suggestions from the WG were split between a format of: "Area of Sky" "Telescope" designation and "Telescope" "Area of Sky". Examples include: COSMOSVLA, HDFSST, GOODS-MUSIC, *vs.* CXOHDF, CXOM31.

Lanie Dickel and Ray Norris submitted several short newspaper articles for the IAU newspaper Nuncio Sidereo III. Thank you, both, for helping spread our important messages to the IAU community.

Marion Schmitz
chair of the Working Group

Transactions IAU, Volume XXVIB
Proc. IAU XXVI General Assembly, August 2006
Karel A. van der Hucht, ed.

doi:10.1017/S1743921308024186

DIVISION XII / COMMISSON 5 / WORKING GROUP FITS

CHAIR **William D. Pence**
VICE-CHAIR **François Ochsenbein**
PAST CHAIR **Donald C. Wells**
MEMBERS **Steven L. Allen, Mark R. Calabretta, Lucio Chiappetti, Daniel Durand, Thierry Forveille, Carlos Gabriel, Eric Greisen, Preben J. Grosbol, Robert J. Hanisch, Walter J. Jaffe, Osamu Kanamitsu, Oleg Yu. Malkov, Clive G. Page, Arnold H. Rots, Richard A. Shaw, Elizabeth Stobie, William T. Thompson, Douglas C. Tody, Andreas Wicenec**

PROCEEDINGS BUSINESS MEETING on 17 August 2006

The business meeting began with a brief review of the current rules and procedures of the WG, which are documented on the WG web page. Four regional FITS committees have been established by the WG, covering North American, Europe, Japan, and Australian/New Zealand, to provide advice to the WG on pending proposals. While it is recognized that this committee structure might need to be revised to provide representation to other regions, the current system is working well, and there were no motions to make any changes at this time.

The chairman then summarized the main accomplishments of the WG during the past three years:

- Re-organized the officers and membership of the WG.
- Established written rules and procedures.
- Approved two new MIME types for use with FITS files.
- Appointed a 4th Regional committee (Australia/New Zealand).
- Approved the variable-length array and TDIMn conventions for use with binary tables.
- Approved World Coordinate System Paper III (on spectral axes).
- Approved support for 64-bit integers in FITS files.
- Released a new version 2.1 of the FITS Standard reflecting the recent approved changes to FITS.
- Approved the HEALPix sky projection WCS convention.
- Created a new Registry of FITS conventions on the Web.

The meeting ended with a discussion of the activities that the WG should undertake during the next 3 years. High priority activities included:

- Add more conventions to the Registry.
- Convene a technical panel to make minor updates and clarifications to the FITS Standard document.
- Update (or retire) the 10 year old FITS User's Guide.
- Transfer material from the NRAO FITS web site to the FITS Support Office web site.
- Support the completion of WCS Paper IV (on distortions).
- Help draft a new WCS Paper V on time coordinates.

William D. Pence
chair of the Working Group

Transactions IAU, Volume XXVIB
Proc. IAU XXVI General Assembly, August 2006
Karel A. van der Hucht, ed.

doi:10.1017/S1743921308024198

DIVISION XII / COMMISSON 5 / TASK FORCE PRESERVATION AND DIGITIZATION OF PHOTOGRAPHIC PLATES

CHAIR	**R. Elizabeth Griffin**
MEMBERS	**88**

PROCEEDINGS BUSINESS MEETING on 17 August 2006

During the GA the Task Force for the Preservation and Digitization of Photographic Plates (PDPP) held an open two-session meeting, during which a formal merger between the WG-AC/CdC (Commission 8) and the TF-PDPP was confirmed. A dominant fraction of the members already belonged to both groups, and the nature of the merger is such that the activities of the WG-AC/CdC are now performed by the TF-PDPP.

It was also agreed to pursue actively a joint European project to digitize collections of direct plates. The project UDAPAC, formed in 2000 at the Royal Observatory Belgium by a consortium of astronomers from nine countries, would become the cradle for the new collaboration, and funding would be sought collectively from the European Union and/or other likely sources.

A short video of the new Harvard scanner, demonstrating the speed with which a new technology plate scanner could operate, convinced the community of the feasibility of digitizing large numbers of plates in a reasonable space of time.

R. Elizabeth Griffin
chair of the Task Force

Transactions IAU, Volume XXVIB
Proc. IAU XXVI General Assembly, August 2006
Karel A. van der Hucht, ed.

doi:10.1017/S1743921308024204

COMMISSION 6 — ASTRONOMICAL TELEGRAMS

TELEGRAMMES ASTRONOMIQUE

PRESIDENT	**Kaare Aksnes**
VICE-PRESIDENT	**Alan C. Gilmore**
ORGANIZING COMMITTEE	**Daniel W. E. Green, Brian G. Marsden Syuichi Nakano, Elizabeth Roemer, Nikolaj N. Samus, Jana Tichá**

COMMISSION 6 WORKING GROUP

Div. XII / Comm. 6 SERVICE	**Central Bureau for Astronomical Telegrams**

PROCEEDINGS BUSINESS MEETING on 17 August 2006

1. Members present

K. Aksnes, A. C. Gilmore, D. W. E. Green, B. G. Marsden, S. Nakano, N. Samus, J. Tichá, H. Yamaoka.

2. Deceased members

Janet Mattei, Director of the AAVSO, had passed away during the triennium. The meeting stood in silence in her memory.

3. Organising Committee 2006-2009

The president proposed the following for the coming triennium: Alan C. Gilmore president, and Nikolai N. Samus vice-president. Members: Brian G. Marsden, Daniel W. E. Green, Syuichi Nakano, Elizabeth Roemer, Jana Tichá, Hitoshi Yamaoka, Kaare Aksnes. Supernova group representative: Hitoshi Yamaoka was invited on to the OC to provide a link with supernova observers.

4. New Member

A new IAU member, S. K. Mattila (United Kingdom) had asked to join the Commission. This was approved by resolution.

5. CBAT Director's report

Green, Director of the Central Bureau for Astronomical Telegrams (CBAT), commented on the triennial report. The report had been circulated to Commission 6 members in October 2005 and will appear in *IAU Transactions XXVIA, Reports on Astronomy 2003-2006* (ed. O. Engvold, 2007, Cambridge: CUP).

Green noted that a large amount of time was spent on emails. With the Bureau's email address well-known, spam emails were an increasing problem. Answering news-media inquiries also took up much time. The Edgar Wilson Award would likely need a change of criteria soon as amateur astronomers cease to discover comets. Problem objects were some comets on the 'NEO Confirmation Page' (NEOCP). A proposed solution is to set up an 'unconfirmed objects' webpage like the Minor Planet Center's 'NEO Confirmation Page'.

Demand for the printed *IAUCs* remains steady. Subscriptions pay for printing and postage. There are many requests for items to be printed in the *Circulars* and not just to appear electronically. Several editorial problems faced by the CBAT Director: use of non-permanent URL addresses (www...) for references; spam email, referred to above; disagreements between authors and referees; weighing automatic publication against editorial involvement. The big issues facing CBAT were: how best to serve the astronomical community via the WWW; how to counter the perception that CBAT is just a supernova announcement service; should the *IAUCs* continue as in both card and electronic form; and the function of CBAT as an archive.

Funding of the bureau is being reviewed. The administration of the Center for Astrophysics has proposed eliminating subscription and line charges. Funding to support the Bureau is being sought from the U.S., Europe, and Japan. Due to subscription lengths, a two-year lead time is needed before phasing them out. Printed *IAUCs* would still need income to cover printing and postal costs. The current link of computer services between CBAT and the MPC is a further complication.

Green has applied to the U.S. National Science Foundation for five-year funding of CBAT, but notes that 80% of NSF applications are rejected. Referee's reviews of the proposal were very supportive. Negative comments centered on the need to provide such things as internet search facilities to the website, the ability of recipients to receive announcements only of the type of objects that they are interested in, and concerns in being able to handle large quantities of information that should result from the pending all-sky surveys (with noted mention of the developing 'Virtual Observatory' concept). However, more automation diminishes the vital requirement to vet and review material. Mike Rudenko proposed bringing XML into CBAT; others noted that, while XML is used in simple Windows web pages, not all browsers can read it. Green is dialoguing with IAU members worldwide to determine how best to improve the services of the CBAT for the astronomical community at large.

There had been suggestions from Division XII outgoing president, Virginia Trimble, that Commission 6 become a Working Group in light of its small membership. Marsden suggested that Trimble had confused the Commission 6 OC list with the Membership list. He also noted that this was the 16th business meeting of Commission 6 that he had attended, showing it was an active and continuing commission. Originally the Commission reported to the Executive Committee. Given that Division XII had largely ignored Commission 6 in the past triennium, perhaps the original arrangement should be reconsidered. Discussion was needed with the incoming Division XII president, Malcolm Smith, to resolve this matter.

Kaare Aksnes
president of the Commission

Transactions IAU, Volume XXVIB
Proc. IAU XXVI General Assembly, August 2006
Karel A. van der Hucht, ed.

doi:10.1017/S1743921308024216

COMMISSION 14 — ATOMIC AND MOLECULAR DATA

DONNÉES ATOMIQUES ET MOLÉCULAIRES

PRESIDENT	**Sveneric Johansson**
VICE-PRESIDENT	**Steven R. Federman**
SECRETARY	**Glenn M. Wahlgren**
ORGANIZING COMMITTEE	**Saul J. Adelman, Emile Biémont, James E. Lawler, Michael E. Mickelson, Donald C. Morton, Tanya A. Ryabchikova, Peter L. Smith, Chantal Stehle**

COMMISSION 14 WORKING GROUPS

Div. XII / Commission 14 WG	**Atomic Spectra and Wavelengths**
Div. XII / Commission 14 WG	**Atomic Transition Probabilities**
Div. XII / Commission 14 WG	**Molecular Structure**
Div. XII / Commission 14 WG	**Gas Phase Reactions**
Div. XII / Commission 14 WG	**Collision Processes**
Div. XII / Commission 14 WG	**Line Broadening**
Div. XII / Commission 14 WG	**Molecular Reactions on Solid Surfaces**
Div. XII / Commission 14 WG	**Optical Properties of Solids**

PROCEEDINGS BUSINESS MEETING on 17 August 2006

1. Attendance

S. R. Federman (co-chair), F. Kupka, G. Nave, G. Peach, R. C. Peterson, T. Ryabchikova, D. Smits, S. Tayal, G. M. Wahlgren (co-chair), G. Zhao.

2. Officers

In our commission the vice-president (VP) becomes the president, and a new VP is chosen from members of the Organizing Committee. The position of secretary was discontinued and its responsibilities incorporated into the VP position. The president announced that the new officers are Steven R. Federman (president) and Glenn M. Wahlgren (vice-president).

3. Organizing Committee

Our commission's usual practice is for a member to serve on the Organizing Committee (OC) for six years, with past presidents serving for three years past their term as president and other officers serving longer than six years if necessary to complete their service as officers. At this time the entire OC was replaced due to the completed service of past OC members S. J. Adelman, E. Biémont, J. E. Lawler, M. E. Mickelson, D. C. Morton, T. A. Ryabchikova, C. Stehle and the promotion of G. M. Wahlgren to VP. The constitution of the new OC is: Steven R. Federman (president), Glenn M. Wahlgren (vice-president); members: Milan Dimitrijevic, Alain Jorissen, Lyudmila I. Mashonkina, Farid Salama, Jonathan Tennyson, and Ewine F. van Dishoeck.

4. Working Groups

A new Working Group (WG) structure will be implemented, which combines the former eight WGs into four. The new WGs and their (parenthetic) former WG compositions will be
- WG *Atomic Data* (Atomic Spectra & Wavelengths; Atomic Transition Probabilities),
- WG *Molecular Data* (Molecular Structure; Gas Phase Reactions),
- WG *Collision Processes* (Collision Processes; Line Broadening),
- WG *Solids and Their Surfaces* (Molecular Reactions on Solid Surfaces; Optical Properties of Solids).

The new structure aims to reduce redundancies and streamline the structure of the tri-ennial report. The chairpersons for these WGs are still being finalized.

5. Membership

Wahlgren reported that the Commission membership has 199 members, including new members to be inducted during the current General Assembly. During the past three year period, 27 members were removed from the Commission roster, and a similar number added. The membership represents a total of 30 countries. Recent efforts at new member recruitment emphasized bringing young members into the Commission. The membership list was updated by the secretary and placed on the Commission's web site.

Communication between the OC and the membership is now fully conducted by electronic mail. A number of news-letter style updates were sent by the Secretary during the past three years. Members remarked that the email updates are appreciated and should continue. In addition, it was suggested that these updates be posted on the Commission's web site. A small fraction of the membership continually does not receive the emails due to various problems with mail servers and discontinued email addresses.

6. Meeting report and future meetings

Important aspects of the work of this Commission are to bring together providers and users of atomic and molecular data and to disseminate data. A number of meetings serve as forums for these discussions. Federman reported on the NASA-sponsored Workshop on Laboratory Astrophysics, held in Las Vegas, Nevada, USA during February 2006. His report highlighted the major points discussed, which culminated in the creation of a 'white-paper' on the status and needs of atomic, molecular, and solid state data in support of NASA missions. Such workshops take place every four years.

Other recent forums emphasizing data for astronomy included a special session of the American Astronomical Society, co-organized by S. Federman, and the triennial Atomic Spectroscopy and Oscillator Strength (ASOS) conference. The most recent ASOS meeting occurred in 2004 in Madison, Wisconsin, USA, and the next in the series is scheduled to take place in Lund, Sweden in 2007. Richard Monier announced a workshop on non-LTE codes and their use during summer 2007; the issue of atomic data requirements is a focus.

New ground- and spaced-based telescopes and instrumentation for infrared (IR) wavelengths will drive the need for new atomic and molecular data. The subject of data for spectrum analysis in the IR is expected to be included in upcoming meetings. In anticipation of this development, the Commission sponsored a science session with more than two dozen attendees following its business meeting. Three invited talks were presented by G.M. Wahlgren (*Atomic Data for IR Spectroscopy*), S. Federman (*Molecular Data for IR Spectroscopy*) and E. van Dishoek (*Solid State and the IR*). Brief contributions of relevance to the work of the Commission were presented by T. Gull (*Spectroscopy of Eta Carinae*), G. Nave (*NIST Data Bases*), G. Peach (*Brown Dwarf Spectral Line Shapes*), and R. Peterson (*Spectroscopy of Metal-Poor Stars*).

Steven R. Federman and Glenn M. Wahlgren
incoming president and vice-president of the Commission

Transactions IAU, Volume XXVIB
Proc. IAU XXVI General Assembly, August 2006
Karel A. van der Hucht, ed.

doi:10.1017/S1743921308024228

COMMISSION 41 — HISTORY OF ASTRONOMY

HISTOIRE DE L'ASTRONOMIE

PRESIDENT	**Alexander A. Gurshtein**
VICE-PRESIDENT	**Il-Seong Nha**
SECRETARY	**Clive L. N. Ruggles**
ORGANIZING COMMITTEE	**David H. DeVorkin, Wolfgang R. Dick, Rajesh Kochhar, Tsuko Nakamura, Luisa Pigatto, F. Richard Stephenson, Brian Warner**

COMMISSION 41 WORKING GROUPS

Div. XII / Commission 41 WG	**Archives**
Div. XII / Commission 41 WG	**Historical Instruments**
Div. XII / Commission 41 WG	**Transits of Venus**

PROCEEDINGS BUSINESS MEETING on 22 August 2006

1. Opening of meeting

On Tuesday 22 August 2006 approximately 40 people attended the Commission 41 *History of Astronomy* Business Meeting at the IAU XXVI General Assembly in Prague. Commission president Alex Gurshtein opened the meeting, welcoming the commission members and calling for a moment of silence for those members who passed away in the last triennium. David DeVorkin was appointed recording secretary for the meeting, with Steven Dick as the scruitineer of the ballot. A moment of silence was then observed in the memory of members departed over the last triennium, including: Jerzy Dobrzycki (Poland), Robert Duncan (Australia), Mohammad Edalati (Iran), Philip Morrison (USA), John Perdix (Australia), Neil Porter (Ireland), Gibson Reaves (USA), Brian Robinson (Australia), and Raymond E. White (USA).

2. Minutes of previous meeting

The minutes of the Sydney meeting were informally approved but not seconded. Clive Ruggles requested that section 4.1, ("4.1 New C41 Members") required alterations that are still in progress since many discrepancies among the various listings are still unreconciled. Clive is doing a heroic job readying this material for the next Secretary.

3. President's report

President Gurshtein noted the activities and accomplishments of the Commission during the previous triennium, including:
(*a*) A proposal to the IAU to declare the year 2009 as the *International Year of Astronomy* in honor of the 400th anniversary of the application of the telescope to astronomy by Galileo;
(*b*) A petition to the European Union to issue common postage stamps with astronomical themes in 2009;
(*c*) a response to the UNESCO-WHC initiative "Astronomy and World Heritage" to include astronomical sites on the "World Heritage List".

(*d*) participation or endorsement in the successful designation of the Struve Arc as a World Heritage Site.
(*e*) Promotion and support of recent publications, including: Proceedings IAU Colloquium No. 196 (*Transits of Venus*, 2005); and *Astronomical Instruments and Archives from the Asia-Pacific Region*, 2004. (*f*) Meetings under Commission 41 auspices: IAU Colloquium No. 196; *Sharing the Celestial Sphere*, Beijing 2005; SEAC06 European Conference on Archaeoastronomy (Greece).
(*g*) Sponsorship of four active Working Groups: Archives, Astronomical Chronology, Historical Instruments, and Transits of Venus.
(*h*) Maintenance of communications with other international bodies, including the IUHPS. President Gurshtein also noted that "Commission 41 is starting to prepare a broad interdisciplinary meeting that could be entitled the First World Congress on the History of Astronomy" in cooperation with the IUHPS. The recording secretary asked for clarification if this was an approved activity of Commission 41, and the matter was referred to "Other Business" (see below).

The President also discussed various difficulties Commission 41 has encountered in the last triennium:
1. The Relations of Commission 41 with the rest of the IAU "not as good as would be desired." The re-structuring of the IAU into Divisions weakened Commission 41's position seriously until Division XII was formed, the "Union wide" division.
2. Commission 41 has not been able to obtain official IAU recognition through officially sanctioned sessions for the history of astronomy. The Commission has not been successful gaining approvals for Joint Discussions, Special Sessions or Symposia under the sanction of the Union either for Sydney or for Prague. The president offered an explanation: that the methodology of astronomy and that practiced by its historians is not in "synchronicity."
3. The President remarked on the fate of the long-awaited "General History of Astronomy" (Michael Hoskin, General Editor) noting that only three volumes of the projected eight have appeared in the past three decades. He noted that if asked by EC members why not more of these volumes have appeared, he would be embarrassed to try and explain the delay. He then declared the series to be "dead" and suggested it be replaced by a "General Chronology of Astronomy."
4. The president suggested that there might be limits on the number of papers any one member of Commission 41 is allowed to author, or deliver, during a General Assembly.

Informal discussion followed the president's report. Various proposals were made and discussed, but none were actually proposed or seconded, or brought to a vote, and therefore do not bear mention in the minutes. The president's report was not moved for acceptance or seconded.
Rajesh Kochhar was invited to provide an informal report on the 2005 Beijing meeting, which he delivered extemporaneously and to the general satisfaction of the membership present.

4. Secretary's report

Clive Ruggles reported that Newsletters #6, #7, and #8 had been issued in April of each year. It had not been possible in this triennium to continue issuing paper copies to Commission members, because of the lack of any funds to cover the cost of this. He noted with appreciation that the distinctive blue program booklet "History of Astronomy: Commission 41 Activities at Prague" was underwritten by the National Observatory of Japan upon Mitsuru Soma's initiative.

The production of definitive membership lists continues to be a serious problem because, as an Inter-Union Commission, Commission 41 membership is managed by bodies that are not in full communication with one another and work under different criteria. This makes it especially hard for the secretary to determine consistently which names are recognized by each of the bodies involved (IAU and the IUHPS/DHST, which does not maintain individuals as members). The suggested solution is to manage the membership list internally in Commission 41, but this does not solve acknowledgment by the parent bodies. Clive Ruggles will reconcile present lists, based upon the names presented in the Sydney "Minutes of the General Business Meeting held on 23 July 2003" and communicate them in due course to the OC.

The secretary then identified 154 names as members of the Commission at the 2003 General Assembly, and 13 elected in 2003 and appointed to Commission 41. A call was made for the acceptance of the secretary's report. It was moved and seconded.

5. Admission of new Commission 41 members

The secretary displayed a set of lists of names of people wishing to be admitted. Commission members approved and seconded that Commission 41 appoint the entire list, in all categories. The following people were approved as members of the Commission: Robert BREINHORST (Germany), Yvan Dutil (Canada), Michael Dworetsky (USA), Anthony Fairall (South Africa), Edward H. Geyer (Germany), Erik Høg (Denmark), Alexander Kosovichev (USA), Ioannis Liritzis (Greece), J. McKim Malville (USA), Gennady Pinigin (Ukraine), Tom Ray (Ireland), Gerhard Ruben* (Germany), Steven Shore (Italy), Michelle Storey (Australia), Peter Usher (USA), Iryna Vavilova (Ukraine), Patrick Woudt (South Africa), and Hitoshi Yamaoka (Japan).

Discussion: a person asked if the IAU membership lists are public knowledge. The secretary answered that the IAU does not allow distribution of name lists. [Secretary's addendum: I've since checked various Commission web pages and many do include name and country lists: it is further details that are prohibited. This does need clarifying with the IAU, since if we are allowed to publish an up-to-date name/country list, we should do so on the website.]

The ICHA admission procedures: "Sections 2 and 3 were not approved or fully discussed. Discussion centered on relative rights and authority of associate and full members.

6. Formal approval of the procedures for admitting non-IAU members to the ICHA

The secretary led the discussion of the new procedures for admitting new non-IAU members to Commission 41 according to guidelines specified in the blue proceedings brochure, pp. 28-29. Approval was sought only for items enumerated in Section 1: "Purpose and status of these procedures." 1.1 was approved; 1.2 was ratified; and 1.3 was approved. Sections 2 and 3 were not approved or fully discussed. Discussion centered on relative rights and authority of associate and full members. No resolutions were framed, none seconded.

7. Election of Commission 41/ICHA Officers for 2003–2006

Elections were duly held, taking into account absentee ballots, resulted in the following Organizing Committee for the 2006-2009 triennium: Nha Il-Seong (Korea, president), Clive Ruggles (UK, vice-president) Alex Gurshtein (USA, past president), Teije de Jong (Netherlands), David DeVorkin (USA), Rajesh Kochhar (India), Tsuko Nakamura (Japan), Wayne Orchiston (Australia), Antonio A. P. Videira (Brazil), and Brian Warner (South Africa). Steve Dick kindly acted as scrutineer of the ballots, ably assisted by John Steele.

8. Working Groups

Each Working group met subsequent to the Commission 41 business meeting, decided on leadership and membership, and if they needed to continue in the next triennium.

Archives: Ileana Chinnici (Italy, Chair), Brenda Corbin (USA), Suzanne Dbarbat (France), Daniel Green (USA), Irakli Simonia (Georgia), Wayne Orchiston (Australia), and Adam Perkins (UK). Contact person: Ileana Chinnici, INAF-Osservatorio Astronomico di Palermo Giuseppe S. Vaiana, Piazza del Parlamento 1, 90134 Palermo, Italy <chinnici@astropa.unipa.it>.

Historical Instruments: Luisa Pigatto (Italy, Chair), Juergen Hamel (Germany), Kevin Johnson (UK), Rajesh Kochhar (India), Tsuko Nakamura (Japan), Nha Il-Seong (Korea), Wayne Orchiston (Australia), Bjorn Pettersen (Norway), Sara Schechner (USA) and Yunli Shi (China). Contact person: Luisa Pigatto, INAF-Osservatorio Astronomico di Padova, Vicolo dell' Osservatorio, 5, I-35122 Padova, Italy <luisa.pigatto@oapd.inaf.it>.

Transits of Venus: Steven Dick (USA, Chair), Hilmar Duerbeck (Germany), Robert van Gent (Netherlands), David Hughes (UK), Willie Koorts (South Africa), Wayne Orchiston (Australia), and Luisa Pigatto (Italy). Contact person: Steven J. Dick, NASA Chief Historian, National Aeronautics and Space Administration, Headquarters, Code IQ, Room CO72, 300 E Street SW, Washington DC 20546-0001, USA <steven.j.dick@nasa.gov>.

Historical Radio Astronomy (joint with Division X): Wayne Orchiston (Australia, Chair), Rod Davies (UK), Suzanne Dbarbat (France), Ken Kellermann (USA), Masaki Morimoto (Japan), Slava Slysh (Russia), Govind Swarup (India), Jasper Wall (Canada), Richard Wielebinski (Germany), and Hugo van Woerden (Netherlands). Contact person: Wayne Orchiston, School of Mathematical and Physical Sciences, James Cook University, Townsville, Queensland 4811, Australia <wayne.orchiston@jcu.edu.au>.

9. Resolutions

No formal resolutions were proposed or discussed. Informally, there were two suggestions: Brenda Corbin proposed that the OC find someone to take over the duties of the web site, since Wolfgang Dick will not be continuing in this valuable function. She offered thanks to W. Dick, and the Commission heartily agreed.

Wayne Orchiston formally proposed that "we congratulate Ileana Chinnici and Clive Ruggles on the excellent job they did producing the C41/ICHA Newsletter during the last triennium". This was loudly seconded and heartily approved.

At a meeting of the new officers and Executive Committee that evening we assigned duties to members for the 2009 OC in Brazil: David DeVorkin - Secretary; Wayne Orchiston - Newsletter Subcommittee - Editor/print Newsletter; Clive Ruggles - Newsletter Subcommittee -Webmaster/website. Since that meeting, once duties were clarified, it was decided that Ileana Chinnici would be the Editor of the print version of the Newsletter. The secretary will act as coordinator and catalyst for the management of the Newsletter and website, which will be the primary responsibilities of the above named. Materials for submission to either form of communication will be the common responsibility of the OC with the approval of the vice-president and/or president.

The balance of the meeting dealt with defining plans for the 2009 General Assembly, which will have as its theme the "International Year of Astronomy 2009" (IYA2009). It is the responsibility of the OC to create a plan that will be acceptable to members of Commission 41 and appealing to the new "All-Union" Division XII. The OC decided that it would aim high, to a full symposium or special session, and given the focus of the IYA on education, we will also attempt to attract joint support by Commission 46. Suggestions for themes centered around the idea of "Instrumentation for Astronomy Through the Ages" which would carry a strong telescope theme but not be limited to optical devices only. We might suggest an invited discourse to the General Assembly, possibly Albert van Helden on Galileo. We strongly agreed to create and appoint a Scientific Organizing Committee, chaired possibly by Brian Warner (who was not at the meeting). We will also need a "Local Organizing Committee" and hope that Antonio A. P. Videira of Brazil will be willing to organize that critical group.

10. Other business

The recording secretary recalled the question of planning for a "First World Congress on the History of Astronomy" reported by the president, asking for someone to read into the record a formal proposal for the Commission's consideration and action. Several voices spoke against the plan, concerned that Commission 41 efforts and resources be wholly engaged for the next General Assembly in 2009. No one spoke for the plan. Hence no resolution was offered and the idea was dropped.

President Gurshtein closed the meeting at about 3:30 pm.

Alexander A. Gurshtein
president of the Commission

Transactions IAU, Volume XXVIB
Proc. IAU XXVI General Assembly, August 2006
Karel A. van der Hucht, ed.

doi:10.1017/S174392130802423X

DIVISION XII / COMMISSION 41 / WORKING GROUP HISTORICAL INSTRUMENTS

CHAIR **Luisa Pigatto**
PAST CHAIR **Nha Il-Seong**
MEMBERS **Jürgen Hamel, Kevin Johnson, Rajesh Kochhar, Tsuko Nakamura, Nha Il-Seong, Wayne Orchiston, Bjrn R. Pettersen, Sara J. Schechner, Shi Yunli**

PROCEEDINGS BUSINESS MEETING on 23 August 2006

1. Introduction

The Working Group Historical Instruments (WG-HI) was founded by the members of Commission 41 at the 2000 Manchester IAU XXIV General Assembly, with the main objectives to assemble a bibliography of existing publications relating to such instruments, and to encourage colleagues to carry out research and publish their results. Membership of the WG-HI has increased from three to nine people since its foundation. This clearly demonstrates the IAU members increasing interest in safeguarding old astronomical instruments and buildings as witness to their own country's cultural heritage and scientific progress.

2. The WG's scientific sessions at Prague IAU GA

The WG-HI held three sessions in Prague, on Wednesday, August 23: the first was devoted to WG business and the last two, to oral and poster papers presentation. The scientific topic focused on the theme The world-wide search for historically significant astronomical instruments. The agenda and presentations for these latter two sessions are available on the WG-HI website.

3. The thesaurus of historical instruments

An important task of the WG-HI is to prepare a Thesaurus of historical instruments used in astronomy and related disciplines such as geography, geodesy, navigation, meteorology and chronology. The Thesaurus is a controlled and structured list of terms defining instruments for observations and measurements in a very precise way. Historical instruments will be defined on the basis of the first printed sources describing the instruments themselves. Identifying correctly old instruments gives an additional useful tool to study and preserve the heritage testifying the progress of astronomy during the centuries.

4. WG Historical Instruments website

From 2006 Prague GA, Luisa Pigatto, has taken the task of making, developing and maintaining the WG-HI website. Presentations and Abstracts at the WGs scientific sessions in Prague as well as members reports, are there available in pdf format. Links to Science Museums and historical instruments collections websites are also given. A list of printed books related to astronomy, geography, geodesy, navigation, meteorology and chronology in which instruments and their use are described, is in progress.

5. Closing remarks

The *International Year of Astronomy 2009* (IYA2009) commemorates the 400th anniversary of Galileos telescope, the main instrument in the history of astronomy. The WG-HI is particularly involved in planning events related to historical instruments, their use and improvement on which the progress in astronomy is based.

Luisa Pigatto
chair of the Working Group

Transactions IAU, Volume XXVIB
Proc. IAU XXVI General Assembly, August 2006
Karel A. van der Hucht, ed.

doi:10.1017/S1743921308024241

COMMISSION 46 — ASTRONOMY EDUCATION AND DEVELOPMENT

ENSEIGNEMENT DE L'ASTRONOMIE

PRESIDENT	**Jay M. Pasachoff**
VICE-PRESIDENT	**Barrie W. Jones**
ORGANIZING COMMITTEE	**John B. Hearnshaw, Michèle Gerbaldi, Lars Lindberg Christensen, Charles R. Tolbert, John R. Percy**

COMMISSION 46 PROGRAM GROUPS

Div. XII / Commission 46 PG	**World-Wide Development of Astronomy**
Div. XII / Commission 46 PG	**Teaching Astronomy Development**
Div. XII / Commission 46 PG	**International Schools for Young Astronomers**
Div. XII / Commission 46 PG	**Exchange of Astronomers**
Div. XII / Commission 46 PG	**National Liaisons on Astronomy Education**
Div. XII / Commission 46 PG	**Collaborative Programs**
Div. XII / Commission 46 PG	**Commission Newsletter**
Div. XII / Commission 46 PG	**Public Information at the Times of Solar Eclipses**
Div. XII / Commission 46 PG	**Exchange of Books, Journals, Materials**

TRIENNIAL REPORT 2003–2006

1. Introduction

The International Astronomical Union (IAU) was founded in 1922 to "promote and safeguard astronomy ... and to develop it through international co-operation". The IAU is funded through its National Members. Almost all of the funds supplied from the dues are used for the development of astronomy.

One of the 40 IAU "Commissions" is Commission 46, formerly called *The Teaching of Astronomy* and more recently, at the 2000 General Assembly, renamed *Astronomy Education and Development*. It is the only Commission that deals exclusively with astronomy education; a previous Commission 38 (*Exchange of Astronomers*), which allocated travel grants to astronomers who need them, and a Working Group on the *Worldwide Development of Astronomy*, have been absorbed by Commission 46.

The Commission's mandate is "to further the development and improvement of astronomy education at all levels, throughout the world".

In general, the Commission works with other scientific and educational organizations to promote astronomy education and development; through the National Liaisons to the Commission, it promotes astronomy education in the countries that adhere to the IAU; and it encourages all programs and projects that can help to fulfill its mandate.

2. Activities

The Commission holds business sessions at each IAU General Assembly. Within the format of the IAU General Assemblies, the Commission organizes or co-sponsors major sessions on education-related topics, such as a Special Session held at the 2003 General Assembly in Sydney, Australia, on which a book was published: *Teaching and Learning Astronomy: Effective Strategies for Educators Worldwide*, edited by John R. Percy and Jay M. Pasachoff (Cambridge: CUP, 2005).

The Commission has also organized two major conferences on astronomy education - in the USA in 1988, and in the UK in 1996:

• Jay M. Pasachoff & John R. Percy, eds., 1990, *The Teaching of Astronomy*, Proc. IAU Colloquium No. 105 (Cambridge: CUP).

• L. Gouguenheim, D. McNally, & J. R. Percy, eds., 1998, *New Trends in Astronomy Teaching*, Proc. IAU Colloquium No. 162 (Cambridge: CUP).

• John Percy, ed., 1996, *Astronomy Education: Current Developments, Future Coordination*, ASP-CS 89 (San Francisco: ASP).

For three decades, the Commission has sponsored one-day workshops for local school teachers, as part of every IAU General Assembly, and as part of several IAU regional meetings. Immediately after the conference that is described in the forthcoming book, a very successful teachers' workshop was held in Sydney, organized by Nicholas Lomb, Sydney Observatory. Because of language difficulties, only local astronomers organized a teachers' conference after the Prague General Assembly.

Until recently, Commission 46 was concerned primarily with tertiary (university-level) education and beyond, but several of its activities have an impact on school-level and public education.

Commission 46 was involved in a series of sessions and a Special Session for the 14–25 August 2006 IAU General Assembly. Special Session 2 on *Innovation in Teaching/Learning Astronomy* was organized by Rosa M. Ros and Jay M. Pasachoff. Special Session 5 on *Astronomy Education in the Developing World* was organized by John Hearnshaw. Both have been published in the *Highlights of Astronomy, Volume 14* (Ed. Karel A. van der Hucht, 2007, CUP).

3. Program Groups of Commission 46

3.1. *PG for the Worldwide Development of Astronomy*

The role of this PG is to visit countries with some astronomical expertise at tertiary (i.e., post high school) level, which are probably not IAU member states, but which would welcome some development of their capabilities in teaching and/or research in astronomy. For example, as a result of a visit last year, Mongolia joined the IAU at the IAU XXVI GA in Prague, and has received advice on broadening their astronomy programs.

OC: John B. Hearnshaw (chair), Athem Alsabti, Alan Batten, Julietta Fierro, Richard Gray, Mary Kay Hemenway, Yoshihide Kozai, Hugo Levato, Peter Martinez, Don Wentzel, Jay White, and Jayant Narliker.

3.2. *PG for Teaching for Astronomy Development*

TAD is intended to assist a country with currently little astronomy which wants to enhance its astronomy education significantly. TAD operates on the basis of a proposal from a professional astronomy organization or on the basis of a contract between the IAU and an academic institution, usually a university.

The capabilities of the TAD program are limited to assistance with university-level activities, such as

(*a*) the creation of university-level astronomy/astrophysics courses and the faculty training and equipment associated with the development and first offering of such courses

(*b*) a basic, largely educationally oriented research capability for faculty and students

(*c*) travel (i.e. transportation) costs of foreign visiting lecturers and of students invited for study at foreign universities, and

(*d*) professional preparations needed as a prerequisite for plans to offer astronomy in schools and for the public. TAD can provide advice about education of school teachers, but not financial support. The training of school teachers and the actual performance of school teaching and public outreach is considered to be part of the national resources.

OC: Jay White, Don Wentzel, Armando Arellano Ferro, Khalil Chamcham, David Clarke, Nguyen Dinh Huan, Nidia Morrell, Derek McNally, John Percy, Maria Cristina Pineda de Carias, and Nguyen Quang Rieu.

3.3. *PG for International Schools for Young Astronomers*

ISYA seeks the participation of young astronomers mainly, but not exclusively, from astronomically developing countries. Participants should generally have finished first-degree studies. ISYA seeks to broaden the participants perspective on astronomy by lectures from an international faculty on selected topics of astronomy, seminars, practical exercises and observations, and exchange of experiences. The most recent ISYAs were in Morocco in 2004 and in Puebla, Mexico in 2005. The next ISYA will be in Malaysia in 2007 (March-April).

OC: Michèle Gerbaldi, Ed Guinan.

3.4. *PG for Exchange of Astronomers*

The PG makes travel grants to qualified individuals in order to enable them to visit institutions abroad where they may interact with the intellectual life and participate in the research of the host institution. It is the objective of the program that astronomy in the home country be enriched after the applicant returns. The PG publishes, both on the IAU web site and in IAU Information Bulletins, all the information needed to apply for a grant under the IAU Exchange of Astronomers program.

Membership 2003-2006: Charles Tolbert (chair), John Percy (vice-chair). The group administers the IAU's program of grants for exchange of astronomers, which enables qualified applicants (usually young astronomers from astronomically-developing countries) to visit institutions abroad for periods of three months or greater.

During the period 2003-2006, 15 grants were awarded, for a total of $US 20,975. Grantees came from 10 different countries, and visited institutions in 9 different countries.

OC: Charles Tolbert, and John Percy.

3.5. *PG for National Liaisons on Astronomy Education*

The main duty of the National Liaison on Astronomy Education is (a) to write the triennial national report, to make it a valuable resource for countries wishing to enhance their astronomy education, and (b) to transmit to the educators of his/her own country the insights that they might glean from the reports and conferences.

OC: Barrie W. Jones.

3.6. *PG for Collaborative Programs*

This Program Group works on activities co-sponsored by UNESCO, COSPAR, UN, ICSU, etc., and carries out interactions with other international organizations. It will be our main link with the 2009 International Year of Astronomy.

OC: Syuzo Isobe, Jonannes Anderson, Christopher Corbally, David Crawford, Julieta Fierro, Hans Haubold, Seigbert Raither, Dale Smith, James C White II, P Willekens, and Peter Willmore.

3.7. *PG for Commission Newsletter*

The Newsletter is published twice a year, and is available (including back issues) on the IAU web site.

OC: Barrie W. Jones, and Tracey Moore.

3.8. *PG for Public Information at the Times of Solar Eclipses*

Timely advice for countries that will experience a solar eclipse, see `<http://www.eclipses.info>` in order to consult with local astronomers and with newspapers.

For the latest transit of Venus, see `<http://www. transitofvenus.info>`.

OC: Jay Pasachoff, Ralph Chou, and Julieta Fierro.

3.9. *PG for the Interchange of Books, Journals, and Materials*

We are re-studying the role of this program group in the context of new electronic document possibilities, but we can still link people needing written material with those for whom the material is surplus.

OC: John Percy

Jay M. Pasachoff
president of the Commission

Transactions IAU, Volume XXVIB
Proc. IAU XXVI General Assembly, August 2006
Karel A. van der Hucht, ed.

doi:10.1017/S1743921308024253

DIVISION XII / COMMISSION 46 / PROGRAM GROUP WORLD-WIDE DEVELOPMENT OF ASTRONOMY

CHAIR **John B. Hearnshaw**
PAST CHAIR **Alan H. Batten**
MEMBERS **A. Athem Alsabti, Alan H. Batten, Julieta Fierro, Richard O. Gray, Mary Kay M. Hemenway, Yoshihide Kozai, Hugo Levato, Hakim L. Malasan, Peter Martinez, Jayant V. Narlikar, Donat G. Wentzel, James C. White**

PROCEEDINGS BUSINESS MEETINGS on 16 August and 24 August 2006. Triennial Report 2003-2006.

1. Introduction

The Program Group for the World-wide Development of Astronomy (PG-WWDA) is one of nine Commission 46 program groups engaged with various aspects of astronomical education or development of astronomy education and research in the developing world. In the case of PG-WWDA, its goals are to promote astronomy education and research in the developing world through a variety of activities, including visiting astronomers in developing countries and interacting with them by way of giving encouragement and support.

2. Aims and objectives of PG-WWDA

The principal aims and objectives of PG-WWDA are: (*a*) To visit developing countries (often IAU non-member states) with some limited astronomical expertise, and which would welcome some development of their capabilities in astronomy; (*b*) To give encouragement, and to explore the possible assistance of the IAU in developing astronomy in these countries; (*c*) To discuss with astronomers in developing countries the available resources for astronomical teaching or research, and to promote international contacts and exchanges between astronomers in these countries and those elsewhere; (*d*) To write reports on the state of astronomy in developing countries for the Commission 46 president and to send these reports to the IAU Executive Committee; and (*e*) If the conditions were deemed favourable, then to follow-up any report with involvement by TAD or other program groups of Commission 46, as may be appropriate.

3. Visit to Mongolia

John Hearnshaw spent a week in Mongolia from 11-18 March 2004 on behalf of PG-WWDA. His visit was hosted by Prof G. Batsukh in the Geophysics Department of the National University of Mongolia (NUM) in Ulaanbaatar. Four academics in this department were teaching undergraduate astronomy. Further astronomers were employed by the Mongolian Technical University as well as by the Mongolian Academy of Sciences at the Research Center of Astronomy and Geophysics (RCAG) and the associated Khurel Togoot Observatory, both being part of the Academy. Visits were made to the observatory and to the Academy, where a meeting with Dr T. Galbaatar, the Acting President of the Mongolian Academy of Sciences, took place on 17 March. About a dozen astronomers are employed in Mongolian universities or at RCAG, and they would benefit greatly if Mongolia were to adhere to the IAU. A subsequent application for membership resulted from these discussions. A series of four lectures was presented at NUM by Hearnshaw during this visit.

4. Visit to Kenya

Peter Martinez (South African Astronomical Observatory) made a visit to Kenya 15-17 June 2004. His visit was hosted by the Physics Department of the University of Nairobi, where Dr Paul Baki is an active astronomer teaching in the department. Four other academics in the department have some interests in astronomy. Work and advice on an undergraduate astronomy syllabus was undertaken during this visit and Dr Martinez gave a series of lectures. Plans for Kenya to acquire a small telescope (about aperture 0.5 m) were discussed and it was proposed that Dr Baki be nominated for individual membership of the IAU. Dr Baki has been invited to make a presentation at the Special Session No. 5 on *Astronomy for the Developing World*) at the IAU XXVI General Assembly in Prague.

5. Contact with Thai astronomers

John Hearnshaw has maintained contact with Thai astronomers, notably at Chiang Mai University (CMU) in northern Thailand. This has partly been through a Thai astronomy PhD student he is supervising in New Zealand, but also through Boonrucksar Soonthornthum, a Thai astronomer who was formerly an MSc student in New Zealand. Boonrucksar was until 2005 Dean of Science at CMU. He is now Director of the new Thai National Astronomical Research Institute (NARIT). He visited New Zealand to meet with this writer in September 2005.

There is substantial astronomical activity in Thailand, both at CMU and at least five other Thai universities. At NARIT a 2.5-m telescope should be installed on Doi Inthanon, Thailand's highest mountain near Chiang Mai, by late 2007 or 2008. Thailand had considered joining the IAU a few years ago, but without a successful outcome. The decision to found and equip NARIT with substantial capital expenditure has now made a re-consideration of this proposal very favourable, and a new application by Thailand to join the IAU in 2006 has been made. This is a logical decision, as of all the IAU non-member countries in the world at the present time (2006), Thailand probably has more professional astronomical activity than any other.

Hearnshaw made plans to visit Chiang Mai University and NARIT in early 2007.

6. Visit to Iraq

Athem Alsabti, a member of PG-WWDA, visited Iraq in April 2004 to investigate the state of astronomy in that country and to explore the possibility of rehabilitating the Mt Korek Observatory in northern Iraq, which had been damaged in the Iran-Iraq war in 1989, before becoming operational. A 3.5-m Zeiss telescope had been installed there, as well as a 1.25-m Ritchey-Chrétien reflector, and a 30-m radio telescope for millimetre wavelengths.

His visit took him to Baghdad University, the Ministry of Higher Education in Baghdad, the Iraqi National Academy of Science, Salahaddin University in Erbil in Kurdish Iraq and then to Mt Korek. He met with vice-chancellors of most Iraqi universities while in Baghdad.

Dr Alsabti reported after this visit that: "There is a very strong support from the Kurdish scientists and authorities (*a*) to rebuild the Observatory, and (*b*) to start astronomical studies in the Kurdish region based at Erbil".

7. Visit to Cuba

John Hearnshaw and Julieta Fierro visited Cuba for a week in January 2005 on behalf of PG-WWDA. The visit was hosted by the Instituto de Geofísica y Astronomía (IGA) in Havana. IGA is a part of the Ministry of Science, Technology and the Environment (CITMA). During their visit they had discussions with Dr Lourdes Palacio Suárez and Prof. Jorge Pérez Doval at IGA. They are respectively director of IGA and head of the astronomy section of IGA. They also had meetings with Dr Lilliam Álvarez Díaz, Director of Science at CITMA and with Dr Oscar Álvarez (CITMA), astronomer attached as specialist in science in that ministry. Between them Hearnshaw and Fierro presented seven talks, seminars or public lectures, all in different venues, and Fierro gave a television interview. They visited all the astronomical facilities of IGA, including Arroyo Naranjo Observatory and 60-cm Cassegrain telescope (in the outskirts of Havana) and the Cacahual solar observatory with its solar telescope and spectrograph.

Cuba used to be an interim member of the IAU, but since the break-up of the Soviet Union its astronomers have been very isolated by political events and it was clear that astronomy is not a high priority for the present Cuban government. Nevertheless, there are a few contacts between Cuban astronomers and those in developed countries. One Cuban astronomer is doing a PhD in Spain in observational astronomy and theoretical cosmologists (led by Dr Rolando Cárdenas Ortiz) from the University Central de Las Villas, Santa Clara, in central Cuba have contacts with those in the U.K. Ways of improving international contacts between astronomers in Cuba and those in the international community were discussed.

8. Visit to Trinidad and Tobago

John Hearnshaw visited the St Augustine campus of the University of the West Indies (UWI) in Trinidad and Tobago for a week in December 2005. His host was Dr Shirin Haque, an astronomer who is the acting head of the Physics Department of UWI. Although she is the only professional astronomer in Trinidad, she heads a small group of active students in the department, and she has established an organization called CARINA (the Caribbean Institute of Astronomy) whose aim is to promote the development of astronomy in the Caribbean region.

During the week in Trinidad, Hearnshaw gave three lectures or seminars, met with astronomy graduate students, visited the National Science Centre, gave radio and television interviews, visited the Trinidad and Tobago Astronomical Society and its observatory, and visited a private observatory on the island of Tobago to which UWI astronomers have regular access. As a result of this visit, Dr Haque, who was trained in astronomy at the University of Virginia in the US, has been nominated for individual membership of the IAU. Trinidad and Tobago does not adhere to the IAU. Such a step could come at a later date, if astronomical activity continues to grow at UWI, as is hoped.

9. Closing remarks

PG-WWDA has had a successful three years since we last met in person at the Sydney IAU XXV General Assembly. We have made contacts with astronomers in a number of developing countries, and as a result of these contacts follow-up work is in progress in many of these places.

We note that of these non-adhering countries, the People's Democratic Republic of Korea has 20 IAU individual members who are virtually isolated politically and hence also scientifically from the rest of the world. In addition there are seven IAU members in Kazakhstan. Neither of these countries adheres to the Union, and both must be future places where PG-WWDA might fruitfully explore contacts.

The highlights for the last three years must be the success in bringing both Mongolia and Thailand to the point of making applications to join the IAU as adhering countries. Significant pockets of astronomers were operating in isolation in both these places.

In the future, apart from North Korea and Kazakhstan mentioned above, we see Colombia (which is a non-member country) as another important place to visit. We are also looking at Jordan, Uzbekistan, Mauritius, Laos, and perhaps several other countries in Latin America, Africa and south-east Asia. We feel it is productive to concentrate on helping countries where a few astronomers are already active and need to make contacts with the international community to grow further.

Often, as in Iraq, the current political situation prevents any major continuing efforts on the ground by PG-WWDA, and the same is true in North Korea. However we note that 19 non-adhering countries have one or more IAU members with presumably few international contacts, a further approximately 20 developing countries do adhere to the union, but the astronomers resident in them still have limited access to international science, and some 115 countries essentially have no professional astronomical activity at all. In these circumstances it is clear that PG-WWDA has plenty of work to do to help astronomers, no matter where they live, to participate in the global international scientific community.

10. New membership of PG-WWDA from August 2006

At the business meeting of Commission 46 on 24 August 2006, the membership of PG-WWDA for the coming triennium 2006-2009 was selected as follows: John Hearnshaw (New Zealand, chair), Athem Alsabti (Iraq/UK), Alan Batten (Canada), Julieta Fierro (Mexico), Ed Guinan (USA, ex officio as co-chair TAD), Yoshihide Kozai (Japan), Hugo Levato (Argentina), Hakim

Malasan (Indonesia), Peter Martinez (South Africa), Larry Marschall (USA, ex officio as co-chair TAD), Jayant Narlikar (India), Pereira Osorio (Portugal), Jay Pasachoff (USA), Karla Perkins (USA), and Jin Zhu (China).

John B. Hearnshaw
chair of the Program Group

Transactions IAU, Volume XXVIB
Proc. IAU XXVI General Assembly, August 2006
Karel A. van der Hucht, ed.

doi:10.1017/S1743921308024265

DIVISION XII / COMMISSION 46 / PROGRAM GROUP INTERNATIONAL SCHOOLS FOR YOUNG ASTRONOMERS

CHAIR **Michèle Gerbaldi**
VICE-CHAIR **Edward F. Guinan**

REPORT ON THE INTERNATIONAL SCHOOLS FOR YOUNG ASTRONOMERS PROGRAM 1967–2006

1. Introduction

The programme *International Schools for Young Astronomers,* hereafter named ISYA, is organized by IAU Commission 46 on *Astronomy Education & Development* since 1967. We present here a brief history of the development of this programme since its creation till 2006. Much more data can be found in the *Transactions of the IAU*, either Vol. A or B, as well as in two papers published in proceedings of meetings; the references of these publications are given at the end of this report.

2. Creation of Commission 46 on The Teaching of Astronomy and of the International Schools for Young Astronomers

In 1964, at the the IAU XII General Assembly in Hamburg, Germany, a Special Meeting was organized on *The Teaching of Astronomy* (*Transactions of the IAU*, vol. XIIB, p. 629, 1964). From this meeting (*Transactions of the IAU*, vol. XIIB, p. 648, 1964) two recommendations were proposed: the organization of an IAU Commission on the *Teaching of Astronomy* and the creation of an *International School for Young Astronomers* in order to re-enforce the "... international cooperation in the domain of astronomy teaching, including the training of the astronomers ...".

The aim of such a School is to give to the young astronomers an intense training in astronomy and astrophysics during three months, rather similar to the one that could be given in an university, and then they would spend one year in an institution to have both a more practical training and a more specialized theoretical training (*Transactions of the IAU*, vol. XIIIA, p. XCV, 1967).

3. From the 1st ISYA in 1967 till the 18th in 1990

The first Summer School for Young Astronomers was organized at Manchester University (UK) during 6.5 weeks. With financial support of UNESCO, the IAU and the host country, four International Schools for Young Astronomers took place consecutively: Manchester (UK) in 1967, already mentioned, then Arcetri (Italy) in 1968, Hyderabad (India) in 1969, and Córdoba (Argentina) in 1970.

Starting in 1969, a new concept for the ISYA emerged: to organize the ISYA always in developing countries and institutions in order to give a concentrated expert instruction and training in special topics of modern astronomy to a number of selected young astronomers, who otherwise would not have such opportunities available to them. This organization had the advantage that the experienced astronomers can help the host institute to plan future teaching and research programs.

Unfortunately, in 1971, co-sponsoring by UNESCO stopped. In view of the importance and usefulness of the ISYA, the IAU Executive Committee decided to allocate funds which would

Table 1. ISYAs 1967–1990

No	Date	Location	Duration (weeks)	Participants
1	1967 March	U.K., Manchester	6.5	12 (12f, 8n)
2	1968 June-July	Italy, Arcetri	8.5	10 (10f, 7n)
3	1969	India, Hyderabad	8	23 (5f, 5n)
4	1970 Oct-Nov	Argentina, Córdoba	8	21 (5n)
5	1973 July-Aug	Indonesia, Lembang	4	8 (3f, 4n)
6	1974 May	Argentina, San Miguel	4	60 (21f, 7n)
7	1975 Sept	Greece, Athens/Thera	4	74 (35f, 16n)
8	1977 Nov	Brazil, Rio	4	29
9	1978 Aug	Nigeria, Nsukka	3	28
10	1979 Sept	Spain, Tenerife	2	36 (7n)
11	1980 Sept-Oct	Yugoslavia, Hvar	3	25
12	1981 Aug-Sept	Egypt, Cairo	3	28 (9n)
13	1983 May-June	Indonesia, Lembang	3	21 (5n)
14	1986 Aug	China, Beijing	3	52 (6n)
15	1986 Sept	Portugal, Espinho	3	30 (19f, 7n)
16	1989 Aug	Cuba, Havana	2	55 (23f, 6n)
17	1990 May-June	Malaysia, Kuala Lumpur and Melaka	2.5	27 (11f)
18	1990 Sept	Morocco, Marrakesh	2.5	53

allow the continuation of the organization of these Schools, albeit that the duration of the Schools had to be reduced by half compared to the previous ones.

Since 1967 the ISYA is a regular programme of the Commission 46. Table 1 gives a list of all 18 ISYAs organized till 1990. The relevant information was taken from the triennial IAU Transactions in Volume A or B (reports from Commission 46).

Table 1 provides information, when available, on the total number of participants (first figure in the last column), the number of foreigners (f) and the number of different nationalities (n). The ISYA in Argentina was on the theme of Physics of Solar Plasmas, the Sun and Interplanetary Medium and Solar Energy; it consists in fact of three parallel schools and it was also granted by the Argentinian Commission Nacional de Estudios Geoheliofisicos.

From 1979 till 1990 the ISYA received a partial financial support from the UNESCO via the ICSU. In 1967, J. Kleczek was nominated General Secretary for the ISYA programme. He did it till 1990.

4. Objectives and organization of ISYA

An ISYA is always oriented toward developing countries, in astronomy, and is taking place there, which makes the IAU ISYA unique among all the Schools which nowadays are organized. Nevertheless, an ISYA is taking place in countries and universities with a reasonable long-term interest in astronomy to sustain further development. During an ISYA there is no donation of piece of research equipment, as example for a telescope.

An ISYA is organized through an agreement signed between the IAU and an university often associated to a project of development (new astronomy department, implementation of a telescope, ...). The main financial conditions are:

- the IAU pays for the travels of the faculty members and all the participants
- the host country pays for the stay of the faculty members and all the participants and provides the meeting facilities.

The duration of an ISYA is now of 3-week, which is needed for the participants to be used to speak and debate in English and in public; the lecturers are asked to stay as long as possible in order that the participants feel at ease to communicate.

The participants' background is that of a M.Sc. degree, but it ranges from just graduated to some PhD on-going. During an ISYA there are lectures but practical activities, computer

Table 2. List of the ISYA since 1992

No	Date	Location	Duration (weeks)	Participants
19	1992 Aug	China, Beijing and Xinglong Observatory	3	30 (17f, 12n, 9w)
20	1994 Jan	India, Pune	3	35 (25f, 13n, 11w)
21	1994 Sept	Egypt, Cairo and Kottamia Observatory	3	41 (12f, 13n, 10w)
22	1995 July	Brazil, Belo Horizonte and Serra Piedade	3	38 (19f, 11n, 15w)
23	1997 July	Iran, Zanjan	3	38 (14f, 8n, 12w)
24	1999 Aug	Romania, Bucharest	3	41 (18f, 9n, 22w)
25	2001 Jan	ChiangMai, Thailand	3	36 (17f, 9n, 6w)
26	2002 Aug	Casleo, Argentina	3	28 (14f, 9n, 10w)
27	2004 July	Al Akhawayn, Morocco	3	29 (18f, 13n, 9w)
28	2005 July-Aug	INAOE, Mexico	3	46 (20f, 10n, 18w)
29	2007 March	Kuala Lumpur, Langkawi, Malaysia	3	35

oriented, are equally important. Participant's talks is another aspect: for most of the students it is the first time that they have the opportunity to give a talk on their research, in English, in public and in front of foreign specialists.

5. From 1992 (19th ISYA) till 2007 (29th ISYA)

Table 2 gives the list of the last ten ISYA and the next one in Malaysia in 2007. It provides information on the number of foreigners (f), the number of different nationalities (n) and of the number of women (w).

The ISYAs were financially co-sponsored by UNESCO (through ICSU) till 2000. Since then the ISYAs are funded only by the IAU.

From 1992 till 1997, Don Wentzel (USA) and Michèle Gerbaldi (France) were respectively the General Secretary and the Assistant General Secretary for these Schools. Since then Michèle Gerbaldi (France) and Ed Guinan (USA) are respectively the chair and the vice-chair for this Programme Group of Commission 46.

Starting in 2007, the next team for the ISYA is : Jean-Pierre De Greve (Vrije Universiteit Brussel, Belgium) as chair and Kam-Ching Leung (USA) as vice-chair.

6. Concluding remarks

More and more "Summer Schools" are being organized by various institutions. But the ISYA are unique: they last longer (three weeks), they are not specialized toward doctoral students and they are fully funded. Even more important, they are taking place where needs have been expressed.

We do not repeat here the detailed analysis done by D. Wentzel in 1996 and in 2006 by M. Gerbaldi on the impact of the ISYA programme. We underline than an ISYA gives to the participants a broad perspective on astronomy and how science works. As a participant quoted "*... we learned not only useful astrophysics, but also had the chance to interact with some of the more advanced researchers in the field...*" Concerning the host institution, light is casted upon its development projects.

Today the *lonely astronomer* is also the one who is not associated to an international project. An ISYA has the perspective to insert more the young researchers in the international domain by, among other, offering them the possibility to start their network of scientific contacts but without cutting them from their roots.

It should be emphasized that no ISYA could have taken place without the enthusiasm of the faculty members who participated to it, giving so freely their time and energy to make a success of these Schools.

Acknowledgements

It is my great pleasure to acknowledge the IAU Executive Committee members and more specially the General Secretaries who have supported continuously this programme. Don Wentzel, with whom I started to work on this programme, is acknowledged for advices which inspired me to continue. Ed Guinan, vice-chairperson of the ISYA programme, is warmly thanks for this partnership during those years as well as for the friendly atmosphere in which we worked for the ISYA programme.

Michèle Gerbaldi
chair of the Programme Group

References

Gerbaldi, M., 2006, *Astronomy for the Developing World, SpS 5, IAU XXVI General Assembly* Prague, 2006, Eds. J. Hearnshaw & P. Martinez (Cambridge: CUP).
Transactions of the IAU 1964, vol. XIIB, p. 629
Transactions of the IAU 1964, vol. XIIB, p. 637
Transactions of the IAU 1964, vol. XIIB, p. 648
Transactions of the IAU 1967, vol. XIIIA, p. XCV
Transactions of the IAU 1967, vol. XIIIB, p. 229
Transactions of the IAU 1970, vol. XIVA, p. 563
Transactions of the IAU 1973 vol. XVA, p. 718
Transactions of the IAU 1973 vol. XVA, p. 719
Wentzel, D., 1996, in: L. Gouguenheim, D. McNally, & J. Percy (eds.) *Proc. IAU Colloquium No. 162, New trends in Astronomy Teaching*, (Cambridge: CUP), p. 27

Transactions IAU, Volume XXVIB
Proc. IAU XXVI General Assembly, August 2006
Karel A. van der Hucht, ed.

doi:10.1017/S1743921308024277

EXECUTIVE COMMITTEE WORKING GROUP YOUNG ASTRONOMERS EVENTS

SOC CHAIR **Michèle Gerbaldi**
MEMBERS **Jean-Pierre De Greve, Michal Dovčiak, Oddbjørn Engvold, Edward F. Guinan, John B. Hearnshaw, Melanie Johnston-Hollitt, Jay M. Pasachoff, John R. Percy, Ignasi Ribas, James C. White**
LOC CHAIR **Michal Dovčiak**
MEMBERS **René Goosmann, Tomáš Pecháček, Ivana Stoklasová**

PROCEEDINGS IAU XXVI GA YOUNG ASTRONOMERS EVENTS on 15–25 August 2006

1. Introduction

At the IAU XXV General Assembly in Sydney, 2003, a questionnaire on the perception of participation of "young astronomers" at IAU meeting was distributed. Following the conclusions from the analysis of this questionnaire, the IAU EC recommended in 2004 that the "young astronomers" concept at the next GA in Prague should be worked out with specific activities.

After extensive discussions with General Secretary Oddbjørn Engvold, with the IAU Officers during their February 2005 meeting, and with several colleagues involved in Commission 46 programmes, two actions were proposed for the IAU XXVI GA in Prague, 2006, in order to cast light on young astronomers and to create links between the "established" astronomers and the new generation of IAU members.

The proposed actions were (*i*) a Young Astronomers Lunch Debate, and (*ii*) a Young Astronomers Consulting Service, as first steps to answer to the concerns expressed by the previous EC:

• to make the IAU better known among the young generation astronomers,
• to identify how the IAU serves and aims to serve the interest of young astronomers,
• to stimulate young astronomers to become actively involved in IAU's programs and activities.

Reciprocally, these actions are a means to alert and educate the IAU itself about its responsibilities for its membership in the broadest sense. These Young Astronomers' Events were primarily organized for astronomers who are enrolled in PhD programs or who have received their Doctorate degrees within three years before the General Assembly.

Registration to these events was done via the Registration Form for the IAU XXVI GA. Dedicated web pages (<`http://astro.cas.cz/yae`>) for these events were all created by Dr. Michal Dovčiak. These pages are now available at the URL: <`http://www.iau.org`> (Commission 46, Astronomy Education & Development).

The organization and venue of the Young Astronomer Lunch Debate and the Young Astronomers Consulting Service are respectively described in Section 2 and 3. In Section 4 and 5 the analysis of the surveys done on each event are presented. In the conclusion, plans for the future are presented.

2. The Young Astronomers Lunch Debate

The purpose of this luncheon was to give to the young astronomers a splendid opportunity to meet with members of the IAU Executive Committee, representatives from ESA, ESO, NASA and with other participating astronomers in order to exchange ideas and to discuss their research, careers, and educational opportunities for Post-Doctoral positions, as well as employment opportunities.

This luncheon was organized through tables of 11 participants (9 young astronomers + 2 guest astronomers), each table discussing one or more topics of interest for young astronomers. The lunch debate took place prior to the Opening Ceremony of the GA on Tuesday August 15th (11h - 13h30). The format selected was a buffet in order to be more flexible according to the different nationalities and diets of the participants.

A questionnaire was prepared in order to select the topics for the table discussions. By the end of June 50 % of the 140 young astronomers, already registered to the lunch debate, answered this questionnaire and a list of topics to be discussed was established. The topics assigned for each table were selected from that list and the young astronomers were asked to register for a table. At the end 177 young astronomers registered and 20 tables were set up, two to three themes being debated at each table.

A difficulty came from the delay between the registration for the IAU XXVI GA, which could have been done as early as March, and the choice of the table and topics associated to it, which was done by the end of June. In July we had to track, by e-mails, the participants to sign up for a table.

A critical point was also the invitation of the guest astronomers, this latter could be done only after a reasonable guess of the number of participants to the lunch. (Annexe 4).

It was planned to record the discussions at each table. Some notes were collected but are difficult to exploit meaningfully. The first feedback of the lunch debate is that the young astronomers were able to start, on the spot, discussions with confirmed astronomers whatever the subject.

Some statistics on the young astronomers who participated to this lunch debate is given in the following tables. Concerning the number of participants per countries, this list is biased because in most of the cases the participants indicated the country where they are doing their studies and not their nationality.

Participants to the Lunch Debate

156 answers of overall 177 participants

Gender: 64 women and 92 men					
Situation:	neither PhD nor Postdoc 11	PhD only 106	PhD and Postdoc 37	Postdoc only 2	
Age group:	below 20 3	20 to 24 30	25 to 29 83	30 to 35 47	above 35 4

Participants to the Lunch Debate

country	number of participants	country	number of participants
Argentina	3	Mexico	2
Australia	3	Morocco	1
Austria	1	Nepal	1
Brazil	7	Netherlands	5
Canada	1	Peru	2
China	7	Poland	4
China (Taiwan)	4	Portugal	1
Czech Republic	13	Romania	1
Democratic Rep. of Congo	1	Russia	14
Denmark	1	Serbia and Montenegro	4
Egypt	1	Slovakia	2
Estonia	2	Slovenia	1
Finland	1	South Africa	1
France	6	South Korea	1
Germany	14	Spain	10
Greece	2	Sweden	2
Hungary	1	Thailand	1
India	5	Turkey	2
Iran	3	United Kingdom	5
Israel	2	Ukraine	8
Italy	11	USA	16
Japan	1	Viet Nam	1
Macedonia	1		

A detailed analysis of the survey on the lunch debate done by Melanie Johnston-Hollitt is presented in Section 4.

3. Young Astronomers Consulting Service

The purpose was to have a clearly identified office during the all GA where young astronomers could meet more advanced astronomers to seek advice on their CV, thesis, jobs, etc. during one-to-one personal contacts.

To launch this programme an e-mail was sent to all the already registered astronomers to the GA, by May 2006. In less than two days, more than 80 volunteers expressed their wish to act as a consultant. A consultant database was created with their domains of expertise, as well as a young astronomers' one with their requests. The registration to the Young Astronomers' Consulting Service continued during the all GA.

The arrangement of the one-to-one discussions started before the venue of the GA through the access to the consultants' and/or young astronomers' database, if permitted by the owner of the data. During the GA, the list of potential consultants with their domains of expertise and their consulting hours was put on the notice-board next to the office of the Consulting Service. The young astronomers were then able to sign in for the discussion with a consultant.

A green tag was sticked on the consultants' badge to emphasize their visibility.

Thus the young astronomers could contact the consultants either by signing at the consulting hours form and meet them at the YACS office at the agreed time, or exchange e-mails, or exchange messages through message-box service, or they could approach the consultant directly seeing their green tag.

The Young Astronomers Office was run efficiently by Czech young astronomers — members of the LOC, during the all GA, but its location prevented its direct visibility.

A report on the run of the Consulting Service during the GA is given by Michal Dovčiak Section 5.

4. Report on the survey from the participants to the Lunch Debate

At the conclusion of the lunch debate, a questionnaire was circulated to all participants and feedback from each table's deliberations was collected. The questions asked and the responses are summarized in Annexes 5 and 6, broken down into responses from young astronomers and guest/senior astronomers respectively.

A total of 149 young astronomers and 18 guests responded to the survey. The responses were overwhelmingly positive and it was clear that the event was a success.

An astounding 98 % of the young astronomers that responded to the question "Do you think the event was a success?" indicated that they did and 92 % indicated that they would like to attend such events in the future (and of the small percentage that did not do it was mostly because they believed they would be "too old" at the next GA).

As a means of facilitating interaction between young and senior astronomers and between young astronomers and their peers the lunch debate was highly successful. 89 % of the young astronomers responded to the question regarding the effectiveness of the event at providing an opportunity to interact with senior astronomers stating it to be effective or very effective and 74 % believed the event to have also provided an effective or very effective opportunity to interact with their fellow young astronomers.

The topics of the lunch debate were also believed to be well or very well covered by 80 % of respondents. The event also served to increase the sense of partipation for young astronomers with 85 agreeing it had served this purpose for them. Given that one of the major complaints identified in the YA survey during the 2003 GA in Sydney was lack of a sense of participation by young astronomers this is a fantastic result.

Other trends in the data demonstrate the importance of the GA website as a means for promotion of young astronomer activities and there was a strong sense in the comments that the creation of a young astronomer mailing list or other permanent web-based contact resource would be of benefit. It would seem that most participants were happy with the lunch format, but not with the timing of the event which both clashed with interesting scientific sessions and was too early in the program for people only attending the second week of the GA. Additionally, a large number of young astronomers commented that there were too many people per table to provide sufficient interaction with the senior astronomer and that it would have increased the benefit for them had the table sizes been reduced. In future it would be better to have the event in the middle of the GA and might be wise to move to an evening format to avoid clashes.

In summary, it is clear that the lunch debate was an overwhelming success and there is a strong demand from young astronomers for this type of activity to continue at future GAs.

5. Report on the survey on the Young Astronomers Consulting Service

Two surveys on the Young Astronomers Consulting Service, hereafter named YACS, were carried out — one assessment questionnaire was for young astronomers and one for consultants.

21 young astronomers filled in the questionnaire out of 145 who have registered for the YACS (Annexe 7). We do not know the exact number of young astronomers that really used this service. This is due to the fact that we tried to make the meetings between young astronomers and consultants as easy to arrange as possible — they could arrange them by exchanging e-mails or messages in their message boxes, i.e. without coming to the YACS office. 14 consultants answered the assessment questionnaire out of 85 registered consultants (Annexe 8). From the answers to the first question of both questionnaires we estimate that up to 30 or 40 young astronomers addressed up to 20 consultants.

Comments on the YACS were given by both the young astronomers and the consultants.

About 81 % of young astronomers addressed 1 or 2 consultants, 19 % addressed 3 of them.

About 71 % of consultants were asked for a discussion by 1 or 2 young astronomers and 21 % by 3 young astronomers. We had one very efficient consultant who talked to 7 young astronomers!

The consulting service was very useful in initiating first contacts between young astronomers and consultants. More than 50 % of young astronomers initiated their contact with consultants through the YACS or they saw the green tag on the consultants' badges (given to them by YACS) and contacted them directly. According to consultants even more contacts were established in this way (i.e., with the help of YACS). It seems that the best way to organize meetings was by exchanging e-mails or messages in message boxes between young astronomers and consultants. On the other hand the consulting hours forms were used only in few cases. It was quite difficult for consultants to know and announce their free time slots in advance.

According to 90 % of young astronomers the discussions with consultants were useful and according to 86 % of them the consultant was a suitable person to provide them with the assistance required. 93 % of consultants had the feeling that the discussions were useful for young astronomer. 95 % of young astronomers found the YACS useful, 76 % would use it in the future again and for 90 % of them the participation in this service enhanced their experience of the IAU General Assembly. 24 % of young astronomers thought there were not enough consultants covering a large range of expertise. Approximately half of the young astronomers and half of the consultants made an ongoing link, i.e., expect further contact with each other. 86 % of consultants expressed readiness to be contacted for advice by young astronomers in the period between General Assemblies and 93 % of them are ready to volunteer as a consultant for this service in the future.

86 % of young astronomers and 64 % of consultants thought it very easy or more-or-less easy to arrange a time and place for the discussions. Most of the discussions took up to 1 hour but there were some of them that lasted even up to 3 hours.

From the above it is quite clear that the service was useful and we think it should be available during next General Assembly as well. Moreover it would be useful to establish some permanent place where young astronomers could find relevant information and ask for advice — this place could be a web page dedicated to young astronomers plus some kind of database of consultants from different countries and research areas who would be willing to help (give a piece of advice via e-mail) even between General Assemblies.

At the end it is needed to emphasize that although only about quarter of the consultants were asked for one-to-one discussion it was very important that so many were accessible. We even advise to encourage more experienced astronomers to register as a consultant during next General Assembly so that there is even better choice in different areas of expertise for young astronomers. There are different opinions on the type of themes that young astronomers wanted to discuss — according to experiences of some of the consultants, young astronomers want to discuss mainly further scientific education and job opportunities. On the other hand some of the young astronomers did not find the right person to talk to because the consultants did not cover enough research fields. This is an interesting question and in future assessment questionnaire it would be good to ask also what types of topics were discussed between young astronomers and consultants.

6. Closing remarks

The Young Astronomers Events during the IAU XXVI GA in Prague were a success as indicated by the surveys and informal discussions. We recall three of the answers given by young astronomers concerning the Lunch debate:

- Has the Young Astronomers Events increased your sense of participation at the IAU GA ? yes : 85 %
- Effectiveness of providing interaction between young astronomers: effective: 74 %
- Effectiveness of providing interaction with senior astronomers: effective : 89 %

These Young Astronomers Events should be maintained for the next GA in 2009 with some modifications, the first one is to have the lunch debate later during the first week of the GA, not necessarily the same day as the Opening Ceremony, but still at the beginning of the GA in order to create this special link between the two generations of astronomers as early as possible.

The lunch debate should be preferably during an evening in order to avoid any overlap with the scientific sessions. The lunch-break is too short to have it in this slot, in the most favorable cases it is one hour.

As many participants are coming to a GA for one week only, several times a second lunch debate, the second week, was mentioned.

If these events are part of the 2009-GA, in Rio, obviously, the team in charge of it should contact Michal Dovčiak (chair of the LOC) as early as possible.

We recommend the creation of a Working Group under the auspice of Commission 46 for a follow-up of the Consulting Service.

The creation of permanent dedicated web pages for the young astronomers with the following suggested topics should be one of the tasks of this Working Group:

- to set up a Consulting Service by e-mail
- to offer selected links to national Young Astronomers (YA) groups, selected links to PhD, PostDoc programmes
- to publicize the offer by Dr. Levato of 5% of the observing time at the Argentina national observatory El Leoncito (2.4 m tel, 0.9 m tel) for YA through applications and selections.

It can be concluded that the Young Astronomers Events during the IAU XXVI GA organized, for the first time, have fulfilled their role: to facilitate the interactions between the young and the established astronomers and to widely increase the sense of participation of the young astronomers to the GA.

Michèle Gerbaldi
chair of the Working Group

Transactions IAU, Volume XXVIB
Proc. IAU XXVI General Assembly, August 2006
Karel A. van der Hucht, ed.

doi:10.1017/S1743921308024289

EXECUTIVE COMMITTEE WORKING GROUP WOMEN IN ASTRONOMY

CO-CHAIRS **Anne J. Green, Sarah T. Maddison**
MEMBERS **Johannes Andersen, Olga B. Dluzhnevskaya, Gloria M. Dubner, Andrea K. Dupree, R. Elizabeth Griffin, W. Miller Goss, Mary Kontizas, Birgitta Nordström, Francesca Primas, Sylvia Torres Peimbert, Yiping Wang, Shahinaz M. Yousef**

BUSINESS MEETING AND WORKING LUNCH on 21 August 2006

1. Introduction

The second Women in Astronomy Lunchtime Meeting was held on Monday 21 August 2006, with more than 250 participants. The meeting was hosted by the EC Working Group for Women in Astronomy, established at the 2003 IAU General Assembly, and was attended by the current President, the Presidents-Elect for this and the next General Assembly, the General Secretary and Vice-Presidents, many senior astronomers, as well as students and young astronomers. It was a particular pleasure to welcome and congratulate the incoming President, Dr Catherine Cesarsky, the first woman to hold the position.

2. Business Meeting report

Lunch was preceded by the Business Meeting attended by the Organising Committee and an overflow audience of interested participants. Following discussion it was agreed that the Organising Committee should hold office for six years, in accordance with practice common among the Divisions. An important agenda item was the reporting on national statistics; one of the primary goals of the Working Group is the collection of global statistics on the status and position of women in astronomy. Several national reports were tabled, including from the United Kingdom, Latin America (organised from Argentina), Russia, Australia and Greece. We plan to survey the wider community with a concise and consistent set of questions relevant to all countries. For this, we need National Representatives who will take responsibility for obtaining the statistics. This will be quite complex to orchestrate, but we are fortunate that many surveys already exist and with the new ones that are planned, a comprehensive demographics database can be established.

It was noted that IAU gender statistics give an incomplete picture, but we cannot be satisfied with recent numbers showing overall only 13 % women members from a total of 8000 in 39 countries, although an encouraging 22 nations recorded an increase since the previous General Assembly. The number of women who hold executive positions on the various IAU Committees is increasing, which is a welcome development.

3. Lunch Meeting report

The meeting theme was "Career Development for Women" with keynote speakers Dr Sunetra Giridhar of the Indian Institute of Astrophysics, Bangalore, and Dr Patricia Knezek, Deputy Director of the WIYN Observatory, Arizona. Every participant received a flyer showing the 1992 Baltimore Charter for Women in Astronomy, current IAU statistics and five suggested topics for discussion in the breakout groups and at the final Plenary session. Some of the issues and comments were:

3.1. *Unequal opportunity*

Has discrimination gone underground? Many participants wanted more flexible criteria for appointments, for more women to be appointed to senior positions and for the visibility of women at conferences to increase. In many instances, talented women are not even submitting applications for positions for which they would be competitive. Sadly, subtle discrimination is still a problem at several institutions.

3.2. *Mentoring and self-confidence*

Do women network effectively? Many young women astronomers expressed the need for role models and effective mentoring, to provide strategies to build self-confidence. Anecdotal evidence suggests women base their job and promotion applications on their achievements rather than on their potential (a more male approach).

3.3. *Family responsibilities*

Is there an easier time for having children? Many noted that the provision of childcare at workplaces and conferences is critical. While maternity leave is now frequently offered, arrangements for childcare at conferences and workplaces are often lacking. Women are still (generally) the primary caregivers, with greater vulnerability for research disruption and mobility limitations. This issue was raised by many of the participants.

3.4. *Dual careers*

Equal advancement of two careers is extremely difficult. Lack of mobility affects women more than men. How can we encourage more options for partners? Can we embrace non-standard career paths as acceptable? The two-body issue is seen as problematic for many women.

Following the meeting, a submission was made to the incoming IAU Executive with two action items for their consideration. The Working Group is keen to assist the Executive to achieve implementation of these items.

4. Items for action

4.1. *Increased visibility for women at conferences*

For all Symposia and scientific meetings, the IAU should strongly encourage the adequate representation of women as invited speakers, members of the Science Organising Committees and Session Chairs. Details of invitees to be provided as part of any meeting proposal.

4.2. *Childcare facilities at conferences*

Many women are constrained in their ability to attend scientific meetings by the lack of childcare provisions. The IAU would require that all future proposals for scientific meetings include some provision for childcare, most likely paid for by the participants, but organised by the hosts.

Finally, the meeting was an excellent if brief opportunity to exchange ideas and experiences, made possible through generous support from the US IAU National Committee for Astronomy and the National Organising Committee of the Prague General Assembly, for which we are greatly appreciative. We look forward to meeting again at the IAU XXVII General Assembly in Rio de Janeiro in 2009.

Ann Green & Sarah Maddison
co-chairs of the Working Group

Transactions IAU, Volume XXVIB
Proc. IAU XXVI General Assembly, August 2006
Karel A. van der Hucht, ed.

doi:10.1017/S1743921308024290

CHAPTER VI

STATUTES, BYE-LAWS AND WORKING RULES

Prague, Czech Republic, 15 August 2006

Introduction.
IAU XXVI General Assembly revisions to the Statutes and Bye-Laws

The current Statutes and Bye-Laws were extensively re-written and approved at the Sydney GA in 2003 in order to streamline many of the processes of the IAU and to bring the IAU more into conformity with the structure of other ICSU unions. These revision have generally been well received, however there have been several changes to the Statutes and Bye-Laws that Individual Members and National Members have deemed desirable in the past three years and which have been brought to the attention of the EC. They are the following:

1. Change the Statutes to allow scientific matters to be decided by a vote of individual rather than national members. National Members pay dues and set the policies of the Union, and are the only voting entities in the IAU under the Sydney revisions, contrary to previous IAU policy and tradition. Many Union Individual Members have argued that questions of science are best settled by individuals who participate in debate and discussion because this is a more accurate reflection of how scientific issues are actually resolved in the community. The EC debated this issue and recommended changing the Statutes to once again allow all scientific questions and resolutions to be decided by a majority vote of individual members at the GA. The EC acknowledges that returning the responsibility for voting to Individual Members could be treated as part of an eventual evolution to voting on Union issues by electronic means.

2. Change the Bye-Laws to allow Division Presidents (DPs) to nominate individuals for membership in the Union. The Sydney Bye-Laws do not permit this. The DPs wish for this authority in order to provide for needed expertise in Commission discussions and in appointments to Working Groups. It often happens that needed expertise is not available in areas related to astronomy, and it can happen that individuals with experience useful to the Union, but not known to the National Committees for Astronomy (NCAs), are sometimes overlooked for Union membership. The EC agrees that the increased flexibility of allowing DPs to nominate Individual Members far outweighs any loss of control on the part of the NCAs to control Individual Membership, and it strongly recommended that the Bye-Laws be revised to permit the nomination of individuals for membership.

3. Change the Bye-Laws to allow Division Organizing Committees (OCs) to exceed 12 persons. The Sydney revisions limited the number of individuals on each OC in order to keep them manageable. However, some DPs have felt that their Divisions are so diverse that adequate representation of the many areas requires more than 12 people. The EC sympathesizes with this argument, but still believes that it is important to keep OCs from proliferating their memberships. Therefore, it believes that occasional exceptions

to the 12-person limit for some OCs is acceptable, and supported such a revision in the Bye-Laws.

4. Change the Bye-Laws to increase the number of dues categories, reflecting the increasing number of Individual Members in many countries and allowing such countries to be assigned an equitable amount of dues. The maximum dues contribution specified in the Sydney Bye-Laws requires a country with a large number of Individual Members to pay a relatively small dues contribution per Individual Member compared to that of some smaller countries. The only adverse effect to assigning some countries to a larger dues category is the increased number of votes this gives to such countries in Union budgetary matters. The EC discussed this issue and felt that the increased resources that would be made available by increased dues far outweighed the increasing influence the larger countries would have in budgetary votes. Therefore, the EC strongly recommended defining two new dues categories that would be appropriate for the largest national members of the IAU.

The above four revisions were published in *Information Bulletin* No. 98, pp. 53-61, brought to the floor in the GA First Session in Prague, and discussed and voted upon by the National Members. All of the proposed revisions passed, and went into effect at the end of that Session.

Robert Williams, Prague, Czech Republic, 15 August 2006

Transactions IAU, Volume XXVIB
Proc. IAU XXVI General Assembly, August 2006
Karel A. van der Hucht, ed.

doi:10.1017/S1743921308024307

STATUTES

Prague, Czech Republic, 15 August 2006

I. OBJECTIVE

1. The International Astronomical Union (referred to as the Union) is an international non-governmental organization. Its objective is to promote the science of astronomy in all its aspects.

II. DOMICILE AND INTERNATIONAL RELATIONS

2. The legal domicile of the Union is Paris.

3. The Union adheres to, and co-operates with the body of international scientific organizations through ICSU: The International Council for Science. It supports and applies the policies on the Freedom, Responsibility, and Ethics in the Conduct of Science defined by ICSU.

III. COMPOSITION OF THE UNION

4. The Union is composed of:

4.a. National Members (adhering organizations)

4.b. Individual Members (adhering persons)

IV. NATIONAL MEMBERS

5. An organization representing a national professional astronomical community, desiring to promote its participation in international astronomy and supporting the objective of the Union, may adhere to the Union as a National Member. Exceptionally, a National Member may represent the community in the territory of more than one nation, provided that no part of that community is represented by another National Member.

6. An organization desiring to join the Union as a National Member while developing professional astronomy in the community it represents may do so on an interim basis, on the same conditions as above, for a period of up to nine years. After that time, it will either become a National Member on a permanent basis, or its membership in the Union will terminate.

7. A National Member is admitted to the Union on a permanent or interim basis by the General Assembly. It may resign from the Union by so informing the General Secretary, in writing.

8. A National Member may be either:

8.a. the organization by which scientists of the corresponding nation or territory adhere to ICSU or:

8.b. an appropriate National Society or Committee for Astronomy, or:

8.c. an appropriate institution of higher learning.

9. The adherence of a National Member is suspended if its dues have not been paid for five years; it resumes, upon the approval of the Executive Committee, when the arrears have been paid. After five years of suspension of a National Member, the Executive Committee may recommend to the General Assembly to terminate the membership.

10. A National Member is admitted to the Union in one of the categories specified in the Bye-Laws.

V. INDIVIDUAL MEMBERS

11. A professional scientist who is active in some branch of astronomy may be admitted to the Union by the Executive Committee as an Individual Member. An Individual Member may resign from the Union by so informing the General Secretary, in writing.

VI. GOVERNANCE

12. The governing bodies of the Union are:

12.a. The General Assembly;

12.b. The Executive Committee; and

12.c. The Officers.

VII. GENERAL ASSEMBLY

13. The General Assembly consists of the National Members and of Individual Members. The General Assembly determines the overall policy of the Union.

13.a. The General Assembly approves the Statutes of the Union, including any changes therein.

13.b. The General Assembly approves Bye-Laws specifying the Rules of Procedure to be used in applying the Statutes.

13.c. The General Assembly elects an Executive Committee to implement its decisions and to direct the affairs of the Union between successive ordinary meetings of the General Assembly. The Executive Committee reports to the General Assembly.

13.d. The General Assembly appoints a Finance Committee, consisting of one representative of each National Member having the right to vote on budgetary matters according to §14.a., to advise it on the approval of the budget and accounts of the Union. The General Assembly also appoints a Finance Sub-Committee to advise the Executive Committee on its behalf on budgetary matters between General Assemblies.

13.e. The General Assembly appoints a Special Nominating Committee to prepare a suitable slate of candidates for election to the incoming Executive Committee.

13.f. The General Assembly appoints a Nominating Committee to advise the Executive Committee on the admission of Individual Members.

14. Voting at the General Assembly on issues of a primarily scientific nature, as determined by the Executive Committee, is by Individual Members. Voting on all other matters is by National Member. Each National Member authorises a representative to vote on its behalf.

14.a. On questions involving the budget of the Union, the number of votes for each National Member is one greater than the number of its category, referred to in article 10. National Members with interim status, or which have not paid their dues for years preceding that of the General Assembly, may not participate in the voting.

14.b. On questions concerning the administration of the Union, but not involving its budget, each National Member has one vote, under the same condition of payment of dues as in §14.a.

14.c. National Members may vote by correspondence on questions concerning the agenda for the General Assembly.

14.d. A vote is valid only if at least two thirds of the National Members having the right to vote by virtue of article §14.a. participate in it.

15. The decisions of the General Assembly are taken by an absolute majority of the votes cast. However, a decision to change the Statutes can only be taken with the approval of at least two thirds of the votes of all National Members having the right to vote by virtue of article §14.a. Where there is an equal division of votes, the President determines the issue.

16. Changes in the Statutes or Bye-Laws can only be considered by the General Assembly if a specific proposal has been duly submitted to the National Members and placed on the Agenda of the General Assembly by the procedure and deadlines specified in the Bye-Laws.

VIII. EXECUTIVE COMMITTEE

17. The Executive Committee consists of the President of the Union, the President-Elect, six Vice-Presidents, the General Secretary, and the Assistant General Secretary, elected by the General Assembly on the proposal of the Special Nominating Committee.

IX. OFFICERS

18. The Officers of the Union are the President, the General Secretary, the President-Elect, and the Assistant General Secretary. The Officers decide short-term policy issues within the general policies of the Union as decided by the General Assembly and interpreted by the Executive Committee.

X.SCIENTIFIC DIVISIONS

19. As an effective means to promote progress in the main areas of astronomy, the scientific work of the Union is structured through its Scientific Divisions. Each Division covers a broad, well-defined area of astronomical science, or deals with international matters of an interdisciplinary nature. As far as practicable, Divisions should include comparable fractions of the Individual Members of the Union.

20. Divisions are created or terminated by the General Assembly on the recommendation of the Executive Committee. The activities of a Division are organized by an Organizing Committee chaired by a Division President. The Division President and a Vice-President are elected by the General Assembly on the proposal of the Executive Committee, and are ex officio members of the Organizing Committee.

XI. SCIENTIFIC COMMISSIONS

21. Within Divisions, the scientific activities in well-defined disciplines within the subject matter of the Division may be organized through scientific Commissions. In special cases, a Commission may cover a subject common to two or more Divisions and then becomes a Commission of all these Divisions.

22. Commissions are created or terminated by the Executive Committee upon the recommendation of the Organizing Committee(s) of the Division(s) desiring to create or terminate them. The activities of a Commission are organized by an Organizing Committee chaired by a Commission President. The Commission President and a Vice-President are appointed by the Organizing Committee(s) of the corresponding Division(s) upon the proposal of the Organizing Committee of the Commission.

XII. BUDGET AND DUES

23. For each ordinary General Assembly the Executive Committee prepares a budget proposal covering the period to the next ordinary General Assembly, together with the accounts of the Union for the preceding period. It submits these, with the advice of the Finance Sub-Committee, to the Finance Committee for consideration before their submission to the vote of the General Assembly.

23.a. The Finance Committee examines the accounts of the Union from the point of view of responsible expenditure within the intent of the previous General Assembly, as interpreted by the Executive Committee. It also considers whether the proposed budget is adequate to implement the policy of the General Assembly. It submits reports on these matters to the General Assembly before its decisions concerning the approval of the accounts and of the budget.

23.b. The amount of the unit of contribution is decided by the General Assembly as part of the budget approval process.

23.c. Each National Member pays annually a number of units of contribution corresponding to its category. The number of units of contribution for each category shall be specified in the Bye-Laws.

23.d. A vote is valid only if at least two thirds of the National Members having the right to vote by virtue of article §14.a. participate in it.

23.e. National Members having interim status pay annually one half unit of contribution.

23.f. The payment of contributions is the responsibility of the National Members. The liability of each National Members in respect of the Union is limited to the amount of contributions due through the current year.

XIII. EMERGENCY POWERS

24. If, through events outside the control of the Union, circumstances arise in which it is impracticable to comply fully with the provisions of the Statutes and Bye-Laws of the Union, the Executive Committee and Officers, in the order specified below, shall take such actions as they deem necessary for the continued operation of the Union. Such action shall be reported to all National Members as soon as this becomes practicable, until an ordinary or extraordinary General Assembly can be convened.

The following is the order of authority: The Executive Committee in meeting or by correspondence; the President of the Union; the General Secretary; or failing the practicability or availability of any of the above, one of the Vice-Presidents.

XIV. DISSOLUTION OF THE UNION

25. A decision to dissolve the Union is only valid if taken by the General Assembly with the approval of three quarters of the National Members having the right to vote by virtue of article §14.a. Such a decision shall specify a procedure for settling any debts and disposing of any assets of the Union.

XV. FINAL CLAUSE

26. These Statutes enter into force on 15 August 2006.

Transactions IAU, Volume XXVIB
Proc. IAU XXVI General Assembly, August 2006
Karel A. van der Hucht, ed.

doi:10.1017/S1743921308024319

BYE-LAWS

Prague, Czech Republic, 15 August 2006

I. MEMBERSHIP

1. An application for admission to the Union as a National Member shall be submitted to the General Secretary by the proposing organization at least eight months before the next ordinary General Assembly.

2. The Executive Committee shall examine the application and resolve any outstanding issues concerning the nature of the proposed National Member and the category of membership. Subsequently, the Executive Committee shall forward the application to the General Assembly for decision, with its recommendation as to its approval or rejection.

3. The Executive Committee shall examine any proposal by a National Member to change its category of adherence to a more appropriate level. If the Executive Committee is unable to approve the request, either party may refer the matter to the next General Assembly.

4. Individual Members are admitted by the Executive Committee upon the nomination of a National Member or the President of a Division. The Executive Committee shall publish the criteria and procedures for membership, and shall consult the Nominating Committee before approving applications for admissions as Individual Members.

II. GENERAL ASSEMBLY

5. The ordinary General Assembly meets, as a rule, once every three years. Unless determined by the previous General Assembly, the place and date of the ordinary General Assembly shall be fixed by the Executive Committee and be communicated to the National Members at least one year in advance.

6. The President may summon an extraordinary General Assembly with the consent of the Executive Committee, and must do so at the request of at least one third of the National Members. The date, place, and agenda of business of an extraordinary General Assembly must be communicated to all National Members at least two months before the first day of the Assembly.

7. Matters to be decided upon by the General Assembly shall be submitted for consideration by those concerned as follows, counting from the first day of the General Assembly:

7.a. A motion to amend the Statutes or Bye-Laws may be submitted by a National Member or by the Executive Committee. Any such motion shall be submitted to the General Secretary at least nine months in advance and be forwarded, with the recommendation of the Executive Committee as to its adoption or rejection, to the National Members at least six months in advance.

7.b. The General Secretary shall distribute the budget prepared by the Executive Committee to the National Members at least eight months in advance. Any motion to modify this budget, or any other matters pertaining to it, shall be submitted to the General Secretary at least six months in advance. Any such motion shall be submitted, with the advice of the Executive Committee as to its adoption or rejection, to the National Members at least four months in advance.

7.c. Any motion or proposal concerning the administration of the Union, and not affecting the budget, by a National Member, or by the Organizing Committee of a Scientific Division of the Union, shall be placed on the Agenda of the General Assembly, provided it is submitted to the General Secretary, in specific terms, at least six months in advance.

7.d. Any motion of a scientific character submitted by a National Member, a Scientific Division of the Union, or by an ICSU Scientific Committee or Program on which the Union is formally represented, shall be placed on the Agenda of the General Assembly, provided it is submitted to the General Secretary, in specific terms, at least six months in advance.

7.e. The complete agenda, including all such motions or proposals, shall be prepared by the Executive Committee and submitted to the National Members at least four months in advance.

8. The President may invite representatives of other organizations, scientists in related fields, and young astronomers to participate in the General Assembly. Subject to the agreement of the Executive Committee, the President may authorise the General Secretary to invite representatives of other organizations, and the National Members or other appropriate IAU bodies to invite scientists in related fields and young astronomers.

III. SPECIAL NOMINATING COMMITTEE

9. The Special Nominating Committee consists of the President and past President of the Union, a member proposed by the retiring Executive Committee, and four members selected by the Nominating Committee from among twelve Members proposed by Presidents of Divisions, with due regard to an appropriate distribution over the major branches of astronomy.

9.a. Except for the President and immediate past President, present and former members of the Executive Committee shall not serve on the Special Nominating Committee. No two members of the Special Nominating Committee shall belong to the same nation or National Member.

9.b. The General Secretary and the Assistant General Secretary participate in the work of the Special Nominating Committee in an advisory capacity.

10. The Special Nominating Committee is appointed by the General Assembly, to which it reports directly. It assumes its duties immediately after the end of the General Assembly and remains in office until the end of the ordinary General Assembly next following that of its appointment, and it may fill any vacancy occurring among its members.

IV. OFFICERS AND EXECUTIVE COMMITTEE

11.a. The President of the Union remains in office until the end of the ordinary General Assembly next following that of election. The President-Elect succeeds the President at that moment.

11.b. The General Secretary and the Assistant General Secretary remain in office until the end of the ordinary General Assembly next following that of their election. Normally the Assistant General Secretary succeeds the General Secretary, but both officers may be re-elected for another term.

11.c. The Vice-Presidents remain in office until the end of the ordinary General Assembly following that of their election. They may be immediately re-elected once to the same office.

11.d. The elections take place at the last session of the General Assembly, the names of the candidates proposed having been announced at a previous session.

12. The Executive Committee may fill any vacancy occurring among its members. Any person so appointed remains in office until the end of the next ordinary General Assembly.

13. The past President and General Secretary become advisers to the Executive Committee until the end of the next ordinary General Assembly. They participate in the work of the Executive Committee and attend its meetings without voting rights.

14. The Executive Committee shall formulate Working Rules to clarify the application of the Statutes and Bye-Laws. Such Working Rules shall include the criteria and procedures by which the Executive Committee will review applications for Individual Membership; standard Terms of Reference for the Scientific Commissions of the Union; rules for the administration of the Union's financial affairs by the General Secretary; and procedures by which the Executive Committee may conduct business by electronic or other means of correspondence. The Working Rules shall be published electronically and in the Transactions of the Union.

15. The Executive Committee appoints the Union's official representatives to other scientific organizations.

16. The Officers and members of the Executive Committee cannot be held individually or personally liable for any legal claims or charges that might be brought against the Union.

V. SCIENTIFIC DIVISIONS

17. The Divisions of the Union shall pursue the scientific objects of the Union within their respective fields of astronomy. Activities by which they do so include the encouragement and organization of collective investigations, and the discussion of questions relating to international agreements, cooperation, or standardization.

They shall report to each General Assembly on the work they have accomplished and such new initiatives as they are undertaking.

18. Each Scientific Division shall consist of:

18.a. An Organizing Committee, normally of 6-12 persons, including the Division President and Vice-President, and a Division Secretary appointed by the Organizing Committee from among its members.

18.b. Members of the Union appointed by the Organizing Committee in recognition of their special experience and interests. The Committee is responsible for conducting the business of the Division.

19. Normally, the Division President is succeeded by the Vice-President at the end of the General Assembly following their election, but both may be re-elected for a second term. Before each General Assembly, the Organizing Committee shall organize an election from among the membership, by electronic or other means suited to the Commission structure of the Division, of a new Organizing Committee to take office for the following term. Election procedures should, as far as possible, be similar among the Divisions and require the approval of the Executive Committee.

20. Each Scientific Division may structure its scientific activities by creating a number of Commissions. In order to monitor and further the progress of its field of astronomy, the Division shall consider, before each General Assembly, whether its Commission structure serves its purpose in an optimum manner. It shall subsequently present its proposals for the creation, continuation or discontinuation of Commissions to the Executive Committee for approval.

21. With the approval of the Executive Committee, a Division may appoint Working Groups to study well-defined scientific issues and report to the Division. Unless specifically re-appointed by the same procedure, such Working Groups cease to exist at the next following General Assembly.

VI. SCIENTIFIC COMMISSIONS

22. A Scientific Commission shall consist of:

22.a. A President and an Organizing Committee consisting of 4-8 persons elected by the Commission membership, subject to the approval of the Organizing Committee of the Division;

22.b. Members of the Union, appointed by the Organizing Committee, in recognition of their special experience and interests, subject to confirmation by the Organizing Committee of the Division.

23. A Commission is initially created for a period of six years. The parent Division may recommend its continuation for additional periods of three years at a time, if sufficient justification for its continued activity is presented to the Division and the Executive Committee. The activities of a Commission is governed by Terms of Reference, which are based on a standard model published by the Executive Committee and are approved by the Division.

24. With the approval of the Division, a Commission may appoint Working Groups to study well-defined scientific issues and report to the Commission. Unless specifically re-appointed by the same procedure, such Working Groups cease to exist at the next following General Assembly.

VII. ADMINISTRATION AND FINANCES

25. Each National Member pays annually to the Union a number of units of contribution corresponding to its category as specified below; National Members with interim status pay annually one half unit of contribution:

Categories as defined in article 10 of the Statutes

I	II	III	IV	V	VI	VII	VIII	IX	X	XI	XII
1	2	4	6	10	14	20	27	35	45	60	80

Number of units of contribution

26. The income of the Union is to be devoted to its objects, including:

26.a. the promotion of scientific initiatives requiring international co-operation;

26.b. the promotion of the education and development of astronomy world-wide;

26.c. the costs of the publications and administration of the Union.

27. Funds derived from donations are reserved for use in accordance with the instructions of the donor(s). Such donations and associated conditions require the approval of the Executive Committee.

28. The General Secretary is the legal representative of the Union. The General Secretary is responsible to the Executive Committee for not incurring expenditure in excess of the amount specified in the budget as approved by the General Assembly.

29. The General Secretary shall consult with the Finance Sub-Committee (cf. Statutes 13.d.) in preparing the accounts and budget proposals of the Union, and on any other matters of major importance for the financial health of the Union. The comments and advice of the Finance Sub-Committee shall be made available to the Officers and Executive Committee as specified in the Working Rules.

30. An Administrative office, under the direction of the General Secretary, conducts the correspondence, administers the funds, and preserves the archives of the Union.

31. The Union has copyright to all materials printed in its publications, unless otherwise arranged.

VIII. FINAL CLAUSE

32. These Bye-Laws enter into force on 15 August 2006.

Transactions IAU, Volume XXVIB
Proc. IAU XXVI General Assembly, August 2006
Karel A. van der Hucht, ed.

doi:10.1017/S1743921308024320

WORKING RULES

Prague, Czech Republic, 25 August 2006

INTRODUCTION AND RATIONALE

The Statutes of the International Astronomical Union (IAU) define the goals and organizational structure of the Union, while the Bye-Laws specify the main tasks of the various bodies of the Union in implementing the provisions of the Statutes. The Working Rules are designed to assist the membership and governing bodies of the Union in carrying out these tasks in an appropriate and effective manner. Each of the sections below is preceded by an introduction outlining the goals to be accomplished by the procedures specified in the succeeding paragraphs. The Executive Committee updates the Working Rules as necessary to reflect current procedures and to optimize the services of the IAU to its membership.

I. NON-DISCRIMINATION

The International Astronomical Union follows the regulations of (ICSU): The International Council for Science and concurs with the actions undertaken by their Standing Committee on Freedom in the Conduct of Science on non-discrimination and universality of science (*cf.* § 21 below).

II. NATIONAL MEMBERSHIP

The aim of the rules for applications for National Membership is to ensure that the proposed National Member adequately represents an astronomical community not already represented by another Member, and that such membership will be of maximum benefit for the community concerned (*cf.* Statutes § IV).

1. Applications for National Membership should therefore clearly describe the following essential conditions:

1.a. the precise definition of the astronomical community to be represented by the proposed Member;

1.b. the present state and expected development of that astronomical community;

1.c. the manner in which the proposed National Member represents this community;

1.d. whether the application is for membership on a permanent or interim basis; and

1.e. the category in which the prospective National Member wishes to be classified (*cf.* Bye-Laws § 25).

2. Applications for National Membership shall be submitted to the General Secretary, who will forward them to the Executive Committee for review as provided in the Statutes.

III. INDIVIDUAL MEMBERSHIP

Professional scientists whose research is directly relevant to some branch of astronomy are eligible for election as Individual Members of the Union (*cf.* Statutes § V). Individual Members are, normally, admitted by the Executive Committee on the proposal of a National Member. However, Presidents of Divisions may also propose individuals for membership in cases when the normal procedure is not applicable or practicable (*cf.* Bye-Laws § 4). The present rules are intended to ensure that all applications for membership are processed on a uniform basis, and that all members are fully integrated in and contributing to the activities of the Union. Thus, the General Assembly appoints a Nominating Committee with a small number of members (*cf.* Statutes § 13.f.) and it, or a Sub-Committee which it appoints, may remain in office until the next General Assembly to allow more time for its work.

3. The term "Professional Scientist" shall normally designate a person with a doctoral degree (Ph.D.) or equivalent experience in astronomy or a related science, and whose professional activities have a substantial component of work related to astronomy.

4. National Members and Division Presidents may propose Individual Members who fall outside the category of professional scientist but who have made major contributions to the science of astronomy, e.g., through education or research related to astronomy. Such proposals should be accompanied by a detailed motivation for what should be seen as exceptions to the rule.

5. Eight months before an ordinary General Assembly, National Members and Presidents of Divisions will be invited to propose new Individual Members; these proposals should reach the General Secretary no later than five months before the General Assembly. Late proposals will normally not be taken into consideration. Proposals from Presidents of Divisions will be submitted by the IAU 3 months before the General Assembly to the relevant National Members, if any, who may add the person(s) in question to their own list of proposals.

6. National Members are urged to propose the deletion of Individual Members who are no longer active in astronomy. Such proposals should be done in consultation with the member concerned and submitted to the General Secretary at the same time as proposals for new Individual Members. National Members shall also promptly inform in writing the General Secretary of the death of any Individual Member represented by them.

7. Proposals for membership shall include the full name, date of birth, and nationality of the candidate, postal and electronic addresses, the University, year, and subject of the M.Sc./Ph.D. or equivalent degree, current affiliation and occupation, the proposing National Member or Division, the Division(s) and/or Commission(s) which the candidate wishes to join, and any further detail that might be relevant.

8. The General Secretary shall submit all proposals for Individual Membership to the Nominating Committee for review, consolidated into two lists:

8.a. one containing all proposals by National Members; and

8.b. one containing all proposals by Presidents of Divisions, in accordance with Bye-Law § 4.

9. The Nominating Committee shall examine all proposals for individual membership and advise the Executive Committee on their approval or rejection.

10. In exceptional cases, the Executive Committee may, on the proposal of a Division, admit an Individual Member between General Assemblies. Such proposals shall be prepared as described above (*cf.*, § 2 and submitted with a justification of the request to bypass the normal procedure. The Executive Committee shall consult the Nominating Committee or relevant National Member, if any, if a standing NC of manageable size has not been appointed, before approving such exceptions to the normal procedure.

11. The General Secretary shall maintain updated lists of all National and Individual Members, and shall make these available to the membership in electronic form. The procedures for dissemination of these lists shall be set by the General Secretary in such a way that the membership directory be properly protected against unintended or inappropriate use.

IV. RESOLUTIONS OF THE UNION

Traditionally, the decisions and recommendations of the Union on scientific and organizational matters of general and significant importance are expressed in the Resolutions of the General Assembly. In order for such Resolutions to carry appropriate weight in the international community, they should address astronomical matters of significant impact on the international society, or matters of international policy of significant importance for the international astronomical community as a whole. Resolutions should be adopted by the General Assembly only after thorough preparation by the relevant bodies of the Union. The proposed resolution text should be essentially complete before the beginning of the General Assembly, to allow Individual and National Members time to study them before the vote by the General Assembly. The following procedures have been designed to accomplish this:

12. Proposals for Resolutions to be adopted by the General Assembly may be submitted by a National Member, by the Executive Committee, a Division, a Commission or a Working Group. They should address specific issues of the nature described above, define the objectives to be achieved, and describe the action(s) to be taken by the Officers, Executive Committee, or Divisions to achieve these objectives.

13. Resolutions proposed for vote by the General Assembly fall in two categories:

13.a. Resolutions with implications for the budget of the Union; or

13.b. Resolutions without financial implications.

Proposals for Resolutions should be submitted on standard forms appropriate for each type, which are available from the IAU Secretariat. They may be submitted in either English or French and will be discussed and voted upon in the original language. Upon submission each proposed Resolution is posted on the Union web site. When the approved Resolutions are published, a translation to the other language will be provided.

14. Resolutions of type A must be submitted to the General Secretary at least nine months before the General Assembly in order to be taken into account in the budget for the impending triennium. Resolutions of type B must be submitted to the General Secretary three months before the beginning of the General Assembly. The Executive Committee may decide to accept late proposals in exceptional circumstances.

15. Before being submitted to the vote of the General Assembly, proposed Resolutions will be examined by the Executive Committee, Division Presidents, and by a Resolutions Committee, which is nominated by the Executive Committee. The Resolutions Committee consists of at least three members of the Union, one of whom should be a member of the Executive Committee, and one of whom should be a continuing member from the previous triennium. It is appointed by the General Assembly during its final session and remains in office until the end of the following General Assembly.

16. The Resolutions Committee will examine the content, wording, and implications of all proposed Resolutions promptly after their submission. In particular, it will address the following points:

i. suitability of the subject for an IAU Resolution;

ii. correct and unambiguous wording;

iii. consistency with previous IAU Resolutions.

The Resolutions Committee may refer a Resolution back to the proposers for revision or withdrawal if it perceives significant problems with the text, but can neither withdraw nor modify its substance on its own initiative. The Resolutions Committee advises the Executive Committee whether the subject of a proposed Resolution is primarily a matter of policy or primarily scientific. The Resolutions Committee will also notify the Executive Committee of any perceived problems with the substance of a proposed Resolution.

17. Proposed Resolutions shall be published in the General Assembly Newspaper before the final session.

18. The Executive Committee will examine the substance and implications of all proposed Resolutions. The Resolutions Committee presents the proposals during the second session of the General Assembly with its own recommendations, and those of the Executive Committee, if any, for their approval or rejection. A representative of the body proposing the Resolution is given the opportunity to defend the Resolution in front of the General Assembly, after which a general discussion takes place before the vote.

V. EXTERNAL RELATIONS

Contacts with other international scientific organizations, national and international public bodies, the media, and the public are increasing in extent and importance. In order to maintain coherent overall policies in matters of international significance, clear delegation of authority is required. Part of this is accomplished by having the Union's representatives in other scientific organization appointed by the Executive Committee (cf. Bye-Laws § 15). Supplementary rules are given in the following.

19. Representatives of the Union in other scientific organizations are appointed by the Executive Committee upon consultation with the Division(s) in the field(s) concerned.

20. In other international organizations, e.g., in the United Nations Organization, the Union is normally represented by the General Secretary or Assistant General Secretary, as decided by the Executive Committee.

21. The Union strongly supports the policies of ICSU: The International Council for Science, as regards the freedom and universality of science. Participants in IAU sponsored activities who feel that they may have been subjected to discrimination are urged, first, to seek clarification of the origin of the incident, which may have been due to misunderstandings or to the cultural differences encountered in an international environment. Should these attempts not prove successful, contact should be made with the General Secretary who will take steps to resolve the issue.

22. Public statements that are attributed to the Union as a whole can be made only by the President, the General Secretary, or the Executive Committee. The General Secretary may, in consultation with the relevant Division, appoint Individual Members of the Union with special expertise in questions that attract the attention of media and the general public as IAU spokespersons on specific matters.

VI. FINANCIAL MATTERS

The great majority of the Union's financial resources are provided by the National Members, as laid out in the Statutes § XII and Bye-Laws § VII. The purpose of the procedures described below is twofold: (i) to provide the best possible advice and guidance to the General Secretary and Executive Committee in planning and managing the Unions financial affairs, and (ii) to provide National Members with a mechanism for continuing input to and oversight over these affairs between and in preparation for the General Assemblies. The procedures adopted to accomplish this are as follows:

23. At the end of each of its final sessions the General Assembly, at the proposal of the Finance Committee, appoints a Finance Sub-Committee of 5-6 members, including a Chair. The Finance Sub-Committee remains in office until the end of the next General Assembly (*cf.* Statutes § 13.d.) and cooperates with the National Members, Finance Committee, Executive Committee and General Secretary in the following manner:

23.a. After the end of each year the General Secretary provides the Finance Sub-Committee with the auditor's report and summary reports covering the financial performance of the Union as compared to the approved budget, together with an analysis of any significant departures, and information on any Executive Committee approvals for budget changes. Upon receipt of the above reports from the General Secretary, the Finance Sub-Committee examines the accounts of the Union in the light of the corresponding budget and any relevant later decisions by the Executive Committee. It reports its findings and recommendations to the Executive Committee at its next meeting. The Finance Sub-Committee may at any time, at the request of the Executive Committee or the General Secretary, or on its own initiative, advise the General Secretary and/or the Executive Committee on any aspect of the Union's financial affairs.

23.b. Early in the year preceding that of a General Assembly, the General Secretary shall submit a preliminary draft of the budget for the next triennium to the Finance Sub-Committee for review. The draft budget, updated as appropriate following the comments and advice of the Finance Sub-Committee, is submitted to the Executive Committee for approval at the EC meeting in the year preceding that of a General Assembly, together with the report of the Finance Sub-Committee. The final budget proposal as approved by the Executive Committee is subsequently submitted to the National Members with a statement of the views of the Finance Sub-Committee on the proposal.

23.c. Before the first session of a General Assembly, the Finance Sub-Committee shall submit a report to the Executive Committee and the Finance Committee on its findings and recommendations concerning the development of the Union's finances over the preceding triennium. The Finance Sub-Committee shall also prepare a slate of candidates for the composition of the Finance Sub-Committee in the next triennium, preferably providing a balance between new and continuing members.

23.d. The report of the Finance Sub-Committee, together with the audited detailed accounts and the earlier comments on the proposed budget for the next triennium, will form a suitable basis for the discussions of the Finance Committee leading to its recommendations to the General Assembly concerning the approval of the accounts for the previous triennium and the budget for the next triennium, as well as the new Finance Sub-Committee to serve during that period.

24. The General Secretary is responsible for managing the Union's financial affairs according to the approved budget (*cf.* Bye-Laws § 28).

24.a. In response to changing circumstances, the Executive Committee may approve such specific changes to the annual budgets as are consistent with the intentions of the General Assembly when the budget was approved.

24.b. Unless authorized by the Executive Committee, the General Secretary shall not approve expenses exceeding the approved budget by more than 10% of any corresponding major budget line or 2% of the total budget in a given year, whichever is larger. This restriction does not apply in cases when external funding has been provided for a specific purpose, e.g. travel grants to a General Assembly.

24.c. Unless specifically identified in the approved budget, contractual commitments in excess of 50,000 CHF or with performance terms in excess of 3 years require the additional approval of the Union President.

25. The National Representatives, in approving the accounts for the preceding triennium, discharge the General Secretary and the Executive Committee of liability for the period in question.

VII. RULES OF PROCEDURE FOR THE EXECUTIVE COMMITTEE

The Executive Committee must respond quickly to events and thus it needs to be able to have discussions and take decisions on a relatively short timescale and without meeting in person. The following rules, as required by Bye-Law 14, are designed to facilitate EC action in a flexible manner, while giving such decisions the same legal status as those taken at actual physical meetings.

26. The Executive Committee should meet in person at least once per year. In years of a General Assembly it should meet in conjunction with and at the venue of the General Assembly. In other years, the Executive Committee decides on the date and venue of its regular meeting. The meetings of the Executive Committee are chaired by the President or, if the President is unavailable, by the President Elect or by one of the Vice-Presidents chosen by the Executive Committee to serve in this capacity.

27. The date and venue of the next regular meeting of the Executive Committee shall be communicated at least six months in advance to all its members and the Advisors, and to all Presidents of Divisions. Any of these persons may then propose items for inclusion in the Draft Agenda of the meeting before the date posted on the IAU Deadlines page.

28. Outgoing and incoming Presidents of Divisions are invited to attend all non-confidential sessions of the outgoing and incoming Executive Committee, respectively, in the years of a General Assembly. The President will invite Presidents of Divisions to attend the meetings of the Executive Committee in the years preceding a General Assembly. Division Presidents attend these sessions with speaking right, but do not participate in any voting.

29. The Executive Committee may take official decisions if at least half of its members participate in the discussion and vote on an issue. Decisions are taken by a simple majority of the votes cast. In case of an equal division of votes, the Chair's vote decides the issue. Members who are unable to attend may, by written or electronic correspondence with the President before the meeting, authorize another member to vote on her/his behalf or submit valid votes on specific issues.

30. If events arise that require action from the Executive Committee between its regular meetings, the Committee may meet by teleconference or by such electronic or other means of correspondence as it may decide. In such cases, the Officers shall submit a clear description of the issue at hand, with a deadline for reactions. If the Officers propose a specific decision on the issue, the decision shall be considered as approved unless a majority of members vote against it by the specified deadline. In case of a delay in communication, or if the available information is considered insufficient for a decision, the deadline shall be extended or the decision deferred until a later meeting at the request of at least two members of the Executive Committee.

31. The Officers of the IAU should, as a rule, meet once a year at the IAU Secretariat in order to discuss all matters of importance to the Union. The other members of the Executive Committee and the Division Presidents shall be invited to submit items for discussion at the Officers' Meetings and shall receive brief minutes of these Meetings.

32. Should any member of the Executive Committee have a conflict of interest on a matter before the Executive Committee that might compromise their ability to act in the best interests of the Union, they shall declare their conflict of interest. The remaining members of the Executive Committee determine the appropriate level of participation in such issues for members with a potential conflict of interest.

VIII. SCIENTIFIC MEETINGS AND PUBLICATIONS

Meetings and their proceedings remain a major part of the activities of the Union. The purpose of scientific meetings is to provide a forum for the development and dissemination of new ideas, and the proceedings are a written record of what transpired.

33. The General Secretary shall publish in the Transactions and on the IAU web site rules for scientific meetings organized or sponsored by the Union.

34. The proceedings of the General Assemblies and other scientific meetings organized or sponsored by the Union shall, as a rule, be published. To ensure prompt publication of Proceedings of IAU Symposia and Colloquia, the Assistant General Secretary is authorized to oversee the production of the material for the Proceedings. The Union shall publish an Information Bulletin at regular intervals to keep Members informed of current and future events in the Union. The Union shall also publish a more informal, periodic Newsletter which it distributes electronically to its members. The Executive Committee decides on the scope, format, and production policies for such publications, with due regard to the need for prompt publication of new scientific results and to the financial implications for the Union. At the present time, publications are in printed and in electronic form.

35. Divisions, Commissions, and Working Groups shall, with the approval of the Executive Committee, be encouraged to issue Newsletters or similar publications addressing issues within the scope of their activity.

IX. TERMS OF REFERENCE FOR DIVISIONS

The Divisions are the scientific backbone of the IAU. They have a main responsibility for monitoring the scientific and international development of astronomy within their subject areas, and for ensuring that the IAU will address the most significant issues of the time with maximum foresight, enterprising spirit, and scientific judgment. To fulfill this role IAU Divisions should maintain a balance between innovation and continuity. The following standard Terms of Reference have been drafted to facilitate that process, within the rules laid down in the Statutes § X and the Bye-Laws § V.

36. As specified in Bye-Law 18, the scientific affairs of the Division are conducted by an Organizing Committee of up to 12 members of the Division, headed by the Division President, Vice-President, and Secretary. Thus, all significant decisions of the Division require the approval of the Organizing Committee, and the President and Vice-President are responsible for organizing the work of the Committee so that its members are consulted in a timely manner. Contact information for the members of the Organizing Committee shall be maintained at the Division web site.

37. Individual Members of the Union are admitted to membership in a Division by its Organizing Committee (cf. Bye-Laws 18). Individual Members active within the field of activity of the Division and interested in contributing to its development should contact the Division Secretary, who will consult the Organizing Committee on the admission of the candidates.

37.a. The Division Secretary shall maintain a list of Division members for ready consultation by the community, including their Commission memberships if any. Updates to the list shall be provided to the IAU Secretariat on a running basis.

37.b. Members may resign from a Division by so informing the Division Secretary.

38. The effectiveness of the Division relies strongly on the scientific stature and dedication of its President and Vice-President to the mission of the Division. The Executive Committee, in proposing new Division Presidents and Vice-Presidents for election by the General Assembly, will rely heavily on the recommendations of the Organizing Committee of the Division. In order to prepare a strong slate of candidates for these positions, and for the succession on the Organizing Committee itself, the following procedures apply:

38.a. Candidates are proposed and selected from the membership of the Division on the basis of their qualifications, experience, and stature in the fields covered by the Division. In addition, the Organizing Committees should have proper gender balance and broad geographical representation.

38.b. At least six months before a General Assembly, the Organizing Committee submits to the membership of the Division a list of candidates for President, Vice-President (for which there should be at least two persons willing to serve), Secretary, and the Organizing Committee for the next triennium. The Organizing Committee requests nominations from the membership in preparing this list, and accepts additional nominations within a month after its submission to the membership. The Vice-President is normally nominated to succeed the President. The outgoing President participates in the deliberations of the new Organizing Committee in an advisory capacity.

38.c. If more names are proposed than there are positions to be filled on the new Organizing Committee, the outgoing Organizing Committee devises the procedure by which the requisite number of candidates is elected by the membership. The resulting list is communicated to the General Secretary at least two months before the General Assembly. The General Secretary may allow any outstanding issues to be resolved at the business meeting of the Division during the General Assembly.

38.d. A member of the Organizing Committee normally serves a maximum of two terms, unless elected Vice-President of the Division in her/his second term. Presidents may serve for only one term.

38.e. The Organizing Committee decides on the procedures for designating the Division Secretary, who maintains the web site, records of the business and membership of the Division, and other rules for conducting its business by physical meetings or by correspondence.

39. A key responsibility of the Organizing Committee is to maintain an internal organization of Commissions and Working Groups in the Division which is conducive to the fulfillment of its mission. The Organizing Committee shall take the following steps to accomplish this task in a timely and effective manner:

39.a. Within the first year after a General Assembly – with the business meeting of the Commission at the General Assembly itself as a natural starting point – the Organizing Committee shall discuss with its Commissions, and within the Organizing Committee itself, if changes in its Commission and Working Group structure may enable it to accomplish its mission better in the future. As a rule, Working Groups should be created (following the rules in Bye-Law § 21 and Bye-Law § 23) for new activities that are either of a known, finite duration or are exploratory in nature. If experience, possibly from an existing Working Group, indicates that a major section of the Division's activities require a coordinating body for a longer period (a decade or more), the creation of a new Commission may be in order.

39.b. Whenever the Organizing Committee is satisfied that the creation of a new Working Group or Commission is well motivated, it may take immediate action as specified in Bye-Law § 21 or Bye-Law § 23. In any case, the Organizing Committee submits its complete proposal for the continuation, discontinuation, or merger of its Commissions and Working Groups to the General Secretary at least three months before the next General Assembly.

39.c. The President and Organizing Committee maintain frequent contacts with the other IAU Divisions to ensure that any newly emerging or interdisciplinary matters are addressed appropriately and effectively.

X. TERMS OF REFERENCE FOR COMMISSIONS

The role of the Commissions is to organize the work of the Union in specialized subsets of the fields of their parent Division(s), when the corresponding activity is judged to be of considerable significance over times of a decade or more. Thus, new Commissions may be created when fields emerge that are clearly in sustained long-term development and where the Union may play a significant role in promoting this development at the international level. Similarly, Commissions may be discontinued when their work can be accomplished effectively by the parent Division. In keeping with the many-sided activities of the Union, Commissions may have purely scientific as well as more organizational and/or interdisciplinary fields. They will normally belong and report to one of the IAU Divisions, but may be common to two or more Divisions. The following rules apply if a Division has more than one Commission.

40. The activities of a Commission are directed by an Organizing Committee of 4-8 members of the Commission, headed by a Commission President and Vice-President (*cf.* Bye-Laws § 22). A member of the Organizing Committee normally serves a maximum of two terms, unless elected Vice-President of the Commission in her/his second term. Presidents may serve for only one term. All members of the Organizing Committee are expected to be active in this task, and are to be consulted on all significant actions of the Commission. The Organizing Committee appoints a Commission Secretary who maintains the records of the membership and activities of the Commission in co-operation with the Division Secretary and the IAU Secretariat. Contact information for the members of the Organizing Committee shall be maintained at the Commission web site.

41. Individual Members of the Union, who are active in the field of the Commission and wish to contribute to its progress, are admitted as members of the Commission by the

Organizing Committee. Interested Members should contact the Commission Secretary, who will bring the request before the Organizing Committee for decision. Members may resign from the Commission by notifying the Commission Secretary. Before each General Assembly, the Organizing Committee may also decide to terminate the Commission membership of persons who have not been active in the work of the Commission; the individuals concerned shall be informed of such planned action before it is put into effect. The Commission Secretary will report all changes in the Commission membership to the Division Secretary and the IAU Secretariat.

42. At least six months before a General Assembly, the Organizing Committee submits to the membership of the Commission a list of candidates for President, Vice-President (for which there should be the names of two persons willing to serve), and the Organizing Committee for the next triennium. The Organizing Committee requests nominations from the membership in preparing this list, and accepts additional nominations within a month after its submission to the membership. The Vice-President is normally nominated to succeed the President. The outgoing President participates in the deliberations of the new Organizing Committee in an advisory capacity. If more names are proposed than available elective positions, the outgoing Organizing Committee devises the procedure by which the requisite number of candidates is elected by the membership. The resulting list is submitted to the Organizing Committee of the parent Division(s) for approval before the end of the General Assembly. Members of the Organizing Committee normally serve a maximum of two terms, unless elected Vice-President of the Commission. Presidents may serve for only one term.

43. At least six months before each General Assembly, the Organizing Committee shall submit to the parent Division(s) a report on its activities during the past triennium, with its recommendation as to whether the Commission should be continued for another three years, or merged with one or more other Commissions, or discontinued. If a continuation is proposed, a plan for the activities of the next triennium should be presented, including those of any Working Groups which the Commission proposes to maintain during that period.

44. The Organizing Committee decides its own rules for the conduct of its business by physical meetings or (electronic) correspondence. Such rules require approval by the Organizing Committee of the parent Division(s).

Transactions IAU, Volume XXVIB
Proc. IAU XXVI General Assembly, August 2006
Karel A. van der Hucht, ed.

doi:10.1017/S1743921308024332

CHAPTER VII

RULES AND GUIDELINES FOR IAU SCIENTIFIC MEETINGS

1. Introduction

The program of IAU scientific meetings is one of the most important means by which the IAU pursues its goal of promoting astronomy through international collaboration. A large fraction of the Union's budget is devoted to the support of the IAU scientific meetings. The Executive Committee (EC) places great emphasis on maintaining high scientific standards, coverage of a balanced spectrum of topics, and an appropriately international flavor for the programme of IAU Meetings. In that respect, the ICSU rules on non-discrimination in the access of qualified scientists from all parts of the world to any IAU meeting apply.

Because of limited available funds, the number of meetings per year that the IAU can sponsor financially is restricted to nine IAU Symposia (CHF 25,000 each); one Regional IAU Meeting (LARIM and APRIM; CHF 25,000 each) in the years between General Assemblies (GAs); and one to two Co-Sponsored Meetings (CHF 3,000 each). Accordingly, not all meeting proposals worthy of support can be awarded IAU sponsorship.

The IAU Colloquium Series has been terminated after IAU Colloquium No. 200 (October 2005), to the benefit of the IAU Symposium Series.

Contacts with meeting organizers during the preparation and conduct of IAU scientific meetings are maintained by the IAU Assistant General Secretary (AGS) and the President of the Coordinating Division (see below).

2. IAU Symposia

The IAU Symposium Series is the scientific flagship of the IAU. Symposia are organized on suitably broad, yet well-defined scientific themes of considerable general interest, and normally last 4 to 5 days. They are intended to significantly advance the field, by seeking answers to current key questions and/or clarify emerging concepts in invited reviews, contributed papers and poster papers. Therefore, their programs should consist of reviews and previews, and should provide ample time for discussion.

IAU Symposium proposals for a certain calendar year, backed by a coordinating IAU Division and supported by a reasonable number of supporting IAU Divisions, IAU Commissions, and IAU Working/Program Groups, are to be submitted to the IAU Proposal Web Server (see below), before December of the year two years before the intended Symposium.

The scientific merit of each IAU Symposium proposal will be evaluated by the twelve IAU Division Presidents (DPs), taking into consideration comments and advice received from the Organizing Committees of IAU Divisions, IAU Commissions and IAU Working/Program Groups. The DPs will provide their recommendations for selection to the IAU EC. The EC will decide on and announce the final selection of the nine IAU Symposia to be held in a certain year, in Spring of the year before.

IAU Symposium Proceedings are published in the IAU Proceedings' Series by the IAU Publisher, since 2004 Cambridge University Press (CUP). To date, CUP has published the Proceedings of IAU Colloquia Nos. 195 through 199, and of IAU Symposia Nos. 222 through 231.

In the year of a GA, six of the nine IAU Symposia of that year are scheduled as GA Symposia, within the scientific program of the GA and held at the GA venue. Each GA Symposium normally lasts 3.5 days. For GA Symposia, the GA Local Organizing Committee (LOC) handles the local organization. The General Secretary (GS), in consultation with the organizers of the individual GA Symposia, coordinates the financial support to be allocated to each of the GA Symposia.

In the year of a GA, the three IAU Symposia outside that GA should, as a rule, be scheduled no closer than three months before or after the dates of that GA.

URL of IAU Proposal Web Server: `<http://solarphys.uio.no/IAU/>` .

2.1. SELECTION CRITERIA FOR IAU SPONSORSHIP OF SYMPOSIA

The following guidelines for the selection of meetings for IAU sponsorship should be observed by prospective proposers:

(*a*) An IAU Symposium should have a well-defined and scientifically relevant theme, be scheduled at a propitious time for significant progress in the field, and be of interest to young researchers as well as senior experts.

(*b*) Where the IAU embraces all fields in astronomy, the proposed IAU Symposium program should maintain a broad and balanced scope and cover the main active fields at appropriate intervals. Accordingly, even scientifically strong proposals in the same or largely overlapping fields can only be approved at some intervals. While some themes have developed series of IAU Symposia with intervals of 3-5 years, approval is not automatically guaranteed, since each proposal will be judged on its own scientific merits.

(*c*) Scientific programmes of proposed IAU Symposia should be well balanced, to be demonstrated by the proposed draft program and the proposed draft list of key speakers.

d) Given the international nature of the Union, IAU Symposia are by definition internationally oriented. This requires a well-balanced geographical and gender distribution of both the proposed Scientific Organizing Committee (SOC) and the proposed key speakers. Normally, substantially less than half of the SOC membership and of the key speakers should come from any single country. The SOC membership should reflect in a balanced way the current activity in the field.

(*e*) The statement that the ICSU rules on non-discrimination in the access to the meeting should be strictly observed, and MUST be explicitly confirmed before a proposal will receive final approval by the EC. A summary of the measures taken to ensure this should be given, and the signatures of both the SOC and LOC Chairpersons are required.

(*f*) Presentation of scientific results in IAU Symposia is by invitation of the SOC chair person. Suitably qualified scientists working in the field may seek invitations. It is the policy of the IAU to promote the full participation of astronomers worldwide in its Symposium program. It is essential that no restriction based on gender, race, color, nationality, religious or political affiliation be imposed on the full participation of all bona fide scientists in any aspect of the organization and conduct of IAU Symposia, either by its organizers, or by the authorities of the host country. Approval of a proposal for an IAU meeting requires explicit guarantees that this principle will be respected.

(*g*) In association with some IAU meetings, educational activities may be organized, like International Schools for Young Astronomers (ISYAs) and Teachers' Workshops. By taking advantage of the presence of many expert national and foreign scientists, one- or two-day events may be organized for the benefit of university and high-school astronomy educators in or near the country hosting the meeting. Such initiatives have generally been well received and successful. While the scientific quality of the proposed Symposium will remain the primary selection criterion for IAU sponsorship, a good parallel educational program will certainly add to the overall merit of a proposal.

(*h*) The scientific merit of each IAU Symposium proposal will be evaluated by the IAU Division Presidents (DPs), taking into consideration comments and advice received from the Organizing Committees of IAU Divisions, IAU Commissions and IAU Working/Program Groups.

In Spring of the year before that of the intended Symposium, the IAU EC, taking into account the recommendations from the DPs, decides on the final selection of the nine Symposia. The decision of the EC on each Symposium, including any conditions to be fulfilled before final approval, are communicated to the proposers by the AGS. A letter of award will be issued to them, accompanied by an official form listing the essential facts of the meeting as approved by the EC. Any revision of the details recorded on this form will require approval by the AGS.

2.2. PROCEDURES FOR PROPOSAL PREPARATION FOR IAU SYMPOSIA

2.2.1. GENERALITIES

Normally, the initiative to propose a scientific meeting for IAU sponsorship originates from a group of scientists in a certain field. In collaboration with colleagues worldwide, they should prepare a draft scientific program and nominations for the members of a SOC, who will be responsible for the scientific aspects of the meeting from its inception to its conclusion. Responsibility for the preparation and timely submission of the final proposal rests with the chair person of the proposed SOC.

Prospective meeting organizers should contact the AGS well in advance of their intended proposal submission by sending a Letter of Intent (LoI, see below).

Application procedures have been designed so as to ensure that the information necessary for the evaluation of the proposals by the IAU Division Presidents and the IAU EC is complete and in an uniform format, that allows direct comparison between proposals as far as possible. Therefore, proposals, with all entries properly answered, have to be submitted electronically to the IAU Proposal Web Server (see above).

2.2.2. LETTER OF INTENT

Before submitting a proposal for an IAU Symposium, proposers must send at their earliest convenience a Letter of Intent (LoI) to the AGS, with a copy to the President of the desired Coordinating IAU Division covering the scientific field of the meeting, stating:
(*a*) the title of the intended IAU Symposium;
(*b*) the full name(s) of the proposed SOC chair person(s);
(*c*) the desired Coordinating IAU Division for the intended IAU Symposium;
(*d*) the venue and the preferred dates; and
(*e*) a short list of topics (up to 10).

A list of received Letters of Intent will be posted and updated on the IAU web site, informing prospective proposers of other existing plans for IAU Symposium proposals. This, in order to avoid unnecessary competition between proposals, and to stimulate possibly collaboration between otherwise competing groups.

2.2.3. TOPIC AND TITLE

The title of a Symposium should state the topic of the meeting as concisely and succinctly as possible. Long and detailed titles do not catch the eye, and are cumbersome for the announcement of a Symposium as well as on the cover of its subsequent Proceedings.

Therefore, as a rule, Symposium titles should be no longer than 10 words (or 70 characters including spaces) in total.

Any change of title after an IAU Symposium has been accepted requires the prior approval of the AGS.

2.2.4. COORDINATING IAU DIVISION; SUPPORTING IAU DIVISIONS, IAU COMMISSIONS, IAU WORKING/PROGRAM GROUPS

An IAU Symposium can be proposed by individual astronomers (preferably members of the IAU), by an IAU Working/Program Group, or by an IAU Commission.

An IAU Division should accept the coordinating responsibilities for an IAU Symposium proposal as Coordinating Division.

When other Divisions, Commissions, and or Working/Program Groups, are listed as supporting a proposal, a report of the communication between the proposers and the above should be submitted together with that proposal.

Proposals must be electronically submitted to the IAU Proposal Web Server (see above) before the posted deadline. Normally, this deadline will be in December, two years before the year of the proposed Symposium.

2.2.5. SCIENTIFIC ORGANIZING COMMITTEE

The composition of the proposed SOC is a key element in assessing the scientific value of a proposal. The SOC of a Symposium has the overall responsibility for its scientific standards and should make sure to cover the principal topics of the field to be covered.

The SOC should normally not be larger than sixteen persons and should represent an optimum scientific, geographic and gender distribution. Therefore, the composition of the SOC should reflect in a positive way the intent of the ICSU Statement on Freedom in the Conduct of Science. Normally, any one institution should not be represented on the SOC by more than one person. It is customary, but not required, that SOC members are members of the IAU.

The SOC chair person(s) and its members are appointed by the IAU EC, as part of the approval process.

The SOC exercises responsibility in five main aspects:

– before the Symposium:

(*a*) The definition of the scientific program of the Symposium, including the choice and distribution of topics for individual sessions, and the selection of invited reviews, invited papers, contributed papers, and poster papers;

(*b*) The choice of key speakers for invited reviews;

(*c*) Provide in the proposal a list of about 10 preliminary scientific programme topics (for announcement in the IAU Information Bulletin);

(*d*) Provide a list of individuals qualifying for IAU Travel Grants, with amounts recommended (for criteria, see below). That list has to be submitted for approval to the AGS at least FIVE months before the start of the Symposium;

– after the Symposium:

(*e*) Within one month after a Symposium, the SOC chair person will send to the AGS the Post Meeting Report of the Symposium, including

(*e*.1) a copy of the final scientific program;

(*e*.2) a list of participants;

(*e*.3) a list of recipients of IAU grants, stating amount and country;

(*e*.4) receipts signed by the recipients of IAU Grants (this does not apply to GA Symposia);

(*e*.5) a report to the IAU EC on the scientific highlights of the meeting (1 - 2 pages).

The Post Meeting Report Form is available at:
`<http://www.iau.org/fileadmin/content/pdfs/PostMeet.pdf>`.

A compilation of the 2005 IAU Post Meeting Reports is available at:
`<http://www.iau.org/fileadmin/content/pdfs/Post_MR05.pdf>`.

Any change of SOC membership after a Symposium has been accepted requires the prior approval of the IAU AGS.

2.2.6. LOCAL ORGANIZING COMMITTEE

The LOC, to be identified in the proposal, is responsible for all aspect of the local arrangements associated with the Symposium. Those tasks include booking and preparation of meeting rooms, provisions for modern audio-visual facilities, for coffee and tea breaks, arranging for necessary transportation for meeting participants, ensuring that accommodation within reasonable price levels is available, and providing assistance to meeting participants with their booking. In addition, the LOC should prepare and schedule social events.

2.2.7. EDITORS OF PROCEEDINGS

It speaks for itself that the success of a Symposium and its Proceedings depends in the first place on arranging for the best possible scientific programme and on selecting the best possible speakers.

It is of paramount importance that the Proceedings of an IAU Symposium will be published timely, as a valuable record of the event for future reference. Arrangements for Authors and Editors for the publication of Proceedings of IAU Symposia are summarized in the `<README.txt>` files of the ftp site
`<ftp://ftp.sron.nl/pub/karelh/UPLOADS/IAU-CUP.dir/Symposium.dir/>`.

Full names of the proposed Editors must be given in the proposal. One of the proposed Editors should be marked as Chief Editor, with prime responsibility for the contacts with the IAU AGS and with the IAU Publisher, Cambridge University Press.

In the contract between the IAU and CUP, it is stipulated that the Proceedings of an IAU Symposium will be published within six months after that Symposium. Since CUP needs three months for its processing and publishing of a complete Proceedings' manuscript, Editors have the first three months after their Symposium to complete their editing task. This requires that all Authors have to deliver their completed manuscripts to the Chief Editor before or during the Symposium. Authors are allowed to submit a revised version of their manuscript to the Chief Editor within four weeks after the Symposium.

Thus, Editors are committed to submit the final complete manuscript of the Proceedings of their IAU Symposium to CUP within three months after their IAU Symposium. Editors should realize that editing an IAU Symposium Proceedings volume can be a full time job for three months.

Any change of Editors, after a Symposium has been accepted, requires the prior approval of the IAU AGS.

2.2.7.1. IAU EDITORIAL BOARD

In order to ensure the quality of the IAU Symposium Proceedings Series, to strengthen the working relation between the IAU Proceedings Editors and the IAU AGS, and to provide a platform for communication, the EC has established the IAU Editorial Board (EB) for the Series.

The EB serves as a communication and support platform for all Editors of IAU Symposium Proceedings, where they can exchange experience and ask for advice, whenever necessary, in their efforts to ensure that all papers published in their Proceedings are of the highest quality and that their Proceedings are published on time, i.e., within six months after their Symposium.

Members of the IAU EB for a certain year are: (a) all Chief-Editors of the Proceedings of IAU Symposia of that year (working members); (b) the IAU AGS (chair); and (c) the IAU GS plus three or four members appointed by the EC for a period of at least three years (advisory members).

The constitution of the EB of a certain year will be listed in all nine IAU Symposium Proceedings of that year. EB members of a certain year will receive copies of all nine IAU Symposium Proceedings of that year.

2.2.8. REGISTRATION FEE

2.2.8.1. REGISTRATION FEE FOR IAU SYMPOSIA OUTSIDE GAs

Keen efforts should be made to keep the Symposium registration fee low, to make the Symposium affordable to all. Such efforts should include the use of low-cost meeting facilities and finding local sponsorship. The acceptable level of a registration fee will depend on local circumstances, and proposers should carefully specify what services the registration fee will cover. Including the price of the Proceedings (2006 participant's price: US$ 75.-), a registration fee of US$ 250.- (~ CHF 320.-) is the current upper limit for IAU Symposia. The EC may reject or withhold approval of otherwise valid proposals if the proposed registration fee is exorbitant.

2.2.8.2. REGISTRATION FEE FOR IAU SYMPOSIA INSIDE GAs

For IAU Symposia held as part of GAs, participants are required to pay the full registration fee for the GA.

2.2.9. VENUE AND ACCOMMODATION

The proposed venue should be reasonably accessible and affordable. The venue should have modern audio-visual facilities, and allow that all poster presentations will be on display during the whole duration of the Symposium, preferably in the tea/coffee break areas.

In order to enable interested and qualified colleagues from all countries around the world to attend a Symposium, affordable accommodation should be available. It is recognized that some hotels offer conference room, board and lodging all in one location, which is most favorable to the all scientific interactions of a Symposium. In case such resorts are expensive, efforts should be made to secure additional financial sponsoring, in order to keep costs down.

2.2.10. SUBMITTING THE PROPOSAL

Once the above requirements are observed, completed proposals for IAU Symposia should be submitted electronically to the IAU Proposal Web Server (see above), before the posted deadline.

3. Travel Grants for IAU Symposia

3.1. TRAVEL GRANTS FOR IAU SYMPOSIA OUTSIDE GAs

IAU Travel Grants are intended to cover in part expenses associated with attendance at the Symposium.

Participants of IAU Symposia may apply for an IAU Travel Grant, using the form available at <`http://www.iau.org/fileadmin/content/pdfs/GrantOutGa.pdf`>.

Proposals for the distribution of IAU Grants to individual participants are provided by the SOC and sent by the SOC chair person to the AGS for approval. IAU priority is to support qualified scientists to whom only limited means of support are available, e.g., colleagues from economically less privileged countries and young scientists. IAU support should carry significant weight in ensuring the participation of the selected beneficiaries, rather than adding comfort for colleagues whose attendance is already assured. In addition, a reasonable gender and geographical distribution is expected; normally, no more than 1/3 of the funds should be allocated to a single country or region.

Within these general guidelines, it is left to the judgment of the SOC how to formulate their proposal for IAU Travel Grant distribution, maintaining the overall scientific standard of the conference as the primary criterion. The recommendation for the distribution of the IAU Travel Grants shall be sent by the SOC Chair to the AGS, specifying for each person: name, nationality, full mailing and e-mail address, amount of proposed grant (in Swiss Francs: CHF), and title and nature of contribution (invited review, invited paper, contributed paper, poster paper). The recommendation of the SOC should reach the AGS no later than FIVE months before the Symposium. This deadline has been found necessary in order to ensure timely notification to beneficiaries and completion of visa formalities. After approval by the AGS of the IAU Travel Grant distribution proposal of the SOC, individual IAU Grant notification letters will be mailed to the recipients by the AGS, with a copy to the SOC Chair and LOC Chair.

The regular administrative procedure is that the LOC opens a bank account in the name of the meeting (or uses a bank account of its institute or university) to which the IAU Secretariat will transfer the allocated IAU Travel Grant. Individual IAU Travel Grants are paid to recipients upon arrival and registration at the Symposium.

3.2. TRAVEL GRANTS FOR IAU SYMPOSIA INSIDE GAs

In case of IAU Symposia inside GAs, IAU Travel Grants are intended to cover in part expenses associated attending the entire GA. For all IAU meetings inside the GAs, electronic applications for Grants are collected by the IAU Secretariat. The Application Form for an IAU Grant for attending a GA is available at
<http://www.iau.org/IAU_Grant_Application_Form_wit.102.0.html>.

Full completion of the form is mandatory, including submission of an Abstract if relevant. The deadline for receiving these applications will be such, that enough time is left to the IAU Secretariat to prepare relevant summaries of applications, to be sent to the SOCs of the different scientific meetings for ranking, before a final selection is made in consultation with the IAU GS and IAU AGS.

IAU Travel Grants for GAs will be allocated by the IAU Secretariat, upon recommendation of the organizers of the events. These grants are paid to recipients upon arrival and registration at the conference.

4. Regional IAU Meetings (RIMs)

The IAU sponsors two series of Regional IAU Meetings (RIMs): a series of triennial meetings in the Latin-American region (LARIM, since 1978), and a series of triennial meetings in the Asian-Pacific region (APRIM, since 1978). A past series of twelve European Regional IAU Meetings (1974-1990) has effectively been succeeded by the series of Joint European and National Astronomy Meetings (JENAM), under the auspices of the European Astronomical Society.

APRIMs and LARIMs are held at the invitation of a national astronomical society in, respectively, the Asian-Pacific region and the Latin-American region in years between GAs. Their purpose, in addition to the discussion of specific scientific topics, is to promote contacts between scientists in the regions concerned, especially young astronomers.

Therefore, both a much wider range of scientific topics, a larger SOC, and a larger total attendance are expected than for IAU Symposia. The Proceedings of RIMs are usually published by a regional publisher or in a regional astronomical publication series.

5. Web site for IAU meetings

As soon as a successful applicant has been informed by the AGS of the approval of her/his proposed IAU meeting, the SOC and LOC are kindly requested to create a web site for that IAU meeting, containing, *inter alia*, those parts of the above information which are essential for the participants of that IAU meeting. The URL of the web site should be communicated to the IAU AGS as soon as available.

Transactions IAU, Volume XXVIB
Proc. IAU XXVI General Assembly, August 2006
Karel A. van der Hucht, ed.

doi:10.1017/S1743921308024344

CHAPTER VIII

EXCUTIVE COMMITTEE WORKING GROUPS

EC Advisory Committee - *Hazards of Near-Earth Objects*
Chair: David Morrison (USA), <david.morrison@arc.nasa.gov>
Members: Richard P. Binzel (USA), Andrea Cusari (Italy), Andrea Milani (Italy), Timothy B. Spahr, Hans Rickman, and Donald K. Yeomans (USA)

EC/WG - *IAU General Assemblies*
Chair: Jan Palouš, <palous@ig.cas.cz>
Members: Richard N. Manchester (XXV GA Australia 2003), Daniela Lazzaro (XXVII GA Brazil 2009), Gang Zhao (XXVIII GA China 2012), and Karel A. van der Hucht (IAU GS, ex officio)

EC/WG - *International Year of Astronomy 2009*
Chair: Catherine J. Cesarsky (France), <catherine.cesarsky@cea.fr>
Members: Yolanda Berenguer (UNESCO, France), Ian F. Corbett (IAU, UK), Dennis Crabtree (Canada), Susana E. Deustua (USA), Kevin Govender (South Africa), Mary Kay M. Hemenway (USA), Robert Hill (UK), Douglas Isbell (USA), Norio Kaifu (Japan), Lars Lindberg Christensen (Denmark, ESA/ESO), Claus Madsen (Denmark, ESO), Ian E. Robson (UK), and Pedro Russo (IAU, Portugal)

EC/WG - *Woman in Astronomy*
Chair: Anne Green
Members: 70

EC/WG - *Publishing*
Chair: Michelle C. Storey (Australia)
Members: 1

EC/WG - *Future Large Scale Facilities*
Chair: Gerard F. Gilmore (UK)
Members: 1

Transactions IAU, Volume XXVIB
Proc. IAU XXVI General Assembly, August 2006
Karel A. van der Hucht, ed.

doi:10.1017/S1743921308024356

CHAPTER IX

IAU DIVISIONS, COMMISSIONS, WORKING GROUPS AND PROGRAM GROUPS 2006 - 2009

1. Introduction

As a result of the triennial reviews by the IAU Division presidents of their Commissions and Working Groups (Bye-Laws § V), the following structural changes have been proposed and approved by the Executive Committee.

2. Discontinued Commissions and Working Groups

Div.I/WG General Relativity in Celestial Mechanics, Astrometry & Metrology
(upgraded to Div.I/Commission 52 Relativity in Fundamental Astronomy)

Div.I/WG - *Re-definition of Universal Time Coordinated*

Div.I/WG - *Precession and the Ecliptic*

Div.I/WG - *Nomenclature for Fundamental Astronomy*

Div.I/WG - *Future Development of Ground-Based Astrometry*

Div.I-III/WG - *Near Earth Objects*
(upgraded to EC - Working Group Hazards of near-Earth Objects)

Div.I/Comm.8/WG - *Astrographic Catologue and Carte du Ciel*

Div.III/WG - *Extrasolar Planets*
(upgraded to Div.III/Commission 53 Extrasolar Planets)

Div.V/WG - *Variable and Binary Stars in Galaxies*

Div.V/Comm.42/WG - *Accretion Physics in Interacting Binaries*

Div.IX/Comm.9 - *Instrumentation and Techniques*

Div.IX/WG - *Optical and Infrared Interferometry*
(upgraded to Div.IX/Commission 54 Optical and Infrared Interferometry)

Div.XII/WG - *Communicating Astronomy with the Public*
(upgraded to Div.XII/Commission 55 Communicating Astronomy with the Public)

Div.XII/Comm.14/WG - *Atomic Spectra and Wavelengths*

Div.XII/Comm.14/WG - *Atomic Transition Probabilities*

Div.XII/Comm.14/WG - *Line Broadening*

Div.XII/Comm.14/WG - *Molecular Structure*

Div.XII/Comm.14/WG - *Molecular Reactions on Solid Surfaces*

Div.XII/Comm.14/WG - *Optical Properties of Solids*

Div.XII/Comm.14/WG - *Gas Phase Reactions*

3. New Commissions and Working Groups

Div.I/Comm.52 - *Relativity in Fundamental Astronomy*
Div.I/WG - *Second Realization of International Celestial Reference Frame*
Div.I/WG - *Numerical Standards in Fundamental Astronomy*
Div.I/WG - *Astrometry by Small Ground-Based Telescopes*
Div.III/Comm.53 - *Extrasolar Planets*
Div.VII/WG - *Galactic Center*
Div.IX/Comm.54 - *Optical and Infrared Interferometry*
Div.IX/WG - *Adaptive Optics*
Div.IX/WG - *Site Testing Instruments*
Div.IX/WG - *Large Telescope Projects*
Div.IX/WG - *Small Telescope Projects*
Div.XII/Comm.55 *Communicating Astronomy with the Public*
Div.XII/Comm.14/WG - *Atomic Data*
Div.XII/Comm.14/WG - *Molecular Data*
Div.XII/Comm.14/WG - *Solids and their Surfaces*

Transactions IAU, Volume XXVIB
Proc. IAU XXVI General Assembly, August 2006
Karel A. van der Hucht, ed.

doi:10.1017/S1743921308024368

DIVISION I - Fundamental Astronomy

URL: <astro.cas.cz/iaudiv1>

President
Jan Vondrák
Astronomical Institute
Academy of Sciences
Bocn II 1401
CZ-141 31 Praha 4
Czech Republic
Phone: +420 267 103043
Fax: +420 272 769023
Email: <vondrak@ig.cas.cz>

Vice-President
Dennis D. McCarthy
US Naval Observatory (USNO)
3450 Massachusetts Avenue NW
Washington, DC 20392-5420
USA
Phone: +1 202 762 1837
Fax: +1 202 762 1563
Email: <dmc@maia.usno.navy.mil>

Organizing Committee

Aleksander Brzezinski, P-C19 (Poland), Joseph A. Burns, P-C7 (USA), Pascale Defraigne, P-C31 (Belgium), Dafydd Wyn Evans, VP-C8 (UK) Toshio Fukushima, P-C4, PP (Japan), George H. Kaplan, VP-C4 (USA), Sergei A. Klioner, P-C52 (Germany), Zoran Knezevic, VP-C7 (Serbia) Irina I. Kumkova, P-C8 (Russia), Chopo Ma, VP-C19 (USA), Richard N. Manchester, VP-C31 (Australia), and Gérard Petit, VP-C52 (France)

PARTICIPATING COMMISSIONS AND COMMISSION WORKING GROUPS

Div.I/Comm.4 - Ephemerides

P: Toshio Fukushima, PP, (Japan), <Toshio.Fukushima@nao.ac.jp>
VP: George H. Kaplan, (USA), <gkaplan@usno.navy.mil>
OC: Jan Vondrák (Czech Rep.), Catherine Hohenkerk (UK), John A. Bangert (USA), Sean E. Urban (USA), Jean-Eudes Arlot (France), Martin Lara (Spain), and Elena V. Pitjeva (Russia)
URL: <http://iau-comm4.jpl.nasa.gov/>

Div.I/Comm.7 - Celestial Mechanics and Dynamical Astronomy

P: Joseph A. Burns (USA), <jab16@cornell.edu>
VP: Zoran Knežević (Serbia), <zoran@aob.bg.ac.yu>
S: David Vokrouhlicky (Czech Rep.), <vokrouhl@mbox.cesnet.cz>
OC: Evangelia Athanassoula (France), C. Beauge (Argentina), B. Erdi (Hungary), A. Maciejewski (Poland), R. Malhotra (USA), Andrea Milani, PP (Italia), A. Morbidelli (France), S.J. Peale (USA), and Ji-Lin Zhou (China)
URL: <http://copernico.dm.unipi.it/comm7>

Div.I/Comm.8 - Astrometry

P: Irina I. Kumkova (Russia), <kumkova@iperas.nw.ru>
VP: Dafydd Wyn Evans (UK), <dwe@ast.cam.ac.uk>
OC: Alexandre H. Andrei (Brazil), Alain Fresneau (France), Imants Platais (USA), Petre P. Popescu (Rumania), Ralf-Dieter Scholz (Germany), Mitsuru Soma (Japan), Norbert Zacharias (USA), and Zi Zhu (China)
URL: <http://www.ast.cam.ac.uk/iau_comm8/>

Div.I/Comm.8/WG - Densification of the Optical Reference Frame
Chair: Norbert Zacharias (USA), <nz@usno.navy.mil>
URL: <http://ad.usno.navy.mil/dens_wg/dens.html>

Div.I/Comm.19 - Rotation of the Earth
P: Aleksander Brzezinski (Poland), <alek@cbk.waw.pl>
VP: Chopo Ma (USA), <cma@virgo.gsfc.nasa.gov>
OC: Patrick Charlot (France), Pascale Defraigne (Belgium), Véronique Dehant (Belgium), Jean O. Dickey (USA), ChengLi Huang (China), Jean Souchay (France), and Jan Vondrák (Chech Republic)
URL: <http://www.astro.oma.be/IAU/>

Div.I/Comm.31 - Time
P: Pascale Defraigne (Belgium), <Pascale.Defraigne@oma.be>
VP: Richard N. Manchester (Australia), <Dick.Manchester@csiro.au>
OC: Mizuhiko Hosokawa (Japan), Sigfrido Leschiutta (Italia), Demetrios Matsakis (USA), Gérard Petit (France), and Zhai ZaoCheng (China)
URL: <http://www.astro.oma.be/IAU/COM31/>

Div.I/Comm.52 - Relativity in Fundamental Astronomy
P: Sergei A. Klioner (Germany), <Sergei.Klioner@tu-dresden.de>
VP: Gérard Petit (France), <gpetit@bipm.org>
OC: Viktor A. Brumberg (Russia), Nicole Capitaine (France), Agnès Fienga (France), Toshio Fukushima (Japan), Bernard R. Guinot (France), Cheng Huang (China), François Mignard (France), Kenneth P. Seidelmann (USA), Michael H. Soffel (Germany), and Patrick T. Wallace (UK)
URL: <http://astro.geo.tu-dresden.de/RIFA>

DIVISION I WORKING GROUPS

Div.I/WG - Second Realization of International Celestial Reference Frame
Chair: Chopo Ma (USA), <cma@gemini.gsfc.nasa.gov>
URL: <http://rorf.usno.navy.mil/ICRF2/>

Div.I/WG - Numerical Standards in Fundamental Astronomy
Chair: Brian J. Luzum (USA), <bjl@maia.usno.navy.mil>
URL: <http://maia.usno.navy.mil/NSFA.html>

Div.I/WG - Astrometry by Small Ground-Based Telescopes
Chair: William Thuillot (France) <thuillot@imcce.fr>
URL: <http://www.imcce.fr/hosted_sites/iau_wgnps/astrom.html>

DIVISION I INTER-DIVISION WORKING GROUPS

Div.I-III/WG - Cartographic Coordinates and Rotational Elements
Chair: Brent A. Archinal (USA), <barchinal@usgs.gov>
URL: <http://astrogeology.usgs.gov/Projects/WGCCRE/>

Div.I-III/WG - Natural Satellites
Chair: Jean-Eudes Arlot (France), <Jean-Eudes.Arlot@obspm.fr>
URL: <http://www.imcce.fr/host/iau_wgnps/iauwg.html>

DIVISION II - Sun and Heliosphere

URL: <http://www.iac.es/proyecto/iau_divii/IAU-DivII/main/index.php>

President
Donald B. Melrose
School of Physics A28
University of Sydney
Sydney, NSW 2006
Australia
Phone: +61 2 9351 4234
Fax: +61 2 9351 7726
Email: <melrose@physics.usyd.edu.au>

Vice-President
Valentin Martínez Pillet
Instituto de Astrofsica de Canarias
Observatorio del Teide
C/ Vía Láctea s/n
E-38200 La Laguna, Tenerife
Spain
Phone: +34 922 605 237
Fax: +34 922 605 210
Email: <vmp@.iac.es>

Secretary
Lidia van Driel-Gesztelyi, VP-C10 (France), <Lidia.vanDriel@obspm.fr>

Organizing Committee
David F. Webb, PP (USA), James A. Klimchuk, P-C10 (USA), Alexander Kosovichev, VP-C12 (USA), Jean-Louis Bougeret, P-C49 (France), and Rudolf von Steiger VP-C49 (Swiss)

PARTICIPATING COMMISSIONS

Div.II/Comm.10 - Solar Activity

P: James A. Klimchuk (USA), <james.klimchuk@nrl.navy.mil>
VP: Lidia van Driel-Gesztelyi (France), <Lidia.vanDriel@obspm.fr>
S: Karel J. Schrijver (USA), <schryver@lmsal.com>
OC: Donald B. Melrose, PP (Australia), Lyndsay Fletcher (UK), Nat Gopalswamy (USA), Richard A. Harrison (UK), Cristina H. Mandrini (Argentina), Hardi Peter (Germany), Saku Tsuneta (Japan), Bojan Vrsnak (Croatia), and Jingxiu Wang (China)
URL: <http://www.mssl.ucl.ac.uk/iau_c10/index.html>

Div.II/Comm.12 - Solar Radiation and Structure

P: Valentin Martínez Pillet (Spain), <vmp@iac.es>
VP: Alexander Kosovichev (USA), <sasha@quake.stanford.edu>
S: John T. Mariska (USA), <mariska@nrl.navy.mil>
OC: Martin Asplund (Australia), Thomas J. Bogdan (USA), Gianna Cauzzi (Italy), Jørgen Christensen-Dalsgaard (Denmark), Lawrence E. Cram Australia), Weiqun Gan (China), Laurent Gizon (Germany), Petr Heinzel (Czech R.), Marta G. Rovira (Argentina), and P. Venkatakrishnan (India)
URL: <http://www.iac.es/proyecto/iau_divii/IAU-DivII/main/index.php>

Div.II/Comm.49 - Interplanetary Plasma and Heliosphere

P: Jean-Louis Bougeret (France), <Jean-Louis.Bougeret@obspm.fr>
VP: Rudolf von Steiger (Swiss), <rudolf.vonsteiger@issibern.ch>
OC: Subramanian Ananthakrishnan (India), Hilary V. Cane (Australia), Nat Gopalswamy (USA), Stephen W. Kahler (USA), Rosine Lallement (France), Blai Sanahuja (Spain), Kazunari Shibata (Japan), Marek Vandas (Czech R.), Frank Verheest (Belgium), and David F. Webb, PP (USA)

DIVISION II WORKING GROUPS

Div.II/WG - Solar Eclipses
Chair: Jay M. Pasachoff (USA), <jay.m.pasachoff@williams.edu>
URL: <http://www.williams.edu/Astronomy/IAU_eclipses/>

Div.II/WG - Solar and Interplanetary Nomenclature
Chair: Edward W. Cliver (USA), <Edward.Cliver@hanscom.af.mil>
URL: <http://www2.bc.edu/ haganmp/Nomenclature.htm>

Div.II/WG - International Solar Data Access
Chair: Robert Bentley (UK), <rdb@mssl.ucl.ac.uk>
URL: <http://www.mssl.ucl.ac.uk/grid/iau/DivII_WG_IntDataAccess.html>

Div.II/WG - International Collaboration on Space Weather
Chair: David F. Webb (USA), <David.Webb@hanscom.af.mil>
URL: <http://www2.bc.edu/ haganmp/MAINpage>

DIVISION III - Planetary Systems Sciences

URL: <http://www.ss.astro.umd.edu/IAU/div3/>

President
Edward L.G. Bowell
Lowell Observatory
Box 1149
1400 West Mars Hill Road
Flagstaff, AZ 86001
USA
Phone: +1 928 774 3358
Fax: +1 928 774 6296
Email: <ebowell@lowell.edu>

Vice-President
Karen J. Meech
Institute for Astronomy
University of Hawai'i Honolulu
2680 Woodlawn Drive
Honolulu, HI 96822
USA
Phone: +1 808 956 6828
Fax: +1 808 956 9580
Email: <meech@ifa.hawaii.edu>

Secretary
Guy J. Consolmagno, (Vatican State City), <brother_guy@mac.com>

Organizing Committee
Alan P. Boss, P-C51 (USA), Régis Courtin, P-C16 (France), Julio A. Fernández, P-C20 (Uruguay), Bo A.S. Gustafson, PP-C21 (USA), Walter F. Huebner, P-C15 (USA), A.-Chantal Levasseur-Regourd (France), Mikhail Ya. Marov (Russia), Michel Mayor, P-C53 (Switzerland), Rita M. Schulz (Germany), Pavel Spurný, P-C22 (Czech Rep.), Giovanni B. Valsecchi, PP-C20 (Italy), Jun-Ichi Watanabe (Japan), Iwan P. Williams, PP (UK), and Adolf N. Witt P-C21 (USA)

PARTICIPATING COMMISSIONS AND COMMISSION WORKING GROUPS

Div.III/Comm.15 - Physical Study of Comets and Minor Planets
P: Walter F. Huebner (USA), <whuebner@swri.edu>
VP: Alberto Cellino (Italy), <Cellino@to.astro.it>
S: Daniel C. Boice (USA), <DBoice@SwRI.edu>
OC: Dominique Bockelee-Morvan (France), Yuehua Ma (China), Harold J. Reitsema (USA), Rita M. Schulz (Netherlands), Petrus M.M. Jenniskens (USA), Dmitrij F. Lupishko (Ukraine), and Gonzalo Tancredi (Uruguay)
URL: <http://iau15.space.swri.edu>

Div.III/Comm.15/WG - Physical Study of Comets
Chair: Tetsuo Yamamoto (Japan), <TY@lowtem.hokudai.ac.jp>
URL: <http://atlas.sr.unh.edu/IAU_Comm15/>

Div.III/Comm.15/WG - Physical Study of Minor Planets
Chair: Ricardo Gil-Hutton (Argentina), <RGilHutton@casleo.gov.ar>
URL: <http://atlas.sr.unh.edu/IAU_Comm15/>

Div.III/Comm.16 - Physical Study of Planets and Satellites
P: Régis Courtin (France), <Regis.Courtin@obspm.fr>
VP: Melissa A. McGrath (USA), <melissa.a.mcgrath@nasa.gov>
S: Luisa M. Lara (Spain), <lara@iaa.es>
OC: Carlo Blanco (Italy), Guy J. Consolmagno (Vatican City State), Leonid V. Ksanfomality (Russia), David Morrison (USA), John R. Spencer (USA), and Viktor G. Tejfel (Kazakhstan)
URL: <http://www.iaa.es/IAUComm16>

Div.III/Comm.20 - Positions and Motions of Minor Planets, Comets and Satellites
P: Julio A. Fernández (Uruguay), <julio@fisica.edu.uy>
VP: Makoto Yoshikawa (Japan), <makoto@pub.isas.ac.jp>
S: Steve Chesley (USA), <steve.chesley@jpl.nasa.gov>
OC: Giovanni B. Valsecchi, PP (Italy), Yulia A. Chernetenko (Russia), Alan C. Gilmore (New Zealand), Daniela Lazzaro (Brazil), Karri Muinonen (Finland), Petr Pravec (Czech Republic), Timothy B. Spahr (USA), David J. Tholen (USA), Jana Tychá (Czech Republic), and Jin Zhu (China)
URL: <http://www.astro.uu.se/IAU/c20/>

Div.III/Comm.20/WG - Motions of Comets
Chair: Julio Fernández (Uruguay, <julio@fisica.edu.uy>
URL: <http://www.astro.uu.se/IAU/c20/wgcomet.html>

Div.III/Comm.20/WG - Distant Objects
Chair: Brian Marsden (USA), <bmarsden@cfa.harvard.edu>
URL: <http://www.astro.uu.se/IAU/c20/>

Div.III/Comm.21 - Light of the Night Sky
P : Adolf N. Witt (USA), <awitt@dusty.astro.utoledo.edu>
VP: Jayant Murthy (India), <jmurthy@yahoo.com>
OC: W. Jack Baggaley (New Zealand), Eli Dwek (USA), Bo A.S. Gustafson (USA), A.-Chantal Levasseur-Regourd (France), Ingrid Mann (Japan), Kalevi Mattila (Finland), and Junichi Watanabe (Japan)
URL: <http://www.astro.ufl.edu/ gustaf/IAUCom21/IAU_Com_21.html>

Div.III/Comm.22 - Meteors, Meteorites and Interplanetary Dust
P: Pavel Spurn (Czech R.), <spurny@asu.cas.cz>
VP: Jun-ichi Watanabe (Japan), <jun.watanabe@nao.ac.jp>
S: Jiri Borovicka (Czech Rep.), <borovic@asu.cas.cz>
OC: William J. Baggaley (New Zealand), Peter G. Brown (Canada), Guy J. Consolmagno (USA), Petrus M.M. Jenniskens (USA), Asta K. Pellinen-Wannberg (Sweden), Vladimir Porubcan (Slovakia), Iwan P. Williams (UK), and Hajime Yano (Japan)
URL: <http://meteor.asu.cas.cz/IAU/>

Div.III/Comm.22/WG - Professional-Amateur Cooperation in Meteors
Chair: Galina O. Ryabova (Russia), <ryabova@niipmm.tsu.ru>
URL: <http://www.asu.cas.cz/english/>

Div.III/Comm.22/TF - Task Force for Meteor Shower Nomenclature
Chair: Petrus M.M. Jenniskens (USA), <pjenniskens@mail.arc.nasa.gov>
URL: <http://meteor.asu.cas.cz/IAU/nomenclature.html>

Div.III/Comm.51 - Bioastronomy
P: Alan P. Boss (USA), <boss@dtm.ciw.edu>
VP: William M. Irvine (USA), <irvine@fcrao1.astro.umass.edu>
OC: Cristiano Cosmovici (Italy), Pascale Ehrenfreund (Netherlands), Karen J. Meech, PP (USA), David W. Latham (USA), David Morrison (USA), and Stephane Udry (Switzerland)
URL: <http://www.dtm.ciw.edu/boss/c51index.html>

Div.III/Comm.53 - Extrasolar Planets
P : Michel Mayor (Switzerland), <michel.mayor@obs.unige.ch>

VP: Alan P. Boss (USA), <boss@dtm.ciw.edu>
OC: Paul Butler (USA), William B. Hubbard (USA), Philip A. Ianna (USA), Martin Kuerster (Germany), Jack J. Lissauer (USA), Karen J. Meech (USA), François Mignard (France), Allan Penny (UK), Andreas Quirrenbach (Germany), Jill C. Tarter (USA), and Alfred Vidal-Madjar (France)

DIVISION III WORKING GROUPS

Div.III/Service - Minor Planet Center (MPC)
Director: Timothy B. Spahr (USA), <tspahr@cfa.harvard.edu>
URL: <http://cfa-www.harvard.edu/iau/mpc.html>

Div.III/Committtee - Minor Planet Center Advisory Committee
Michael F. A'Hearn (USA), Steven R. Chesley (USA), Hans Rickman (Sweden), and Giovanni B. Valsecchi (Italy)

Div.III/WG - Committee on Small Bodies Nomenclature (CSBN)
Chair: Jana Tichá (Czech Republic), <jticha@klet.cz>
S: Brian Marsden (USA), <bmarsden@cfa.harvard.edu>
URL: <http://www.ss.astro.umd.edu/IAU/csbn/>

Div.III/WG - Planetary System Nomenclature (WG-PSN)
Chair: Rita M. Schulz (Netherlands), <rschulz@rssd.esa.int>
URL: <http://planetarynames.wr.usgs.gov/append1.html>

DIVISION III INTER-DIVISION WORKING GROUPS

Div.III-I/WG - Cartographic Coordinates and Rotational Elements
Chair: Brett A. Archinal (USA), <barchinal@usgs.gov>
URL: <http://astrogeology.usgs.gov/Projects/WGCCRE/#wgm>

Div.III-I/WG - Natural Satellites (formerly Div.III- C20 WG)
Chair: Jean-Eudes Arlot (France), <Jean-Eudes.Arlot@obspm.fr>
URL: <http://www.imcce.fr/host/iau_wgnps/iauwg.html>

DIVISION IV - Stars

URL: <http://www.galax.obspm.fr/IAUdiv4>

President
Monique Spite
GEPI
Observatoire de Paris-Meudon
5 Place J. Janssen
F-92195 Meudon Cedex
France
Phone: +33 1 45 077 839
Fax: +33 1 45 077 878
Email: <monique.spite@obspm.fr>

Vice-President
Christopher J. Corbally
Vatican Observatory Research Group
Steward Observatory
University of Arizona
933 North Cherry Avenue
Tucson, AZ 85721
USA
Phone: +1 520 621 3225
Fax: +1 520 621 1532
Email: <ccorbally@as.arizona.edu>

Organizing Committee

Christine Allen, P-C26 (Mexico), Francesca d'Antona, P-C35 (Italy), Dainis Dravins, PP (Sweden), Sunetra Giridhar, P-C45 (India), John D. Landstreet, P-C36 (Canada), and Mudumba Parthasarathy, P-C29 (India)

PARTICIPATING COMMISSIONS AND COMMISSION WORKING GROUPS

Div.IV/Comm.26 - Double and Multiple Stars

P: Christine Allen (Mexico), <chris@astroscu.unam.mx>
VP: Jose A.D. Docobo (Spain), <oadoco@usc.es>
OC: Yuri Y. Balega (Russia) John Davis (Australia), William I. Hartkopf, PP (USA), Brian D. Mason (USA), Edouard Oblak (France), Terry D. Oswalt (USA), Dimitri Pourbaix (Belgium), and Colin D. Scarfe (Canada)
URL: <http://ad.usno.navy.mil/wds/dsl.html>

Div.IV/Comm.26/WG - Binary and Multiple System Nomenclature

Chair: Brian D. Mason (USA), <wih@usno.navy.mil>
URL: <http://ad.usno.navy.mil/wds/newwds.html>

Div.IV/Comm.29 - Stellar Spectra

P: Mudumba Parthasarathy (India), <partha@iiap.res.in>
VP: Nikolai E. Piskunov (Sweden), <piskunov@astro.uu.se>
OC: Fiorella Castelli (Italy), Kenneth G. Carpenter (USA), Katia Cunha (Brazil), Philippe R.J. Eenens (Mexico), Ivan Hubeny (USA), Sylvia C.F. Rossi (Brazil), Chris Sneden, PP (USA), Masahide Takada-Hidai (Japan), Glenn M. Wahlgren (Sweden), and Werner W. Weiss (Austria)
URL: <http://www.iiap.res.in/personnel/partha/IAUcom29.html>

Div.IV/Comm.35 - Stellar Constitution

P: Francesca d'Antona (Italy), <dantona@mporzio.astro.it>
VP: Corinne Charbonnel (Switzerland), <Corinne.Charbonnel@obs.unige.ch>
OC: Gilles Fontaine (Canada), Richard B. Larson (USA), John Lattanzio (Australia), James W. Liebert (USA), Ewald Mueller (Germany), Achim Weiss (Germany), and Lev R. Yungelson (Russia)
URL: <http://iau-c35.stsci.edu>

Div.IV/Comm.36 - Theory of Stellar Atmospheres
P : John D. Landstreet (Canada), <jlandstr@uwo.ca>
VP: Martin Asplund (Australia), <martin@mso.anu.edu.au>
OC: Suchitra C. Balachandran (USA), Svetlana V. Berdyugina (Switzerland), Peter H. Hauschildt (Germany), Hans G. Ludwig (France), Lyudmila I. Mashonkina (Russia), K.N. Nagendra (India), Joachim Puls (Germany), Sofia Randich (Italia), Monique Spite, PP (France), and Grazina Tautvaisiene (Lithuania)
URL: <http://www.galax.obspm.fr/IAU36/>

Div.IV/Comm.45 - Stellar Classification
P: Sunetra Giridhar (India), <giridhar@iiap.res.in>
VP: Richard O. Gray (USA), <grayro@appstate.edu>
OC: Coryn A.L. Bailer-Jones (Germany), Christopher J. Corbally, PP (USA), Laurent Eyer (Switserland), Michael J. Irwin (UK), Joseph D. Kirkpatrick (USA), Steven Majewski (USA), Dante Minniti (Chile), and Birgitta Nordström (Denmark)
URL: <http://www.iap.fr/SitesHeberges/com45uai/index.html>

DIVISION IV WORKING GROUPS

Div.IV/WG - Massive Stars
Chair: Stanley P. Owocki (USA), <owocki@bartol.udel.edu>
URL: <http://www.astroscu.unam.mx/massive_stars/index.php>

Div.IV/WG - Abundances in Red Giants
Chair: John Lattanzio (Australia), <john.lattanzio@sci.monash.edu.au>
URL: <http://www.maths.monash.edu.au/ johnl/wgarg/>

DIVISION IV INTER-DIVISION WORKING GROUPS

Div.IV-V/WG - Active OB Stars
co-Chair: Juan Fabregat (Spain), <Juan.Fabregat@uv.es>
co-Chair: Geraldine J. Peters (USA), <gjpeters@mucen.usc.edu>
URL: <http://www.astro.virginia.edu/ dam3ma/benews/iauwg_abs.html>

Div.IV-V/WG - Ap and Related Stars
Chair: Margarida Cunha (Portugal), <mcunha@astro.up.pt>
URL: <http://www.eso.org/gen-fac/pubs/apn/apwg/>

Div.IV-V-IX/WG - Standard Stars
Chair: Christopher J. Corbally (USA), <ccorbally@as.arizona.edu>
URL: <http://stellar.phys.appstate.edu/ssn/>

DIVISION V - Variable Stars

URL: <http://www.konkoly.hu/IAUDV/>

President
Alvaro Giménez
Centro de Astrobiologia
INTA / CSIC
Carretera de Torrejon a Ajalvir
Torreon de Ardoz
E-28850 Madrid
Spain
Phone/Fax: +34 91 520 1111
Email: <agimenez@rssd.esa.int>

Vice-President
Steven D. Kawaler
Department of Physics and Astronomy
Iowa State University
A323 Zaffarano Hall
Ames, IA 50011-3160
USA
Phone: +1 515 294 9728
Fax: +1 515 294 6027
Email: <sdk@iastate.edu>

Secretary
Conny Aerts (Belgium), <conny@ster.kuleuven.ac.be>

Organizing Committee
Jørgen Christensen-Dalsgaard,PP (Denmark), Edward F. Guinan (USA), Donald W. Kurtz (UK), and Slavek M. Rucinski, P-C42 (Canada)

PARTICIPATING COMMISSIONS

Div.V/Comm.27 - Variable Stars
P: Steven D. Kawaler (USA), <sdk@iastate.edu>
VP: Gerald Handler (Austria), <handler@astro.univie.ac.at>
OC: Conny Aerts (Belgium), Timothy R. Bedding (Australia), Márcio Catelán (Chile), Margarida Cunha (Portugal), Laurent Eyer (Switzerland), C. Simon Jeffery (N. Ireland), Peter Martinez (South Africa), Katalin Olah (Hungary), Karen Pollard (New Zealand), and Seetha Somasundaram (India)
URL: <http://www.konkoly.hu/IAUC27>

Div.V/Comm.42 - Close Binary Stars
P: Slavek Rucinski (Canada), <rucinski@astro.utoronto.ca>
VP: Ignasi Ribas (Spain), <iribas@ieec.uab.es>
OC: Alvaro Giménez, PP (Spain), Petr Harmanec (Czech Rep.), Ronald W. Hilditch (UK), Janusz Kaluzny (Poland), Panayiotis Niarchos (Greece), Birgitta Nordström (Denmark), Katalin Olah (Hungary), Mercedes T. Richards (USA), Colin D. Scarfe, PP (Canada), Edward M. Sion (USA), Guillermo Torres (USA), and Sonja Vrielmann (Germany)
URL: <http://www.konkoly.hu/IAUDV/>

DIVISION V WORKING GROUP

Div.V/WG - Spectroscopic Data Archiving
Chair: Elizabeth Griffin (Canada), <Elizabeth.Griffin@nrc-cnrc.gc.ca>
URL: <http://www.konkoly.hu/SVO/>

INTER-DIVISION WORKING GROUPS

Div.IV-V/WG - Active OB Stars
co-Chair: Juan Fabregat (Spain), <Juan.Fabregat@uv.es>

co-Chair: Geraldine J. Peters (USA), <gjpeters@mucen.usc.edu>
URL: <http://www.astro.virginia.edu/ dam3ma/benews/iauwg_abs.html>

Div.IV-V/WG - Ap and Related Stars
Chair: Margarida Cunha (Portugal) <mcunha@astro.up.pt>
URL: <http://www.eso.org/gen-fac/pubs/apn/apwg/>

Div.IV-V-IX/WG - Standard Stars
Chair: Christopher J. Corbally (USA), <ccorbally@as.arizona.edu>
URL: <http://stellar.phys.appstate.edu/ssn/>

DIVISION VI - Interstellar Matter
URL: <www.div6.qub.ac.uk>

President
School of Mathematics and Physics
Faculty of Engineering & Physical Sciences
Queen's University Belfast
13 Stranmillis Road
Belfast BT9 5AF
Northern Ireland
Phone: +44 2890 976523
Fax: +44 2890 974536
E-mail: <Tom.Millar@qub.ac.uk>

Vice-President
Astronomy Department
University of Illinois
103 Astronomy Building
1002 West Green Street
Urbana, IL 61801
USA
Phone: +1 217 333 5535
Fax: +1 217 244 7638
Email: <chu@astro.uiuc.edu>

Organizing Committee
Dieter Breitschwerdt (Germany), Michael G. Burton (Australia), Sylvie Cabrit (France), Paola Caselli (Italia), John E. Dyson, PP (UK), Gary J. Ferland (USA), Elisabete M.de Gouveia Dal Pino (Brazil), Mika J. Juvela (Finland), Bon-Chul Koo (S. Korea), Sun Kwok (Hong Kong), Susana Lizano (Mexico), Michael Rozyczka (Poland), L. Viktor Toth (Hungary), Masato Tsuboi (Japan), and Ji Yang (China Nanjing)

PARTICIPATING COMMISSION

Div.VI/Comm.34 - Interstellar Matter
P = P Division VI, VP = VP Division VI, OC = OC Division VI

DIVISION VI WORKING GROUPS

Div.VI/WG - Star Formation
Chair: Franceso Palla (Italy), <palla@arcetri.astro.it>
URL: <http://www.arcetri.astro.it/sfwg/>

Div.VI/WG - Astrochemistry
Chair: Ewine F. van Dishoeck (Netherlands), <ewine@strw.leidenuniv.nl>
URL: <http://www.strw.leidenuniv.nl/iau34/>

Div.VI/WG - Planetary Nebulae Chair: Arturo Manchado (Spain), <amt@iac.es>
URL: <http://www.iac.es/proyect/PNgroup/wg/>

DIVISION VII - Galactic System

URL: <http://www.ari.uni-heidelberg.de/interessantes/iaudivisionVII/IAUDivVII.html>

President
Ortwin Gerhard
MPI für Extraterrestrische Physik
Giessenbachstrasse 1
D-85748 Garching bei München
Germany
Phone: +49 89 30000 3539/3503
Fax: +49 89 30000 3351
Email: <gerhard@mpe.mpg.de>

Vice-President
Despina Hatzidimitriou
Department of Physics
University of Crete
P.O. Box 2208
GR-710 03 Heraklion
Greece
Phone: +30 2 810 394 212
Fax: +30 2 810 394 201
Email: <dh@physics.uoc.gr>

Organizing Committee
Charles J. Lada, VP-C37 (USA), Ata Sarajedini (USA), Patricia A. Whitelock, PP (South Africa), Rosemary F. Wyse, VP-C33 (USA), and Joseph Lazio (USA)

PARTICIPATING COMMISSIONS

Div.VII/Comm.33 - Structure and Dynamics of the Galactic System
P: Ortwin Gerhard (Germany), <gerhard@mpe.mpg.de>
VP: Rosemary F. Wyse (USA), <wyse@pha.jhu.edu>
OC: Yuri N. Efremov (Russia), Wyn Evans (UK), Chris Flynn (Finland), Jonathan E. Grindlay (USA), Joseph Lazio (USA), Birgitta Nordström (Denmark), Patricia A. Whitelock (South Africa), and Chi Yuan (South Korea)
URL: <http://www.saao.ac.za/IAU/IAUComm33.html>

Div.VII/Comm. 37 Star - Clusters and Associations
P: Despina Hatzidimitriou (Greece), <dh@physics.uoc.gr>
VP: Charles J. Lada (USA), <clada@cfa.harvard.edu>
OC: Russell D. Cannon (Australia), Kyle McC. Cudworth (USA), Gary S. Da Costa (Australia), LiCai Deng (China), Young-Wook Lee (South Korea), Ata Sarajedini (USA), and Monica Tosi (Italy)
URL: <http://www.ari.uni-heidelberg.de/interessantes/iaucommission37/IAUComm37.html>

DIVISION VII WORKING GROUP

Div.VII/WG - Galactic Center
Chair: Joseph Lazio (USA), <Lazio@nrl.navy.mil>
URL: <http://www.aoc.nrao.edu/ gcnews/>

DIVISION VIII - Galaxies and the Universe

URL: <http://www.star.bris.ac.uk/iau/>

President
Sadanori Okamura
Department of Astronomy
School of Science
University of Tokyo
7-3-1 Hongo Bunkyo-ku
Tokyo 113-0033
Japan
Phone: +81 3 5841 4257
Fax: +81 3 5841 7644
Email: <okamura@astron.s.u-tokyo.ac.jp>

Vice-President
Elaine M. Sadler
School of Physics A28
University of Sydney
Sydney, NSW 2006
Australia
Phone: +61 2 9351 2622
Fax: +61 2 9351 7726
Email: <ems@physics.usyd.edu.au>

Organizing Committee
Francesco Bertola, PP (Italy), Françoise Combes, P-C28 (France), Roger L. Davies, VP-C28 (UK), Thanu Padmanabhan, VP-C47 (India), and Rachel L. Webster, P-C47 (Australia)

Webmaster
Mark Birkinshaw (UK), <Mark.Birkinshaw@bristol.ac.uk>

PARTICIPATING COMMISSIONS

Div.VIII/Comm.28 - Galaxies

P: Franoise Combes (France), <francoise.combes(@)obspm.fr>
VP: Roger L. Davies (UK), <roger.davies@durham.ac.uk>
OC: Avishai Dekel (Israel), Marijn Franx (Netherlands), John S. Gallagher (USA), Naomasa Nakai (Japan), Monica Rubio (Chile), Valentina Karachentseva (Ukraine), Gillian R. Knapp (USA), Renée C. Kraan-Korteweg (South Africa), Bruno Leibundgut (Germany), Jayant V. Narlikar (India), and Elaine M. Sadler (Australia)
URL: <http://aramis.obspm.fr/IAU28/index.php>

Div.VIII/Comm.47 - Cosmology

P: Rachel L. Webster (Australia), <rwebster@physics.unimelb.edu.au>
VP: Thanu Padmanabhan (India), <paddy@iucaa.ernet.in>
OC: (tbd)
URL: <tbd>

DIVISION VIII WORKING GROUP

Div.VIII/WG - Supernovae

co-Chair: Brian Schmidt (Australia), <brian@mso.anu.edu.au>
co-Chair: Wolfgang Hillebrandt (Germany), <wfh@MPA-Garching.MPG.de>
URL: <http://www.star.bris.ac.uk/iau/news/sn_wg_ga_26.pdf>

DIVISION IX - Optical and Infrared Techniques

URL: <http://www.ifa.hawaii.edu/users/kud/iau/DivIX.htm>

President
Rolf-Peter Kudritzki
Institute for Astronomy
2680 Woodlawn Drive
Honolulu, HI 96822
USA
Phone: +1 808 956 8566
Fax: +1 808 946 3467
Email: <kud@ifa.hawaii.edu>

Vice-President
Andreas Quirrenbach
Landessternwarte, Univerity of Heidelberg
Koenigstuhl 12
D-69117 Heidelberg
Germany
Phone: +49 6221 54 1792
Fax: +49 6221 54 1702
Email: <A.Quirrenbach@lsw.uni-heidelberg.de>

Organizing Committee
Michael G. Burton WG A (Australia), Xiangqun Cui (China), Martin Cullum (Germany), Peter Martinez P-C25 (South Africa), Guy S. Perrin P-C54 (France), Andrei A. Tokovinin (Chile), Guillermo Torres VP-C30 (USA), and Stephane Udry P-C30 (Switzerland)

Consultants
Christiaan L. Sterken (Belgium) and Michel Dennefeld (France)

PARTICIPATING COMMISSIONS AND COMMISSION WORKING GROUPS

Div.IX/Comm.25 - Stellar Photometry and Polarimetry
P: Peter Martinez (South Africa), <peter@saao.ac.za>.
VP: Eugene F. Milone (Canada), <milone@ucalgary.ca>.
OC: Arlo U. Landolt, PP (USA), Edward G. Schmidt (USA), Carme Jordi (Spain), Christiaan L. Sterken (Belgium), Alexei Mironov (Russia), and Qian Sheng-Bang (China)
URL: <http://www.vub.ac.be/STER/IAU/IAUComm25.html>

Div.IX/Comm.25 - WG Infrared Astronomy
Chair: Eugene F. Milone (Canada), <milone@ucalgary.ca>
URL: <http://www.ucalgary.ca/ milone/IRWG/>

Div.IX/Comm.30 - Radial Velocities
P: Stephane Udry (Switzerland), <stephane.udry@obs.unige.ch>
VP: Guillermo Torres (USA), <torres@cfa.harvard.edu>
OC: Francis C. Fekel (USA), Kenneth C. Freeman (Australia), Elena V. Glushkova (Russia), Goeffrey W. Marcy (USA), Robert D. Mathieu (USA), Birgitta Nordström, PP (Denmark), Dimitri Pourbaix (Belgium), Catherine Turon (France), and Tomaz Zwitter (Slovenia)
URL: <http://www.ctio.noao.edu/science/iauc30/iauc30.html>

Div.IX/Comm.30 WG - Radial-Velocity Standard Stars
Chair: Stephane Udry (Switzerland), <stephane.udry@obs.unige.ch>
URL: <http://obswww.unige.ch/ udry/std/std.html>

Div.IX/Comm.30 WG - Stellar Radial Velocity Bibliography
Chair: Hugo Levato (Argentina), <hlevato@casleo.gov.ar>
URL: <www.casleo.gov.ar/indexingles.htm>

Div.IX/Comm.30 WG - Catolog of Orbital Elements of Spectroscopic Binary Systems
Chair: Dimitri Pourbaix (Belgium), <pourbaix@astro.ulb.ac.be>
URL: <http://sb9.astro.ulb.ac.be/>

Div.IX/Comm.54 - Optical and Infrared Interferometry
P: Guy S. Perrin (France), <guy.perrin@obspm.fr>
VP: Stephen T. Ridgway (USA), <ridgway@noao.edu>
S : Gerard T. van Belle (USA), <gerard(@)ipac.caltech.edu>
OC: Gilles Duvert (France), Reinhard Genzel (Germany), Christopher Haniff (UK), Christian A. Hummel (Germany), Peter R. Lawson (USA), John D. Monnier (USA), Didier Queloz (Switzerland), Peter G. Tuthill (Australia), and Farrokh Vakili (France)
URL: <http://olbin.jpl.nasa.gov/iau/2006/commission.html>

DIVISION IX WORKING GROUPS

Div.IX/WG - Adaptive Optics
Chair: (tbd)

Div.IX/WG - Site Testing Instruments
Chair: Andrei A. Tokovinin (Chile), <atokovinin@ctio.noao.edu>

Div.IX/WG - Large Telescope Projects
Chair: (tbd)

Div.IX/WG - Small Telescope Projects
Chair: (tbd)>

Div.IX/WG - Detectors
Chair: Tim Abbott (USA), <tabbott@ctio.noao.edu>
URL: <http://www.ctio.noao.edu/CCD-world/>

Div.IX/WG - Sky Surveys
Chair: Quentin Parker (Australia), <qap@ics.mq.edu.au>
URL: <http://alba.stsci.edu/wgss/>

INTER-DIVISION WORKING GROUPS

Div.IV-V-IX/WG - Standard Stars
Chair: Christopher J. Corbally (USA), <ccorbally@as.arizona.edu>
URL: <http://stellar.phys.appstate.edu/ssn/>

Div.IX-X/WG Encouraging the International Development of Antarctic Astronomy
Chair: Michael G. Burton (Australia), <mgb@phys.unsw.edu.au>
URL: <www.phys.unsw.edu.au/jacara/iau>

Div.IX-X-XI/WG - Astronomy from the Moon
Chair: Sallie L. Baliunas (USA), <baliunas@cfa.harvard.edu>
vice-Chair: Yoji Kondo (USA), <kondo@stars.gsfc.nasa.gov>
URL: <http://www.cfa.harvard.edu/moon/>

DIVISION X - Radio Astronomy
URL: <http://www.bao.ac.cn/IAU_COM40/>

President
Rendong Nan
National Astronomical Observatories
Chinese Academy of Sciences
A20 Datun Road, Chaoyang District
Beijing 100012
P.R. China
Phone: +86 10 6487 7280
Fax: +86 10 6485 2055
Email: <nrd@bao.ac.cn>

Vice-President
Russell A. Taylor
Department of Physics and Astronomy
University of Calgary
2500 University Drive NW
Calgary AB T2N 1N4
Canada
Phone: +1 403 220 5385
Fax: +1 403 289 3331
Email: <russ@ras.ucalgary.ca>

Organizing Committee
Christopher L. Carilli (USA), Jessica Chapman (Australia), Gloria M. Dubner (Argentina), Michael Garrett (Netherlands), W. Miller Goss (USA), Richard E. Hills (UK), Hisashi Hirabayashi (Japan), Luis F. Rodriguez, PP (Mexico), Prajval Shastri (India), and Jose M. Torrelles (Spain)

PARTICIPATING COMMISSION

Div.X/Comm.40 - Radio Astronomy
P = P Division X, VP = VP Division X, OC = OC Division X

DIVISION X WORKING GROUPS

Div.X/WG - Global VLBI
Chair: Jonathan D. Romney (USA), <jromney@nrao.edu>
URL: <http://www.bao.ac.cn/IAU_COM40/WG/WgVLBI.html>

Div.X/WG - Interference Mitigation
Chair: Anastasios Tzioumis (Australia), <atzioumi@atnf.csiro.au>
URL: <http://www.bao.ac.cn/IAU_COM40/WG/WgRF.html>

Div.X/WG - Astrophysically Important Spectral Lines
Chair: Masatoshi Ohishi (Japan), <ohishi@nao.ac.jp>
URL: <http://www.bao.ac.cn/IAU_COM40/WG/WgSL.html>

DIVISION X INTER-DIVISION WORKING GROUPS

Div.IX-X/WG - Encouraging the International Development of Antarctic Astronomy
Chair: Michael G. Burton (Australia), <mgb@phys.unsw.edu.au>
URL: <http://www.bao.ac.cn/IAU_COM40/WG/WgAntarc.html>

Div.IX-X-XI WG - Astronomy from the Moon
Chair: Sallie Baliunas (USA), <baliunas@cfa.harvard.edu>
vice-Chair: Yoji Kondo (USA), <kondo@stars.gsfc.nasa.gov>
URL: <http://www.cfa.harvard.edu/moon/>

Div.X-XII/WG - Historic Radio Astronomy
Chair: Wayne Orchiston (Australia), <Wayne.Orchiston@jcu.edu.au>
URL: <http://www.bao.ac.cn/IAU_COM40/WG/WgHistRA.html>

DIVISION XI - Space and High Energy Astrophysics

URL: < http://www.mpe.mpg.de/IAU_DivXI/

President
Günther Hasinger
MPI für Extraterrestrische Physik
Giessenbachstrasse 1
D-85748 Garching-bei-München
Germany
Phone: +49 893 0000 3401
Fax: +49 893 0000 3404
Email: <ghasinger@mpe.mpg.de>

Vice-President
Christine Jones
Harvard-Smithonian Center for Astrophysics
60 Garden Street
Cambridge, MA 02138
USA
Phone: +1 617 495 7137
Fax: +1 617 495 7356
Email: <cjf@cfa.harvard.edu>

Organizing Committee
João Braga (Brazil), Noah Brosch (Israel), Thijs de Graauw (Chile), Leonid I. Gurvits (Netherlands), George Helou (USA), Ian D. Howarth (UK), Hideyo Kunieda (Japan), Thierry Montmerle (France), Haruyuki Okuda (Japan), Marco Salvati (Italy), and Kulinder Pal Singh (India)

PARTICIPATING COMMISSION

Div.XI/Comm.44 - Space and High Energy Astrophysics
P = P Division XI, VP = VP Division XI, OC = OC Division XI

DIVISION XI WORKING GROUP

Div.XI/WG - Particle Astrophysics
Chair: Reinhard Schlickeiser (Germany), <office@tp4.ruhr-uni-bochum.de>
URL: < http://iau.physik.rub.de/ >

DIVISION XI INTER-DIVISION WORKING GROUP

Div.IX-X-XI/WG - Astronomy from the Moon
Chair: Sallie Baliunas (USA), < baliunas@cfa.harvard.edu >.
Vice-Chair: Yoji Kondo (USA), < kondo@stars.gsfc.nasa.gov >
OC: Oddbjorn Engvold (Norway), Karel A. van der Hucht (Netherlands), Norio Kaifu (Japan), Haruyuki Okuda (Japan), and Yervant Terzian (USA)
URL: < http://www.cfa.harvard.edu/moon/ >

DIVISION XII - Union-Wide Activities

URL: <http://www.iaudivisionxii.org/>

President
Malcolm G. Smith
AURA / CTIO / NOAO
Casilla 603
La Serena
Chile
Phone: +56 51 205 217
Fax: +56 51 205 356
Email: <msmith@ctio.noao.edu>

Vice-President
Françoise Genova
Observatoire Astronomique
Université de Strasbourg
11 rue de l'Université
F-67000 Strasbourg
France
Phone: +33 3 9024 2476
Fax: +33 3 9024 2432
Email: <genova@astro.u-strasbg.fr>

Secretary / Webmaster
Lars Lindberg Christensen (ESO/ESA), <lars@eso.org>

Organizing Committee
Johannes Andersen, PP (Denmark), Steven R. Federman, P-C14 (USA), Alan C. Gilmore, P-C6 (New Zealand), Raymond P. Norris, P-C5 (Australia), Ian E. Robson, P-C55 (UK), Il-Seong Nha, P-C41 (South Korea), Magda G. Stavinschi, P-C46 (Romania), Richard J. Wainscoat, VP-C50 (USA), and Virginia L. Trimble, PP (USA)

PARTICIPATING COMMISSIONS AND COMMISSION WORKING GROUPS

Div.XII/Comm.5 - Documentation and Astronomical Data
P : Raymond P. Norris, WG AD (Australia), <Ray.Norris@csiro.au>
VP: Masatoshi Ohishi (Japan), <ohishi@nao.ac.jp>
OC: Françoise Genova, PP (France), Uta Grothkopf, WG Libraries (Germany), Oleg Yu. Malkov (Russia), Masatoshi Ohishi (Japan), William D. Pence, WG FITS (USA), Marion Schmitz, WG Nomen. (USA), Robert J. Hanisch, WG VO (USA), and Xu Zhou (China)
URL: <http://www.atnf.csiro.au/people/rnorris/IAUC5/>

Div.XII/Comm.5/WG - Astronomical Data
Chair: Raymond P. Norris (Australia), <Ray.Norris@csiro.au>
URL: <http://www.atnf.csiro.au/people/rnorris/WGAD/>

Div.XII/Comm.5/WG - Nomenclature
Chair: Marion Schmitz (USA), <zb4ms@ipac.caltech.edu>
URL: <http://vizier.u-strasbg.fr/Dic/iau-spec.htx>

Div.XII/Comm.5/WG - Libraries
Chair: Uta Grothkopf (Germany), <ugrothko@eso.org>
URL: <http://www.eso.org/gen-fac/libraries/IAU-WGLib/>

Div.XII/Comm.5/WG - FITS Data Format
Chair: William D. Pence (USA), <William.D.Pence@nasa.gov>
vice-Chair: François Ochsenbein (France), <francois@astro.u-strasbg.fr>
URL: < http://fits.gsfc.nasa.gov/iaufwg/ >

Div.XII/Comm.5/WG - Virtual Observatories, Data Centers and Networks
Chair: Robert J. Hanisch (USA), <hanisch@stsci.edu>
URL: < http://cdsweb.u-strasbg.fr/IAU/wgvo.html >

Div.XII/Comm.5/TF - Preservation and Digitization of Photographic Plates
Chair: Elizabeth Griffin (Canada), <Elizabeth.Griffin@hia-iha.nrc-cnrc.gc.ca>
URL: < http://www.lizardhollow.net/PDPP.htm >

Div.XII/Comm.6 - Astronomical Telegrams
P: Alan C. Gilmore (New Zealand), <alan.gilmore@canterbury.ac.nz>
VP: Nicolay Samus (Russia), <samus@lnfm1.sai.msu.ru>
OC: Brian G. Marsden (USA), Daniel W.E. Green (USA), Syuichi Nakano (Japan), Elizabeth Roemer (USA), Jana Ticha (Czech Rep.), Hitoshi Yamaoka (Japan), and Kaare Aksnes (Norway)
URL: <http://cfa-www.harvard.edu/iau/Commission6.html>

Div.XII/Comm.6/Service - Central Bureau for Astronomical Telegrams
Chair: Daniel W. E. Green (USA), <dgreen@cfa.harvard.edu>
URL: <http://cfa-www.harvard.edu/iau/cbat.html>

Div.XII/Comm.14 - Atomic and Molecular Data
P: Steven R. Federman (USA), <steven.federman@utoledo.edu>
VP: Glenn M. Wahlgren (Sweden), <glenn.wahlgren@astro.lu.se>
OC: Sveneric Johansson (Sweden), Milan Dimitrijevic (Serbia), Alain Jorissen (Belgium), Lyudmila I. Mashonkina (Russia), Farid Salama (USA), Jonathan Tennyson (UK), and Ewine F. van Dishoeck (Netherlands)
URL: <http://www.astro.lu.se/Research/astrophys/iau/>

Div.XII/Comm.14/WG - Atomic Data
co-Chair: Gillian Nave, (USA), <gnave@nist.gov>
co-Chair: Glenn M. Wahlgren (Sweden), <glenn.wahlgren@astro.lu.se>
co-Chair: Jeffrey R. Fuhr (USA), <jeffrey.fuhr@nist.gov>
URL: <http://iacs.cua.edu/IAUC14>

Div.XII/Comm.14/WG - Molecular Data
Chair: John H. Black (Sweden), <John.Black@chalmers.se>
URL: <http://iacs.cua.edu/IAUC14>

Div.XII/Comm.14/WG - Collision Processes
Chair: Philip Stancil (USA), <stancil@physast.uga.edu>
co-Chair: Gillian Peach (UK), < ucap22g@ucl.ac.uk>
URL: <http://iacs.cua.edu/IAUC14/>

Div.XII/Comm.14/WG - Solids and their Surfaces
Chair: G. Vidali
URL: <http://iacs.cua.edu/IAUC14/>

Div.XII/Comm.41 - History of Astronomy
P: Il-Seong Nha (South Korea), <slisnha@chol.com>
VP: Clive L.N. Ruggles (UK), <rug@le.ac.uk>
OC: Alexander A. Gurshtein (Russia/USA), Teije de Jong (Netherlands), Rajesh K. Kochhar (India), Tsuko Nakamura (Japan), Wayne Orchiston (Australia), Antonio A.P. Videira (Brazil), and Brian Warner (South Africa)
URL: <http://www.le.ac.uk/has/c41/>

Div.XII/Comm.41/WG - Archives
Chair: Ileana Chinnici (Italy), <chinnici@astropa.unipa.it>
URL: <http://www.le.ac.uk/has/c41/wgarc.html>

Div.XII/Comm.41/WG - Historical Instruments
Chair: Luisa Pigatto (Italy), <luisa.pigatto@oapd.inaf.it>
URL: <http://www.oapd.inaf.it/museo/PagineInglesi/History

Div.XII/Comm.41/WG - Transits of Venus
Chair: Steven J. Dick (USA), <steven.j.dick@nasa.gov>
URL: <http://www.le.ac.uk/has/c41/wgtov.html>

Div.XII/Comm.46 - Astronomy Education and Development
P: Magdalena G. Stavinschi (Romania), <magda@aira.astro.ro>
VP: Rosa M. Ros (Spain), <ros@mat.upc.es>
OC: Michèle Gerbaldi (France), Edward F. Guinan (USA), John B. Hearnshaw (New Zealand), Margarita Metaxa (Greece), Nidia Morrell (Argentina), Mazlan Othman (Malaysia), John R. Percy (Canada), Charles R. Tolbert (USA), Silvia Torres-Peimbert (Mexico), and James C. White (USA)
URL: <http://physics.open.ac.uk/IAU46/>

Div.XII/Comm.46/PG - World Wide Development of Astronomy
Chair: John B. Hearnshaw (New Zealand), <john.hearnshaw@canterbury.ac.nz>
URL: <http://physics.open.ac.uk/IAU46/programme%20groups.html#PGWWDA>

Div.XII/Comm.46/PG - Teaching for Astronomy Development
co-Chair: Laurence A. Marshall (USA), <marshall@gettysburg.edu>
co-Chair: Edward F. Guinan (USA), <edward.guinan@villanova.edu>
URL: <http://physics.open.ac.uk/IAU46/programme%20groups.html#TAD>
<http://physics.open.ac.uk/IAU46/guidelines.html>

Div.XII/Comm.46/PG - International Schools for Young Astronomers
Chair: Jean-Pierre de Greve (Belgium), <jpdgreve@vub.ac.be>
vice-Chair: Kam-Ching Leung (USA), <kleung@unlserve.unl.edu>
URL: <http://physics.open.ac.uk/IAU46/programme%20groups.html#ISYA>

Div.XII/Comm.46/PG - Collaborative Programs
Chair: Hans J. Haubold (Austria), <haubold@kph.tuwien.ac.at>
URL: <http://physics.open.ac.uk/IAU46/programme%20groups.html#CP>

Div.XII/Comm.46/PG - Exchange of Astronomers
Chair: John R. Percy (Canada), <jpercy@credit.erin.utoronto.ca>
URL: <http://physics.open.ac.uk/IAU46/programme%20groups.html#EoA>

Div.XII/Comm.46/PG - National Liaison on Astronomy Education
Chair: Barrie W. Jones (UK), <b.w.jones@open.ac.uk>
URL: <http://physics.open.ac.uk/IAU46/programme%20groups.html#NL>

Div.XII/Comm.46/PG - Commission Newsletter
Chair: Barrie W. Jones (UK), <b.w.jones@open.ac.uk>
URL: <http://physics.open.ac.uk/IAU46/programme%20groups.html#News>

Div.XII/Comm.46/PG - Public Education on the Occasions of Solar Eclipses
Chair: Jay M. Pasachoff (USA), <jay.m.pasachoff@williams.edu>
URL: <http://physics.open.ac.uk/IAU46/programme%20groups.html#PGSE>

Div.XII/Comm.46/PG - Exchange of Books and Journals
Chair: Susana E. Deustua (USA), <deustua@aas.org>
URL: <http://physics.open.ac.uk/IAU46/programme%20groups.html#Ex>

Div.XII/Comm.50 - Protection of Existing and Potential Observatory Sites
P : Hugo E. Schwarz (Chile, † 2006)
VP: Richard J. Wainscoat, (USA), acting president ,<rjw@ifa.hawaii.edu>
OC: Carlo Blanco (Italy), Jim Cohen, PP (UK, † 2006), David L. Crawford (USA), Magarita Metaxa (Greece), and Woodruff T. Sullivan, PP (USA)
URL: <http://www.ctio.noao.edu/cgi-bin/iau50.pl>

Div.XII/Comm.50/WG - Controlling Light Pollution
Chair: Richard J. Wainscoat (USA), <rjw@ifa.hawaii.edu>
URL: <http://www.ctio.noao.edu/light_pollution/iau50/>

Div.XII/Comm.55 - Communicating Astronomy with the Public
P : Ian E. Robson (UK), <eir@roe.ac.uk>
VP: Dennis Crabtree (Canada), <Dennis.Crabtree@nrc-cnrc.gc.ca>
S : Lars Lindberg Christensen (ESO/ESA), <lars@eso.org>
OC: Augusto Damineli Neto (Brazil), Richard T. Fienberg (USA), Anne Green (Australia), Ajit K. Kembhavi (India), Birgitta Nordström (Denmark), Oscar Alvarez-Pomares (Cuba), Kazuhiro Sekiguchi (Japan), Patricia A. Whitelock (South Africa), and Jin Zhu (China)
URL: <http://www.communicatingastronomy.org/>

Div.XII/Comm.55/WG - Washington Charter
Chair: Dennis Crabtree (Canada), <Dennis.Crabtree@nrc.ca>
URL: <http://www.communicatingastronomy.org/washington_charter/index.html>

Div.XII/Comm.55/WG - Virtual Astronomy Multimedia Project
Chair: Adrienne Gaulthier (USA), <gauthier@as.arizona.edu>
URL: <http://www.communicatingastronomy.org/repository/index.html>

Div.XII/Comm.55/WG - Best practices
Chair: Lars Lindberg Christensen (Germany), <lars@eso.org>
URL: <http://www.communicatingastronomy.org/bestpractices/index.html>

Div.XII/Comm.55/WG - Communicating Astronomy Journal
Chair: Pedro Russo (Germany), <prusso@eso.org>
URL: <http://www.communicatingastronomy.org/journal/index.html>

Div.XII/Comm.55/WG - New Ways of Communicating Astronomy with the Public
Chair: Michael West (USA), <mwest@gemini.edu>
URL: <http://www.communicatingastronomy.org/newways/index.html>

Div.XII/Comm.55/WG - Communicating Astronomy with the Public conferences
Chair: Ian E. Robson (UK), <eir@roe.ac.uk>
URL: <http://www.communicatingastronomy.org/capconferences/index.html>

DIVISION XII INTER-DIVISION WORKING GROUP

Div.X-XII/WG - Historic Radio Astronomy
Chair: Wayne Orchiston (Australia), <Wayne.Orchiston@jcu.edu.au>
URL: <http://www.le.ac.uk/has/c41/wghra.html>

Transactions IAU, Volume XXVIB
Proc. IAU XXVI General Assembly, August 2006
Karel A. van der Hucht, ed.

doi:10.1017/S174392130802437X

CHAPTER X

NATIONAL MEMBERSHIP

Introduction

This chapter lists the Adhering Organizations of the 63 National Members of the International Astronomical Union. As of September 2008, the IAU had 7 members with *Interim* status, 21 members in Category I, 10 members in Category II, 11 members in Category III, 3 members in Category IV, 4 members in Category V, 1 member in Category VI, 5 members in Category VII, and 1 member in Category X.

Transactions IAU, Volume XXVIB
Proc. IAU XXVI General Assembly, August 2006
Karel A. van der Hucht, ed.

doi:10.1017/S1743921308024381

NATIONAL MEMBERS, ADHERING ORGANIZATIONS
(September 2008)

Year of Adherence	Country	Category	Units	Individual Members
1927	ARGENTINA CONICET Rivadavia 1917 AR - 1033 Buenos Aires	II	2	110
1935, 1994	ARMENIA Armenian National Academy of Sciences Marshal Baghramian av 24 AM - 375019 Yerevan	I	1	25
1939	AUSTRALIA Australian Academy of Sciences Ian Potter House Gordon St. G.P.O. Box 783 AU Canberra, ACT 2601	IV	6	241
1955	AUSTRIA Die österreichische Akademie der Wissenschaften Dr Ignaz Seipel Platz 2 AT - 1010 Vienna	I	1	36
1920	BELGIUM Koninklijke Vlaamse Academie voor Wetenschap en Kunsten Académie royale des Sciences, des Lettres et des Beaux-Arts Palais des Académies Rue Ducale 1 BE - 1000 Bruxelles	IV	6	106
1998	BOLIVIA Academia Nacional de Ciencias de Bolivia Av 16 de Julio No. 1732 BO - La Paz	*Interim*	0.5	0
1961	BRAZIL CNPq Av W3 Norte Quadra 507 B Caixa Postal 11-1142 BR - Brasilia, DF 70740	II	2	160

1957	BULGARIA Bulgarian Academy of Sciences 1 15 Noemvri Street BG-1040 Sofia	I	1	48
1920	CANADA National Research Council of Canada Director, International Relations Office Strategy and Development Branch 100 Sussex Drive CA - Ottawa ON K1A 0R6	V	10	224 65
1947	CHILE CONICYT Chilean Academy of Sciences Casilla 297 V CL - Santiago	II	2	65
1935	CHINA NANJING Purple Mountain Observatory Chinese Academy of Sciences 2 West Beijing Rd CN - 210008 Nanjing	VI	14	315
1959	CHINA TAIPEI Institute of Astronomy and Astrophysics Academia Sinica 128 Academia Rd Sec. 2 Nankang TW - 115 Taipei	II	2	40
1935, 1994	CROATIA Hrvatsko Astronomsko Društvo Croatian Academy of Sciences & Arts Kaciceva 26 HR - 41000 Zagreb Hrvatska	I	1	14
1922, 1993	CZECH REPUBLIC Academy of Sciences Narodni 3 CZ - 117 20 Praha 1	III	4	81
1922	DENMARK Kongelige Danske Videnskabernes Selskab H. C. Andersens Bvd 35 DK - 1553 Copenhagen V	III	4	64

1925	EGYPT Academy of Scientific Reseach & Technology 101 Kasr El-Eini Str EG - 11516 Cairo	III	4	56
1935, 1992	ESTONIA Estonian Academy of Sciences Kohtu 6 EE - 10130 Tallinn	I	1	21
1948	FINLAND The Finnish Academy of Science and Letters Mariankatu 5 A, FI - 00170 Helsinki	II	2	62
1920	FRANCE COFUSI Académie des Sciences 23 Quai Conti FR - 75006 Paris	VII	20	661
1951	GERMANY Rat Deutscher Sternwarten MPI für Extraterrestrische Physik Giessenbachstrasse 1 DE - 85748 Garching bei München	VII	20	522
1920	GREECE Hellenic Ministry of Development General Secretariat for Research & Technology 14-18 Messogeion Ave GR - 115 27 Athens	III	4	104
1947	HUNGARY Hungarian Academy of Sciences Nador ulice 7 HU - 1051 Budapest	II	2	47
1988	ICELAND Ministry of Education, Sciences and Culture Office of Financial Affairs Solvholsgata 4 Dunhaga 3 IS -150 Reykjavik	I	1	5
1964	INDIA Indian National Science Academy 2 Bahadur Shah Zafar Marg IN - 110 002 New Delhi	V (2009)	10	214

1964, 1979	INDONESIA Indonesian Institute of Sciences Lembaga Ilmu Pengetahuan Indonesia - LIPI Gedung Widya Graha Jl. Jenderal Gatot Subroto 10 ID - 12710 Jakarta 12710	I	1	16
1969	IRAN University of Tehran Office International Relations Enghlab Ave IR - Tehran	I	1	24
1947	IRELAND The Royal Irish Academy 19 Dawson Street IE - 2 Dublin	I	1	40
1954	ISRAEL Israel Academy of Sciences & Humanities Albert Einstein Square 43 Jabotinsky Square PO Box 4040 IL - 91040 Jerusalem	III (2009)	4	66
1920	ITALY Istituto Nazionale di AstroFisica - INAF Via del Parco Mellini 84 IT - 00136 Roma	VII	20	527
1920	JAPAN Science Council of Japan 7-22-34 Roppongi Minato-ku JP - 106-8555 Tokyo	VII	20	547
1973	KOREA, RP Korean Astronomical Society 61 Hwa-am Dong, Yusung Ku KR - 305-348 Taejon	II	8	98
1996	LATVIA University of Latvia Raina blvd 19 LV - 1586 Riga	I	1	15

2006	LEBANON National Council for Scientific Research Riad El Solh 1107 PO Box 11-8281 LB - 2260 Beirout	*Interim*	0.5	5
1935, 1993	LITHUANIA Lithuanian Academy of Sciences MTP-1 Gedimino Prospekt 3 LT - 2600 Vilnius	I	1	15
1988	MALAYSIA Kementerian Sains Teknologi dan Inovasi Aras 5, Block 2, Menara PjH, Presint 2 MY - Putrajaya 62100	*Interim*	0.5	7
1921	MEXICO Instituto de Astronomía Universidad Nacional Autonoma de Mexico Circuito exterior s/n Ciudad Universitaria MX - Mexico DF 04510	III	4	98
2006	MONGOLIA Mongolian Academy of Sciences Sukhbaatar Sq.3 A. Amar Street 1 MN - 210620a Ulaanbaatar 11	*Interim*	0.5	0
1988, 2001	MOROCCO CNCPRST 52 bd Omar Ibn Khattab BP 8027 MA - 10102 Agdal-Rabat	*Interim*	0.5	7
1922	NETHERLANDS Koninklijke Nederlandse Akademie van Wetenschappen Kloveniersburgwal 29 Postbus 19121 NL - 1000 GC Amsterdam	V	10	187
1964	NEW ZEALAND Royal Society of New Zealand PO Box 598 NZ - Wellington	II	2	28

2003	NIGERIA Nigerian Academy of Science University of Lagos PMB 1004 Univ of Lagos Post Office Akoka-Yaba NG - Lagos	I	1	10
1922	NORWAY Det Norske Videnskaps-Akademi Drammensveien 78 NO - 0271 Oslo	II	2	29
1988	PERU Consejo Nacional de Ciencia y Tecnologia Apartado Postal 1984 PE - 100 Lima	*Interim*	0.5	4
2001	PHILIPPINES c/o Chief, AGSSB Science Garden Complex Agham Rd, Diliman PH - 1101 Quezon City	*Interim*	0.5	4
1922	POLAND Polish Academy of Sciences Palac Kultury i Nauki Plac Defilad 1 PL - 00-901 Warsaw	III	4	140
1924	PORTUGAL Secção Portuguesa das Uniões Internacionais Astronómica e Geodésica - SPUIAGG Rua da Artilharia Um 107 PT - 1099-052 Lisboa	II	2	37
1922	ROMANIA Romanian Academy of Sciences Calea Victoriei 125 Sector 1, Cod 010071 RO - 71102 Bucharest	I	1	34
1935, 1992	RUSSIAN FEDERATION Russian Academy of Sciences Foreign Relations Department Leninskij Prospekt 14 RU - 117901 Moscow	V	10	383

1988	SAUDI ARABIA King Abdulaziz City for Science & Technology KACST Box 6086 SA - 11442 Riyadh	I	1	11
1935, 2003	SERBIA Drustvo astronoma Srbije Astronomska opservatorija Volgina 7 RS - 11160 Belgrade 74	I	1	32
1922, 1993	SLOVAKIA Slovak Academy of Sciences Scientific Secretary Stefanikova 49 SK - 814 38 Bratislava	I	1	31
1938	SOUTH AFRICA National Research Foundation P.O. Box 2600 ZA - 0001 Pretoria	III	4	62
1922	SPAIN Comision Nacional de Astronomia General Ibanez de Ibero 3 ES - 28003 Madrid	IV	6	254
1925	SWEDEN Royal Swedish Academy of Sciences Box 50005 SE - 104 05 Stockholm	III	4	110
1923	SWITZERLAND Swiss Academy of Sciences Schwarztorstrasse 9 CH - 3007 Bern	III	4	94
1935, 1993	TAJIKISTAN Institute of Astronomy Tajik Academy of Sciences Bukhoro Str 22 TJ - 734042 Dushanbe	I	1	6
2006	THAILAND National Astronomical Research Institute of Thailand Ministry of Science and Technology Yothee Rd., Ratchathewi TH - 10400 Bangkok	I	1	10

1961 TURKEY I 1 31
Turkish Astronomical Society
Sabanci Ünivrsitesi Lojmanlari
Orhanli-Tuzla
TR - 34956 Istanbul

1935, 1993 UKRAINE III 4 171
National Academy of Science of Ukraine
Ulitza Volodimirska 54
UA - 04053 Kiev

1920 UNITED KINGDOM VII 20 664
Royal Astronomical Society
Burlington House
Piccadilly
GB - London W1J OBQ

1920 UNITED STATES X 35 2497
The National Academies
Board on International Scientific Organizations
500 Fifth St. NW
US - Washington DC 20001

1932 VATICAN CITY STATE I 1 5
Governatorato Citta del Vaticano
VA - 00120 Citta del Vaticano

1953 VENEZUELA I 1 20
Centre de Investigaciones de Astronomía
Apdo Postal 264
VE - Mérida 5101 A

Transactions IAU, Volume XXVIB
Proc. IAU XXVI General Assembly, August 2006
Karel A. van der Hucht, ed.

doi:10.1017/S1743921308024393

CHAPTER XI

INDIVIDUAL MEMBERSHIP

INDIVIDUAL MEMBERSHIP BY ADHERING COUNTRY

Argentina

Abadi Mario
Aguero Estela L.
Ahumada Andrea V.
Ahumada Javier Alejandro
Alonso Maria V.
Aquilano Roberto Oscar
Arnal Edmundo Marcelo
Azcarate Diana E.
Bagala Liria G.
Bajaja Esteban
Barba Rodolfo H.
Bassino Lilia P.
Baume Gustavo L.
Beauge Christian
Benaglia Paula
Benvenuto Omar
Bosch Guillermo L.
Brandi Elisande E.
Branham Richard L.
Brunini Adrian
Calderon Jesus
Cappa de Nicolau Cristina
Carpintero Daniel Diego
Carranza Gustavo J.
Castagnino Mario
Castelletti Gabriela
Cellone Sergio Aldo
Cidale Lydia S.
Cincotta Pablo M.
Cionco Rodolfo G.
Claria Juan
Colomb Fernando R.
Combi Jorge A.
Cora Sofia A.
Costa Andrea
Cruzado Alicia
Dasso Sergio
De Biasi Maria S.
Di Sisto Romina P.
Donzelli Carlos J.
Dubner Gloria M.
Feinstein Carlos
Feinstein Alejandro
Fernandez Laura I.
Fernandez Silvia M.
Ferrer Osvaldo E.
Filloy Emilio M. E. E.
Forte Juan C.
Garcia Lambas D.
Garcia Beatriz E.
Garcia Lia G.
Giacani Elsa B.
Gil-Hutton Ricardo A.
Giordano Claudia M.
Goldes Guillermo V.
Gomez Mercedes
Gomez Daniel O.
Gonzalez Jorge F.
Grosso Monica Gladys
Hernandez Carlos A.
Hol Pedro E.
Iannini Gualberto
Lapasset Emilio
Levato Hugo
Lopez Garcia Z. l.
Lopez Carlos
Lopez Jose A.
Lopez Garcia Francisco
Luna Homero G.
Machado Marcos
Malaroda Stella M.
Mandrini Cristina H.
Manrique Walter T.
Marabini Rodolfo J.
Marraco Hugo G.
Martin Maria C.
Mauas Pablo
Melita Mario Daniel
Milone Luis A.
Morras Ricardo
Muriel Hernan
Muzzio Juan C.
Nunez Josue A.
Olano Carlos A.
Orellana Rosa B.
Orsatti Ana María
Perdomo Raul
Piacentini Ruben
Piatti Andrés E.
Pintado Olga I.
Plastino Angel R.
Poeppel Wolfgang G. L.
Rabolli Monica
Reynoso Estela M.
Ringuelet Adela E.
Romero Gustavo E.
Rovero Adrián C.
Rovira Marta G.
Sahade Jorge
Semenzato Roberto
Sistero Roberto F.
Solivella Gladys R.
Tissera Patricia B.
Valotto Carlos A.
Vazquez Ruben A.
Vega E. I.
Velazquez Pablo F.
Vergne María Marcela
Villada Monica M.
Vucetich Hector
Wachlin Felipe C.

Armenia

Abrahamian Hamlet V.
Gigoyan Kamo S.
Gurzadyan Grigor A.
Gurzadyan Vahagn G.
Gyulbudaghian Armen L.
Hambaryan Valeri V.
Harutyunian Haik A.
Hovhannessian Rafik Kh.
Kalloglian Arsen T.
Kandalyan Rafik A.
Khachikian Edward Yerem
Magakian Tigran Y.
Mahtessian Abraham P.
Malumian Vigen H.
Melikian Norair D.

Mickaelian Areg M.
Nikoghossian Arthur G.
Parsamyan Elma S.
Petrosian Artaches R.
Pikichian Hovhannes Vahram
Sanamian V. A.
Sedrakian David
Shakhbazian Romelia K.
Shakhbazyan Yurij L.
Yengibarian Norair

Australia

Argast Dominik
Ashley Michael
Asplund Martin
Bailes Matthew
Bailey Jeremy A.
Bailin Jeremy
Ball Lewis
Barnes David G.
Bedding Timothy R.
Bessell Michael S.
Bhat Ramesh N D
Bicknell Geoffrey V.
Biggs James
Birch Peter Vaughan
Blair David Gerald
Bland-Hawthorn Jonathan
Boyle Brian J.
Bray Robert J.
Bridges Terry J.
Briggs Franklin
Brooks Kate J.
Brown Ronald D.
Bruntt Hans
Burman Ronald R.
Burton Michael G.
Cairns Iver H.
Calabretta Mark R.
Cally Paul S.
Campbell-Wilson Duncan
Cane Hilary V.
Cannon Russell D.
Carter Bradley Darren
Caswell James L.
Chapman Jessica
Chapman Jacqueline F.
Clay Roger
Colless Matthew
Corbett Elizabeth A.
Costa Marco E.
Couch Warrick
Cram Lawrence Edward
Cramer Neil F.
Crocker Roland M.
Croom Scott M.
Cunningham Maria R.
Curran Stephen J.
Da Costa Gary Stewart
Davis John
Dawson Bruce
De Blok Erwin
De Kool Marthijn
De Propris Roberto
Donea Alina C.
Dopita Michael A.
Drinkwater Michael J.
Duldig Marcus L.
Durrant Christopher J.
Edwards Paul J.
Ekers Ronald D.
Elford William Graham
Ellingsen Simon P.
Ellis G. R. A.
Ellis Simon C.
Erickson William C.
Evans Robert
Fenton K. B.
Ferrario Lilia
Filipovic Miroslav D.
Fluke Christopher J.
Forbes Duncan Alan
Forbes J. E.
Francis Paul
Frater Robert H.
Freeman Kenneth C.
Fux Roger M.
Galloway David
Gascoigne S. C. B.
George Martin
Germany Lisa M.
Gingold Robert Arthur
Godfrey Peter Douglas
Graham Alister W. McK.
Green Anne
Greenhill John
Hall Peter J.
Haynes Roslynn
Haynes Raymond F.
Hopkins Andrew M.
Horiuchi Shinji
Horton Anthony J.
Hosking Roger J.
Hughes Stephen W.
Humble John Edmund
Hunstead Richard W.
Hurley Jarrod R.
Hyland Harry R. Harry
Jackson Carole A.
Jauncey David L.
Jenkins Charles R.
Jerjen Helmut
Jones Paul
Jones Heath D.
Kalnajs Agris J.
Keay Colin S. l.
Kedziora-Chudozer Lucyna L.
Keller Stefan C.
Kennedy Hans Daniel
Kesteven Michael J. l.
Kilborn Virginia A.
Killeen Neil
Kiss Laszlo L.
Koribalski Baerbel Silvia
Kuncic Zdenka
Lambeck Kurt
Lattanzio John
Lawrence Jon S.
Lawson Warrick
Lewis Geraint F.
Li Bo
Liffman Kurt
Lineweaver Charles H.
Lomb Nicholas Ralph
Lovell James E.
Luck John M.
Luo Qinghuan

Mackie Glen
Maddison Sarah T.
Madsen Gregory J.
Malin David F.
Manchester Richard N.
Mardling Rosemary A.
Mathewson Donald S.
McAdam Bruce W. B.
McClure-Griffiths Naomi M.
McConnell David
McCulloch Peter M.
McGee Richard Xavier
McGregor Peter John
McKinnon David H.
McLean Donald J.
McNaught Robert H.
McSaveney Jennifer A.
Melatos Andrew
Melrose Donald B.
Middelberg Enno
Mills Franklin P.
Mills Bernard Y.
Milne Douglas K.
Monaghan Joseph J.
Moreton G. E.
Morgan Peter
Muller Erik M.
Murdoch Hugh S.
Murphy Tara
Murray James R.
Nelson Graham John
Newell Edward B.
Nikoloff Ivan
Norris John
Norris Raymond P.
O'Byrne John W
Orchiston Wayne
Ord Stephen M.
O'Sullivan John David
O'Toole Simon J.
Ott Juergen A.
Page Arthur
Pandey Bierndra P.
Parker Quentin A.
Peterson Bruce A.
Phillips Christopher J.
Pimbblet Kevin A.
Pongracic Helen
Power Chris B.
Pracy Michael B.
Prentice Andrew J. R.
Proctor Robert N.
Protheroe Raymond J.
Rees David E.
Reynolds John
Robertson James Gordon
Robinson Peter A.
Robinson Garry
Russell Kenneth S.
Ryder Stuart
Sackett Penny
Sadler Elaine M.
Sault Robert J
Savage Ann
Schmidt Brian P.
Sharma Dharma P.
Sharp Robert
Sheridan K. V.
Shimmins Albert John
Shobbrook Robert R.
Shortridge Keith
Slee O. B.
Smith Robert G.
Smith Craig H.
Sood Ravi
Sparrow James G.
Stathakis Raylee A.
Staveley-Smith Lister
Stewart Ronald T.
Stibbs Douglas W. N.
Storey Michelle C.
Storey John W. V.
Subrahmanyan Ravi
Sutherland Ralph S.
Tango William J.
Taylor Kenneth N. R.
Tingay Steven J.
Tinney Christopher G.
Titov Oleg A.
Trampedach Regner
Tuohy Ian R.
Turtle A. J.
Tuthill Peter G.
Tzioumis Anastasios
Van der Borght Rene
Vaughan Alan
Walker Mark Andrew
Walsh Wilfred M.
Walsh Andrew J.
Wardle Mark J.
Waterworth Michael
Watson Robert
Watson Frederick Garnett
Webb John
Webster Rachel L.
Wellington Kelvin
White Graeme Lindsay
Whiteoak John B.
Wickramasinghe D. T.
Wild John Paul
Willes Andrew J.
Wilson Peter R.
Wood Peter R.
Wright Alan E.
Wyithe Stuart
Zealey William J.

Austria

Auner Gerhard
Breger Michel
Dorfi Ernst Anton
Dvorak Rudolf
Firneis Maria G.
Firneis Friedrich J.
Goebel Ernst
Handler Gerald
Hanslmeier Arnold
Hartl Herbert
Haubold Hans J.
Haupt Hermann F.
Hensler Gerhard
Hron Josef
Jackson Paul
Kerschbaum Franz
Koeberl Christian
Lebzelter Thomas
Leubner Manfred P.
Maitzen Hans M.
Nittmann Johann
Paunzen Ernst
Pfleiderer Jorg
Pilat-Lohinger Elke E.

Rakos Karl D.
Schindler Sabine
Schnell Anneliese
Schober Hans J.
Schroll Alfred
Schuh Harald
Stift Martin Johannes
Weber Robert
Weinberger Ronald
Weiss Werner W.
Zeilinger Werner W

Belgium

Aerts Conny
Alvarez Rodrigo
Arnould Marcel L.
Arpigny Claude
Baes Maarten
Berghmans David
Biemont Emile
Blommaert Joris A. D. L.
Blomme Ronny
Boffin Henri M. J.
Borkowski Virginie
Bruyninx Carine
Burger Marijke
Cadez Vladimir
Callebaut Dirk K.
Claeskens Jean-François
Clette Frederic
Cuypers Jan
David Marc
De Cat Peter
De Cuyper Jean-Pierre M.
De Greve Jean-Pierre
De Groof Anik
De Keyser Johan
De Loore Camiel
De Ridder Joris
De Rijcke Sven
De Rop Yves
De Viron Olivier
Debehogne Henri Sc.
Decin Leen K. E.
Defraigne Pascale
Dehant Véronique
Dejaiffe Rene J.
Dejonghe Herwig B.
Denis Carlo
Denoyelle Jozef K.
Dommanget Jean
Elst Eric Walter
Fremat Yves
Fu Jianning
Gabriel Maurice R.
Gerard Jean-Claude M. C.
Goossens Marcel
Goriely Stephane
Gosset Eric
Grevesse Nicolas
Groenewegen Martin
Henrard Jacques
Hensberge Herman
Heynderickx Daniel
Hochedez Jean-François E.
Houziaux Leo
Hutsemekers Damien
Huygen Eric
Jamar Claude A. J.
Jorissen Alain
Lampens Patricia
Lemaire Joseph F.
Lemaitre Anne
Magain Pierre
Manfroid Jean
Noels Arlette
Palmeri Patrick
Paquet Paul
Pauwels Thierry
Perdang Jean M.
Poedts Stefaan
Pourbaix Dimitri
Quinet Pascal
Rauw Gregor
Rayet Marc
Renson P. F. M.
Reyniers Maarten
Robe Henri H. G.
Roosbeek Fabian
Roth Michel A.
Royer Pierre
Runacres Mark C.
Sauval A. Jacques
Scuflaire Richard
Siess Lionel
Simon Paul C.
Sinachopoulos Dimitris
Siopis Christos
Smette Alain
Smeyers Paul
Sterken Christiaan L.
Surdej Jean M. G.
Swings Jean-Pierre
Thoul Anne A.
Van de Steene Griet C.
Van der Linden Ronald
Van Dessel Edwin Ludo
Van Eck Sophie
Van Hoolst Tim
Van Rensbergen Walter
Van Santvoort Jacques
Van Winckel Hans
Vandenbussche Bart
Verheest Frank
Verschueren Werner
Vreux Jean Marie
Waelkens Christoffel
Zander Rodolphe

Brazil

Abraham Zulema
Alcaniz Jailson S.
Aldrovandi Ruben
Andrei Alexandre H.
Angeli Claudia A.
Arany-Prado Lilia I.
Assafin Marcelo
Baptista Raymundo
Barbosa Cassio L.
Barbuy Beatriz
Barroso Jr Jair
Batalha Celso Correa
Benevides Soares Paulo
Berman Marcelo S.
Bica Eduardo L. D.

Boechat-Roberty
Heloisa M.
Bonatto Charles J.
Braga João
Bruch Albert
Canalle Joao B. G.
Capelato Hugo Vicente
Carvalho Joel C.
Carvano Jorge M. F.
Chan Roberto
Chian Abraham
Chian-Long
Corradi Wagner J. B.
Correia Emilia
Costa Joaquim E. R.
Cuisinier Francois C.
Cunha Katia
Da Costa Luiz A. N.
Da Costa Jose Marques
Da Rocha Vieira E.
Da Silva Licio
Daflon Simone
D'Amico Flavio
Damineli Neto Augusto
Dantas Christine C.
De Aguiar Odylio Denys
De Almeida Amaury A.
De Araujo Francisco X.
De Carvalho Reinaldo
De Freitas Mourao
Ronaldo R.
De Gouveia Dal Pino
Elisabete M.
De la Reza Ramiro
De Medeiros Jose Renan
De Souza Ronaldo
Del Peloso Eduardo F.
Dias da Costa Roberto D.
Diaz Marcos P.
Do Nascimento Jose D.
Dottori Horacio A.
Drake Natalia
Ducati Jorge R.
Emilio Marcelo
Faundez-Abans Max
Ferraz Mello Sylvio
Foryta Dietmar William
Franco Gabriel Armando P.
Freitas Mourao R.
Friaca Amancio C. S.
Giacaglia Giorgio E.
Gimenez de Castro Carlos
Guillermo
Giuliatti Winter Silvia M.
Gomes Alercio M.
Gomes Rodney D. S.
Gomide Fernando de Mello
Gonzales'a Walter D.
Gregorio-Hetem Jane C.
Gruenwald Ruth B.
Hetem Jr. Annibal
Idiart Thais E.
Jablonski Francisco J.
Jafelice Luiz C.
Janot Pacheco Eduardo
Jatenco-Pereira Vera
Jayanthi Udaya B.
Kaufmann Pierre
Kepler S. O.
Kotanyi Christophe
Lazzaro Daniela
Leister Nelson Vani
Leite Scheid P.
Lepine Jacques R. D.
Lima Jose A. S.
Lima Botti Luiz Claudio
Lima Neto Gastao B.
Lopes Dalton De faria
Lorenz-Martins Silvia
Machado Maria A. D.
Maciel Walter J.
Magalhaes Antonio Mario
Maia Marcio R. G. S.
Maia Marcio A. G.
Marques Dos Santos P.
Martin Inacio Malmonge
Martin Vera A. F.
Matsuura Oscar T.
Medina Tanco Gustavo A.
Meliani Mara T.
Mendes Luiz T. S.
Mendes Da Costa Aracy
Mendes de Oliveira
Cláudia L.
Miranda Oswaldo D.
Mothe-Diniz Thais
Neves de Araujo Jose Carlos
Novello Mario
Oliveira Grijo A. K.
Opher Reuven
Ortiz Roberto
Palmeira Ricardo A. R.
Pastoriza Miriani G.
Pellegrini Paulo S. S.
Penna Jucira L.
Pereira Claudio B.
Piazza Liliana Rizzo
Plana Henri M.
Pompeia Luciana
Poppe Paulo C. d. R.
Porto de Mello Gustavo F.
Quarta Maria Lucia
Quast Germano Rodrigo
Rao K. Ramanuja
Raulin Jean-Pierre
Ribeiro Marcelo B.
Rocha-Pinto Helio J.
Roig Fernando V.
Rosa Reinaldo R.
Rossi Silvia C. F.
Santos Nilton Oscar
Santos Jr Joao F.
Sawant Hanumant S.
Scalise Jr Eugenio
Schaal Ricardo E.
Schuch Nelson Jorge
Silva Adriana V. R.
Soares Domingos S. L.
Sodre Laerte
Steiner Joao E.
Storchi-Bergmann Thaisa
Takagi Shigetsugu
Tateyama Claudio Eiichi
Teixeira Ramachrisna
Telles Eduardo
Tello Bohorquez Camilo
Torres Carlos Alberto O.
Tsuchida Masayoshi
Vaz Luiz Paulo Ribeiro
Veiga Carlos Henrique
Videira Antonio A.
Viegas Sueli M. M.
Vieira Martins Roberto
Vieira Neto Ernesto
Vilas-Boas José W. S.
Vilhena Rodolpho
Moraes R.
Villas da Rocha Jaime F.
Villela Neto Thyrso
Willmer Christopher N. A.
Winter Othon Cabo
Yokoyama Tadashi

Bulgaria

Antov Alexandar
Bachev Rumen S.
Bonev Tanyu
Borissova Jordanka
Duchlev Peter I.
Filipov Latchezar Georgiev
Georgiev Leonid
Georgiev Tsvetan
Golev Valery K.
Iliev Ilian
Ivanov Georgi R.
Ivanova Violeta
Kalinkov Marin P.
Kirilova Daniela
Kjurkchieva Diana
Kolev Dimitar Zdravkov
Komitov Boris
Konstantinova-Antova Renada K.
Kovachev Bogomil Jivkov
Kraicheva Zdravka
Kunchev Peter
Kurtev Radostin G.
Marchev Dragomir V.
Markov Haralambi S.
Markova Nevjana
Mihov Boyko M.
Nedialkov Petko L.
Nikolov Nikola S.
Nikolov Andrej
Panov Kiril
Petrov Georgi Trendafilov
Peykov Zvezdelin I.
Popov Vasil Nikolov
Raikova Donka
Russev Ruscho Minchev
Russeva Tatjana
Sbirkova-Natcheva T.
Semkov Evgeni Hristov
Shkodrov Vladimir G.
Spassova Nedka Marinova
Stanishev Vallery D.
Stateva Ivanka K.
Stavrev Konstantin Y.
Stoev Alexey D.
Strigatchev Anton
Tomov Nikolai A.
Tsvetkov Tsvetan
Tsvetkov Milcho K.
Tsvetkova Katja
Umlenski Vasil
Yankulova Ivanka
Zhekov Svetozar A.

Canada

Abraham Roberto G.
Aikman G. Chris L.
Auman Jason R.
Avery Lorne W.
Balogh Michael L.
Bartel Norbert Harald
Bastien Pierre
Basu Dipak
Basu Shantanu
Batten Alan H.
Beaudet Gilles
Beaulieu Sylvie F.
Bell Morley B.
Bennett Philip D.
Bergeron Pierre
Bishop Roy L.
Bochonko D. Richard
Bohlender David
Bolton Charles Thomas
Bond John Richard
Borra Ermanno F.
Brassard Pierre
Brooks Randall C.
Broten Norman W.
Brown Peter Gordon
Burke J. Anthony
Caldwell John James
Campbell-Brown Margaret D.
Cannon Wayne H.
Carignan Claude
Carlberg Raymond Gary
Charbonneau Paul
Clark Thomas A.
Clark Thomas Alan
Clarke Thomas R.
Clarke David A.
Claude Stephane M.
Clement Maurice J.
Climenhaga John L.
Clutton-Brock Martin
Connors Martin G.
Cote Stéphanie
Cote Patrick
Couchman Hugh M. P.
Courteau Stephane J.
Coutts-Clement Christine
Crabtree Dennis
Crampton David
Davidge Timothy J.
Dawson Peter
De Robertis Michael M.
Demers Serge
Dewdney Peter E. F.
Di Francesco James
Dobbs Matt A.
Dougherty Sean M.
Douglas R. J.
Doyon Rene
Drissen Laurent
Duley Walter W.
Duncan Martin J.
Dupuis Jean
Durand Daniel
Dutil Yvan
Dyer Charles Chester
English Jayanne
Fahlman Gregory G.
Feldman Paul A.
Feldman Paul Donald
Fernie J. Donald
Ferrarese Laura
Fich Michel
Fletcher J. Murray
Fontaine Gilles
Forbes Douglas
Fort David N.
Gaizauskas Victor

Galt John A.
Garrison Robert F.
Gower Ann C.
Gray David F.
Gregory Philip C.
Griffin R. Elizabeth
Guenther David Bruce
Gulliver Austin Fraser
Halliday Ian
Hanes David A.
Harris Gretchen L. H.
Harris William E.
Hartwick F. David A.
Hawkes Robert Lewis
Henriksen Richard N.
Hesser James E.
Hickson Paul
Higgs Lloyd A.
Holmgren David E.
Houde Martin
Hube Douglas P.
Hudson Michael J.
Hutchings John B.
Innanen Kimmo A.
Irwin Alan W.
Irwin Judith
Israel Werner
Jarrell Richard A.
Jayawardhana Ray
Jeffers Stanley
Johnson Jennifer A.
Johnstone Douglas I.
Joncas Gilles
Jones James
Kaspi Victoria M.
Kavelaars JJ. Matthew
Kenny Harold
Knee Lewis
Koehler James A.
Kouba Jan
Kronberg Philipp
Lake Kayll William
Lamontagne Robert
Landecker Thomas L.
Landstreet John D.
Leahy Denis A.
Legg Thomas H.
Lester John B.
Locke Jack L.
Lowe Robert P.
MacLeod John M.
Marlborough J. Michael
Martel Hugo
Martin Peter G.
Matthews Jaymie
Matthews Brenda C.
McCall Marshall Lester
McCutcheon William H.
McIntosh Bruce A.
Menon T. K.
Merriam James B.
Michaud Georges J.
Milone Eugene F.
Mitalas Romas Assoc
Mitchell George F.
Mochnacki Stephan W.
Moffat John W.
Moffat Anthony F. J.
Monin Dmitry
Moorhead James M.
Morbey Christopher L.
Morton Donald C.
Nadeau Daniel
Navarro Julio Fernando
Naylor David A.
Nemec James
Nicholls Ralph W.
Odgers Graham J.
Page Don Nelson
Pathria Raj K.
Pen Ue-Li
Percy John R.
Pineault Serge
Plume Rene
Popelar Josef
Pritchet Christopher J.
Pryce Maurice H. l.
Purton Christopher R.
Racine Rene
Rice John B.
Richardson Eric Harvey
Richer Harvey B.
Robb Russell M.
Robert Carmelle
Roberts Scott C.
Roger Robert S.
Rogers Christopher
Rosvick Joanne M.
Routledge David
Rucinski Slavek M.
Rutledge Robert E.
Safi-Harb Samar
Scarfe Colin David
Schade David J.
Scott Douglas
Scrimger J. Norman
Seaquist Ernest R.
Sher David
Sigut Aaron T. A.
Sills Alison I.
Smylie Douglas E.
Sreenivasan S. Ranga
Stagg Christopher
Stairs Ingrid H.
Steinbring Eric
Stetson Peter B.
St-Louis Nicole
Sutherland Peter G.
Talon Suzanne
Tapping Kenneth F.
Tassoul Jean-Louis
Tatum Jeremy B.
Taylor Russell A.
Tikhomolov Evgeniy
Turner David G.
Vallee Jacques P.
Van den Bergh Sidney
Van Kerkwijk Marten H.
VandenBerg Don
Veran Jean-Pierre
Vorobyov Eduard I.
Wade Gregg A.
Walker Gordon A. H.
Webb Tracy M. A.
Webster Alan R.
Wehlau Amelia F.
Welch Gary A.
Welch Douglas L.
Wesemael Francois
Wesson Paul S.
Widrow Larry M.
Wiegert Paul A.
Willis Anthony G.
Wilson Christine
Woodsworth Andrew W.
Wu Yanqin
Yang Stephenson L. S.
Yee Howard K. C.

Chile

Alcaino Gonzalo
Alloin Danielle
Alvarez Hector
Aparici Juan
Bagnulo Stefano
Beasley Anthony James
Boehnhardt Hermann
Bouchet Patrice
Bronfman Leonardo
Campusano Luis E.
Carrasco Guillermo
Catelan Márcio
Celis Leopoldo
Costa Edgardo
De Buizer James M.
Fouque Pascal
Garay Guido
Geisler Douglas P.
Gieren Wolfgang P.
Gutierrez-Moreno A.
Hamuy Mario
Hardy Eduardo
Hubrig Swetlana
Hummel Christian Aurel
Infante Leopoldo
Jehin Emmanuel
Kaufer Andreas
Krisciunas Kevin
Krzeminski Wojciech
Kunkel William E.
Liller William
Lindgren Harri
Loyola Patricio
Mathys Gautier
Maury Alain J.
May J.
Maza Jose
Melnick Jorge
Melo Claudio H.
Méndez René A.
Minniti Dante
Morrell Nidia
Naef Dominique
Noel Fernando
Nurnberger Dieter E. A.
Nyman Lars-Aake
Pedreros Mario
Phillips Mark M.
Quintana Hernan
Richtler Tom
Roth Miguel R.
Rubio Monica
Ruiz Maria Teresa
Schmidtobreick Linda
Schoeller Markus
Smith Malcolm G.
Suntzeff Nicholas B.
Szeifert Thomas
Tokovinin Andrei A.
Torres Carlos
Vogt Nikolaus
Walker Alistair Robin
West Michael J.

China Nanjing

Ai Guoxiang
Bao Shudong
Bi Shao Lan
Bian Yulin
Cao Huilai
Cao Xinwu
Chan Kwing Lam
Chang Jin
Chang Ruixiag
Chao Zhang Zhen
Chen DaMing
Chen Li
Chen Peng Fei
Chen Yongjun
Chen Meidong
Chen Peisheng
Chen Daohan
Chen Jiansheng
Chen Yang
Chen Dong
Chen Yafeng
Chen Li
Chen Yang
Cheng Kwongsang
Cheng Fuzhen
Chu Yaoquan
Chun Sun Y.
Cui Xiangqun
Cui Shizhu
Cui Zhenhua
Dai Zigao
Deng Zugan
Deng YuanYong
Deng LiCai
Di Xiaohua
Ding Mingde
Dong Xiaojun
Fan Zuhui
Fan Yu
Fan Junhui
Fang Cheng
Fang Li Li
Fang Liu Bi
Feng Long Long
Fong Chugang
Fu Yanning
Fu Jian-Ning
Gan Weiqun
Gang Zhang Shou
Gao Yuping
Gao Buxi
Gao Bilie
Gu Qiusheng
Gu Sheng-hong
Guo Hongfang
Han Zhanwen
Han Yanben
Han JinLin
Han Tianqi
Hao Jinxin
He Miao-fu
He XiangTao
Hong Xiaoyu
Hong Zhang
Hong Wu
Hong Sun Cai
Hou Jinliang
Hu Xiaogong
Hu Wenrui
Hu Yonghui
Hu Tiezhu
Hu Fuxing

Huang Cheng-Li
Huang Tianyi
Huang Jiehao
Huang Keliang
Huang YongFeng
Huang Guangli
Huang He
Huang Runqian
Huang Cheng
Ji Haisheng
Ji Jianghui
Jiang Xiaojun
Jiang Yun Chun
Jiang Xiaoyuan
Jiang Biwei
Jiang Dongrong
Jiang Chongguo
Jiang Zhaoji
Jin Shenzeng
Jin WenJing
Jin Biaoren
Jing Hairong
Jing Yipeng
Kwok Sun
Lei Chengming
Li Guangyu
Li Xinnan
Li Zongwei
Li Zhongyuan
Li Tipei
Li Zhigang
Li Xiaoqing
Li Qibin
Li Xiangdong
Li Guoping
Li Jinling
Li Li Qingkang
Li Yuanjie
Li Wei
Li Hui
Li Qi
Li Yan
Li Zhiping
Li Zhengxing
Li Kejun
Lian Luo Xin
Liang Shiguang
Liao Dechun
Liao Xinhao
Lin Xuan-bin
Lin Qing
Liu Fukun
Liu Ciyuan
Liu Xinping
Liu Caipin
Liu Yuying
Liu Lin
Liu Yongzhen
Liu Wenzhong
Liu Xiang
Liu Zhong
Liu Qingzhong
Lou Yu-Qing
Lu BenKui
Lu Ruwei
Lu Jufu
Lu Tan
Lu Chunlin
Luo Xianhan
Luo Dingchang
Luo Shaoguang
Ma Yuehua
Ma Xingyuan
Ma YuQian
Ma Jingyuan
Ma Zhenguo
Ma Wenzhang
Mao Rui-Qing
Mao Weijun
Meng Xinmin
Min Wang Yu
Min Yuan Wei
Ming Bai Jin
Nakashima Jun-ichi
Nan Ren-Dong
Pan Liande
Pei Chunchuan
Peng Bo
Peng Qiuhe
Qi Guanrong
Qian Shengbang
Qiao Guojun
Qiao Rongchuan
Qiming Wang
Qin Yi-Ping
Qin Bo
Qin Zhihai
Qiu Yulei
Qiu Yaohui
Qu Zhong Quan
Qu Qinyue
Rong Jianxiang
Rui Qi
Shao Zhengyi
Shen Kaixian
Sheng Wan Xiao
Shi Shengcai
Shu Chenggang
Song Jinan
Su Ding-qiang
Sun Yisui
Sun Jin
Sun Xiaochun
Tan Huisong
Tang Yuhua
Tang Zheng Hong
Tao Jin-he
Tao Jun
Wang Hongchi
Wang Yiping
Wang Shouguan
Wang Gang
Wang Tinggui
Wang Yong
Wang Huaning
Wang Zhenru
Wang Shui
Wang Jingxiu
Wang Zhengming
Wang Jiaji
Wang Rongbin
Wang Min
Wang jiancheng
Wang Na
Wang Ding-Xiong
Wang Yanan
Wang Junjie
Wang Kemin
Wei Jianyan
Wei Liu Xiao
Wei Daming
Wei Wenren
Wenlei Shan
Wu Dong Shao
Wu Xuebing
Wu Haitao
Wu Shaoping
Wu Xuejun
Wu De Jin
Wu Bin

Wu Xiangping
Wu Yuefang
Wu Shouxian
Wu Guichen
Wu Xinji
Wu Lianda
Xi Zezong
Xia Yi
Xia Yifei
Xia Xiao-Yang
Xiang Shouping
Xiao Naiyuan
Xie Guangzhong
Xiong Yaoheng
Xiong Da Run
Xu Jin
Xu Aoao
Xu Jiayan
Xu Chongming
Xu Renxin
Xu Jun
Xu Weibiao
Xue Suijian
Yan Yihua
Yan Jun
Yang Ji
Yang Shijie
Yang Fumin
Yang Zhiliang
Yang Shimo
Yang Tinggao
Yang Zhigen
Yao Zhangqiu
Yao Zhi Da
Yao Yongqiang
Yao Qijun
Ye Shuhua
Ye Binxun
Ye Wenwei
Yi Meiliang
Yi Zhaohua
Yin Xinhui
Yong Zhou Li
You Junhan
Yu Dai
Yu Wang Xiang
Yu Nanhua
Yu Zhiyao
Yuan Ye-fei
Zeng Qin
Zhang Xiuzhong
Zhang Jian
Zhang Yang
Zhang Jin
Zhang Xizhen
Zhang Shuang Nan
Zhang Huawei
Zhang Li
Zhang Hongbo
Zhang Mei
Zhang Jialu
Zhang Jiaxiang
Zhang Baozhou
Zhang Tongjie
Zhang Qiang
Zhang Zhongping
Zhao Gang
Zhao Yongheng
Zhao Jun Liang
Zhao Juan
Zhao You
Zhao Donghai
Zhao Zhaowang
Zhao Changyin
Zheng Xinwu
Zheng Xiaoping
Zheng Xuetang
Zheng Weimin
Zhong Gu Bo
Zhong Min
Zhou Xu
Zhou Hongnan
Zhou Yonghong
Zhou Ji-Lin
Zhou Daoqi
Zhou Youyuan
Zhu Xingfeng
Zhu Nenghong
Zhu Wenyao
Zhu Yaozhong
Zhu Zi
Zhu Jin
Zhu Yongtian
Zhu Zhenxi
Zhu Zheng
Zou Zhenlong
Zuo Yingxi

China Taipei

Cai Michael
Chang Hsiang-Kuang
Chao Benjamin F.
Chen Wen Ping
Chin Yi-nan
Chiueh Tzihong
Chou Dean-Yi
Chou Chih-Kang
Chou Yi
Fu Hsieh-Hai
Fu-Shong Kuo
Gir Be Young
Hasegawa Tatsuhiko
Hirano Naomi
Hsiang-Kuang Tseng
Huang Yinn-Nien
Huang Yi-Long
Hwang Woei-yann P.
Hwang Chorng-Yuan
Ip Wing-Huen
Jiang Ing-Guey
Ko Chung-Ming
Kuan Yi-Jehng
Lee Jong Truenliang
Lee Thyphoon
Lim Jeremy
Ling Chih-Bing
Matsushita Satoki
Nee Tsu-Wei
Ng Kin-Wang
Sawada-Satoh Satoko
Shen Chun-Shan
Shu Frank H.
Sun Wei-Hsin
Ting Yeou-Tswen
Tsai Chang-Hsien
Tsao Mo
Tsay Wean-Shun
Wu Hsin-Heng
Yuan Chi
Yuan Kuo-Chuan

Croatia (Republic of)

Andreic Zeljko
Bozic Hrvoje
Brajsa Roman
Dadic Zarko
Jurdana-Sepic Rajka
Kotnik-Karuza Dubravka
Martinis Mladen E.
Pavlovski Kresimir
Rosa Dragan
Ruzdjak Vladimir
Solaric Nikola
Spoljaric Drago
Vrsnak Bojan
Vujnovic Vladis

Cuba

Alvarez-Pomares Oscar
Boytel Jorge del Pino
Cardenas Rolando P.
Doval Jorge M. Pérez
Garcia Eduardo del Pozo
Quiros Israel
Taboada Ramon Rodriguez

Czech Republic

Ambroz Pavel
Barta Miroslav
Bicak Jiri
Borovicka Jiri
Bouska Jiri
Bumba Vaclav
Bursa Milan
Ceplecha Zdenek
Dovciak Michal
Durech Josef
Ehlerova Sona
Farnik Frantisek
Fischer Stanislav
Grygar Jiri
Hadrava Petr
Harmanec Petr
Heinzel Petr
Hejna Ladislav
Horacek Jiri
Horsky Jan
Hudec Rene
Jungwiert Bruno
Karas Vladimir
Karlicky Marian
Kasparova Jana
Kleczek Josip
Klokocnik Jaroslav
Klvana Miroslav
Korcakova Daniela
Kostelecky Jan
Koten Pavel
Kotrc Pavel
Koubsky Pavel
Kraus Michaela
Krticka Jiri
Kubat Jiri
Lala Petr
Markova Eva
Mayer Pavel
Meszaros Attila
Mikulasek Zdenek
Moravec Zdenek
Nickeler Dieter H.
Novotny Jan
Odstrcil Dusan
Ouhrabka Miroslav
Palous Jan
Pecina Petr
Perek Lubos
Pesek Ivan
Pokorny Zdenek
Polechova Pavla
Pravec Petr
Ron Cyril
Ruzickova-Topolova B.
Schwartz Pavol
Sehnal Ladislav
Semerak Oldrich
Sidlichovsky Milos
Sima Zdislav
Simek Milos
Simon Vojtech
Skoda Petr
Sobotka Michal
Solc Martin
Spurny Pavel
Stefl Stanislav
Stefl Vladimir
Stuchlik Zdenek
Subr Ladislav
Svestka Jiri
Ticha Jana
Tlamicha Antonin
Valnicek Boris
Vandas Marek
Vetesnik Miroslav
Vokrouhlicky David
Vondrak Jan
Vykutilova Marie
Wolf Marek
Wunsch Richard

Denmark

Andersen Anja C.
Andersen Johannes
Arentoft Torben
Baerentzen Jorn
Bazot Michael
Borysow Aleksandra
Brandenburg Axel
Brandt Soeren K.
Budiz-Jorgensen Carl
Chenevez Jerome
Christensen Per R.

Christensen-Dalsgaard Jørgen
Clausen Jens Viggo
Davis Tamara M
Dorch Søren Bertil F.
Einicke Ole H.
Frandsen Soeren
Fynbo Johan P. U.
Galsgaard Klaus
Gammelgaard Peter
Grundahl Frank
Hannestad Steen
Hansen Leif
Helmer Leif
Helt Bodil E.
Hjorth Jens
Hoeg Erik
Hornstrup Allan
Jensen Brian L.
Johansen Karen T.
Jorgensen Henning E.
Jorgensen Uffe Graae
Kjaergaard Per
Kjeldsen Hans
Knude Jens Kirkeskov
Kristensen Leif Kahl
Linden-Voernle Michael J. D.
Lund Niels
Madsen Jes
Moesgaard Kristian P.
Naselsky Pavel D.
Nissen Poul Erik
Nordlund Aake
Nordstrom Birgitta
Norgaard-Nielsen Hans U.
Novikov Igor D.
Olsen Fogh H. J.
Olsen Lisbeth F.
Olsen Erik H.
Pedersen Holger
Pedersen Kristian
Petersen J. Otzen
Pethick Christopher J.
Rasmussen Ib L.
Sommer-Larsen Jesper
Stritzinger Maximilian D.
Tauris Thomas M.
Teixeira Teresa C. V. S.
Thejll Peter Andreas
Thomsen Bjarne B.
Ulfbeck Ole C.
Vedel Henrik
Watson Darach J.
Westergaard Niels J.

Egypt

Abd El Hamid Rabab
Abdelkawi M. Abubakr
Abou'el-ella Mohamed S.
Abulazm Mohamed Samir
Ahmed Mostafa Kamal
Ahmed Abdel-aziz Bakry
Ahmed Imam Ibrahim
Aiad A. Zaki
Amer Morsi M.
Awad Mervat El-Said
Awadalla Nabil Shoukry
Beheary Mohamed Mohamed
El Basuny Ahmed Alawy
El Nawaway Mohamed Saleh
El Raey Mohamed E.
El Shahawy Mohamad
El-Saftawy Magdy
El-Sharawy Mohamed Bahgat
Gaber Ali Eid
Galal A. A.
Gamaleldin Abdulla I.
Ghobros Roshdy Azer
Hady Ahmed Abdel
Hamdy M. A. M.
Hamid S. El Din
Hanna Magdy Abd El-Malek
Hassan Inal A.
Helali Yhya E.
Ismail Hamed Abdel-Hamid
Ismail Mohamed Nader
Issa Issa Aly
Kahil Magd Elias
Kamal Fouad Youssef
Kamel Osman M.
Khalil Khalil Ibrahim
Khalil Nisreen Madg-Eldin
Mahmoud Farouk M. A. B.
Marie Mohamed Abdelsalam
Mikhail Joseph Sidky
Morcos Abd El Fady B.
Nawar Samir
Osman Anas Mohamed
Rassem Mohamed Abdel-Aziz
Saad Nadia Ahmed
Saad Somaya Mohamed
Saad Abdel-naby S.
Selim Hadia Hassan
Shalabiea Osama M. A.
Shaltout Mosalam A. M.
Sharaf Mohamed Adel
Soliman Mohamed Ahmed
Tadross Ashraf Latif
Tawadros Maher Jacoub
Wanas Mamdouh I.
Yousef Shahinaz M.
Youssef Nahed H.

Estonia

Annuk Kalju
Einasto Maret
Einasto Jaan
Gramann Mirt
Haud Urmas
Kipper Tonu
Kolka Indrek
Leedjarv Laurits
Malyuto Valeri
Nugis Tiit
Pelt Jaan
Pustylnik Izold

Saar Enn
Sapar Lili
Sapar Arved
Suhhonenko Ivan
Tago Erik
Tenjes Peeter
Traat Peeter
Veismann Uno
Vennik Jaan
Viik Tõnu

Finland

Aittola Marko
Berdyugin Andrei V.
Berdyugina Svetlana V.
Donner Karl Johan
Duemmler Rudolf
Flynn Chris
Haikala Lauri K.
Hakala Pasi J.
Hannikainen Diana C.
Hanninen Jyrki
Harju Jorma Sakari
Heinamaki Pekka S.
Huovelin Juhani
Jetsu Lauri J.
Juvela Mika J.
Kaasalainen Mikko K.
Karttunen Hannu
Katajainen Seppo J.
Kocharov Leon G.
Korpi Maarit J.
Kotilainen Jari
Kultima Johannes
Lahteenmaki Anne M.
Lainela Markku J.
Laurikainen Eija
Lehtinen Kimmo K.
Lehto Harry J.
Lumme Kari A.
Markkanen Tapio
Mattila Kalevi
Mikkola Seppo
Muinonen Karri
Nevalainen Jukka H.
Niemi Aimo
Nilsson Kari J.
Nurmi Pasi T.
Oja Heikki
Piirola Vilppu E.
Piironen Jukka O.
Pohjolainen Silja H.
Portinari Laura
Poutanen Juri
Rahunen Timo
Raitala Jouko T.
Rautiainen Pertti T.
Riehokainen Aleksandr
Riihimaa Jorma J.
Roos Matts
Salo Heikki
Sillanpaa Aimo Kalevi
Tahtinen Leena
Takalo Leo O.
Teerikorpi Veli Pekka
Terasranta Harri T.
Tiuri Martti
Tornikoski Merja T.
Tuominen Ilkka V.
Valtaoja Esko
Valtonen Mauri J.
Vilhu Osmi
Wiik Kaj J.
Zheng Jia-Qing

France

Abada-Simon Meil
Abgrall Herve
Aboudarham Jean
Acker Agnes
Adami Christophe
Aime Claude
Alard Christophe L.
Alecian Georges
Alimi Jean-Michel A.
Allard Nicole
Allegre Claude
Aly Jean-Jacques
Amari Tahar
Amram Philippe
Andrillat Yvette
Arduini-Malinovsky Monique
Arenou Frederic
Arias Elisa Felicitas
Arlot Jean-Eudes
Arnaud Jean-Paul
Arnaud Monique D.
Artru Marie-Christine
Artzner Guy
Asseo Estelle
Assus Pierre
Athanassoula Evangelia
Aubier Monique G.
Audouze Jean
Auriere Michel
Auvergne Michel
Avignon Yvette
Azzopardi Marc
Babusiaux Carine
Bacchus Pierre
Baglin Annie
Balanca Christian
Balkowski-Mauger Chantal
Ballereau Dominique
Balmino Georges G.
Baluteau Jean-Paul
Barge Pierre
Barlier Francois E.
Barriot Jean-Pierre
Barucci Maria A.
Basa Stephane
Baudry Alain
Beaulieu Jean-Philippe R.
Bely-Dubau Francoise
Benaydoun Jean-Jacques
Benest Daniel
Ben-Jaffel Lofti
Bensammar Slimane
Bergeat Jacques Georges
Berger Jacques G.
Bergeron Jacqueline A.
Berruyer-Desirotte Nicole

Bertaux Jean-Loup
Berthier Jerôme
Berthomieu Gabrielle
Bertout Claude
Beuzit Jean-Luc
Bezard Bruno G.
Bienayme Olivier
Bijaoui Albert
Billebaud Francoise
Binetruy Pierre
Biraud François
Birlan Mirel I.
Biver Nicolas
Bizouard Christian
Blamont Jacques-Emile
Blanchard Alain
Blazit Alain
Bocchia Romeo
Bockelee-Morvan Dominique
Boehm Torsten C.
Boer Michel
Boily Christian M.
Bois Eric
Boisse Patrick
Boissier Samuel
Boisson Catherine
Bommier Veronique
Bonazzola Silvano
Bonifacio Piercarlo
Bonnarel Francois
Bonneau Daniel
Bonnet Roger M.
Bonnet-Bidaud Jean-Marc
Bontemps Sylvain
Borde Suzanne
Borgnino Julien
Bosma Albert
Bottinelli Lucette
Boucher Claude
Bouchet François R.
Bougeard Mireille L.
Bougeret Jean-Louis
Bouigue Roger
Boulanger Francois
Boulesteix Jacques
Boulon Jacques J.
Bouvier Jerôme
Boyer Rene
Brahic André
Braine Jonathan
Briot Danielle
Brouillet Nathalie
Brunet Jean-Pierre
Bryant John
Buat Véronique
Buecher Alain
Burgarella Denis
Burkhart Claude
Cabrit Sylvie
Calame Odile
Cambresy Laurent
Capitaine Nicole
Caplan James
Casandjian Jean-Marc
Casoli Fabienne
Casse Michel
Castets Alain
Catala Claude
Cayatte Veronique
Cayrel Roger
Cayrel de Strobel Giusa
Cazenave Anny
Celnikier Ludwik
Chabrier Gilles
Chadid-Vernin Merieme
Chalabaev Almas
Chamaraux Pierre
Chambe Gilbert
Chapront Jean
Chapront-Touze Michelle
Charlot Patrick
Charlot Stephane
Charpinet Stéphane
Chassefiere Eric
Chaty Sylvain
Chauvineau Bertrand
Chelli Alain
Chevrel Serge
Chollet Fernand
Chopinet Marguerite
Clairemidi Jacques
Colas François
Colin Jacques
Collin Suzy
Colom Pierre
Colombi Stephane
Combes Françoise
Combes Michel
Comte Georges
Connes Janine
Connes Pierre
Cornille Marguerite
Coude du Foresto Vincent
Coupinot Gerard
Courtes Georges
Courtin Régis
Couteau Paul
Couturier Pierre A
Cox Pierre
Creze Michel
Crifo Francoise
Crovisier Jacques
Cruvellier Paul E.
Cruzalebes Pierre
Cuny Yvette J.
Daigne Gerard
Daigne Frederic
Daniel Jean-Yves
Dauphole Bertrand
Davoust Emmanuel
De Bergh Catherine
De la Noe Jerome
De Lapparent-Gurriet Valérie
De Laverny Patrick
Debarbat Suzanne V.
Decourchelle Anne C.
Deharveng Jean-Michel
Deharveng Lise
Delaboudiniere Jean-Pierre
Delannoy Jean
Deleflie Florent
Delmas Christian
Demoulin Pascal
Denisse Jean-Francois
Dennefeld Michel
Descamps Pascal
Desesquelles Jean
Despois Didier
d'Hendecourt Louis
Divan Lucienne
Doazan Vera
Dohlen Kjetil
Dolez Noel
Dollfus Audouin
Donas Jose
Donati Jean-Francois
Dourneau Gerard
Downes Dennis

Drossart Pierre
Dubau Jacques
Dubois Marc A.
Dubois Pascal
Dubout Renee
Duc Pierre-Alain
Duchesne Maurice
Ducourant Christine
Dufay Maurice
Dulieu Francois
Dulk George A.
Dumont Anne-Marie
Dumont Rene
Dumont Simone
Duriez Luc
Durouchoux Philippe
Durret Florence
Duval Marie-France
Duvert Gilles
Egret Daniel
Eidelsberg Michele
Enard Daniel
Encrenaz Pierre J.
Encrenaz Therese
Erard Stéphane
Falgarone Edith
Faurobert-Scholl Marianne
Feautrier Nicole
Fehrenbach Charles
Felenbok Paul
Ferlet Roger
Ferrando Philippe Robert
Ferrari Marc
Ferrari Cecile
Ferriere Katia M.
Fienga Agnès
Fillion Jean-Hugues
Floquet Michele
Florsch Alphonse
Forestini Manuel
Fort Bernard P.
Forveille Thierry
Fossat Eric G.
Fox Andrew J.
Foy Renaud
Fraix-Burnet Didier
Francois Patrick
Freire Ferrero Rubens G.
Fresneau Alain
Friedjung Michael
Fringant Anne-Marie
Frisch Helene
Frisch Uriel
Froeschle Claude
Froeschle Christiane D.
Froeschle Michel
Fulchignoni Marcello
Gabriel Alan H.
Gallouet Louis
Gambis Daniel
Garcia Rafael A.
Gargaud Muriel
Garnier Robert
Gautier Daniel
Gay Jean
Genova Françoise
Georgelin Yvon P.
Gerard Eric
Gerbaldi Michèle
Gerin Maryvonne
Gillet Denis
Giraud Edmond
Goldbach Claudine
Goldwurm Andrea
Gomez Ana E.
Gonczi Georges
Gontier Anne-Marie
Gonzalez Jean-Francois
Gordon Charlotte
Goret Philippe
Gouguenheim Lucienne
Goupil Marie-Jose
Gouttebroze Pierre
Granveaud Michel
Grec Gerard
Grenier Isabelle
Greve Albert
Grewing Michael
Gry Cecile
Guelin Michel
Guibert Jean
Guiderdoni Bruno
Guinot Bernard R.
Hadamcik Edith
Halbwachs Jean-Louis
Hameury Jean-Marie
Hammer Francois
Harvey Christopher C.
Hayli Abraham
Hebrard Guillaume
Heck Andre
Henon Michel C.
Henoux Jean-Claude
Herpin Fabrice
Hestroffer Daniel
Heudier Jean-Louis
Heydari-Malayeri Mohammad
Heyvaerts Jean
Hill Vanessa M.
Hoang Binh Dy
Hua Chon Trung
Hubert-Delplace Anne-Marie
Hui bon Hoa Alain
Imbert Maurice
Irigoyen Maylis
Israel Guy Marcel
Jablonka Pascale
Jacq Thierry
Jasniewicz Gerard
Jedamzik Karsten
Joly Monique
Joly Francois
Jorda Laurent
Joubert Martine
Jung Jean
Kahane Claudine
Kandel Robert S.
Katz David A.
Kazes Ilya
Klein Karl L.
Kneib Jean-Paul
Koch-Miramond Lydie
Koechlin Laurent
Koechlin Laurent
Koutchmy Serge
Kovalevsky Jean
Krikorian Ralph
Kunth Daniel
Labeyrie Jacques
Labeyrie Antoine
Lacey Cédric
Lachieze-Rey Marc
Lafon Jean-Pierre J.
Lagage Pierre-Olivier
Lagrange Anne-Marie
Lallement Rosine
Laloe Suzanne
Lamy Philippe

Lancon Ariane
Lannes Andre
Laques Pierre
Laskar Jacques
Lasota Jean-Pierre
Latour Jean J.
Launay Jean-Michel
Launay Françoise
Laurent Claudine
Laval Annie
Lazareff Bernard
Le Bertre Thibaut R.
Le Borgne Jean-Francois
Le Bourlot Jacques
Le Contel Jean-Michel
Le Coroller Hervé
Le Floch André
Le Guet Tully Françoise
Le Squeren Anne-Marie
Leach Sydney
Lebre Agnes
Lebreton Yveline
Lecavelier des Etangs Alain
Lefebvre Michel
LeFevre Jean
Lefevre Olivier
Lega Elena
Leger Alain
Lelievre Gerard
Lemaire Philippe
Lemaire Jean-louis
Lemaitre Gerard R.
Lena Pierre J.
Leorat Jacques
Lequeux James
Lerner Michel-Pierre
Leroy Bernard
Leroy Jean-Louis
Lesteven Soizick
Lestrade Jean-Francois
Levasseur-Regourd Anny-Chantal
Lignieres François
Lortet Marie-Claire
Losco Lucette
Loucif Mohammed Lakhdar
Louise Raymond
Loulergue Michelle
Loup Cecile
Louys Mireille Y.
Lucas Robert
Ludwig Hans G.
Luminet Jean-Pierre
Madden Suzanne
Magnan Christian
Maillard Jean-Pierre
Malbet Fabien
Malherbe Jean-Marie
Mamon Gary A.
Mangeney André
Marcelin Michel
Marchal Christian
Marco Olivier
Martin Francois
Martin Jean-Michel P.
Martres Marie-Josephe
Masnou Francoise
Masnou Jean-Louis
Mathez Guy
Mathias Philippe
Maucherat J.
Maurice Eric N.
Maurogordato Sophie
Mauron Nicolas
Mavrides Stamatia
Mazure Alain
McCarroll Ronald
Megessier Claude
Mein Pierre
Mein Nicole
Mekarnia Djamel
Mellier Yannick
Meneguzzi Maurice M.
Merat Parviz
Mercier Claude
Metris Gilles
Meyer Jean-Paul
Mianes Pierre
Michard Raymond
Michel Patrick
Michel Laurent
Michel Eric
Mignard François
Millet Jean
Milliard Bruno
Mirabel Igor Felix
Mochkovitch Robert
Monier Richard
Monin Jean-Louis
Monnet Guy J.
Montmerle Thierry
Montmessin Franck
Morbidelli Alessandro
Moreau Olivier
Moreels Guy
Morel Pierre-Jacques
Mosser Benoît
Motch Christian
Mouchet Martine
Mouradian Zadig M.
Mourard Denis
Mulholland John Derral
Muller Richard
Muratorio Gerard
Nadal Robert
Namouni Fathi
Neiner Coralie
Nguyen-Quang Rieu
Noens Jacques-Clair
Nollez Gerard
Nottale Laurent
Oblak Edouard
Ochsenbein François
Omnes Roland
Omont Alain
Pagani Laurent P.
Paletou Frederic
Parcelier Pierre
Parisot Jean-Paul
Paturel Georges
Paul Jacques
Pecker Jean-Claude
Pedersen Bent M.
Pedoussaut Andre
Pellas Paul
Pelletier Guy
Pello Roser Descayre
Pequignot Daniel
Perault Michel
Perrier-Bellet Christian
Perrin Marie-Noel
Perrin Jean-Marie
Perrin Guy S.
Petit Pascal
Petit Jean-Marc
Petit Gérard
Petitjean Patrick
Petrini Daniel
Picat Jean-Pierre
Pick Monique

Pierre Marguerite
Pineau des Forets Guillaume
Plez Bertrand
Pollas Christian
Poquerusse Michel
Poulle Emmanuel
Pouquet Annick
Poyet Jean-Pierre
Praderie Françoise
Prantzos Nikos
Prevot-Burnichon Marie-Louise
Prieur Jean-Louis
Proisy Paul E.
Proust Dominique
Provost Janine
Puget Jean-Loup
Puy Denis
Quemerais Eric
Querci Francois R.
Querci Monique
Rabbia Yves
Raoult Antoinette
Rapaport Michel
Rayrole Jean R.
Reeves Hubert
Reinisch Gilbert
Renard Jean-Baptiste
Requieme Yves
Revaz Yves
Reyle Céline
Riazuelo Alain
Ricort Gilbert
Rieutord Michel
Robillot Jean-Maurice
Robin Annie C. R
Robley Robert
Rocca-Volmerange Brigitte
Roques Françoise
Roques Sylvie
Rostas Francois
Rothenflug Robert
Rouan Daniel
Roudier Thierry
Roueff Evelyne M. A.
Rousseau Jeanine
Rousseau Jean-Michel
Rousselot Philippe
Royer Frédéric
Rozelot Jean-Pierre
Sadat Rachida
Sahal-Brechot Sylvie
Saissac Joseph
Salez Morvan C.
Samain Denys
Sanchez Norma G.
Sareyan Jean-Pierre
Sauty Christophe
Schatzman Evry
Scheidecker Jean-Paul
Schmieder Brigitte
Schneider Jean
Scholl Hans
Schumacher Gerard
Seconds Alain-Philippe
Semel Meir
Semelin Benoit
Sibille Francois
Sicardy Bruno
Signore Monique
Simon Guy
Simon Jean-Louis
Simonneau Eduardo
Sirousse Zia Haydeh
Sivan Jean-Pierre
Slezak Eric
Sol Helene
Sotirovski Pascal
Soubiran Caroline
Soucail Geneviève
Souchay Jean
Souriau Jean-Marie
Spallicci Alessandro D. A. M.
Spielfiedel Annie
Spite François M.
Spite Monique
Stasinska Grazyna
Stee Philippe
Stehle Chantal
Steinberg Jean-Louis
Stellmacher Irène
Stellmacher Götz
Sygnet Jean-Francois
Tagger Michel
Talon Raoul
Tanga Paolo
Tarrab Irene
Taton Rene
Tchang-Brillet Lydia
Terquem Caroline E.
Terzan Agop
Theureau Gilles
Thevenin Frederic
Thiry Yves R.
Thomas Claudine
Thuillot William
Thum Clemens
Tobin William
TranMinh Nguyet
Tran-Minh Francoise
Trottet Gerard
Truong Bach
Tuckey Philip A.
Tully John A.
Turck-Chieze Sylvaine
Turon Catherine
Vakili Farrokh
Valiron Pierre
Valls-Gabaud David
Valtier Jean-Claude
van Driel Wim
van Driel-Gesztelyi Lidia
van't Veer-Menneret Claude
van't-Veer Frans
Vapillon Loic J.
Vaubaillon Jérémie J.
Vauclair Gérard P.
Vauclair Sylvie D.
Vauglin Isabelle
Verdet Jean-Pierre
Vergez Madeleine
Vernin Jean
Vernotte François
Veron Marie-Paule
Veron Philippe
Vial Jean-Claude
Viala Yves
Viallefond Francois
Vidal Jean-Louis
Vidal-Madjar Alfred
Vienne Alain
Vigroux Laurent
Vilkki Erkki U.
Vilmer Nicole
Vinet Jean-Yves
Viton Maurice
Vollmer Bernd
Volonte Sergio

Vuillemin Andre
Walch Jean-Jacques
Wenger Marc
Widemann Thomas
Wilson Brian G.
Wink Joern Erhard
Wlerick Gerard
Woltjer Lodewijk
Wozniak Herve
Zahn Jean-Paul
Zamkotsian Frederic
Zavagno Annie
Zeippen Claude
Zorec Juan

Georgia

Bartaya R. A.
Borchkhadze Tengiz M.
Chkhikvadze Iakob N.
Dolidze Madona V.
Dzhapiashvili Victor P.
Dzigvashvili R. M.
Kalandadze N. B.
Khatisashvili Alfez Sh.
Khetsuriani Tsiala S.
Kiladze R. I.
Kogoshvili Natela G.
Kumsiashvily Mzia I.
Kurtanidze Omar M.
Lominadze Jumber
Salukvadze G. N.
Shvelidze Teimuraz D.
Toroshlidze Teimuraz I.

Germany

Abraham Peter
Albrecht Rudolf
Altenhoff Wilhelm J.
Alves Joao F.
Andersen Michael I.
Anzer Ulrich
Appenzeller Immo
Arp Halton
Arshakian Tigran G.
Aschenbach Bernd
Aurass Henry
Axford W. Ian
Aznar Cuadrado Regina
Baade Robert
Baade Dietrich
Baars Jacob W. M.
Baessgen Martin
Bahner Klaus
Baier Frank W.
Bailer-Jones Coryn A. L.
Ballester Pascal
Balthasar Horst
Banday Anthony J.
Barrow Colin H.
Bartelmann Matthias
Barwig Heinz
Baschek Bodo
Bastian Ulrich
Beck Rainer
Becker Werner
Behr Alfred
Bender Ralf
Benvenuti Piero
Berkefeld Thomas
Berkhuijsen Elly M.
Bernstein Hans-Heinrich
Beuermann Klaus P.
Bien Reinhold
Biermann Peter L.
Birkle Kurt
Bleyer Ulrich
Bludman Sidney A.
Boehringer Hans
Boerner Gerhard
Boerngen Freimut
Boller Thomas
Bomans Dominik J.
Bonaccini Domenico
Boschan Peter
Bothmer Volker
Brandt Peter N.
Brauninger Heinrich
Breitschwerdt Dieter
Breysacher Jacques
Brinkmann Wolfgang
Britzen Silke H.
Brosche Peter
Bruggen Marcus
Bruls Jo H.
Buechner Joerg
Bues Irmela D.
Burkert Andreas M.
Burwitz Vadim
Butler Keith
Camenzind Max
Cameron Robert
Carlson Arthur W.
Carsenty Uri
Cesarsky Diego A.
Cesarsky Catherine J.
Chini Rolf
Christensen Lars Lindberg
Ciardi Benedetta
Corbett Ian F.
Cullum Martin
Curdt Werner
Dachs Joachim
de Boer Klaas Sjoerds
Degenhardt Detlev
Deinzer W.
Deiss Bruno M.
Dennerl Konrad
Dettmar Ralf-Juergen
Deubner Franz-Ludwig
Dick Wolfgang R.
Diehl Roland L.
Diercksen Geerd H. F.
Dobler Wolfgang
Dobrzycka Danuta
Dobrzycki Adam
d'Odorico Sandro
Dorschner Johann
Drechsel Horst
Dreizler Stefan
Duerbeck Hilmar W.
Dumke Michael
Duschl Wolfgang J.
Ehlers Jürgen
Eisenhauer Frank

Eisloeffel Jochen
Engels Dieter
Ensslin Torsten A.
Fahr Hans Joerg
Falcke Heino D.
Feitzinger Johannes
Fendt Christian
Fichtner Horst
Fiebig Dirk
Fosbury Robert A. E.
Fraenz Markus
Freudling Wolfram
Fricke Klaus
Fried Josef Wilhelm
Fritze Klaus
Fritze-von Alvensleben Uta
Fuchs Burkhard
Fuerst Ernst
Gail Hans-Peter
Gandorfer Achim
Geffert Michael
Gehren Thomas
Genzel Reinhard
Geppert Ulrich R. M. E.
Gerhard Ortwin
Geyer Edward H.
Giacconi Riccardo
Gilmozzi Roberto
Gizon Laurent
Glatzel Wolfgang
Glindemann Andreas
Gottloeber Stefan
Gouliermis Dimitrios
Graham David A.
Grahl Bernd H.
Grebel Eva K.
Gredel Roland
Groote Detlef
Grosbol Preben Johnson
Grossmann-Doerth U.
Groten Erwin
Grothkopf Uta
Gruen Eberhard
Guenther Eike
Guertler Joachin
Guesten Rolf
Gussmann Ernst August
Gutcke Dietrich
Haefner Reinhold
Haenel Andreus
Haerendel Gerhard
Hagen Hans-Juergen
Hahn Gerhard J.
Hamann Wolf-Rainer
Hammer Reiner
Hanuschik Reinhard
Harris Alan William
Hartogh Paul
Hasinger Günther
Hatchell Jennifer
Hauschildt Peter H.
Hazlehurst John
Heber Ulrich
Hefele Herbert
Hegmann Michael
Heidt Jochen
Hempelmann Alexander M.
Henkel Christian
Henning Thomas
Herbstmeier Uwe
Hering Roland
Herrmann Dieter
Hessman Frederic Victor
Hildebrandt Joachim
Hilker Michael
Hillebrandt Wolfgang
Hippelein Hans H.
Hirth Wolfgang Ernst
Hofmann Wilfried
Holmberg Johan
Hopp Ulrich
Horedt Georg Paul
Huchtmeier Walter K.
Huensch Matthias
Huettemeister Susanne
Hummel Wolfgang
Ilyin Ilya V.
Isserstedt Joerg
Jahreiss Hartmut
Janka Hans Thomas
Jockers Klaus
Jordan Stefan
Junkes Norbert
Just Andreas
Kaehler Helmuth
Kaeufl Hans Ulrich
Kahabka Peter
Kalberla Peter
Kanbach Gottfried
Kauffmann Guinevere A. M.
Kaufmann Jens Peter
Kegel Wilhelm H.
Keller Hans-Ulrich F.
Keller Horst U.
Kelz Andreas
Kendziorra Eckhard
Kentischer Thomas
Kerber Florian
Khanna Ramon
King Lindsay J.
Kippenhahn Rudolf
Kirk John
Kissler-Patig Markus
Kitsionas Spyridon
Klare Gerhard
Klein Ulrich
Klessen Ralf S.
Kley Wilhelm
Kliem Bernhard
Klinkhamer Frans
Klioner Sergei A.
Klose Sylvio
Kneer Franz
Knudsen Kirsten K.
Kobayashi Chiaki
Koehler Peter
Koester Detlev
Kohoutek Lubos
Kollatschny Wolfram
Korhonen Heidi H.
Kraus Alexander
Krause Marita
Krautter Joachim
Kreysa Ernst
Krichbaum Thomas P.
Krivov Alexander V.
Krivova Natalie A.
Kroll Peter
Kroupa Pavel
Krueger Harald
Krugel Endrik
Kuehne Christoph F.
Kundt Wolfgang
Kuntschner Harald
Kupka Friedrich
Labs Dietrich
Lee Henry
Lehman Holger
Leibundgut Bruno
Leinert Christoph

Lemke Dietrich
Lemke Michael
Lenhardt Helmut
Lenzen Rainer
Lesch Harald
Li Li-Xin
Liebscher Dierck-E
Lobanov Andrei
Lombardi Marco
Luest Reimar
Luks Thomas
Lutz Dieter
Mac Low Mordecai-Mark
Mandel Holger
Mann Gottfried
Mannheim Karl
Marsch Eckart
Matas Vladimir R.
Matthews Owen M.
Mattig W.
Mauder Horst
Mebold Ulrich
Meeus Gwendolyn
Mehlert Dörte
Meinig Manfred
Meisenheimer Klaus
Meister Claudia Veronika
Menten Karl M.
Merkle Fritz
Meusinger Helmut
Meyer Friedrich
Meyer-Hofmeister Eva
Mezger Peter G.
Milvang-Jensen Bo
Miralles Joan-Marc
Moehler Sabine
Moehlmann Diedrich
Moellenhoff Claus
Moeller Palle
Moorwood Alan F. M.
Muecket Jan P.
Mueller Volker
Mueller Ewald
Mueller Thomas G.
Mundt Reinhard
Nagel Thorsten
Neckel Th.
Neckel Heinz
Neeser Mark J.
Nesis Anastasios
Neuhaeuser Ralph
Neukum G.
Nothnagel Axel
Notni Peter
Odenkirchen Michael
Oestreicher Roland
Oetken L.
Oleak H.
Ossendrijver Mathieu
Ott Heinz-Albert
Padovani Paolo
Paresce Francesco
Patat Ferdinando
Patnaik Alok Ranjan
Patzer A. Beate C.
Pauldrach Adalbert W. A.
Peter Hardi
Petr-Gotzens Monika G.
Pflug Klaus
Philipp Sabine D.
Pinkau K.
Pitz Eckhart
Polatidis Antonios
Popescu Cristina Carmen
Porcas Richard
Preuss Eugen
Preuss Oliver
Primas Francesca
Przybilla Norbert
Puls Joachim
Quinn Peter
Quirrenbach Andreas
Raedler K. H.
Rauch Thomas
Rauer Heike
Refsdal Sjur
Reich Wolfgang
Reif Klaus
Reimers Dieter
Reinsch Klaus
Reiprich Thomas H.
Rejkuba Marina
Rendtel Juergen
Renzini Alvio
Richichi Andrea
Richter Gotthard
Richter Bernd
Ripken Hartmut W.
Ritter Hans
Rix Hans-Walter
Roemer Max
Roepke Friedrich K.
Roeser Hans-peter
Roeser Hermann-Josef
Roeser Siegfried
Rohlfs Kristen
Romaniello Martino
Ros Ibarra Eduardo
Rosa Dorothea
Rosa Michael Richard
Rosswog Stephan H. K.
Roth Markus
Rothacher Markus
Roy Alan L.
Ruder Hanns
Ruediger Guenther
Rupprecht Gero
Saglia Roberto Philip
Sakelliou Irini
Sams Bruce Jones III
Schaefer Gerhard
Schaffner-Bielich Jurgen
Schaifers Karl
Scheffler Helmut
Schilbach Elena
Schilke Peter
Schindler Karl
Schinnerer Eva
Schleicher Helmold
Schlemmer Stephan
Schlichenmaier Roef
Schlickeiser Reinhard
Schlosser Wolfhard
Schlueter A.
Schmadel Lutz D.
Schmeidler F.
Schmid-Burgk J.
Schmidt H. U.
Schmidt Wolfgang
Schmidt Wolfram
Schmidt Robert W.
Schmidt-Kaler Theodor
Schmitt Juergen H. M. M.
Schmitt Dieter
Schneider Peter
Schoenberner Detlef
Schoenfelder Volker
Scholz M.
Scholz Ralf-Dieter
Schramm Thomas

Schroeder Rolf
Schubart Joachim
Schuecker Peter
Schuessler Manfred
Schuh Sonja
Schulz R. Andreas
Schutz Bernard F.
Schwan Heiner
Schwartz Rolf
Schwekendiek Peter
Schwenn Rainer W.
Schwope Axel
Scorza de Appl Cecilia
Sedlmayer Erwin
Seggewiss Wilhelm
Seifert Walter
Seitter Waltraut C.
Seitz Stella
Sengbusch Kurt V.
Shaver Peter A.
Sherwood William A.
Siebenmorgen Ralf
Sieber Wolfgang
Sigwarth Michael
Simon Klaus Peter
Soffel Michael H.
Solanki Sami K.
Solf Josef
Sollazzo Claudio
Soltau Dirk
Springel Volker
Springer Tim
Spruit Henk C.
Spurzem Rainer
Stahl Otmar Richard
Staubert Ruediger
Staude Juergen
Staude Hans Jakob
Stecklum Bringfried
Steffen Matthias
Steiner Oskar
Steinert Klaus Guenter
Steinle Helmut R.
Steinmetz Matthias
Stix Michael
Stoecker Horst
Storm Jesper
Strassmeier Klaus G.
Strong Andrew W.
Stumpff Peter
Stutzi Juergen
Szostak Roland
Tacconi Linda J.
Tacconi-Garman Lowell E.
Tanaka Yasuo
Tarenghi Massimo
Teriaca Luca
Theis Christian
Thielheim Klaus O.
Thomas Hans Christoph
Tiersch Heinz
Traving Gerhard
Treder H. J.
Trefftz Eleonore E.
Treumann Rudolf A.
Truemper Joachim
Tscharnuter Werner M.
Tueg Helmut
Tuffs Richard J.
Ulmschneider Peter
Ulrich Marie-Helene D.
van den Ancker Mario E.
Voelk Heinrich J.
Voges Wolfgang H.
Voigt Hans H.
Volkmer Reiner
von Appen-Schnur Gerhard F. O.
von Borzeszkowski H. H.
von der Luehe Oskar
von Weizsaecker C. F.
Vrielmann Sonja
Wagner Stefan
Walter Hans G.
Wambsganss Joachim
Warmels Rein Herm
Wedemeyer-Boehm Sven
Wehrse Rainer
Weidemann Volker
Weigelt Gerd
Weis Kerstin
Weiss Achim
Weiss Axel
Wendker Heinrich J.
Werner Klaus
West Richard M.
White Simon David Manion
Wiedemann Günter R.
Wiegelmann Thomas
Wiehr Eberhard
Wielebinski Richard
Wielen Roland
Wilms Jörn
Wilson P.
Winnewisser Gisbert
Wisotzki Lutz
Wittkowski Markus
Wittmann Axel D.
Witzel Arno
Woehl Hubertus
Wolf Bernhard
Wolf Rainer E. A.
Wolfschmidt Gudrun
Wuensch Johann Jakob
Wunner Guenter
Wyrowski Friedrich W.
Yamada Shoichi
Yonehara Atsunori
Yorke Harold W.
Zekl Hans Wilhelm
Zensus J-Anton
Zerull Reiner H.
Zickgraf Franz Josef
Ziegler Harald
Ziegler Bodo L.
Zimmermann Helmut
Zinnecker Hans
Zylka Robert

Greece

Alissandrakis Costas
Anastasiadis Anastasios
Antonacopoulos Gregory
Antonopoulou Evgenia
Arabelos Dimitrios
Asteriadis Georgios
Avgoloupis Stavros
Banos George J.
Bellas-Velidis Ioannis
Boumis Panayotis
Bozis George
Caranicolas Nicholas

Caroubalos Constantinos A.
Charmandaris Vassilis
Chryssovergis Michael
Contadakis Michael E.
Contopoulos Ioannis
Contopoulos George
Daglis Ioannis A.
Danezis Emmanuel
Dapergolas Anastasios
Dara Helen
Deliyannis Jean
Dialetis Dimitris
Dionysiou Demetrios
Efthymiopoulos Christos
Evangelidis E. A.
Georgantopoulos Ioannis
Geroyannis Vassilis S.
Gontikakis Constantin
Goudas Constantine L.
Goudis Christos D.
Hadjidemetriou John D.
Hantzios Panayiotis
Harsoula Mirella K.
Hatzidimitriou Despina
Hilaris Alexander E.
Hiotelis Nicolaos
Isliker Heinz
Kalvouridis Tilemahos
Kazantzis Panayotis
Kokkotas Konstantinos
Kontizas Mary
Kontizas Evangelos
Korakitis Romylos
Krimigis Stamatios M.
Kylafis Nikolaos D.
Laskarides Paul G.
Liritzis Ioannis
Manimanis Vassilios
Mastichiadis Apostolos
Mavraganis Anastasios G.
Mavridis Lyssimachos N.
Mavromichalaki Helen
Merzanides Constantinos
Metaxa Margarita
Moussas Xenophon
Niarchos Panayiotis
Nicolaidis Efthymios
Nindos Alexander
Papadakis Iossif
Papaelias Philip M.
Papathanasoglou Dimitrios
Papayannopoulos Theodoros
Patsis Panos
Persides Sotirios C.
Petropoulos Basil Ch.
Pinotsis Antonis D.
Plionis Manolis
Poulakos Constantine
Preka-Papadema Panagiota
Prokakis Theodore J.
Rovithis Peter
Rovithis-Livaniou Helen
Sarris Emmanuel T.
Sarris Eleftherios
Seimenis John
Seiradakis John H.
Skokos Charalambos
Spyrou Nicolaos
Stathopoulou Maria
Svolopoulos Sotirios
Terzides Charalambos
Theodossiou Efstratios
Tritakis Basil P.
Tsamparlis Michael
Tsikoudi Vassiliki
Tsinganos Kanaris
Tsiropoula Georgia
Tziotziou Konstantinos
Vardavas Ilias Mihail
Varvoglis Harry
Veis George
Ventura Joseph
Vlachos Demetrius G.
Vlahakis Nektarios
Vlahos Loukas
Voglis Nikos
Xilouris Emmanouel
Xiradaki Evangelia
Zachariadis Theodosios
Zachilas Loukas
Zafiropoulos Basil
Zikides Michael C.

Hungary

Almar Ivan
Bagoly Zsolt
Balazs Bela A.
Balazs Lajos G.
Baranyi Tuende
Barcza Szabolcs
Barlai Katalin
Benkoe Jozsef M.
Erdi Bálint
Fejes Istvan
Forgacs-Dajka Emese
Frey Sandor
Gerlei Otto
Grandpierre Attila
Gyori Lajos
Hegedues Tibor
Horvath Andras
Ill Marton J.
Illes Almar Erzsebet
Jankovics Istvan
Jurcsik Johanna
Kalman Bela
Kanyo Sandor
Kelemen Janos
Koevari Zsolt
Kollath Zoltan
Kovacs Geza
Kovacs Agnes
Kun Maria
Lovas Miklos
Ludmany Andras
Olah Katalin
Paparo Margit
Patkos Laszlo
Petrovay Kristof
Szabados Laszlo
Szatmary Karoly
Szecsenyi-Nagy Gábor A.
Szego Karoly
Szeidl Bela
Toth Laszlo V.
Toth Imre
Vargha Magda
Veres Ferenc
Vinko Jozsef
Wynn-Williams C. G.
Zsoldos Endre

Iceland

Bjoernsson Gunnlaugur
Gudmundsson Einar H.
Juliusson Einar
Saemundsson Thorsteinn

India

Abhyankar Krishna D.
Acharya Bannanje S.
Agrawal P. C.
Ahmad Farooq
Alladin Saleh Mohamed
Ambastha Ashok K.
Anandaram Mandayam N.
Ananthakrishnan Subramanian
Ansari S. M. Razaullah
Antia H. M.
Anupama G. C.
Apparao K. M. V.
Ashok N. M.
Babu G. S. D.
Bagare S. P.
Bagla Jasjeet S.
Balasubramanian V.
Balasubramanyam Ramesh
Baliyan Kiran S.
Ballabh Goswami Mohan
Bandyopadhyay A.
Banerjee Dipankar P. K.
Banerji Sriranjan
Banhatti Dilip G.
Basu Baidyanath
Bhandari Rajendra
Bhandari N.
Bharadwaj Somnath
Bhat Narayana P.
Bhatia Prem K.
Bhatia Vishnu B.
Bhatnagar K. B.
Bhatt H. C.
Bhattacharjee Pijushpani
Bhattacharya Dipankar
Bhattacharyya J. C.
Biswas Sukumar
Boddapati Anandarao G.
Bondal Krishna Raj
Chakrabarti Sandip Kumar
Chakraborty Deo K.
Chandra Suresh
Chandrasekhar Thyagarajan
Chaoudhuri Arnab R.
Chaubey Uma Shankar
Chengalur Jayaram N.
Chitre Shashikumar M.
Chokshi Arati
Cowsik Ramanath
Dadhich Naresh
Damle S. V.
Das Mrinal Kanti
Das P. K.
Degaonkar S. S.
Desai Jyotindra N.
Deshpande Avinash
Deshpande M. R.
Dhurandhar Sanjeev
Duari Debiprosad
Duorah Hira Lal
Dwarakanath K. S.
Dwivedi Bhola N.
Gaur V. P.
Ghosh S. K.
Ghosh P.
Giridhar Sunetra
Gokhale Moreshwar H.
Gopala Rao U. V.
Goswami J. N.
Goyal Ashok Kumar
Goyal A. N.
Gupta Ranjan
Gupta Surendra S.
Gupta Yashwant
Gupta Sunil K.
Hallan Prem P.
Hasan S. Sirajul
Iyengar Srinivasan Rama
Iyer B. R.
Jain Rajmal
Jog Chanda J.
Joshi Umesh C.
Kantharia Nimisha G.
Kapoor Ramesh Chander
Kasturirangan K.
Kaul Chaman
Kembhavi Ajit K.
Khare Pushpa
Kilambi G. C.
Kochhar Rajesh K.
Konar Sushan
Krishna Gopal
Krishna Swamy K. S.
Krishnan Thiruvenkata
Kulkarni Prabhakar V.
Kulkarni Vasant K.
Lal Devendra
Mahra H. S.
Manchanda R. K.
Mangalam Arun
Manoharan P. K.
Marar T. M. k.
Mitra Abhas Kumar
Mohan Chander
Mohan Anita
Mohan Vijay
Murthy Jayant
Nagendra K. N.
Nair Sunita
Namboodiri P. M. S.
Narain Udit
Naranan S.
Narasimha Delampady
Narayana J. V.
Narlikar Jayant V.
Nath Biman B.
Nath Mishra Kameshwar
Nayar S. R. Prabhakaran
Nityananda Ram
Ojha Devendra K.
Padalia T. D.
Padmanabhan Thanu
Padmanabhan Janardhan
Pande Girish Chandra
Pandey A. K.
Pandey S. K.
Pandey Uma Shankar

Parthasarathy Mudumba
Pati A. K.
Paul Biswajit
Peraiah Annamaneni
Prabhu Tushar P.
Prasanna A. R.
Radhakrishnan V.
Raghavan Nirupama
Rajamohan R.
Raju P. K.
Raju Vasundhara
Ramadurai Souriraja
Ramamurthy Swaminathan
Ramana Murthy P. V.
Rangarajan K. E.
Rao M. N.
Rao Pasagada Vivekananda
Rao N. Kameswara
Rao A. Pramesh
Rao Ramachandra V.
Rao D. Mohan
Rao Arikkala Raghurama
Rautela B. S.
Raveendran A. V.
Ray Alak
Raychaudhuri Amalkumar
Rengarajan T. N.
Sagar Ram
Saha Swapan Kumar
Sahni Varun
Saikia Dhruba Jyoti
Sanwal Basant Ballabh
Sapre Ashok Kumar
Sarma N. V. G.
Sastri Hanumath J.
Sastry Shankara K.
Sastry Ch. V.
Saxena P. P.
Saxena A. K.
Sengupta Sujan K.
Seshadri Sridhar
Shahul Hameed Mohin
Shastri Prajval
Shevgaonkar R. K.
Shukla K.
Shukre C. S.
Singal Ashok K.
Singh Mahendra
Singh Jagdev
Singh Kulinder Pal
Singh Harinder P.
Sinha Krishnanand
Sivaram C.
Sivaraman K. R.
Somasundaram Seetha
Sonti Sreedhar Rao
Sreekantan B. V.
Sreekumar Parameswaran
Srianand Roghunathan
Srinivasan G.
Srivastava Ram K.
Srivastava J. B.
Srivastava Dhruwa Chand
Stephens S. A.
Subrahmanya C. R.
Subramanian Kandaswamy
Subramanian K. R.
Subramanian Prasad
Swarup Govind
Talwar Satya P.
Tandon S. N.
Tonwar Suresh C.
Tripathi B. M.
Tripathy Sushanta C.
Udaya Shankar N.
Uddin Wahab
Vahia Mayank N.
Vaidya P. C.
Vardya M. S.
Varma Ram Kumar
Vasu-Mallik Sushma
Vats Hari OM
Venkatakrishnan P.
Verma R. P.
Verma V. K.
Vinod S. Krishan
Vishveshwara C. V.
Vivekanand M.
Vivekananda Rao
Yadav Jagdish Singh

Indonesia

Ardi Eliani
Dawanas Djoni N.
Djamaluddin Thomas
Herdiwijaya Dhani
Hidayat Bambang
Hidayat Taufiq
Ibrahim Jorga
Kunjaya Chatief
Malasan Hakim Luthfi
Premadi Premana W.
Radiman Iratio
Raharto Moedji
Ratag Mezak Arnold
Siregar Suryadi
Sukartadiredja Darsa
Wiramihardja Suhardja

Iran (Islamic Republic of)

Adjabshirizadeh Ali
Ardebili M. Reza
Asareh Habibolah
Bordbar Gholam H.
Dehghani Mohammad Hossein
Edalati Sharbaf Mohammad Taghi
Ghanbari Jamshid
Hamedivafa Hashem
Jalali Mir Abbas
Jassur Davoud MZ
Kalafi Manoucher
Khalesseh Bahram
Khesali Ali R.
Kiasatpour Ahmad
Malakpur Iradj
Mansouri Reza
Nasiri Sadollah G.
Nasr-Esfahani Bahram
Nozari Kourosh
Rahvar Sohrab
Riazi Nematollah
Samimi Jalal
Shadmehri Mohsen
Sobouti Yousef

Ireland

Breslin Ann
Butler Raymond F.
Callanan Paul
Carroll P. Kevin
Cawley Michael
Cunniffe John
Devaney Martin N.
Downes Turlough
Drury Luke O'Connor
Elliott Ian
Espey Brian Russell
Fegan David J.
Florides Petros S.
Gillanders Gerard
Golden Aaron
Hanlon Lorraine O.
Haywood J.
Hoey Michael J.
Kennedy Eugene T.
Kiang Tao
Lang Mark
McBreen Brian Philip
McKeith Niall Enda
McKenna Lawlor S.
Meurs Evert J. A.
Murphy John A.
Norci Laura
O'Connor Seamus L.
O'Mongain Eon
O'Sullivan Creidhe
O'Sullivan Denis F.
Quinn John
Ray Thomas P.
Redfern Michael R.
Redman Matthew P.
Shearer Andrew
Smith Niall
Trappe Neil A.
Wrixon Gerard T.

Israel

Alexander Tal E.
Almoznino Elhanan
Barkana Rennan
Barkat Zalman
Bar-Nun Akiva
Bekenstein Jacob D.
Braun Arie
Brosch Noah
Cuperman Sami
Dekel Avishai
Eichler David
Eppelbaum Lev V.
Ershkovich Alexander
Eviatar Aharon
Finzi Arrigo
Formiggini Lilliana
Gedalin Michael
Glasner Shimon Ami
Goldman Itzhak
Goldsmith S.
Griv Evgeny
Harpaz Amos
Heller Ana B.
Horwitz Gerald
Ibbetson Peter Aaron
Israelevich Peter
Joseph Joachim H.
Kaspi Shai
Katz Joseph
Kovetz Attay
Kozlovsky Ben Z.
Leibowitz Elia M.
Levinson Amir
Lyubarsky Yury E.
Maoz Dan
Marco Shmulik
Mazeh Tsevi
Meerson Baruch
Meidav Meir
Mekler Yuri
Merman Hirsh G. A.
Neeman Yuval
Netzer Nathan
Netzer Hagai
Nusser Adi
Ohring George
Piran Tsvi
Podolak Morris
Prialnik-Kovetz Dina
Price Colin
Rakavy Gideon
Regev Oded
Rephaeli Yoel
Ribak Erez N.
Sack Noam
Segaluvitz Alexander
Seidov Zakir F.
Shaviv Giora
Shiryaev Alexander A.
Steinitz Raphael
Tuchman Ytzhak
Usov Vladimir V.
Vager Zeev
Vidal Nissim V.
Waxman Eli
Yeivin Y.

Italy

Aiello Santi
Albanese Lara
Alcala Juan Manuel
Altamore Aldo
Ambrosini Roberto
Amendola Luca
Andreani Paola Michela
Andretta Vincenzo
Andreuzzi Gloria
Angeletti Lucio
Anile Angelo M.
Antonello Elio
Antonucci Ester
Antonuccio-Delogu Vincenzo
Arnaboldi Magda
Auriemma Giulio
Badiali Massimo
Baffa Carlo
Baldinelli Luigi

Bandiera Rino
Banni Aldo
Baratta Giovanni Battista
Baratta Giuseppe Antonio
Barbaro Guido
Barberis Bruno
Barbieri Cesare
Barbon Roberto
Bardelli Sandro
Barletti Raffaele
Barone Fabrizio
Bartolini Corrado
Battinelli Paolo
Battistini Pierluigi
Becciani Ugo
Bedogni Roberto
Belinski Vladimir A.
Bellazzini Michele
Belloni Tomaso
Belvedere Gaetano
Bemporad Alessandro
Benacchio Leopoldo
Benetti Stefano
Bernacca Pierluigi
Berrilli Francesco
Berta Stefano
Bertelli Gianpaolo
Bertin Giuseppe
Bertola Francesco
Bettoni Daniela
Bianchi Simone
Bianchini Antonio
Bignami Giovanni F.
Biviano Andrea
Blanco Carlo
Blanco Armando
Boccaletti Dino
Bocchino Fabrizio
Bodo Gianluigi
Bolzonella Micol
Bombaci Ignazio
Bonanno Giovanni
Bondi Marco
Bonifazi Angelo
Bono Giuseppe
Bonoli Fabrizio
Bonometto Silvio A.
Borgani Stefano
Braccesi Alessandro
Bragaglia Angela
Brand Jan
Brescia Massimo
Bressan Alessandro
Brocato Enzo
Broglia Pietro
Brunetti Gianfranco
Bucciarelli Beatrice
Buonanno Roberto
Burderi Luciano
Busa Innocenza
Busarello Giovanni
Buson Lucio M.
Busso Maurizio
Buzzoni Alberto
Cacciani Alessandro
Caccianiga Alessandro
Cacciari Carla
Calamai Giovanni
Calamida Annalisa
Caloi Vittoria
Calura Francesco
Calvani Massimo
Capaccioli Massimo
Capaccioni Fabrizio
Cappellaro Enrico
Cappi Alberto
Cappi Massimo
Capria Maria Teresa
Caputo Filippina
Capuzzo Dolcetta Roberto
Caraveo Patrizia
Cardini Daniela
Carollo Daniela
Carpino Mario
Carraro Giovanni
Carretta Eugenio
Carretti Ettore
Carusi Andrea
Caselli Paola
Cassatella Angelo
Castelli Fiorella
Catalano Franco A.
Catalano Santo
Catanzaro Giovanni
Cauzzi Gianna
Cavaliere Alfonso G.
Cavallini Fabio
Cazzola Paolo
Cecchi-Pellini Cesare
Celletti Alessandra
Cellino Alberto
Centurion Martin Miriam
Ceppatelli Guido
Cerroni Priscilla
Cerruti Sola Monica
Cester Bruno
Cevolani Giordano
Chiappetti Lucio
Chiappini Moraes Leite
Cristina
Chincarini Guido L.
Chinnici Ileana
Chiosi Cesare S.
Chiuderi Claudio
Chiuderi-Drago Franca Pr.
Chlistovsky Franca
Cinzano Pierantonio
Ciotti Luca
Citterio Oberto
Claudi Riccardo U.
Clementini Gisella
Codella Claudio
Colafrancesco Sergio
Colangeli Luigi
Coluzzi Regina
Comastri Andrea
Comoretto Giovanni
Conconi Paolo
Coradini Angioletta
Corbelli Edvige
Corsini Enrico M.
Cosentino Rosario
Cosmovici Cristiano Batalli
Costa Enrico
Covino Elvira
Covone Giovanni
Cremonese Gabriele
Cristiani Stefano V.
Crivellari Lucio
Cugusi Leonino
Curir Anna
Cusumano Giancarlo
Cutispoto Giuseppe
D' Odorico Valentina
Dallacasa Daniele
Dall-Oglio Giorgio
Dall'Ora Massimo
D'Amico Nicolo'
Danese Luigi
D'Antona Francesca

Danziger I. John
De Biase Giuseppe A.
De Felice Fernando
De Martino Domitilla
De Petris Marco
De Ruiter Hans Rudolf
De Sanctis M. Cristina
De Sanctis Giovanni
De Zotti Gianfranco
Delbo Marco
Dell' Oro Aldo
Della Ceca Roberto
Della Valle Massimo
Di Cocco Guido
Di Fazio Alberto
Di Martino Mario
Di Mauro Maria Pia
Di Serego Alighieri Sperello
Diaferio Antonaldo
D'Onofrio Mauro
Dotto Elisabetta
Drimmel Ronald E.
Einaudi Giorgio
Elia Davide
Emanuele Alessandro
Ermolli Ilaria
Ettori Stefano
Facondi Silvia Rosa
Falchi Ambretta
Falciani Roberto
Falomo Renato
Fanti Carla Giovannini
Fanti Roberto
Faraggiana Rosanna
Fasano Giovanni
Federici Luciana
Felli Marcello
Feretti Luigina
Ferluga Steno
Ferrari Attilio
Ferreri Walter
Ferrini Federico
Focardi Paola
Fofi Massimo
Forti Giuseppe
Franceschini Alberto
Franchini Mariagrazia
Frasca Antonio
Fulle Marco
Fusco-Femiano Roberto
Fusi-Pecci Flavio
Gai Mario
Galeotti Piero
Galletta Giuseppe
Galletto Dionigi
Galli Daniele
Gallino Roberto
Garilli Bianca
Gaudenzi Silvia
Gavazzi Giuseppe
Gemmo Alessandra
Gervasi Massimo
Ghia Piera Luisa
Ghirlanda Giancarlo
Ghisellini Gabriele
Giallongo Emanuele
Giannone Pietro
Giannuzzi Maria A.
Gioia Isabella M.
Giordano Silvio
Giovanardi Carlo
Giovannelli Franco
Giovannini Gabriele
Girardi Marisa
Girardi Leo A.
Godoli Giovanni
Gomez Maria Teresa
Granato Gian Luigi
Gratton Raffaele G.
Greggio Laura
Gregorini Loretta
Gronchi Giovanni Federico
Grueff Gavril
Guarnieri Adriano
Guerrero Gianantonio
Guzzo Luigi
Hack Margherita
Held Enrico V.
Hunt Leslie
Iijima Takashi
Iovino Angela
Janssen Katja
Kawakatu Nozomu
La Barbera Francesco
La Padula Cesare
La Spina Alessandra
Lai Sebastiana
Lanciano Nicoletta
Landi Simone
Landi Degli Innocenti Egidio
Landini Massimo
Landolfi Marco
Lanza Antonino F.
Lanzafame Alessandro C.
Lari Carlo
Lattanzi Mario G.
Lazzarin Monica
Leone Franco
Leschiutta S.
Leto Giuseppe
Ligori Sebastiano
Limongi Marco
Lisi Franco
Londrillo Pasquale
Longo Giuseppe
Lucchesi David M.
Maccacaro Tommaso
Maccagni Dario
Maccarone Maria Concetta
Maceroni Carla
Mack Karl-Heinz
Maffei Paolo
Magazzu Antonio
Maggio Antonio
Magni Gianfranco
Maiolino Roberto
Malagnini Maria Lucia
Manara Alessandro A.
Mancuso Santi
Mandolesi Nazzareno
Mannucci Filippo
Mantegazza Luciano
Mantovani Franco
Marano Bruno
Maraschi Laura
Marconi Alessandro
Marconi Marcella
Mardirossian Fabio
Marilli Ettore
Maris Michele
Marmolino Ciro
Marzari Francesco
Marziani Paola
Marzo Giuseppe A.
Masani A.
Massaglia Silvano
Matt Giorgio
Matteucci Francesca

Mazzali Paolo A.
Mazzei Paola
Mazzitelli Italo
Mazzoni Massimo
Mazzucconi Fabrizio
Meneghetti Massimo
Mennella Vito
Mercurio Amata
Mereghetti Sandro
Merighi Roberto
Merluzzi Paola
Messerotti Mauro
Messina Sergio
Messina Antonio
Mezzetti Marino
Micela Giuseppina
Milani Andrea
Milano Leopoldo
Mineo Teresa
Missana Marco
Molaro Paolo
Molinari Emilio
Monaco Pierluigi
Morbidelli Roberto
Morbidelli Lorenzo
Morossi Carlo
Moscadelli Luca
Moscardini Lauro
Motta Santo
Mucciarelli Paola
Mulas Giacomo
Munari Ulisse
Mureddu Leonardo
Napolitano Nicola R.
Natali Giuliano
Natta Antonella
Nesci Roberto
Nicastro Luciano
Nipoti Carlo
Nobili L.
Nobili Anna M.
Nocera Luigi
Noci Giancarlo
Nucita Achille A.
Oliva Ernesto
Origlia Livia
Orio Marina
Orlandini Mauro
Orlando Salvatore
Ortolani Sergio
Pacini Franco
Pagano Isabella
Palagi Francesco
Palla Francesco
Pallavicini Roberto
Palumbo Maria Elisabetta
Palumbo Giorgio G. C.
Pannunzio Renato
Paolicchi Paolo
Parma Paola
Pasian Fabio
Pasinetti Laura E.
Pastori Livio
Paterno Lucio
Patriarchi Patrizio
Pellegrini Silvia
Peres Giovanni
Perola Giuseppe C.
Persi Paolo
Pettini Marco
Pian Elena
Picca Domenico
Piccioni Adalberto
Pigatto Luisa
Piotto Giampaolo
Pipino Antonio
Piro Luigi
Pirronello Valerio
Pizzella Guido
Pizzella Alessandro
Pizzichini Graziella
Poggianti Bianca M.
Polcaro V. F.
Poletto Giannina
Poma Angelo
Porceddu Ignazio E. P.
Poretti Ennio
Prandoni Isabella
Preite Martinez Andrea
Proverbio Edoardo
Pucillo Mauro
Rafanelli Piero
Raimondo Gabriella
Raiteri Claudia M.
Ramella Massimo
Rampazzo Roberto
Randich Sofia
Ranieri Marcello
Reale Fabio
Reardon Kevin
Righini Alberto
Romano Giuliano
Romano Patrizia
Romoli Marco
Rossi Corinne
Rossi Alessandro
Rossi Lucio
Ruffini Remo
Russo Guido
Sabbadin Franco
Saggion Antonio
Salinari Piero
Salvati Marco
Sancisi Renzo
Santin Paolo
Saracco Paolo
Sarasso Maria
Sbordone Luca
Scappini Flavio
Scaramella Roberto
Scardia Marco
Schneider Raffaella
Sciortino Salvatore
Scodeggio Marco
Scuderi Salvo
Secco Luigi
Sedmak Giorgio
Selvelli Pierluigi
Serio Salvatore
Setti Giancarlo
Severino Giuseppe
Shore Steven N.
Silva Laura
Silvestro Giovanni
Silvotti Roberto
Simone Zaggia
Sironi Giorgio
Smaldone Luigi A.
Smart Richard L.
Smriglio Filippo
Spadaro Daniele
Spagna Alessandro
Spinoglio Luigi
Stalio Roberto
Stanga Ruggero
Stanghellini Carlo
Stefano Andreon
Stirpe Giovanna M.
Strafella Francesco
Straus Thomas

Strazzulla Giovanni
Tagliaferri Gianpiero
Tanzella-Nitti Giuseppe
Tavecchio Fabrizio
Tempesti Piero
Ternullo Maurizio
Terranegra Luciano
Testi Leonardo
Tofani Gianni
Tomasi Paolo
Torelli M.
Tormen Giuseppe
Tornambe Amedeo
Torricelli Guidetta
Tosi Monica
Tozzi Gian Paolo
Tozzi Paolo
Tozzi Andrea
Trevese Dario
Trigilio Corrado
Trinchieri Ginevra
Trussoni Edoardo
Tuccari Gino
Turatto Massimo
Turolla Roberto
Ubertini Pietro
Umana Grazia
Uras Silvano
Vagnetti Fausto
Vallenari Antonella
Valsecchi Giovanni B.
Velli Marco
Ventura Rita
Ventura Paolo
Venturi Tiziana
Vercellone Stefano
Vettolani Giampaolo
Vietri Mario
Vignali Cristian
Vigotti Mario
Villata Massimo
Viotti Roberto
Virgopia Nicola
Vittone Alberto Angelo
Vittorio Nicola
Vladilo Giovanni
Walmsley C. Malcolm
Wolter Anna
Zamorani Giovanni
Zampieri Luca
Zanichelli Alessandra
Zaninetti Lorenzo
Zanini Valeria
Zannoni Mario
Zappala Rosario Aldo
Zappala Vincenzo
Zavatti Franco
Zitelli Valentina
Zlobec Paolo
Zucca Elena
Zuccarello Francesca

Japan

Agata Hidehiko
Aikawa Yuri
Aikawa Toshiki
Akabane Kenji A.
Akabane Kenji A.
Akita Kyo
Ando Hiroyasu
Aoki Shinko
Aoki Wako
Arafune Jiro
Arai Kenzo
Arimoto Nobuo
Asai Ayumi
Awaki Hisamitsu
Ayani Kazuya
Azuma Takahiro
Baba Naoshi
Baba Hajime
Chiba Takeshi
Chiba Masashi
Chikada Yoshihiro
Chikawa Michiyuki
Daishido Tsuneaki
Deguchi Shuji
Dobashi Kazuhito
Doi Mamoru
Dotani Tadayasu
Ebisuzaki Toshikazu
Ebizuka Noboru
Enome Shinzo
Eriguchi Yoshiharu
Ezawa Hajime
Fharo Toshihiro
Fujimoto Masa-Katsu
Fujimoto Masayuki
Fujimoto Shin-ichiro
Fujishita Mitsumi
Fujita Yoshio
Fujita Yutaka
Fujita Mitsutaka
Fujiwara Takao
Fujiwara Tomoko
Fujiwara Akira
Fukuda Ichiro
Fukuda Naoya
Fukue Jun
Fukugita Masataka
Fukui Takao
Fukui Yasuo
Fukushige Toshiyuki
Fukushima Toshio
Funato Yoko
Furusho Reiko
Goto Tomotsugu
Gouda Naoteru
Gunji Shuichi
Habe Asao
Hachisu Izumi
Hamabe Masaru
Hamajima Kiyotoshi
Hanami Hitoshi
Hanaoka Yoichiro
Hanawa Tomoyuki
Handa Toshihiro
Hara Tetsuya
Hara Hirohisa
Hasegawa Ichiro
Hasegawa Tetsuo
Hashimoto Masa-aki
Hashimoto Osamu
Hatsukade Isamu
Hattori Makoto
Hayasaki Kimitake
Hayashi Chushiro
Hiei Eijiro
Hirabayashi Hisashi
Hirai Masanori
Hirao Takanori

Hirashita Hiroyuki
Hirata Ryuko
Hirayama Tadashi
Hiromoto Norihisa
Hirota Tomoya
Honma Mareki
Horaguchi Toshihiro
Hori Genichiro
Horiuchi Ritoku
Hosokawa Mizuhiko
Hozumi Shunsuke
Hurukawa Kiitiro
Ichikawa Takashi
Ichikawa Shin-ichi
Ichimaru Setsuo
Iguchi Satoru
Iguchi Osamu
Iijima Shigetaka
Ikeuchi Satoru
Imai Hiroshi
Imanishi Masatoshi
Inada Naohisa
Inagaki Shogo
Inatani Junji
Inoue Hajime
Inoue Makoto
Inoue Takeshi
Inutsuka Shu-ichiro
Iriyama Jun
Ishida Keiichi
Ishida Toshihito
Ishida Manabu
Ishiguro Masato
Ishihara Hideki
Ishii Takako T.
Ishimaru Yuhri
Ishizaka Chiharu
Ishizawa Toshiaki A.
Ishizuka Toshihisa
Ita Yoshifusa
Ito Takashi
Ito Yutaka
Ito Kensai A.
Itoh Hiroshi
Itoh Naoki
Itoh Masayuki
Iwamuro Fumihide
Iwasaki Kyosuke
Iwata Takahiro
Iye Masanori
Izumiura Hideyuki
Jeong Woong-Seob
Jugaku Jun
Kaburaki Osamu
Kaifu Norio
Kajino Toshitaka
Kakinuma Takakiyo T.
Kakuta Chuichi
Kamaya Hideyuki
Kambe Eiji
Kameya Osamu
Kamijo Fumio
Kanamitsu Osamu
Kaneda Hidehiro
Kaneko Noboru
Karoji Hiroshi
Kashikawa Nobunari
Kasuga Takashi
Kato Takako
Kato Mariko
Kato Taichi
Kato Shoji
Kato Tsunehiko
Kato Yoshiaki
Kato Ken-ichi
Kawabata Shusaku
Kawabata Kinaki
Kawabata Kiyoshi
Kawabata Koji S.
Kawabe Ryohei
Kawaguchi Ichiro
Kawaguchi Kentarou
Kawai Nobuyuki
Kawakita Hideyo
Kawamura Akiko
Kawara Kimiaki
Kawasaki Masahiro
Kawata Yoshiyuki
Kiguchi Masayoshi
Kimura Toshiya
Kinoshita Hiroshi
Kinugasa Kenzo
Kitai Reizaburo
Kitamoto Shunji
Kitamura Masatoshi
Kobayashi Shiho
Kobayashi Eisuke
Kobayashi Yukiyasu
Kobayashi Hideyuki
Koda Jin
Kodaira Keiichi
Kodama Tadayuki
Kodama Hideo
Kogure Tomokazu
Kohno Kotaro
Koide Shinji
Koike Chiyoe
Kojima Yasufumi
Kojima Masayoshi
Kokubo Eiichiro
Komiyama Yutaka
Kondo Masayuki
Kondo Masaaki
Kosai Hiroki
Koshiba Masatoshi
Koshiishi Hideki
Koyama Katsuji
Kozai Yoshihide
Kozasa Takashi
Kubota Jun
Kudoh Takahiro
Kumagai Shiomi
Kumai Yasuki
Kunieda Hideyo
Kuno Nario
Kurokawa Hiroki
Kusano Kanya
Kusunose Masaaki
Machida Mami
Maeda Keiichi
Maeda Kei-ichi
Maeda Koitiro
Maehara Hideo
Maihara Toshinori
Makino Junichiro
Makino Fumiyoshi
Makishima Kazuo
Makita Mitsugu
Makiuti Sin'itirou
Manabe Seiji
Mann Ingrid
Masai Kuniaki
Masuda Satoshi
Matsuda Takuya
Matsuhara Hideo
Matsui Takafumi
Matsumoto Ryoji
Matsumoto Toshio
Matsumoto Tomoaki
Matsumoto Katsura

Matsumoto Hironori
Matsumura Masafumi
Matsuo Hiroshi
Matsuoka Masaru
Matsuura Shuji
Matsuura Mikako
Mikami Takao
Mineshige Shin
Mitsuda Kazuhisa
Miyaji Shigeki
Miyama Syoken
Miyamoto Masanori
Miyata Emi
Miyawaki Ryosuke
Miyazaki Atsushi
Miyoshi Shigeru
Miyoshi Makoto
Mizumoto Yoshihiko
Mizuno Norikazu
Mizuno Akira
Mizuno Takao
Mizuno Shun
Mizutani Kohei
Momose Muntake
Mori Masao
Morimoto Masaki
Morita Kazuhiko
Morita Koh-ichiro
Moriyama Fumio
Motizuki Yuko
Motohara Kentaro
Mukai Tadashi
Murakami Hiroshi
Murakami Toshio
Murakami Hiroshi
Murakami Izumi
Murayama Takashi
Nagase Fumiaki
Nagata Tetsuya
Nagataki Shigehiro
Nakada Yoshikazu
Nakagawa Naoya
Nakagawa Yoshinari
Nakagawa Yoshitsugu
Nakagawa Takao
Nakai Naomasa
Nakajima Koichi
Nakajima Hiroshi
Nakajima Junichi
Nakajima Tadashi
Nakamichi Akika
Nakamoto Taishi
Nakamura Akiko M.
Nakamura Fumitaka
Nakamura Takuji
Nakamura Yasuhisa
Nakamura Tsuko
Nakamura Takashi
Nakanishi Kouichiro
Nakanishi Hiroyuki
Nakano Syuichi
Nakano Makoto
Nakano Takenori
Nakao Yasushi
Nakasato Naohito
Nakayama Kunji
Nakazawa Kiyoshi
Nambu Yasusada
Nariai Kyoji
Narita Shinji
Nishi Keizo
Nishi Ryoichi
Nishida Minoru
Nishida Mitsugu
Nishikawa Jun
Nishimura Jun
Nishimura Shiro E.
Nishimura Masayoshi
Nishimura Osamu
Nishio Masanori
Nitta Shin-ya
Noguchi Masafumi
Noguchi Kunio
Nomoto Ken'ichi
Obi Shinya
Ogawa Hideo
Ogawara Yoshiaki
Ogura Katsuo
Ohashi Takaya
Ohashi Yukio
Ohishi Masatoshi
Ohki Kenichiro
Ohnishi Kouji
Ohta Kouji
Ohtani Hiroshi
Ohtsubo Junji
Ohyama Noboru
Okamoto Isao
Okamura Sadanori
Okazaki Akira
Okazaki Atsuo T.
Okuda Toru
Okuda Haruyuki
Okumura Sachiko
Okumura Shin-ichiro
Omukai Kazuyuki
Onaka Takashi
Onishi Toshikazu
Oohara Ken-ichi
Oowaki Naoaki
Osaki Yoji
Oyabu Shinki
Ozaki Masanobu
Ozeki Hiroyuki
Sabano Yutaka
Sadakane Kozo
Saigo Kazuya
Saijo Keiichi
Saio Hideyuki
Saito Takao
Saito Mamoru
Saito Kuniji
Saito Yoshihiko
Saito Sumisaburo
Saitoh Takayuki
Sakai Junichi
Sakamoto Seiichi
Sakao Taro
Sakashita Shiro
Sakurai Kunitomo
Sakurai Takashi
Sasaki Misao
Sasaki Minoru
Sasaki Shin
Sasao Tetsuo
Sato Isao
Sato Katsuhiko
Sato Koichi
Sato Shinji
Sato Fumio
Sato Shuji
Sato Jun'ichi
Sato Humitaka
Sawa Takeyasu
Seki Munezo
Sekido Mamoru
Sekiguchi Maki
Sekiguchi Naosuke
Sekiguchi Tomohiko
Sekimoto Yutaro

Shibahashi Hiromoto
Shibai Hiroshi
Shibasaki Kiyoto
Shibata Yukio
Shibata Kazunari
Shibata Shinpei
Shibata Katsunori M.
Shibata Masaru
Shibazaki Noriaki
Shigeyama Toshikazu
Shimasaku Kazuhiro
Shimojo Masumi
Shimura Toshiya
Shioya Yasuhiro
Simoda Mahiro
Sofue Yoshiaki
Soma Mitsuru
Sorai Kazuo
Suda Takuma
Suda Kazuo
Suematsu Yoshinori
Sugai Hajime
Suganuma Masahiro
Sugawa Chikara
Sugimoto Daiichiro
Suginohara Tatsushi
Sugitani Koji
Sugiyama Naoshi
Sunada Kazuyoshi
Suto Yasushi
Suzuki Hideyuki
Suzuki Tomoharu
Suzuki Takeru
Tachihara Kengo
Tagoshi Hideyuki
Takaba Hiroshi
Takada-Hidai Masahide
Takahara Fumio
Takahara Mariko
Takahashi Junko
Takahashi Masaaki
Takahashi Koji
Takahashi Tadayuki
Takakubo Keiya
Takami Hideki
Takano Toshiaki
Takano Shuro
Takase Bunshiro
Takata Masao
Takato Naruhisa
Takayanagi Kazuo
Takeda Hidenori
Takeda Yoichi
Takenouchi Tadao
Takeuchi Tsutomu T.
Takeuti Mine
Takizawa Motokazu
Tamenaga Tatsuo
Tamura Shin'ichi
Tamura Motohide
Tanabe Hiroyoshi
Tanabe Kenji
Tanabe Toshihiko
Tanaka Riichiro
Tanaka Yasuo
Tanaka Yutaka D.
Tanaka Masuo
Tanaka Wataru
Taniguchi Yoshiaki
Tanikawa Kiyotaka
Taruya Atsushi
Tashiro Makoto
Tatekawa Takayuki
Tatematsu Ken'ichi
Tawara Yuzuru
Terashima Yuichi
Terashita Yoichi
Tomimatsu Akira
Tomisaka Kohji
Tomita Akihiko
Tomita Kenji
Tomita Koichiro
Tosa Makoto
Tosaki Tomoka
Totani Tomonori
Totsuka Yoji
Toyama Kiyotaka
Tsubaki Tokio
Tsuboi Masato
Tsuchiya Toshio
Tsuji Takashi
Tsujimoto Takuji
Tsunemi Hiroshi
Tsuneta Saku
Tsuru Takeshi
Uchida Juichi
Ueda Yoshihiro
Uemura Makoto
Ueno Sueo
Ueno Munetaka
Uesugi Akira
Ukita Nobuharu
Umeda Hideyuki
Umemoto Tomofumi
Umemura Masayuki
Unno Wasaburo
Utsumi Kazuhiko
Wada Takehiko
Wada Keiichi
Wakamatsu Ken-ichi
Wako Kojiro
Wanajo Shinya
Washimi Haruichi
Watanabe Noriaki
Watanabe Takashi
Watanabe Tetsuya
Watanabe Jun-ichi
Watarai Hidneori
Watarai Kenya
Watari Shinichi
Yabushita Shin A.
Yagi Masafumi
Yahagi Hideki
Yamada Masako
Yamada Toru
Yamada Yoshiyuki
Yamagata Tomohiko
Yamamoto Tetsuo
Yamamoto Satoshi
Yamamoto Yoshiaki
Yamamoto Masayuki
Yamamura Issei
Yamaoka Hitoshi
Yamasaki Noriko Y.
Yamasaki Tatsuya
Yamasaki Atsuma
Yamashita Yasumasa
Yamashita Kojun
Yamauchi Shigeo
Yamauchi Makoto
Yanagisawa Masahisa
Yano Taihei
Yano Hajime
Yasuda Naoki
Yasuda Haruo
Yokosawa Masayoshi
Yokoyama Koichi
Yokoyama Jun-ichi
Yonekura Yoshinori
Yoneyama Tadaoki

Yoshida Haruo
Yoshida Shigeomi
Yoshida Atsumasa
Yoshida Junzo
Yoshida Michitoshi
Yoshida Fumi
Yoshida Takashi
Yoshii Yuzuru
Yoshikawa Makoto
Yoshimura Hirokazu
Yoshioka Satoshi
Yoshioka Kazuo
Yoshizawa Masanori
Yuasa Manabu
Yui Yukari Y.

Korea (Republic of)

Ahn Youngsook
Ann Hong Bae
Byun Yong-Ik
Cha Seung-Hoon
Chae Jongchul
Chang Heon-Young
Chang Kyongae
Cho Se-Hyung
Cho Kyung Suk
Choe Seung Urn
Choi Kyu-Hong
Choi Minho
Choi Chul-Sung
Chou Kyong Chol
Chun Mun-suk
Chung Hyun Soo
Han Cheongho
Han Wonyong
Hong Seung Soo
Hwang Jai-chan
Hyung Siek
Im Myungshin
Ishiguro Masateru
Jang Minwhan
Jeong Jang Hae
Jung Jae Hoon
Kang Hyesung
Kang Young-Woon
Kang Yong Hee
Kim Yongha
Kim Yonggi
Kim Kwang-tae
Kim Sug-Whan
Kim Jongsoo
Kim Sungsoo S.
Kim Chun Hwey
Kim Ho-il
Kim Hyun-Goo
Kim Yong-Cheol
Kim Tu-Whan
Kim Chulhee
Kim Kap-sung
Kim Seung-Lee
Kim Bong Gyu
Kim Sungeun
Koo Bon-Chul
Lee Myung Gyoon
Lee Young Wook
Lee Eun-Hee
Lee Youngung
Lee Woo-baik
Lee Sang Gak
Lee Chang-Won
Lee Hee-Won
Lee Jae-Woo
Lee Dae-Hee
Lee Hyung Mok
Lee Dong-Hun
Lee Yong-Sam
Lee Sungho
Lee Seong-Jae
Lee Yong Bok
Lee Kang Hwan
Lee Dong-Wook
Minh Young Chol
Minn Young Key
Moon Shin Haeng
Moon Yong-Jae
Nha Il-Seong
Noh Hyerim
Noh Hyerim
Oh Kap-Soo
Oh Kyu Dong
Park Hong Suh
Park Myeong-gu
Park Seok Jae
Park Jang-Hyun
Park Changbom
Park Young-Deuk
Park Pil-Ho P.
Park Il-Hung
Rey Soo-Chang
Rhee Myung-Hyun
Ryu Dongsu
Sohn Young-Jong
Song In-Ok
Song Doo Jong
Suh Kyung-Won
Sung Hwankyung
Yang Jongmann
Yang Hong-Jin
Yi Sukyoung Ken
Yi Yu
Yim Hong-Suh
Yoon Suk-Jin
Yoon Tae S.
Yun Hong-Sik

Latvia

Abele Maris K.
Alksnis Andrejs
Balklavs-Grinhofs Arturs E.
Daube-Kurzemniece Ilga A.
Dzervitis Uldis
Eglitis Ilgmars
Freimanis Juris
Grasberg Ernest K.
Lapushka Kazimirs K.
Ryabov Boris I.
Salitis Antonijs
Shmeld Ivar
Vilks Ilgonis
Zacs Laimons
Zhagars Youris H.

Lebanon

Bittar Jamal
El Eid Mounib
Plassard J.
Sabra Bassem Mohamad
Touma Jihad Rachid

Lithuania

Bartasiute Stanislava
Bartkevicius Antanas
Bogdanovicius Pavelas
Cernis Kazimieras
Kazlauskas Algirdas S.
Kupliauskiene Alicija
Lazauskaite Romualda
Meistas Edmundas G.
Rudzikas Zenonas R.
Sperauskas Julius
Straizys Vytautas P.
Sudzius Jokubas
Tautvaisiene Grazina
Vansevicius Vladas
Zdanavicius Kazimer

Malaysia

Abdul Aziz Ahmad Zaharim
Abu Kassim Hasan B.
Ilyas Mohammad
Mahat Rosli H.
Majid Abdul Bin A H
Mohd Zambri Zainuddin
Othman Mazlan

Mexico

Aguilar Luis A. C.
Allen Christine
Andernach Heinz J
Arellano Ferro Armando
Aretxaga Itziar
Arthur Jane
Avila Foucault Remy F.
Avila-Reese Vladimir
Ballesteros-Paredes Javier
Binette Luc
Bisiacchi Gianfranco
Bravo-Alfaro Hector
Canto Jorge
Cardona Octavio
Carigi Leticia
Carraminana Alberto
Carrasco Bertha Esperanza
Carrasco Luis
Carrillo Moreno Rene
Chatterjee Tapan Kumar
Chavarria-K Carlos
Chavez-Dagostino Miguel
Chavushyan Vahram
Colin Pedro
Contreras Maria E.
Cornejo Alejandro A.
Costero Rafael
Coziol Roger
Cruz-Gonzalez Irene
Cuevas Salvador C.
Curiel Salvador
Daltabuit Enrique
De La Herran Jose V.
Dultzin-Hacyan Deborah
Echevarria Juan Roman
Eenens Philippe
Escalante Vladimir
Fierro Julieta
Franco José
Galindo-Trejo Jesus
Garcia-Barreto José Antonio
Gomez Yolanda
Gonzalez Alejando
Gonzalez J. Jesus
Guichard Jose
Henney William John
Hiriart David
Hughes David H.
Kemp Simon N.
Klapp Jaime
Koenigsberger Gloria
Kurtz Stanley E.
Lekth Evgeni
Lizano Susana
Loinard Laurent R.
Lopez Jose Alberto
Lopez Cruz Omar
Malacara Daniel
Martinez Mario
Martos Marco A.
Mayya Divakara
Mendez Emmanuel
Mendoza V. Eugenio E.
Mendoza-Torress Jose-Eduardo
Migenes Victor
Moreno Edmundo
Moreno-Corral Marco A.
Obregon Octavio
Page Dany
Peimbert Manuel
Pena José
Pena Miriam
Perez de Tejada Hector A.
Perez-Peraza Jorge A.
Phillips John Peter
Poveda Arcadio
Puerari Ivanio
Recillas-Cruz Elsa
Rodriguez Luis F.
Rodriguez Monica
Rohrmann Rene D.
Rosado Margarita
Ruelas-Mayorga R. A.
Santillan Alfredo J.

Sarmiento-Galan Antonio F.
Schuster William John
Serrano Alfonso
Tapia Mauricio
Tenorio-Tagle Guillermo
Torres-Peimbert Silvia
Tovmassian Gaghik
Tovmassian Hrant Mushegi
Valdes-Sada Pedro A.
Vazquez-Semadeni Enrique
Wall William F.
Warman Josef
Wilkin Francis P.

Mongolia

Morocco

Benkhaldoun Zouhair
Chamcham Khalil
El Bakkali Larbi
Kadiri Samir
Lazrek Mohamed
Najid Nour-Eddine
Touma Hamid

Netherlands

Achterberg Abraham
Baan Willem A.
Barthel Peter
Baud Boudewijn
Begeman Kor G.
Beintema Douwe A.
Bennett Kevin
Bijleveld Willem
Blaauw Adriaan
Bleeker Johan A. M. Ir.
Bloemen Hans (J. B. G. M.)
Boeker Torsten
Boland Wilfried
Bontekoe Romke
Borgman Jan
Bos Albert
Bosma Pieter B.
Brandl Bernhard R.
Braun Robert
Bregman Jacob D. Ir.
Brinkman Bert C.
Brouw Willem N.
Brown Anthony G. A.
Butcher Harvey R.
Campbell Robert M.
Clavel Jean
De Bruijne Jos H. J.
De Bruyn A. Ger
De Geus Eugene
De Graauw Thijs
De Jager Cornelis
De Jong Teije
De Korte Pieter A. J.
De Koter Alex
De Vries Cornelis
De Zeeuw Pieter T.
Dekker E.
Den Herder Jan-Willem
Deul Erik R.
Dewi Jasinta Dini Maria
Dominik Carsten
Douglas Nigel
Ehrenfreund Pascale
Fender Robert P.
Fitton Brian
Foing Bernard H.
Foley Anthony
Franx Marijn
Fridlund Malcolm
Fritzova-Svestka L.
Gabuzda Denise C.
Garrett Michael
Gathier Roel
Gimenez Alvaro
Goedbloed Johan P.
Gondoin Philippe A. D.
Groot Paul J.
Gurvits Leonid I.
Habing Harm J.
Hammerschlag Robert H.
Hammerschlag-Hensberge G.
Heintze J. R. W.
Heise John
Helmi Amina
Helmich Frank P.
Henrichs Hubertus F.
Heras Ana M.
Hermsen Willem
Heske Astrid
Hoekstra Roel
Houdebine Eric
Hovenier J. W.
Hoyng Peter
Hubenet Henri
Hulsbosch A. N. M.
Icke Vincent
In't Zand Johannes J. M.
Israel Frank P.
Ives John Christopher
Jaffe Walter Joseph
Jakobsen Peter
Jonker Peter G.
Jourdain de Muizon Marie
Kaastra Jelle S.
Kahlmann Hans C.
Kaper Lex
Katgert Peter
Katgert-Merkelijn J. K.
Keller Christoph U.
Koopmans Leon V. E.
Koornneef Jan F.
Kuijken Koen H.
Kuijpers H. Jan M. E.
Kuiper Lucien
Kuperus Max

Kwee K. K.
Lamers Henny J. G. L. M.
Langer Norbert
Larsen Soeren S.
Laureijs Rene J.
Le Poole Rudolf S.
Levin Yuri
Linnartz Harold
Lub Jan
Markoff Sera B.
Martens Petrus C.
Mendez Mariano
Miley George K.
Morales Rueda Luisa
Morganti Raffaella
Namba Osamu
Nelemans Gijs
Nieuwenhuijzen Hans
Ollongren A.
Olnon Friso M.
Olthof Henk
Oosterloo Thomas
Parmar Arvind Nicholas
Peacock Anthony
Pel Jan Willem
Peletier Reynier Frans
Perryman Michael A. C.
Pogrebenko Sergei
Pols Onno R.
Portegies Zwart Simon F.
Pottasch Stuart R.
Prusti Timo
Raassen Ion
Raimond Ernst
Roelfsema Peter
Roettgering Huub
Roos Nicolaas
Rutten Robert J.
Sanders Robert
Savonije Gerrit Jan
Schaye Joop
Scheepmaker Anton
Schilizzi Richard T.
Schrijver Johannes
Schulz Rita M.
Schutte Willem Albert
Schwarz Ulrich J.
Schwehm Gerhard
Shipman Russell F.
Smit J. A.
Snellen Ignas A. G.
Spaans Marco
Spoelstra T. A. Th.
Stappers Benjamin W.
Stark Ronald
Strom Richard G.
Svestka Zdenek
Tauber Jan
te Lintel Hekkert Peter
The Pik-Sin
Tinbergen Jaap
Tjin-a-Djie Herman R. E.
Tolstoy Eline
Trager Scott C.
Valentijn Edwin A.
van Albada Tjeerd S.
van Bueren Hendrik G.
van de Stadt Herman
van de Weygaert Rien
van den Heuvel Edward P. J.
van den Oord Bert H. J.
van der Hucht Karel A.
van der Hulst Jan M.
van der Klis Michiel
van der Kruit Pieter C.
van der Laan Harry
van der Tak Floris F. S.
van der Werf Paul P.
van Dishoeck Ewine F.
van Duinen R. J.
van Genderen Arnoud M.
van Gent Robert H.
van Haarlem Michiel
van Houten-Groeneveld Ingrid
van Langevelde Huib Jan
van Woerden Hugo
Verbunt Franciscus
Verheijen Marc A. W.
Vermeulen Rene Cornelis
Vink Jacco
Waters Laurens B. F. M.
Wesselius Paul R.
Wijers Ralph A. M. J.
Wijnands Rudy
Wijnbergen Jan
Wild Wolfgang
Winkler Christoph
Wise Michael W.
Zaroubi Saleem

New Zealand

Adams Jenni
Albrow Michael
Allen William
Andrews Frank
Arnold Richard A.
Baggaley William J.
Blow Graham L.
Bond Ian A.
Budding Edwin
Carter Brian
Christie Grant W.
Cottrell Peter L.
Craig Ian
Dodd Richard J.
Gilmore Alan C.
Gulyaev Sergei A.
Hearnshaw John B.
Hill Graham
Jones Albert F.
Kerr Roy P.
Kilmartin Pamela
Kitaeff Vyacheslav V.
Pollard Karen
Rumsey Norman J.
Skuljan Jovan
Sullivan Denis John
Sweatman Winston L.
Taylor Andrew
Walker William S. G.
Yock Philip

Nigeria

Akujor Chidi E.
Ejikeme Chukwude A.
Eze Romanus N.
Okeke Pius N.
Okeke Francisca N.
Okoye Samuel E.
Okpala Kingsley C.
Rabiu Akeem B.
Schmitter Edward F.
Urama Johnson O.

Norway

Aksnes Kaare
Andersen Bo Nyborg
Brahde Rolf
Brekke Pål O. L.
Brynildsen Nils
Carlsson Mats
Dahle Haakon
Elgaroy Oystein
Engvold Oddbjørn
Esser Ruth
Gudiksen Boris V.
Hansteen Viggo
Haugan Stein Vidar H.
Hauge Oivind
Havnes Ove
Jaunsen Andreas O.
Kjeldseth-Moe Olav
Leer Egil
Lie-Svendsen Oystein
Lilje Per Vidar Barth
Lin Yong
Pettersen Bjørn R.
Puetzfeld Dirk
Rouppe van der Voort Luc H. M.
Skjaeraasen Olaf
Solheim Jan Erik
Stabell Rolf
Trulsen Jan K.
Wiik Toutain Jun E.
Wikstol Oivind

Peru

Aguilar Maria Luisa
Ishitsuka Mutsumi
Melendez Jorge
Reyes Rafael Carlos

Philippines

Celebre Cynthia P.
Soriano Bernardo M.
Torres Jesus Rodrigo F.
Vinluan Renato A.

Poland

Bajtlik Stanislaw
Bem Jerzy
Berlicki Arkadiusz
Borkowski Kazimierz
Breiter Slawomir
Brzezinski Aleksander
Bulik Tomasz
Bzowski Maciej
Chodorowski Michal
Choloniewsski Jacek
Chyzy Krzysztof Tadeusz
Ciurla Tadeusz
Cugier Henryk
Czern Bozena
Daszynska-Daszkiewicz Jadwiga
Demianski Marek
Drozyner Andrzej
Dybczynski Piotr A.
Dziembowski Wojciech A
Essai Essai
Falewicz Robert K.
Flin Piotr
Gaska Stanislaw
Gesicki Krzysztof
Giersz Miroslav
Gil Janusz A.
Godlowski Wlodzimierz
Gorgolewski Stanislaw Pr.
Gozdziewski Krzysztof
Grabowski Boleslaw
Gronkowski Piotr M.
Grzedzielski Stanislaw Pr.
Haensel Pawel
Hanasz Michal B.
Hanasz Jan
Heller Michael
Hurnik Hieronim
Iwaniszewska Cecylia
Jahn Krzysztof
Jakimiec Jerzy
Janiuk Agnieszka
Jaranowski Piotr
Jaroszynski Michal
Jerzykiewicz Mikolaj
Jopek Tadeusz Jan
Juszkiewicz Roman
Kaluzny Janusz
Kijak Jaroslaw
Kiraga Marcin M.
Kluzniak Wlodzimiere

Kolaczek Barbara
Kopacki Grzegorz
Kosek Wieslaw
Kozlowski Maciej
Krasinski Andrzej
Kreiner Jerzy M.
Krelowski Jacek
Krempec-Krygier Janina
Krolikowska-Soltan Malgorzata
Kruszewski Andrzej
Krygier Bernard
Kryszczynska Agnieszka
Krzesinski Jerzy H.
Kubiak Marcin A.
Kurpinska-Winiarska M.
Kurzynska Krystyna
Kus Andrzej Jan
Kwiatkowski Tomasz
Lehmann Marek
Machalski Jerzy
Maciejewski Andrzej J.
Madej Jerzy
Marecki Andrzej
Maslowski Jozef
Michalec Adam
Michalek Grzegorz
Michalowski Tadeusz
Mietelski Jan S.
Mikolajewska Joanna
Mikolajewski Maciej P.
Moderski Rafal
Molenda-Zakowicz Joanna C.
Moskalik Pawel
Nastula Jolanta
Niedzielski Andrzej
Ogloza Waldemar
Opolski Antoni
Ostrowski Michal
Otmianowska-Mazur Katarzyna
Pietrzynski Grzegorz
Pigulski Andrzej
Pojmanski Grzegorz
Pokrzywka Bartlomiej
Pres Pawek T.
Rompolt Bogdan
Rozyczka Michal
Rudak Bronislaw
Rudawy Pawel
Rudnicki Konrad
Rys Stanislaw
Sarna Marek Jacek
Schillak Stanislaw
Schreiber Roman
Schwarzenberg-Czerny A.
Semeniuk Irena
Sienkiewicz Ryszard
Sikora Marek
Sikorski Jerzy
Sitarski Grzegorz
Smak Jozef I.
Sokolowski Lech
Soltan Andrzej Maria
Soszynski Igor
Stawikowski Antoni
Stepien Kazimierz
Strobel Andrzej
Sylwester Barbara
Sylwester Janusz A.
Szczerba Ryszard
Szutowicz Slawomira E.
Szydlowski Marek W.
Szymanski Michal
Szymczak Marian
Tomczak Michal
Turlo Zygmunt
Tylenda Romuald
Udalski Andrzej
Urbanik Marek
Usowics Jerzy Bogdan
Winiarski Maciej
Wnuk Edwin
Woszczyk Andrzej
Woszczyna Andrzej
Wytrzyszczak Iwona
Zdunik Julian
Zdziarski Andrzej
Zieba Stanislaw
Ziolkowski Janusz
Ziolkowski Krzysztof
Zola Stanislaw

Portugal

Anton Sonia
Augusto Pedro
Avelino Pedro
Braga da Costa Campos L. M.
Costa Vitor
Cunha Margarida
da Costa Antonio A.
De Avillez Miguel P.
Fernandes Joao
Fernandes Amadeu
Ferreira Joao M.
Gameiro Jorge
Garcia Paulo J. V.
Gizani Nectaria
Lago Maria T. V. T.
Lima Joao J.
Lobo Catarina
Luz David
Machado Folha Daniel F.
Magalhaes Antonio A. S.
Marques Manuel N.
Martins Carlos J. A. P.
Mendes Virgilio B.
Moitinho André
Monteiro Mario Joao P. F. G.
Moreira Morais Maria Helena
Osorio Isabel Maria T. V. P.
Osorio Jose Pereira
Pascoal Antonio J. B.
Roos-Serote Maarten C.
Santos Nuno Miguel C.
Santos Filipe D.
Santos Agostinho Rui J.
Serote Roos Margarida
Viana Pedro
Vicente Raimundo O.
Yun Joao L.

Romania

Barbosu Mihail
Blaga Christina O.
Botez Elvira
Breahna Iulian
Burs Lucian
Carsmaru Maria M.
Dinulescu Simona
Dumitrache Cristiana
Dumitrescu Alexandru
Ghizaru Mihai
Imbroane Alexandru
Lungu Nicolaie
Marilena Mierla
Maris Georgeta
Mihaila Ieronim
Mioc Vasile
Oprescu Gabriela
Oproiu Tiberiu
Parv Bazil
Pop Vasile
Pop Alexandru V.
Popescu Nedelia A.
Popescu Petre P.
Predeanu Irina
Roman Rodica
Rusu I.
Stanila George
Stavinschi Magda G.
Suran Marian Doru
Szenkovits Ferenc
Tifrea Emilia
Todoran Ioan
Toro Tibor
Ureche Vasile
Vass Gheorghe

Russian Federation

Abalakin Viktor K.
Afanas'ev Viktor L.
Afanasjeva Praskovja M.
Akim Efraim L.
Altyntsev Alexandre T.
Andronova Anna A.
Antipova Lyudmila I.
Antokhin Igor I.
Antonov Vadim A.
Arkharov Arkadij A.
Arkhipova Vera P.
Artamonov Boris P.
Artyukh Vadim S.
Babadzhahianc Mikhail K.
Bagrov Alexander V.
Bajkova Anisa T.
Bakhtigaraev Nail S.
Balega Yurij Yu
Barabanov Sergey I.
Baranov Alexander S.
Barkin Yuri V.
Baryshev Yurij V.
Batrakov Yurij V.
Baturin Vladimir A.
Belkovich Oleg I.
Berdnikov Leonid N.
Beskin Gregory M.
Beskin Vasily S.
Bikmaev Ilfan F.
Bisikalo Dmitrij V.
Bisnovatyi-Kogan Gennadij S.
Blinnikov Sergey I.
Bobylev Vadim V.
Bochkarev Nikolai G.
Bogod Vladimir M.
Bondarenko Lyudmila N.
Borisov Nikolay V.
Borovik Valery N.
Borozdin Konstantin N.
Boyarchuk Alexander A.
Brumberg Victor A.
Burba George A.
Burdyuzha Vladimir V.
Burenin Rodion A.
Bykov Andrei M.
Bystrova Natalja V.
Chashei Igor V.
Chechetkin Valerij M.
Cherepashchuk Anatolij M.
Chernetenko Yulia A.
Chernin Artur D.
Chernov Gennadij P.
Chertok Ilya M.
Chertoprud Vadim E.
Chugai Nikolaj N.
Churazov Evgenij M.
Dadaev Aleksandr N.
Dagkesamansky Rustam D.
Dambis Andrei K.
Danilov Vladimir M.
Devyatkin Aleksandr V.
Dluzhnevskaya Olga B.
Dodonov Sergej N.
Dokuchaev Vyacheslav I.
Dokuchaeva Olga D.
Doubinskij Boris A.
Dravskikh Aleksandr F.
Dudorov Alexander E.
Efremov Yurij N.
Emelianov Nikolaj V.
Eroshkin Georgij I.
Esipov Valentin F.
Fabrika Sergei Nikolaevich
Fadeyev Yurij A.
Fedotov Leonid V.
Finkelstein Andrej M.
Fokin Andrei
Fomichev Valerij V.
Fomin Valerij A.
Fominov Aleksandr M.
Fridman Aleksej M.
Galeev Albert A.
Garaimov Vladimir I.
Gayazov Iskander S.
Gelfreikh Georgij B.
Gilfanov Marat R.
Ginzburg Vitalij L.
Glagolevskij Yurij V.
Glebova Nina I.
Glushkova Elena V.
Glushneva Irina N.
Gnedin Yurij N.
Goncharov Georgij A.
Gorschkov Aleksandr G.
Gosachinskij Igor V.
Grachev Stanislav I.
Grebenev Sergej A.
Grebenikov Evgenij A.

Grechnev Victor V.
Grib Sergey A.
Grigorjev Viktor M.
Grishchuk Leonid P.
Gubanov Vadim S.
Gubchenko Vladimir M.
Gulyaev Rudolf A.
Gusev Alexander V.
Guseva Irina S.
Hagen-Thorn Vladimir A.
Idlis Grigorij M.
Ikhsanov Robert N.
Ikhsanova Vera N.
Ilin Alexey E.
Illarionov Andrei F.
Ilyasov Yuri P.
Imshennik Vladimir S.
Ioshpa Boris A.
Ipatov Aleksandr V.
Ivanov Vsevolod V.
Ivanov Evgenij V.
Ivanov Dmitrij V.
Ivanov-Kholodny Gor S.
Kaidanovski Mikhail N.
Kaigorodov Pavel V.
Kalenskii Sergei V.
Kaltman Tatyana I.
Kanayev Ivan I.
Karachentsev Igor D.
Kardashev Nicolay S.
Karitskaya Evgeniya A.
Kascheev Rafael A.
Kastel' Galina R.
Katsova Maria M.
Kazarovets Elena V.
Khaikin Vladimir B.
Khaliullin Khabibrachman F.
Kholshevnikov Konstantin V.
Kholtygin Aleksandr F.
Khromov Gavriil S.
Kim Iraida S.
Kiselev Aleksej A.
Kislyakov Albert G.
Kisseleva Tamara P.
Kitchatinov Leonid
Klochkova Valentina G.
Kocharov Grant E.
Kocharovsky Vitalij V.
Kokurin Yurij L.
Kolesnik Yuri B
Kolesov Aleksandr K.
Komberg Boris V.
Kompaneets Dmitrij A.
Kondratiev Vladislav I.
Kononovich Edvard V.
Kopylov Aleksandr I.
Kornilov Viktor G.
Korzhavin Anatolij N.
Koshelyaevsky Nikolay B.
Kostina Lidiya D.
Kostyakova Elena B.
Kovalev Yuri A.
Kovaleva Dana A.
Krasinsky George A.
Ksanfomality Leonid V.
Kudryavtsev Dmitry
Kuimov Konstantin V.
Kulikova Nelli V.
Kumajgorodskaya Raisa N.
Kumkova Irina I.
Kuril'chik Vladimir N.
Kurt Vladimir G.
Kutuzov Sergej A.
Kuzmin Arkadij D.
Kuzmin Andrej V.
Kuznetsov Eduard D.
Kuznetsov Vladimir D.
Lamzin Sergei A.
Larionov Mikhail G.
Larionov Valeri M.
Lavrov Mikhail I.
Leushin Valerij V.
Lipunov Vladimir M.
Livshits Mikhail A.
Loktin Alexhander V.
Loskutov Viktor M.
Lotova Natalja A.
Lozinskaya Tatjana A.
Lukash Vladimir N.
Lukashova Marina V.
Lutovinov Alexander A.
Makalkin Andrei B.
Makarov Dmitry I.
Makarov Valentin I.
Makarova Lidia N.
Malkin Zinovy M.
Malkov Oleg Yu.
Malofeev Valery Mikhailovich
Mardyshkin Vyacheslav V.
Marov Mikhail Ya
Mashonkina Lyudmila I.
Maslennikov Kirill L.
Massevich Alla G.
Matveenko Leonid I.
Medvedev Yurij D.
Mikisha Anatoly
Miletsky Eugeny V.
Mingaliev Marat G.
Mironov Aleksey V.
Mishurov Yuri N.
Mitrofanova Lyudmila A.
Mogilevskij Emmanuil I.
Moiseenko Sergey G.
Moiseev Alexei V.
Myachin Vladimir F.
Nadyozhin Dmitrij K.
Nagirner Dmitrij I.
Nagnibeda Valerij G.
Nagovitsyn Yurij A.
Naumov Vitalij A.
Nefedeva Antonina I.
Nikiforov Igor I.
Nikitin Aleksej A.
Novoselov Viktor S.
Obridko Vladimir N.
Oraevsky Victor N.
Orlov Victor V.
Pakhomov Yuri V.
Pamyatnykh Aleksej A.
Panchuk Vladimir E.
Papushev Pavel G.
Parfinenko Leonid D.
Parijskij Yurij N.
Pashchenko Mikhail I.
Pavlinsky Mikhail N.
Petrov Gennadij M.
Petrov Sergei D.
Petrovskaya Margarita S.
Piskunov Anatolij E.
Pitjev Nikolaj P.
Pitjeva Elena V.
Pluzhnik Eugene
Pogodin Mikhail A.
Poliakov Evgenij V.
Polozhentsev Dmitrij D.
Polyachenko Evgeny V.

Polyakhova Elena N.
Popov Mikhail V.
Popov Viktor S.
Porfir'ev Vladimir V.
Postnov Konstantin A.
Potter Heino I.
Prodan Yurij I.
Prokhorov Mikhail E.
Pushkin Sergej B.
Pustilnik Simon A.
Rahimov Ismail A.
Rastorguev Aleksej S.
Razin Vladimir A.
Reshetnikov Vladimir P.
Revnivtsev Mikhail G.
Rizvanov Naufal G.
Rodin Alexander E.
Rodionova Zhanna F.
Romanov Andrey M.
Romanyuk Iosif I.
Rudenko Valentin N.
Rudnitskij Georgij M.
Ruskol Evgeniya L.
Ryabchikova Tatiana A.
Ryabov Yurij A.
Ryabova Galina O.
Rykhlova Lidiya V.
Rzhiga Oleg N.
Sachkov Mikhail E.
Sagdeev Roald Z.
Sakhibullin Nail A.
Samodurov Vladimir A.
Samus Nikolaj N.
Sazhin Mikhail V.
Sazonov Sergey
Serber Alexander V.
Shakht Natalia A.
Shakura Nikolaj I.
Shaposhnikov Vladimir E.
Shapovalova Alla I.
Sharina Margarita E.
Shchekinov Yuri A.
Sheffer Evgenij K.
Shefov Nikolaj N.
Shematovich Valerij I.
Shenavrin Victor I.
Shevchenko Ivan I.
Shevchenko Vladislav V.
Shibanov Yuri A.
Shishov Vladimir I.
Shokin Yurij A.
Sholomitsky Gennady B.
Sholukhova Olga
Shor Viktor A.
Shulov Oleg S.
Shustov Boris M.
Sidorenkov Nikolaj S.
Silant'ev Nikolaj A.
Sil'chenko Olga K.
Skripnichenko Vladimir I.
Skulachov Dmitrij P.
Slysh Vyacheslav I.
Smirnov Grigorij T.
Smol'kov Gennadij Ya
Smolentsev Sergej G.
Snegirev Sergey D.
Snezhko Leonid I.
Sobolev Andrej M.
Soboleva Natalja S.
Sokolov Leonid L.
Sokolov Vladimir V.
Sokolov Viktor G.
Solovaya Nina A.
Soloviev Alexandr A.
Somov Boris V.
Sorochenko Roman L.
Sorokin Nikolaj A.
Sotnikova Natalia Y.
Stankevich Kazimir S.
Stepanov Aleksander V.
Strukov Igor A.
Struminsky Alexei
Suleimanov Valery F.
Sunyaev Rashid A.
Surdin Vladimir G.
Svechnikov Marij A.
Sveshnikov Mikhail L.
Tapia-Perez Santiago
Taranova Olga G.
Tatevyan Suriya K.
Tchouikova Nadezhda A.
Teplitskaya Raisa B.
Terekhov Oleg V.
Terentjeva Aleksandra K.
Tlatov Andrej G.
Tokarev Yurij V.
Trushkin Sergej A.
Tsvetkov Dmitry Yu.
Tsygan Anatolij I.
Tutukov Aleksandr V.
Tyul'bashev Sergei A.
Udal'tsov Vyacheslav A.
Utrobin Victor P.
Vainshtein Leonid A.
Vaisanen Petri S. M.
Valeev Sultan G.
Val'tts Irina E.
Valyaev Valerij I.
Valyavin Gennady G.
Varshalovich Dmitrij A.
Vashkovyak Sofja N.
Vashkov'yak Michael A.
Vassiliev Nikolaj N.
Vereshchagin Sergej V.
Verkhodanov Oleg V.
Vikhlinin Alexey A.
Vityazev Andrej V.
Vityazev Veniamin V.
Vlasyuk Valerij
Volkov Evgeni V.
Voloshina Irina B.
Voshchinnikov Nikolai V.
Wiebe Dmitri S.
Yakovlev Dmitrij G.
Yakovleva Valerija A.
Yarov-Yarovoj Mikhail S.
Yudin Ruslan V.
Yungelson Lev R.
Yushkin Maxim V.
Zabolotny Vladimir F.
Zagretdinov Renat V.
Zaitsev Valerij V.
Zakharova Polina E.
Zasov Anatolij V.
Zharkov Vladimir N.
Zharov Vladimir E.
Zheleznyakov Vladimir V.
Zhitnik Igor A.
Zhugzhda Yuzef D.
Zhukov Vladimir I.
Zinchenko Igor
Zlotnik Elena Ya

Saudi Arabia

Al-Malki M. B.
Almleaky Yasseen
Al-Mostafa Zaki A.
Basurah Hassan
Bukhari Fadel A. A.
Eker Zeki
Goharji Adan
Kordi Ayman S.
Malawi Abdulrahman
Niazy Adnan Mohammad
Saleh Magdy Y.

Serbia (Republic of)

Angelov Trajko
Arsenijevic Jelisaveta
Atanackovic-Vukmanovic Olga M.
Cirkovic Milan M.
Cvetkovic Zorica D.
Dacic Miodrag D.
Dimitrijevic Milan
Djurasevic Gojko
Djurovic Dragutin M.
Ignjatovic Ljubinko M.
Jankov Slobodan S.
Jovanovic Predrag
Jovanovic Bozidar
Knezevic Zoran
Kovacevic Andjelka B.
Kubicela Aleksandar
Kuzmanoski Mike
Lazovic Jovan P.
Lukacevic Ilija S.
Mihajlov Anatolij A.
Milogradov-Turin Jelena
Nikolic Silvana
Ninkovic Slobodan
Pakvor Ivan
Pejovic Nadezda R.
Popovic Luka C.
Popovic Georgije
Sadzakov Sofija
Samurovic Srdjan S.
Segan Stevo
Simovljevitch Jovan L.
Urosevic Dejan V.
Vince Istvan
Vukicevic-Karabin Mirjana M.

Slovakia

Budaj Jan
Chochol Drahomir
Dorotovic Ivan
Dzifcakova Elena
Galad Adrian
Galis Rudolf
Hajdukova Maria
Hefty Jan
Hric Ladislav
Kapisinsky Igor
Klacka Jozef
Klocok Lubomir
Kocifaj Miroslav
Komzik Richard
Kucera Ales
Minarovjech Milan
Neslusan Lubos
Palus Pavel
Pittich Eduard M.
Pittichova Jana
Porubcan Vladimir
Pribulla Theodor
Rusin Vojtech
Rybak Jan
Rybansky Milan
Saniga Metod
Skopal Augustin
Svoren Jan
Sykora Julius
Tremko Jozef
Ziznovsky Jozef
Zverko Juraj

South Africa

Baart Edward E.
Balona Luis A.
Bassett Bruce A.
Beesham Aroonkumar
Block David Lazar
Booth Roy S.
Buckley David A. H.
Charles Philip Allan
Cooper Timothy
Cress Catherine M.
Cunow Barbara
De Jager Ocker C.
De Jager Gerhard
Dunsby Peter
Ellis George F. R.
Fairall Anthony P.
Fanaroff Bernard L.
Feast Michael W.
Flanagan Claire Susan
Fraser Brian D.
Gaylard Michael John
Glass Ian S.
Govinder Keshlan S.
Hashimoto Yasuhiro
Hellaby Charles William
Hers Jan
Hughes Arthur R. W.
Jonas Justin L.
Kilkenny David
Kniazev Alexei
Koen Marthinus
Kraan-Korteweg Renée C.
Laney Clifton D.
Leeuw Lerothodi L.
Maharaj Sunil Dutt
Martinez Peter

Medupe Rodney T.
Meintjes Petrus J.
Menzies John W.
Mokhele Khotso D.
Nicolson George D.
O'Donoghue Darragh
Poole Graham
Potter Stephen B.
Raubenheimer Barend C.
Rijsdijk Case
Roche David N.
Romero-Colmenero Encarnacion
Sefako Ramotholo R.
Smit Jan A.
Smits Derck P.
Soltynski Maciej G
Still Martin D.
Stoker Pieter H.
Van der Walt Diederick J.
Viollier Raoul D.
Wainwright John
Warner Brian
Whitelock Patricia Ann
Williams William T.
Winkler Hartmut
Woudt Patrick A.

Spain

Abad Medina Alberto J.
Abia Carlos A.
Acosta Pulido Jose A.
Alberdi Antonio
Alcobe Santiago
Alcolea Javier
Aldaya Victor
Alfaro Emilio Javier
Alvarez Pedro
Amado Gonzalez Pedro
Anglada Guillem
Aparicio Antonio
Argueso Francisco
Arribas Santiago
Atrio Barandela Fernando
Bachiller Rafael
Balcells Marc
Ballester Jose L.
Ballesteros Ezequiel
Barcia Alberto
Barcons Xavier
Barrado Navascues David
Battaner Eduardo
Beckman John E.
Belizon Fernando
Bellot Rubio Luis Ramon
Belmonte Aviles Juan Antonio
Benn Chris R.
Bernabeu Guillermo
Betancor Rijo Juan
Boloix Rafael
Bonet Jose A.
Bravo Eduardo
Bujarrabal Valentin
Calvo Manuel
Camarena Badia Vicente
Canal Ramon M.
Carbonell Marc
Cardus Almeda J. O.
Casares Jorge
Castaneda Héctor F.
Castro-Tirado Alberto J.
Catala Poch M. A.
Cepa Jordi
Cernicharo Jose
Clement Rosa Maria
Codina Vidal J. M.
Colina Luis
Collados Manuel
Colomer Francisco
Coma Juan Carlos
Cornide Manuel
Costa Victor
Cubarsi Rafael
De Castro Elisa
De Vicente Pablo
Deeg Hans
Del Olmo Orozco A.
Del Rio Gerardo
Del Toro Iniesta Jose C.
Delgado Antonio J.
Diaz Angeles Isabel
Djupvik Anlaug A.
Docobo Jose A. Durantez
Domingo Vicente
Dominguez Inma
Eiroa Carlos
Elipe Sánchez Antonio
Elizalde Emilio
Estalella Robert
Esteban César
Fabregat Juan
Fabricius Claus V.
Fernandez David
Fernandez-Figueroa M. J.
Ferrandiz Jose Manuel
Ferrer Martinez Sebastian
Ferriz Mas Antonio
Figueras Francesca
Floria Peralta Luis Mario
Folgueira Marta
Fonseca Gonzalez Maria Victoria
Fuensalida Jiménez Jesús
Fuente Asuncion
Galadi Enriquez David
Galan Maximino J.
Gallart Carme
Gallego Jesús
Gallego Juan Daniel
Garcia Domingo
Garcia Lopez Ramon J.
Garcia de la Rosa Ignacio
Garcia-Berro Enrique
Garcia-Burillo Santiago
Garcia-Pelayo Jose
Garrido Rafael
Garzon Francisco L.
Gomez Gonzalez Jesus
Gomez Jose F.
Gomez Monique
Gomez de Castro Ana I.
Gonzales-Alfonso Eduardo
Gonzalez Serrano J. I.
Gonzalez de Buitrago Jesús
Gonzalez Delgado Rosa M.
Gonzalez Martinez Pais Ignacio
Gonzalez-Riestra R.
Gorgas Garcia Javier
Grieger Bjoern
Hernandez-Pajares Manuel
Hernanz Margarita

Herrero Davó Artemio
Hidalgo Miguel A.
Isern Jorge
Israelian Garik
Jimenez Mancebo A. J.
Jordi Carme
Jose Jordi
Kessler Martin F.
Kidger Mark R.
Labay Javier
Lahulla J. Fornies
Lara Luisa M.
Lara Martin
Lazaro Carlos
Ling Josefina F.
Loiseau Nora
Lopez De Coca M. D. P.
Lopez Arroyo M.
Lopez Corredoira Martin
Lopez Gonzalez Maria J.
Lopez Hermoso Maria Rosario
Lopez Moratalla Teodoro
Lopez Moreno Jose Juan
Lopez Puertas Manuel
Lopez Valverde M. A.
Luri Xavier
Mahoney Terence J.
Mampaso Antonio
Manchado Arturo T.
Manrique Alberto
Manteiga Outeiro Minia
Marcaide Juan-Maria
Marco Amparo
Maria de Garcia J. M.
Marquez Isabel
Marti Josep
Martin Eduardo L.
Martin Diaz Carlos
Martinez Garcia Vicent J.
Martinez Pillet Valentin
Martinez Roger Carlos
Martinez-Gonzalez Enrique
Martin-Pintado Jesus
Masegosa Gallego J.
Massaguer Josep
Mauersberger Rainer
Mediavilla Evencio
Medina Jose
Metcalfe Leo
Miralda-Escude Jordi
Moles Mariano J.
Molina Antonio
Montes David
Montesinos Benjamin
Morales-Duran Carmen
Moreno Fernando
Moreno Insertis Fernando
Moreno Lupianez Manuel
Muinos Haro Jose L.
Munoz Tunon Casiana
Najarro de la Parra Francisco
Negueruela Ignacio
Nunez Jorge
Oliver Ramón
Oscoz Alejandro
Ostensen Roy H.
Palle Pere Lluís
Paredes Poy Josep M.
Perea-Duarte Jaime D.
Perez Enrique
Perez Fournon Ismael
Perez Hernandez Fernando J.
Planesas Pere
Prieto Mercedes M.
Prieto Cristina
Quintana Jose M.
Rebolo Rafael
Reglero-Velasco Victor
Rego Fernandez M.
Regulo Clara R.
Ribas Ignasi
Ribo Marc
Rioja Maria J.
Rizzo Jose Ricardo
Roca Cortes Teodoro
Rodrigo Rafael
Rodriguez Eloy
Rodriguez Espinosa Jose M.
Rodriguez Hidalgo Inés
Rodriguez-Eillamil R.
Rodriguez-Franco Arturo
Rolland Angel
Romero Perez M. Pilar
Ros Rosa M.
Rossello Gaspar
Rozas Maite
Ruiz Cobo Basilio
Ruiz-Lapuente María P.
Rutten Renee G. M.
Sabau-Graziati Lola
Sala Ferran
Salazar Antonio
Salvador-Sole Eduardo
Sanahuja Blai
Sanchez Francisco M.
Sanchez Filomeno
Sanchez Almeida J.
Sanchez-Lavega Agustin
Sanchez-Saavedra M. Luisa
Sanroma Manuel
Sansaturio Maria Eugenia
Sanz Jaume
Sanz Jose L.
Sein-Echaluce M. Luisa
Sempere Maria J.
Sequeiros Juan
Serra Ricart Miquel
Sevilla Miguel J.
Simo Carles
Skillen Ian
Solanes Josep M.
Tafalla Mario
Talavera Antonio
Tamazian Vakhtang S.
Toffolatti Luigi
Torra Jordi
Torrejon Jose Miguel
Torrelles Jose M.
Trigo-Rodriguez Josep M.
Trujillo Bueno Javier
Trullols I. Farreny Enric
Unger Stephen
Vallejo Miguel
Varela Perez Antonia M.
Vazdekis Alexandre
Vazquez Manuel
Verdes-Montenegro Lourdes
Verdugo Eva
Vila Samuel C.
Vilchez Jose M.
Vinuales Gavin Ederlinda
Vives Teodoro Jose
Watson Robert A.
Zamorano Jaime

Sweden

Aalto Suzanne E.
Abramowicz Marek
Andersen Torben E.
Ardeberg Arne L.
Artymowicz Pawel
Baath Lars B.
Barklem Paul S
Bergman Per-Goeran
Bergstrom Lars
Bergvall Nils Ake Sigvard
Bjornsson Claes-Ingvar
Black John Harry
Carlqvist Per A.
Carlson Per
Cato B. Torgny
Conway John
Davidsson Björn J.
De Mello Duilia F.
Dravins Dainis
Edsjo Joakim
Edvardsson Bengt
Ellder Joel
Elvius Aina M.
Eriksson Kjell
Faelthammar Carl Gunne
Feltzing Sofia
Fransson Claes
Fredga Kerstin
Freytag Bernd
Frisk Urban
Gahm Goesta F.
Gleisner Hans
Goobar Ariel M.
Gustafsson Bengt
Hartman Henrik
Heikkila Arto J.
Heiter Ulrike
Hjalmarson Ake G.
Hoefner Susanne
Hoeglund Bertil
Hogbom Jan A.
Holm Nils G.
Holmberg Gustav
Horellou Cathy
Hulth Per Olof
Johansson Lennart
Johansson Lars Erik B.
Johansson Sveneric
Justtanont-Liseau Kay
Kiselman Dan
Kochukhov Oleg
Kollberg Erik L.
Korn Andreas J.
Kozma Cecilia A.
Kristiansson Krister
Lagerkvist Claes-Ingvar
Larsson Stefan
Larsson-Leander Gunnar
Lauberts Andris
Lehnert B. P.
Lindblad Per O.
Linde Peter
Lindegren Lennart
Lindqvist Michael
Liseau Rene
Loden Kerstin R.
Lofdahl Mats G.
Lundqvist Peter
Lundstedt Henrik
Lundstrom Ingemar
Magnusson Per
Mellema Garrelt
Nilsson Hampus
Nordh Lennart H.
Nummelin Albert F.
Oja Tarmo
Olberg Michael
Olofsson Kjell
Olofsson Goeran S.
Olofsson Hans
Ostlin Göran T.
Owner-Petersen Mette M.
Pearce Mark
Pellinen-Wannberg Asta K.
Piskunov Nikolai E.
Raadu Michael A.
Rickman Hans
Roennaeng Bernt O.
Romeo Alessandro B.
Roslund Curt
Rosquist Kjell
Rydbeck Gustaf H. B.
Ryde Felix M.
Ryde Nils A. E.
Sandqvist Aage
Scharmer Goeran Bjarne
Schoier Fredrik L.
Sinnerstad Ulf E.
Soderhjelm Staffan
Sollerman Jesper
Stenholm Lars
Stenholm Björn
Sundin Maria I.
Sundman Anita
Thomasson Magnus
Torkelsson Ulf J.
Van Groningen Ernst
Wahlgren Glenn Michael
Wallinder Fredrik
Westerlund Bengt E.
Wiedling Tor
Winnberg Anders
Wramdemark Stig S.

Switzerland

Altwegg Kathrin
Arzner Kaspar J.
Audard Marc
Babel Jacques
Bartholdi Paul
Behrend Raoul
Benz Arnold O.
Benz Willy
Berthet Stephane
Beutler Gerhard
Bianda Michele
Binggeli Bruno
Blecha Andre Boris G.
Bochsler Peter
Burki Gilbert
Buser Roland
Carollo Marcella
Carrier Fabien
Charbonnel Corinne
Courbin Frederic Y. M.
Courvoisier Thierry J.-L.

Cramer Noel
Csillaghy Andre
Dessauges-Zavadsky Miroslava
Diethelm Roger
Dubath Pierre
Duerst Johannes
Fenkart Rolf P.
Finsterle Wolfgang
Fluri Dominique M.
Folini Doris
Froehlich Claus
Gautschy Alfred
Geiss Johannes
Golay Marcel
Goy Gerald
Grenon Michel
Guedel Manuel
Hauck Bernard
Huber Martin C. E.
Hugentobler Urs
Jetzer Philippe F.
Kallenbach Reinald
Labhardt Lukas
Liebendoerfer Matthias
Lilly Simon J.
Locher Kurt H.
Maeder Andre
Maetzler Christian
Magun Andreas
Marsden Stephen C.
Martinet Louis
Mayor Michel
Megevand Denis
Mermilliod Jean-Claude
Meylan Georges
Meynet Georges
Moore Ben
Mowlavi Nami
Nicolet Bernard
North Pierre
Nussbaumer Harry
Paltani Stéphane
Pauluhn Anuschka
Pepe Francesco A.
Pfenniger Daniel
Pont Frederic
Povel Hanspeter
Queloz Didier
Raboud Didier
Ramelli Renzo G. E.
Rauscher Thomas
Rufener Fredy G.
Samland Markus
Schaerer Daniel
Schild Hansruedi
Schildknecht Thomas
Schmid Hans Martin
Schmutz Werner
Segransan Damien
Steinlin Uli
Stenflo Jan O.
Straumann Norbert
Tammann Gustav A.
Thielemann Friedrich-Karl
Thomas Nicolas
Trefzger Charles F.
Tuerler Marc A.
Udry Stephane
Verdun Andreas K.
Von Steiger Rudolf
Walder Rolf
Walter Roland
Wild Paul
Wurz Peter
Zelenka Antoine

Tajikistan

Babadzhanov Pulat B.
Ibadinov Khursand I.
Kokhirova Gulchehra I.
Minnikulov Nasriddin K.
Sahibov Firuz H.
Subhon Ibadov

Thailand

Channok Chanruangrit
Eungwanichayapant Anant
Gasiprong Nipon
Kramer Busaba H.
Muanwong Orrarujee
Pacharin-Tanakun P.
Ruffolo David
Saiz Alejandro
Songsathaporn Ruangsak
Soonthornthum Boonrucksar

Turkey

Akan Mustafa Can
Akcayli Melek M. A.
Akyol Mustafa Unal
Alpar Ali
Aslan Zeki
Atac Tamer
Aydin Cemal
Bolcal Cetin
Boydag-Yildizdogdu F. S.
Bozkurt Sukru
Demircan Osman
Derman I. Ethem
Dogan Nadir
Engin Semanur
Enginol Turan B.
Ercan E. Nihal
Ertan A. Yener
Eskioglu A. Nihat
Evren Serdar
Ezer-Eryurt Dilhan
Gudur N.
Gulmen Omur
Gulsecen Hulusi
Hamzaoglu Esat E. H.
Hazer S.
Hotinli Metin
Ibanoglu Cafir

Kandemir Guelcin
Karaali Salih
Keskin Varol
Kiral Adnan
Kirbiyik Halil
Kiziloglu Nilguen
Kiziloglu Umit
Kocer Durcun
Marsoglu A.
Mentese Huseyin
Oekten Adnan
Oezel Mehmet Emin
Oezkan Mustafa Tuerker
Ozguc Atila
Pekuenlue E. Rennan
Saygac A. Talat
Sezer Cengiz
Tektunali H. Gokmen
Topaktas Latif A.
Tufekcioglu Zeki
Tunca Zeynel
Yilmaz Fatma
Yilmaz Nihal

Ukraine

Akimov Leonid
Alexandrov Yuri V.
Alexandrov Alexander N.
Andrienko Dmitry A.
Andrievsky Sergei
Andronov Ivan
Antonyuk Kirill
Babin Arthur
Baranovsky Edward A.
Belskaya Irina N.
Berczik Peter
Bolotin Sergei
Bolotina Olga
Borysenko Sergiy
Bruns Andrey V.
Danylevsky Vassyl
Denishchik Yurii
Dlugach Zhanna M.
Dorokhova Tetyana
Doroshenko Valentina T.
Dragunova Alina V.
Dudinov Volodymyr N.
Duma Dmitrij P.
Efimenko Volodymyr M.
Efimov Yuri
Epishev Vitali P.
Fedorov Petro
Fedorova Elena
Fomin Valery
Fomin Piotr Ivanovich
Gershberg R. E.
Glazunova Ljudmila
Golovatyj Volodymyr
Gopasyuk Olga S.
Gopka Vera F.
Gorbanev Jury
Gor'kavyi Nikolai
Gozhy Adam
Grinin Vladimir P.
Guseva Natalia G.
Halevin Alexander V.
Hnatyk Bohdan Ivanovych
Hudkova Ludmila
Ivanchuk Victor I.
Ivanova Aleksandra V.
Ivchenko Vasily
Izotov Yuri
Izotova Iryna Yu.
Kalenichenko Valentin
Karachentseva Valentina
Karetnikov Valentin G. R.
Karpov Nikolai V.
Kazantsev Anatolii M.
Kazantseva Liliya
Kharchenko Nina
Kharin Arkadiy S.
Khmil Sergiy V.
Khoda Oleg
Kiselev Nikolai N.
Kislyuk Vitalij S.
Klymyshyn I. A.
Kolesnikov Sergey
Kondrashova Nina N.
Konovalenko Alexander A.
Kontorovich Victor M.
Korotin Sergey
Korsun Pavlo P.
Korsun Alla
Kostik Roman I.
Kotov Valery
Koval I. K.
Koval'chuk George U.
Kovtyukh Valery V.
Kravchuk Sergei
Kruchinenko Vitaliy G.
Krugly Yurij N.
Kryshtal Alexander N.
Kryvdyk Volodymyr G.
Kryvodubskyj Valery N
Kudashkina Larisa S.
Kudrya Yury M.
Kurochka Evgenia Vasilevna
Kuz'kov Vladimir P.
Lazorenko Peter F.
Leiko Uliana
Litvinenko Leonid N.
Lozitskij Vsevolod
Lukyanyk Igor V.
Lupishko Dmitrij F.
Lyubchik Yuri
Lyubimkov Leonid S.
Lyuty Victor M.
Makarenko Ekaterina N.
Marsakova Vladislava
Melnik Valentin N.
Men A. V.
Mishenina Tamara
Mkrtichian David E.
Morozhenko A. V.
Nazarenko Victor
Novosyadlyj Bohdan
Parnovsky Sergei
Pavlenko Yakov V.
Pavlenko Elena
Petrov Peter P.
Petrov G. M.
Petrova Svetlana
Pilyugin Leonid
Pinigin Gennadij I.
Plachinda Sergei I.
Polosukhina-Chuvaeva Nina
Pozhalova Zhanna
Prokof'eva Valentina V.
Pronik Iraida I.
Pronik Vladimir I.
Protsyuk Yuri

Psaryov Volodymyr A.
Pugach Alexander F.
Rachkovsky D. N.
Rashkovskij Sergey L.
Romanchuk Pavel R.
Romanov Yuri S.
Romanyuk Yaroslav O.
Rosenbush Alexander E.
Rostopchina Alla
Ryabov Michael I.
Savanov Igor S.
Sergeev Aleksandr V.
Sergeev Sergey G.
Shakhov Boris A.
Shakhovskaya Nadejda I.
Shakhovskoj Nikolay M.
Shchukina Nataliya
Sheminova Valentina A.
Shevchenko Vasilij
Shkuratov Yurii
Shulga Valery
Shulga Oleksandr V.
Shulman Leonid M.
Silich Sergey
Sizonenko Yuri V.
Skulskyj Mychajlo Y.
Sobolev Yakov M.
Sodin Leonid
Sokolov Konstantin
Stepanian Natali N.
Steshenko N. V.
Stodilka Myroslav
Tarady Vladimir K.
Tarasov Anatolii E.
Tarasova Taya
Terebizh Valery Yu
Tsap Teodor T.
Tsap Yuri
Udovichenko Sergei N.
Usenko Igor
Vavilova Iryna B.
Vidmachenko Anatoliy P.
Voitenko Yuriy M.
Voitsekhovska Anna
Voloschuk Yuri I.
Volvach Alexander
Volyanskaya Margarita Yu
Voroshilov Volodymyr I.
Yanovitskij Edgard G.
Yatsenko Anatolij I.
Yatskiv Yaroslav S.
Yukhimuk Adam K.
Yushchenko Alexander V.
Zakhozhaj Volodimir
Zhdanov Valery
Zheleznyak Alexander P.
Zhilyaev Boris

United Kingdom

Aarseth Sverre J.
Adamson Andrew
Ade Peter A. R.
Aggarwal Kanti Mal
Aitken David K.
Albinson James
Alexander Paul
Allan Peter M.
Allen Anthony John
Allington-Smith Jeremy R.
Almaini Omar
Alsabti Abdul Athem
Anderson Bryan
Andrews David A.
Aragon-Salamanca Alfonso
Archontis Vasilis
Ardavan Houshang
Argyle Robert William
Asher David J.
Atherton Paul David
Axon David
Bailey Mark Edward
Baldry Ivan K.
Baldwin John E.
Balikhin Michael
Barclay Charles E.
Barlow Michael J.
Barocas Vinicio
Barr Jordi McGregor
Barrow John David
Barstow Martin Adrian
Baruch John
Baskill Darren S.
Bastian Nathan J.
Bastin John A.
Bates Brian
Bath Geoffrey T.
Battye Richard A.
Beale John S.
Beggs Denis W.
Bell Steven A.
Bell Burnell Jocelyn Susan
Bennett Jim A.
Bentley Robert D.
Berger Mitchell
Berrington Keith Adrian
Bersier David
Beurle Kevin
Bewsher Danielle
Bingham Robert
Binney James J.
Birkinshaw Mark
Blackman Clinton Paul
Blackwell Donald E.
Blundell Katherine M.
Bode Michael F.
Boksenberg Alec
Boles Thomas
Bonnel Ian A.
Bonnor W. B.
Botha Gert J. J.
Boyd David R.
Boyd Robert L. F.
Boyle Stephen
Brand Peter W. J. l.
Branduardi-Raymont Graziella
Branson Nicholas J. B. A.
Bremer Malcolm N.
Bridgeland Michael
Bridle Sarah L.
Brinks Elias
Bromage Gordon E.
Bromage Barbara J. I.
Brooke John M.
Brookes Clive J.
Brown Paul James Frank
Brown John C.
Browne Ian W. A.
Browning Philippa K.
Browning Philippa
Bruck Mary T.
Bryce Myfanwy
Bunclark Peter Stephen

Bunker Andrew J.
Bureau Martin G.
Burgess Alan
Burgess David D.
Burton William M.
Butchins Sydney Adair
Butler Christopher John
Cameron Andrew Collier
Campbell Alison
Canavezes Alexandre G. R.
Cappellari Michele
Cargill Peter J.
Carr Bernard John
Carson T. R.
Carswell Robert F.
Carter David
Catchpole Robin M.
Cawthorne Timothy V.
Chakrabarty Dalia
Chan Siu Kuen Josphine
Chapman Allan
Chapman Sandra C.
Christou Apostolos
Cioni Maria-Rosa L.
Clark David H.
Clarke David
Clarke Catherine
Clegg Peter E.
Clegg Robin E. S.
Clifton Gloria
Clowes Roger G.
Clube S. V. M.
Coe Malcolm
Cole Shaun M.
Coles Peter
Conway Robin G.
Conway Andrew J.
Cooke B. A.
Cooke John Alan
Cooper Nicholas J.
Couper Heather
Crawford Ian Andrew
Crawford Carolin
Cropper Mark
Crowther Paul
Cruise Adrian Michael
Culhane Leonard
Czerny Michal
Daintree Edward J.
Dalla Silvia C.
Damian Audley M.
Davidson William
Davies Rodney D.
Davies John K.
Davies Jonathan Ivor
Davies Melvyn B.
Davies Roger L.
Davis Richard J.
De Grijs Richard
De Groot Mart
Dennison P. A.
Dent William
Dewhirst David W.
Dhillon Vikram Singh
Diamond Philip John
Dickens Robert J.
Diego Francisco
Dipper Nigel A.
Disney Michael J.
Diver Declan Andrew
Donnison John Richard
Dormand John Richard
Downes Ann Juliet B
Doyle John Gerard
Drew Janet
Driver Simon P.
Duffett-Smith Peter James
Dufton Philip L.
Dunkin Sarah K.
Dunlop Storm
Dunlop James
Dworetsky Michael M.
Dyson John E.
Eales Stephen A.
Eccles Michael J.
Edge Alastair
Edmunds Michael Geoffrey
Edwin Roger P.
Efstathiou George
Eggleton Peter P.
Elkin Vladimir
Elliott Kenneth Harrison
Ellis Richard S.
Elsworth Yvonne P.
Emerson James P.
Erdelyi Robert
Evans Dafydd Wyn
Evans Wyn
Evans Kenton Dower
Evans Roger G.
Evans Aneurin
Evans Christopher J.
Evans Michael W.
Eyres Stewart P. S.
Fabian Andrew C.
Falle Samuel A.
Faulkner Andrew J.
Fawell Derek R.
Ferguson Annette M.
Ferreras Ignacio P.
Fiedler Russell
Field David
Fielder Gilbert
Fitzsimmons Alan
Fletcher Lyndsay
Flower David R.
Fludra Andrzej
Fong Richard
Fox W. E.
Fraser Helen J.
Frenk Carlos S.
Freyhammer Lars Michael
Fuller Gary A.
Furniss Ian
Fyfe Duncan J.
Gadsden Michael
Gaensicke Boris T.
Garrington Simon T.
Garton W. R. S.
Gibson Brad K.
Gietzen Joseph W.
Gill Peter B. J.
Gilmore Gerard F.
Gledhill Timothy M.
Glencross William M.
Godwin Jon Gunnar
Gondhalekar Prabhakar
Gonzalez-Solares Eduardo
Goodwin Simon P.
Gough Douglas O.
Grady Monica M.
Graffagnino Vito G.
Grainge Keith J. B.
Grainger John F.
Grant Ian P.
Gray Norman
Gray Meghan
Green Simon F.
Green David A
Green Anne M.

Green Robin M.
Griffin Matthew J.
Griffin Roger F.
Griffiths William K.
Guest John E.
Gull Stephen F.
Gunn Alastair G.
Guthrie Bruce N. G.
Hadley Brian W.
Haehnelt Martin G.
Hall Andrew Norman
Haniff Christopher
Hardcastle Martin J.
Harper David
Harra Louise K.
Harris-Law Stella
Harrison Richard A.
Hartquist Thomas Wilbur
Hassall Barbara J. M.
Haswell Carole A.
Hawarden Timothy G.
Hawking Stephen W.
Hawkins Michael R. S.
Hayward John
Hazard Cyril
Heavens Alan
Heggie Douglas C.
Hellier Coel
Hendry Martin A.
Hermans Dirk
Hewett Paul
Hewish Antony
Hey James Stanley
Hide Raymond
Hilditch Ronald W.
Hills Richard E.
Hingley Peter D.
Hirschi Raphael
Hoare Melvin G.
Hohenkerk Catherine Y.
Holloway Nigel J.
Hood Alan W.
Horne Keith
Houdek Gunter
Hough James
Howarth Ian Donald
Hughes David W.
Hughes David W.
Humphries Colin M.
Hunt G. E.
Hysom Edmund J.
Ireland John G.
Ireland Jack
Irwin Michael John
Irwin Patrick G. J.
Isaak Kate G.
Ivison Robert J.
Jackson John Charles
Jackson Neal J.
Jakobsson Pall
James John F.
James Richard A.
Jameson Richard F.
Jardine Moira M.
Jeffery Christopher S.
Jenkins Charles R.
Johnstone Roderick
Jones Bernard J. T.
Jones Janet E.
Jones Michael
Jones Laurence R.
Jones Derek H. P.
Jones Barrie W.
Jordan Carole
Jorden Paul Richard
Jupp Alan H.
Kaiser Christian R.
Kanbur Shashi
Khan Josef I.
Khosroshahi Habib G.
King David Leonard
King Andrew R.
King-Hele Desmond G.
Kingston Arthur E.
Kitchin Christopher R.
Knapen Johan Hendrik
Knapp Johannes
Kolb Ulrich
Kollerstrom Nicholas
Kontar Eduard P.
Kramer Michael
Kroto Harold
Kurtz Donald W.
Labrosse Nicolas
Lahav Ofer
Laing Robert
Lancaster Brown Peter
Lang James
Lapington Jonathan S.
Lasenby Anthony
Lawrence Andrew
Leahy J. Patrick
Lee Terence J.
Liddle Andrew
Liu Xiaowei
Lloyd Christopher
Lloyd Huw
Lloyd Evans Thomas Harry
Longair Malcolm S.
Longbottom Aaron
Love Gordon D.
Loveday Jon N.
Lovell Sir Bernard
Lucey John
Lucy Leon B.
Lynas-Gray Anthony E.
Lynden-Bell Donald
Lyne Andrew G.
Lyon Ian C.
Maartens Roy
MacCallum Malcolm A. H.
Maccarone Thomas J.
MacDonald Geoffrey H.
MacGillivray Harvey T.
Maciejewski Witold B.
MacKay Craig D.
Mackay Duncan H.
MacKinnon Alexander L.
Maddison Ronald Ch.
Maddox Stephen
Magorrian Stephen J.
Major John
Mann Robert G.
Mao Shude
Marek John
Marsh Julian C. D.
Marsh Thomas R.
Marshall Kevin P.
Martin Anthony R.
Masheder Michael
Mason Keith Owen
Mason Helen E.
Mason John William
Massey Robert M.
Matheson David Nicholas
Mathioudakis Mihalis
Mattila Seppo K.
Mccombie June
McDonnell J. A. M.
McHardy Ian Michael

McMahon Richard
McMullan Dennis
McNally Derek
McWhirter R. W. Peter
Meaburn John
Meadows A. Jack
Meikle William P. S.
Meiksin Avery Abraham
Merrifield Michael R.
Message Philip J.
Mestel Leon
Miles Howard G.
Millar Thomas J.
Miller John C.
Miller Steven
Mills Allan A.
Mitton Simon
Mitton Jacqueline
Moffatt Henry Keith
Monteiro Tania S.
Moore Daniel R.
Moore Patrick
Morgan David H.
Morgan Brian Lealan
Morison Ian
Morris Simon
Morris David
Morris Michael C.
Morris Rhys H.
Morrison Leslie V.
Moss David L.
Murdin Paul G.
Murphy Michael T.
Murray Carl D.
Murray Andrew C.
Murray John B.
Murtagh Fionn
Muxlow Thomas
Nagashima Masahiro
Nakariakov Valery M.
Napier William M.
Napiwotzki Ralf
Naylor Tim
Nelson Richard P.
Nelson Alistair H.
Neukirch Thomas
New Roger
Newton Gavin
Nicolson Iain
North John David
Norton Andrew J.
O'Brien Tim
O'Brien Paul Thomas
Okamoto Takashi
Oliveira Joana M.
Oliver Sebastian J.
Onuora Lesley Irene
Orford Keith J.
Osborne Julian P.
Osborne John L.
O'Shea Eoghan
Padman Rachael
Page Mathew J.
Page Clive G.
Palmer Philip
Papaloizou John C. B.
Parker Edward A.
Parker Neil
Parkinson John H.
Parmentier Genevieve A.
Parnell Clare E.
Paxton Harold J. B. R.
Peach John V.
Peach Gillian
Peacock John Andrew
Pearce Gillian
Pearce Frazer R.
Pedlar Alan
Penny Alan John
Penston Margaret
Perry Judith J.
Petford A. David
Petkaki Panagiota
Pettini Max
Phillipps Steven
Phillips Kenneth J. H.
Pijpers Frank Peter
Pike Christopher David
Pilkington John D. H.
Pillinger Colin
Pittard Julian M.
Podsiadlowski Philipp
Pollacco Don
Pollacco Don
Ponman Trevor
Ponsonby John E. B.
Pooley Guy
Pounds Kenneth A.
Priest Eric R.
Pringle James E.
Prinja Raman
Proctor Michael R. E.
Pye John P.
Quenby John J.
Raine Derek J.
Rapley Christopher G.
Rasmussen Jesper
Rawlings Jonathan
Rawlings Steven
Raychaudhury Somak
Read Andrew M.
Reay Newrick K.
Rees Martin J.
Regnier Stephane
Regoes Enikoe
Reig Pablo
Richardson Kevin J.
Richer John
Rijnbeek Richard Philip
Riley Julia M.
Roberts Timothy P.
Robertson Norna
Robertson John Alistair
Robinson Andrew
Robinson William J.
Robson Ian E.
Roche Patrick F.
Rowan-Robinson Michael
Rowson Barrie
Roxburgh Ian W.
Roy Archie E.
Ruderman Michael A.
Ruderman Michael S.
Ruffert Maximilian
Ruggles Clive L. N.
Russel Sara S.
Ryan Sean Gerard
Sakano Masaaki
Sanford Peter William
Sansom Anne E.
Sarre Peter J.
Satterthwaite Gilbert E.
Saunders Richard D. E.
Scarrott Stanley M.
Schroder Anja C.
Schroeder Klaus Peter
Schwartz Steven Jay
Scott Paul F.
Serjeant Stephen
Seymour P. A. H.

Shakeshaft John R.
Shallis Michael J.
Shanklin Jonathan D.
Shanks Thomas
Sharples Ray
Shaw Simon E.
Shone David
Shukurov Anvar Muhamadovich
Silk Joseph I.
Simmons John Francis l
Simnett George M.
Simons Stuart
Simpson Allen D. C.
Sims Mark R.
Skilling John
Skinner Gerald K.
Sleath John
Smail Ian
Smalley Barry
Smith Francis Graham
Smith Robert Connon
Smith Nigel J. T.
Smith Rodney M.
Smith Michael
Smith Linda J.
Smyth Michael J.
Somerville William B.
Sorensen Soren-Aksel
Speer R. J.
Spencer Ralph E.
Stannard David
Steel Duncan I.
Steele John
Steele Colin D. C.
Stephenson F. Richard
Stevens Ian R.
Steves Bonita Alice
Stewart John Malcolm
Stewart Paul
Stickland David J.
Stolyarov Vladislav A.
Summers Hugh P.
Sutherland William
Sylvester Roger
Taub Liba
Tavakol Reza
Taylor Gordon D.
Taylor Donald Boggia
Taylor Angela C.
Taylor Fredric W.
Taylor John K.
Tedds Jonathan A.
Tennyson Jonathan
Terlevich Elena
Terlevich Roberto Juan
Thoburn Christine
Thomas Peter A.
Thomas David V.
Thomasson Peter
Thomson Robert
Thorne Anne P.
Tobias Steven M.
Tout Christopher
Tozer David C.
Tritton Susan Barbara
Tritton Keith Peter
Trotta Roberto
Tsiklauri David
Turner Martin J. l.
Tworkowski Andrzej S.
Van der Raay Herman B.
Van Leeuwen Floor
Van Loon Jacobus Th.
Vaughan Simon A.
Veck Nicholas
Vekstein Gregory
Venn Kimberly A.
Verwichte Erwin A. O.
Vink Jorick S.
Viti Serena
Volonteri Marta
Waddington Ian
Walker Simon N.
Walker David Douglas
Walker Helen J.
Walker Ian Walter
Walker Edward N.
Wall Jasper V.
Wallace Patrick T.
Wallis Max K.
Walsh Robert
Walton Nicholas A.
Ward Henry
Ward Martin John
Ward-Thompson Derek
Warner Peter J.
Warwick Robert S.
Watson Micheal G.
Watt Graeme David
Webster Adrian S.
Weiss Nigel O.
Wellgate G. Bernard
Wheatley Peter J.
White Glenn J.
Whiting Alan B.
Whitrow Gerald James
Whitworth Anthony Peter
Wickramasinghe N. Chandra
Wilkins George A.
Wilkinson Althea
Wilkinson Peter N.
Williams David A.
Williams Iwan P.
Williams Peredur M.
Williams Robin
Willis Allan J.
Willmore A. Peter
Willstrop Roderick V.
Wilson Lionel
Wilson Michael John
Woan Graham
Wolfendale Arnold W.
Wolstencroft Ramon D.
Wood Roger
Wood Janet H.
Woolfson Michael M.
Worrall Gordon
Worrall Diana Mary
Worswick Susan
Wright Andrew N.
Wright Ian P.
Yallop Bernard D.
Zane Silvia
Zarnecki John Charles
Zharkova Valentina
Zijlstra Albert
Zuiderwijk Edwardus J.

United States

Aannestad Per Arne
Abbett William P.
Abbott W.
Abbott David C.
Ables Harold D.
Abramenko Valentina
Abt Helmut A.
Acton Loren W.
Adams Thomas F.
Adams Fred
Adams Mark T.
Adams Jr James H.
Adelman Saul J.
Adler David Scott
A'Hearn Michael F.
Ahluwalia Harjit S.
Ahmad Imad Aldean
Aizenman Morris L.
Ake III Thomas Bellis
Akeson Rachel
Akiyama Masayuki
Albers Henry
Alexander Joseph K.
Alexander David R.
Allan David W.
Allen Lori E.
Allen Ronald J.
Allen Jr John E.
Allende Prieto Carlos
Aller Margo F.
Aller Hugh D.
Alley Carrol O.
Allison Michael D.
Aloisi Alessandra
Alpher Ralph Asher
Altrock Richard C.
Altschuler Daniel R.
Altschuler Martin D.
Altunin Valery I.
Alvarez Manuel
Alvarez del Castillo Elizabeth M.
Ambruster Carol
Anand S. P. S.
Andersen Morten
Anderson Kurt S. J.
Anderson Kinsey A.
Anderson Christopher M.
Andersson Bengt Goeran
Angel J. Roger P.
Angione Ronald J.
Anosova Joanna
Anthony-Twarog Barbara J.
Antiochos Spiro K.
Aoki Kentaro
Appleby John F.
Appleton Philip Noel
Archinal Brent A.
Armandroff Taft E.
Armstrong John Thomas
Arnett W. David
Arnold James R.
Arny Thomas T.
Arons Jonathan
Arthur David W. G.
Aschwanden Markus
Aspin Colin
Assousa George Elias
Athay R. Grant
Atkinson David H.
Atreya Sushil K.
Auer Lawrence H.
Augason Gordon C.
Avrett Eugene H.
Ayres Thomas R.
Backer Donald Ch.
Bagnuolo Jr William G.
Bagri Durgadas S.
Bahcall Neta A.
Bailyn Charles D.
Baird Scott R.
Baker Andrew J.
Baker Joanne
Balachandran Suchitra C.
Balasubramaniam K. S.
Balbus Steven A.
Baldwin Ralph B.
Baldwin Jack A.
Balick Bruce
Baliunas Sallie L.
Ball John A.
Bally John
Balonek Thomas J.
Balser Dana S.
Bandermann L. W.
Bangert John A.
Bania Thomas Michael
Bardeen James M.
Barden Samuel Charles
Barger Amy J.
Baring Matthew G.
Barker Timothy
Barker Edwin S.
Barnbaum Cecilia
Barnes Sydney A.
Barnes Aaron
Barnes Graham
Barnes III Thomas G.
Baron Edward
Barrett Paul Everett
Barsony Mary
Barth Charles A.
Barth Aaron J.
Barton Elizabeth J.
Barvainis Richard
Basart John P.
Bash Frank N.
Basri Gibor B.
Bastian Timothy Stephen
Basu Sarbani
Batchelor David A.
Batson Raymond Milner
Bauer Wendy H.
Bauer Carl A.
Baum Stefi Alison
Baum William A.
Baustian W. W.
Bautz Laura P.
Baym Gordon Alan
Beavers Willet I.
Bechtold Jill
Becker Stephen A.
Becker Robert Howard
Becker Peter Adam
Becker Robert A.
Beckers Jacques M.
Becklin Eric E.
Beckwith Steven V. W.
Beebe Herbert A.
Beebe Reta Faye
Beer Reinhard
Beers Timothy C.
Begelman Mitchell Craig

Beiersdorfer Peter
Bell Roger A.
Bell Jeffrey F.
Bell Barbara
Bell III James F.
Belserene Emilia P.
Belton Michael J. S.
Bender Peter L.
Benedict George F.
Benevolenskaya Elena
Benford Gregory
Bennett Charles L.
Bensby Thomas L.
Benson Priscilla J.
Berendzen Richard
Berg Richard A.
Berge Glenn L.
Bergin Edwin A.
Bergstralh Jay T.
Berman Robert Hiram
Berman Vladimir
Bernat Andrew Plous
Bertschinger Edmund
Bettis Dale G.
Bhavsar Suketu P.
Bianchi Luciana
Bidelman William P.
Bieging John Harold
Bignell R. Carl
Billingham John
Binzel Richard P.
Biretta John Anthony
Bjorkman Karen S.
Bjorkman Jon E.
Black Adam Robert S.
Blades John Chris
Blair Guy Norman
Blair William P.
Blakeslee John P.
Blanco Victor M.
Blandford Roger David
Blasius Karl Richard
Bless Robert C.
Blitz Leo
Bloemhof Eric E.
Blondin John M.
Blum Robert D.
Blumenthal George R.
Boboltz David A.
Bobrowsky Matthew
Bock Douglas C.-J.
Boden Andrew F.
Bodenheimer Peter
Boesgaard Ann M.
Boeshaar Gregory Orth
Bogdan Thomas J.
Boggess Albert
Boggess Nancy W.
Bohannan Bruce Edward
Bohlin Ralph C.
Bohlin J. David
Bohm Karl-Heinz
Bohm-Vitense Erika
Boice Daniel Craig
Bolatto Alberto D.
Boldt Elihu
Boley Forrest I.
Bonanos Alceste Z.
Bond Howard E.
Bonsack Walter K.
Book David L.
Bookbinder Jay A.
Bookmyer Beverly B.
Booth Andrew J.
Bopp Bernard W.
Bord Donald John
Borderies Nicole
Borkowski Kazimierz J.
Borne Kirk D.
Bornmann Patricia L.
Boss Alan P.
Bourke Tyler L.
Bowell Edward L. G.
Bowen George H.
Bower Geoffrey C.
Bower Gary Allen
Bowers Phillip F.
Bowyer C. Stuart
Boyce Peter B.
Boyle Richard P.
Boynton Paul Edward
Bracewell Ronald N.
Bradley Paul A.
Bradley Arthur J.
Bradstreet David H.
Branch David R.
Brandt John C.
Brandt William N.
Branscomb L. M.
Brault James W.
Braun Douglas C.
Breakiron Lee Allen
Brecher Kenneth
Brecher Aviva
Breckinridge James B.
Bregman Joel N.
Brickhouse Nancy S.
Bridle Alan H.
Britt Daniel T.
Broadfoot A. Lyle
Broderick John
Brodie Jean P.
Brosius Jeffrey W.
Broucke Roger
Brown Robert L.
Brown Robert Hamilton
Brown Alexander
Brown Douglas N.
Brown Thomas M.
Brown Warren R.
Brownlee Robert R.
Brownlee Donald E.
Brucato Robert J.
Bruenn Stephen W.
Brugel Edward W.
Bruhweiler Frederick C.
Bruner Marilyn E.
Bruning David H.
Brunk William E.
Buchler J. R.
Buff James S.
Buhl David
Buie Marc W.
Bunner Alan N.
Buote David A.
Buratti Bonnie J.
Burbidge Eleanor Margaret
Burbidge Geoffrey R.
Burgasser Adam J.
Burke Bernard F.
Burkhead Martin S.
Burlaga Leonard F.
Burns Joseph A.
Burns Jr Jack O'Neal
Burrows Adam Seth
Burrows David Nelson
Burstein David
Burton W. Butler
Busko Ivo C.
Buta Ronald J.

Buti Bimla
Butler Paul R.
Butler Dennis
Butler Bryan J.
Butterworth Paul S
Buzasi Derek
Byard Paul L.
Byrd Gene G.
Cahn Julius H.
Caillault Jean Pierre
Caldwell John A. R.
Calvin William H.
Calzetti Daniela
Cameron Winifred S.
Campbell Donald B.
Campbell Murray F.
Campbell Belva G. S.
Campins Humberto
Canfield Richard C.
Canizares Claude R.
Cannon John M.
Canzian Blaise
Capen Charles F.
Capriotti Eugene R.
Carbon Duane F.
Carilli Christopher L.
Carleton Nathaniel P.
Carlson John B.
Carney Bruce William
Caroff Lawrence J.
Carpenter Lloyd
Carpenter Kenneth G.
Carr Thomas D.
Carr John Sherman
Carruthers George R.
Carter William Eugene
Casertano Stefano
Cash Jr Webster C.
Cassinelli Joseph P.
Castelaz Micheal W.
Castelli John P.
Castor John I.
Caton Daniel B.
Catura Richard C.
Cefola Paul J.
Centrella Joan M.
Cersosimo Juan Carlos
Chaboyer Brian C.
Chaffee Frederic H.
Chaisson Eric J.
Chamberlain Joseph M.
Chambers John E.
Chambliss Carlson R.
Chance Kelly V.
Chandler Claire
Chandra Subhash
Chapman Gary A.
Chapman Clark R.
Chapman Robert D.
Chartas George
Chatzichristou Eleni T.
Chen Kwan-yu
Chen Hsiao-Wen
Cheng Kwang Ping
Chesley Steven R.
Cheung Cynthia Y.
Chevalier Roger A.
Chitre Dattakumar M.
Chiu Liang-Tai George
Chiu Hong Yee
Chodas Paul Winchester
Choudhary Debi Prasad
Christian Carol Ann
Christiansen Wayne A.
Christodoulou Dmitris
Christy Robert F.
Christy James Walter
Chu You-Hua
Chubb Talbot A.
Chupp Edward L.
Churchwell Edward B.
Ciardi David R.
Ciardullo Robin
Clark Barry G.
Clark Frank Oliver
Clark George W.
Clark Jr Alfred
Clarke Tracy E.
Clarke John T.
Claussen Mark J.
Clayton Geoffrey C.
Clayton Donald D.
Clem James L.
Clemens Dan P.
Clifton Kenneth St.
Cline Thomas L.
Cliver Edward W.
Clowe Douglas I.
Cochran Anita L.
Cochran William David
Cocke William John
Code Arthur D.
Coffeen David L.
Coffey Helen E.
Cohen Martin
Cohen Jeffrey M.
Cohen Leon
Cohen Marshall H.
Cohen Judith
Cohen Ross D.
Cohen Richard S.
Cohn Haldan N.
Colbert Edward J. M.
Colburn David S.
Cole David M.
Coleman Paul Henry
Coletti Donna J.
Colgate Stirling A.
Collins George W. II
Combi Michael R.
Comins Neil Francis
Condon James J.
Conklin Edward K.
Connolly Leo P.
Conselice Christopher J.
Conti Peter S.
Cook Kem H.
Cook John W.
Cooray Asantha R.
Corbally Christopher
Corbet Robin Henry D.
Corbin Thomas Elbert
Corbin Brenda G.
Corbin Michael R.
Corcoran Michael Francis
Cordes James M.
Cordova France A. D.
Corliss C. H.
Corwin Jr Harold G.
Cotton Jr William D.
Coulson Iain M.
Counselman Charles C.
Cowan John J.
Cowie Lennox Lauchlan
Cowley Anne P.
Cowley Charles R.
Cox Arthur N.
Cox Donald P.
Craine Eric R.
Crane Philippe

Crane Patrick C.
Crannell Carol Jo
Crawford David L.
Crawford Fronefield
Crenshaw Daniel Michael
Crocker Deborah Ann
Crotts Arlin Pink
Cruikshank Dale P.
Crutcher Richard M.
Cudaback David D.
Cudworth Kyle McC.
Culver Roger Bruce
Cuntz Manfred
Czyzak Stanley J.
Dahn Conard Curtis
Dalgarno Alexander
Daly Ruth Agnes
Danby J. M. Anthony
Danford Stephen C.
Danks Anthony C.
Danly Laura
Danner Rolf
Dappen Werner
Datlowe Dayton
David Laurence P.
Davidson Kris
Davies Paul Charles W.
Davies Ashley Gerard
Davila Joseph M.
Davis Christopher J.
Davis Gary R.
Davis Sumner P.
Davis Marc
Davis Michael M.
Davis Donald R.
Davis Morris S.
Davis Robert J.
Davis Jr Cecil G.
de Frees Douglas J.
de Jong Roelof S.
de Jonge J. K.
De Marchi Guido
De Marco Orsola
de Pater Imke
de Vincenzi Donald
de Young David S.
Dearborn David Paul K.
Deeming Terence J.
DeGioia-Eastwood Kathleen
Deliyannis Constantine P.
Dell'Antonio P. Ian
Delsemme Armand H.
Demarque Pierre
Deming Leo Drake
Dempsey Robert C.
Dennis Brian R.
Dennison Edwin W.
Dent William A.
Deprit Andre
Dere Kenneth P.
Dermer Charles Dennison
Dermott Stanley F.
Despain Keith Howard
Deupree Robert G.
Deustua Susana E.
Deutschman William A.
Devinney Edward J.
Devorkin David H.
Dewey Rachel J.
Dewitt Bryce S.
Dewitt-Morette Cecile Pr
Dhawan Vivek
Dick Steven J.
Dickel Helene R.
Dickel John R.
Dickey Jean O'Brien
Dickey John M.
Dickinson Dale F.
Dickman Steven R.
Dickman Robert L.
Didkovsky Leonid
Dieter Conklin Nannielou H.
Dietrich Matthias
Digel Seth William
Dikova Smilyana D.
Dinerstein Harriet L.
Dinescu-Casetti Dana Ioana
Dixon Robert S.
Djorgovski Stanislav
Doherty Lowell R.
Dolan Joseph F.
Donahue Robert A.
Donn Bertram D.
Doppmann Gregory W.
Doschek George A.
Douglas James N.
Downes Ronald A.
Downs George S.
Doyle Laurance R.
Draine Bruce T.
Drake Stephen A.
Drake Frank D.
Drake Jeremy
Dreher John W.
Dressel Linda L.
Dressler Alan
Drever Ronald W. P.
Drilling John S.
Dryer Murray
Dubin Maurice
Dufour Reginald James
Dukes Jr. Robert
Duncan Douglas Kevin
Duncombe Raynor L.
Dunham David W.
Dupree Andrea K.
Dupuy David L.
Durisen Richard H.
Durney Bernard
Durrance Samuel T.
Durrell Patrick R.
Duthie Joseph G.
Duvall Jr Thomas L.
Dwarkadas Vikram V.
Dwek Eli
Dyck Melvin
Dyson Freeman J.
Eaton Joel A.
Eddy John A.
Edelson Rick
Edmondson Frank K.
Edwards Alan Ch.
Egan Michael P.
Eicher David
Eilek Jean
El Baz Farouk
Elias Nicholas
Elitzur Moshe
Elliot James L.
Elmegreen Bruce Gordon
Elmegreen Debra M.
Elste Gunther H.
Elston Wolfgang E.
Elvis Martin S.
Emerson Darrel Trevor
Emslie A. Gordon
Endal Andrew S.
Epps Harland Warren

Epstein Gabriel Leo
Epstein Eugene E.
Epstein Richard I.
Eshleman Von R.
Eskridge Paul B.
Esposito Larry W.
Esposito F. Paul
Etzel Paul B.
Eubanks Thomas Marshall
Evans J. V.
Evans Neal J.
Evans Ian Nigel
Evans W. Doyle
Evans Nancy R.
Ewen Harold I.
Ewing Martin S.
Eyer Laurent
Fabbiano Giuseppina
Faber Sandra M.
Fabricant Daniel G.
Fahey Richard P.
Falco Emilio E.
Falk Jr Sydney W.
Fall S. Michael
Faller James E.
Fallon Frederick W.
Fang Li-zhi
Fanselow John Lyman
Farnham Tony L.
Fassnacht Christopher D.
Faulkner John
Fay Theodore D.
Fazio Giovanni G.
Federman Steven Robert
Feigelson Eric D.
Fekel Francis C.
Feldman Uri
Felten James E.
Ferland Gary J.
Fesen Robert A.
Fey Alan Lee
Feynman Joan
Fiala Alan D.
Fichtel Carl E.
Fiedler Ralph L.
Field George B.
Fienberg Richard T.
Filippenko Alexei V.
Fink Uwe
Finn Lee Samuel
Firor John W.
Fischel David
Fischer Jacqueline
Fisher George H.
Fisher Richard R.
Fisher Philip C.
Fisher J. Richard
Fishman Gerald J.
Fitch Walter S.
Fitzpatrick Edward L.
Fix John D.
Flannery Brian Paul
Fleck Bernhard
Fleck Robert Charles
Fleischer Robert
Fleming Thomas Anthony
Fliegel Henry F.
Florkowski David R.
Fogarty William G.
Foltz Craig B.
Fomalont Edward B.
Fontenla Juan M.
Forbes Terry G.
Ford Holland C.
Ford Jr W. Kent
Forman William Richard
Forrest William John
Forster James Richard
Foster Roger S.
Foukal Peter V.
Fox Kenneth
Frail Dale Andrew
Frank Juhan
Franklin Fred A.
Franz Otto G.
Frazier Edward N.
Fredrick Laurence W.
Freedman Wendy L.
French Richard G.
Friberg Per
Friedlander Michael
Friedman Scott David
Friel Eileen D.
Friend David B.
Frisch Priscilla
Frogel Jay Albert
Frost Kenneth J.
Fruchter Andrew S.
Fruscione Antonella
Frye Glenn M.
Ftaclas Christ
Fuhr Jeffrey Robert
Fulbright Jon P.
Fullerton Alexander W.
Funes José G.
Furuya Ray S.
Fuse Tetsuharu
Gaensler Bryan M.
Gaetz Terrance J.
Gaisser Thomas K.
Gallagher III John S.
Gallet Roger M.
Gallimore Jack F.
Galloway Duncan K.
Galvin Antoinette B.
Gaposchkin Edward M.
Garcia Michael R.
Gardner Jonathan P.
Garlick George F.
Garmany Catherine D.
Garmire Gordon P.
Garnett Donald Roy
Garstang Roy H.
Gary Gilmer Allen
Gary Dale E.
Gaskell C. Martin
Gatewood George
Gatley Ian
Gaume Ralph A.
Gauss Stephen F.
Gaustad John E.
Geballe Thomas R.
Gebbie Katharine B.
Gehrels Tom
Gehrels Neil
Gehrz Robert Douglas
Gelderman Richard
Geldzahler Bernard J.
Geller Margaret Joan
Genet Russel M.
Gerakines Perry A.
Gergely Tomas E.
Germain Marvin E.
Gerola Humberto
Gezari Daniel Ysa
Ghez Andrea
Ghigo Francis D.
Ghosh Tapashi
Ghosh Kajal Kumar
Giampapa Mark S.

Gibson James
Gibson Steven J.
Gibson David M.
Giclas Henry L.
Gierasch Peter J.
Gies Douglas R.
Gigas Detlef
Gilliland Ronald L.
Gillingham Peter
Gilman Peter A.
Gilra Daya P.
Gingerich Owen
Giovane Frank
Giovanelli Riccardo
Gizis John E.
Glaser Harold
Glaspey John W.
Glass Billy Price
Glassgold Alfred E.
Glatzmaier Gary A.
Glinski Robert J.
Goebel John H.
Goldman Martin V.
Goldreich Peter
Goldsmith Donald W.
Goldsmith Paul F.
Goldstein Richard M.
Goldwire Jr Henry C.
Golub Leon
Gonzalez Guillermo
Goode Philip R.
Goodman Alyssa Ann
Goodrich Robert W.
Goody R. M.
Gopalswamy Nat
Gordon Courtney P.
Gordon Kurtiss J.
Gordon Mark A.
Gorenstein Paul
Gorenstein Marc V.
Gosling John T.
Goss W. Miller
Gott J. Richard
Gottesman Stephen T.
Gotthelf Eric
Gottlieb Carl A.
Gould Robert J.
Graboske Jr Harold C.
Gradie Jonathan Carey
Grady Carol Anne
Graham Eric
Graham John A.
Grandi Steven Aldridge
Grasdalen Gary L.
Grauer Albert D.
Grav Tommy
Gray Peter Murray
Gray Richard O.
Grayzeck Edwin J.
Green Jack
Green Elizabeth M.
Green Richard F.
Green Daniel William Edward
Greenberg Richard
Greenhill Lincoln J.
Greenhouse Matthew A.
Greenstein George
Gregg Michael David
Gregory Stephen Albert
Greisen Eric
Greve Thomas R.
Greyber Howard D.
Griest Kim
Griffin Ian Paul
Griffiths Richard E.
Grindlay Jonathan E.
Grinspoon David Harry
Gronwall Caryl A.
Gross Richard Sewart
Gross Peter G.
Grossman Allen S.
Grossman Lawrence
Grundy William M.
Grupe Dirk
Gudehus Donald Henry
Guetter Harry Hendrik
Guhathakurta Madhulika
Guidice Donald A.
Guinan Edward F.
Gulkis Samuel
Gull Theodore R.
Gunn James E.
Gurman Joseph B.
Gurshtein Alexander A.
Gursky Herbert
Gustafson Bo A. S.
Guzik Joyce A.
Gwinn Carl R.
Habbal Shadia Rifai
Hackwell John A.
Haddock Fred T.
Haghighipour Nader
Hagyard Mona J.
Haisch Bernhard Michael
Haisch Jr Karl E.
Hajian Arsen R.
Hakkila Jon E.
Hale Alan
Hall Patrick B.
Hall Donald N.
Hall Douglas S.
Hallam Kenneth L.
Hamilton Andrew J. S.
Hamilton Douglas P.
Hammel Heidi B.
Hammond Gordon L.
Hanisch Robert J.
Hankins Timothy Hamilton
Hanner Martha S.
Hansen Carl J.
Hansen Richard T.
Hanson Robert B.
Hapke Bruce W.
Hardebeck Ellen G.
Hardee Philip
Harmer Charles F. W.
Harmer Dianne L.
Harms Richard James
Harnden Jr Frank R.
Harnett Julienne
Harper Graham M.
Harrington J. Patrick
Harris Daniel E.
Harris Alan William
Harris Hugh C.
Hart Michael H.
Harten Ronald H.
Hartigan Patrick M.
Hartkopf William I.
Hartmann Lee William
Hartmann William K.
Hartmann Dieter H.
Hartoog Mark Richard
Harvel Christopher Alvin
Harvey Paul Michael
Harvey Gale A.
Harvey John W.
Harwit Martin
Hasan Hashima

Haschick Aubrey
Hathaway David H.
Hatzes Artie P.
Hauser Michael G.
Havlen Robert J.
Hawkins Isabel
Hawley Suzanne Louise
Hayama Kazuhiro
Hayashi Saeko S.
Hayashi Masahiko
Hayes Donald S.
Haymes Robert C.
Haynes Martha P.
Hazen Martha L.
Heacox William D.
Heap Sara R.
Heasley James Norton
Hecht James H.
Heckathorn Harry M.
Heckman Timothy M.
Heeschen David S.
Hegyi Dennis J.
Heiles Carl
Hein Righini Giovanna
Heiser Arnold M.
Helfand David John
Helfer H. Lawrence
Helin Eleanor Francis
Heller Clayton
Helmken Henry F.
Helou George
Hemenway Mary Kay M.
Hemenway Paul D.
Henden Arne A.
Henning Patricia A.
Henriksen Mark Jeffrey
Henry Richard B. C.
Henry Richard Conn
Herbig George H.
Herbst William
Herbst Eric
Herczeg Tibor J.
Hernquist Lars Eric
Hershey John L.
Hertz Paul L.
Hewitt Adelaide
Hewitt Anthony V.
Heyer Mark
Hibbard John E.
Hibbs Albert R.
Hildebrand Roger H.
Hildner Ernest
Hill Henry A.
Hill Frank
Hill Grant
Hillenbrand Lynne A.
Hilliard Ron
Hills Jack G.
Hillwig Todd C.
Hilton James Lindsay
Hindsley Robert Bruce
Hinkle Kenneth H.
Hinners Noel W.
Hintz Eric G.
Hintzen Paul Michael N.
Ho Luis Chi
Ho Paul T. P.
Hoard Donald W.
Hobbs Robert W.
Hobbs Lewis M.
Hockey Thomas Arnold
Hodapp Klaus-Werner
Hodge Paul W.
Hoeflich Peter
Hoeksema Jon Todd
Hoessel John Greg
Hoff Darrel Barton
Hoffman Jeffrey Alan
Hofner Peter
Hogan Craig J.
Hogg David E.
Holberg Jay B.
Hollenbach David John
Hollis Jan Michael
Hollowell David Earl
Hollweg Joseph V.
Holman Gordon D.
Holt Stephen S.
Holzer Thomas E.
Honeycutt R. Kent
Horch Elliott P.
Horner Scott D.
Horowitz Paul
Houck James R.
Houdashelt Mark L.
Houk Nancy
House Lewis L.
Howard William E.
Howard W. Michael
Howard Robert F.
Howard Sethanne
Howell Ellen S.
Howell Steve B.
Hrivnak Bruce J.
Hu Esther M.
Huang Jiasheng
Hubbard William B.
Hubeny Ivan
Huchra John Peter
Hudson Hugh S.
Hudson Reggie L.
Huebner Walter F.
Huenemoerder David P.
Huggins Patrick J.
Hughes John P.
Hughes Philip
Huguenin G. Richard
Humphreys Roberta M.
Hundhausen Arthur
Hunten Donald M.
Hunter Christopher
Hunter James H.
Hunter Deidre Ann
Hunter Todd R.
Huntress Wesley T.
Hurford Gordon J.
Hurley Kevin C.
Hurwitz Mark V.
Hut Piet
Hutter Donald John
Hyun Jong-June
Ianna Philip A.
Iben Jr Icko
Ignace Richard
Illing Rainer M. E.
Illingworth Garth D.
Imamura James
Imhoff Catherine L.
Impey Christopher D.
Ingerson Thomas
Inglis Michael
Ipatov Sergei I.
Ipser James R.
Irvine William M.
Ivans Inese
Ivezic Zeljko
Jackson James M.
Jackson Bernard V.
Jackson Peter D.
Jackson William M.

Jacobs Kenneth C.
Jacobson Robert A.
Jacoby George H.
Jaffe Daniel T.
Janes Kenneth A.
Janiczek Paul M.
Jannuzi Buell Tomasson
Janssen Michael Allen
Jarrett Thomas H.
Jastrow Robert
Jedicke Robert
Jefferies Stuart M
Jefferys William H.
Jenkins Edward B.
Jenkner Helmut
Jenner David C.
Jenniskens Petrus Matheus Marie
Jevremovic Darko
Jewell Philip R.
Johnson Thomas James
Johnson Hugh M.
Johnson Hollis R.
Johnson Donald R.
Johnson Torrence V.
Johnson Fred M.
Johnston Kenneth J.
Jokipii J. R.
Joner Michael D.
Jones Christine
Jones Barbara
Jones Dayton L.
Jones Frank Culver
Jones Harrison P.
Jones Thomas Walter
Jones Burton
Jones Eric M.
Jordan Stuart D.
Jorgensen Inger
Jørgensen Jes K.
Joselyn Jo Ann c
Joseph Charles Lynn
Joseph Robert D.
Joss Paul Christopher
Joy Marshall J.
Judge Philip
Junkkarinen Vesa T.
Junor William
Jura Michael
Jurgens Raymond F.
Kafatos Minas
Kaftan May A.
Kahler Stephen W.
Kaiser Mary E.
Kaitchuck Ronald H.
Kaler James B.
Kalkofen Wolfgang
Kammeyer Peter C.
Kamp Lucas Willem
Kane Sharad R.
Kanekar Nissim
Kaplan George H.
Kaplan Lewis D.
Karovska Margarita
Karp Alan H.
Karpen Judith T.
Kassim Namir E.
Katz Jonathan I.
Kaufman Michele
Kawaler Steven D.
Kaye Anthony B.
Keel William C.
Keene Jocelyn Betty
Keil Stephen L.
Keil Klaus
Keller Geoffrey
Keller Charles F.
Kellermann Kenneth I.
Kellogg Edwin M.
Kemball Athol
Kennicutt Robert C.
Kent Stephen M.
Kenyon Scott J.
Khare Bishun N.
Kielkopf John F.
Killen Rosemary M.
Kilston Steven D.
Kim Dong-Woo
Kimble Randy A.
King David S.
King Ivan R.
Kinman Thomas D.
Kinney Anne L.
Kiplinger Alan L.
Kirby Kate P.
Kirkpatrick Joseph D.
Kirkpatrick Ronald C.
Kirshner Robert Paul
Kisseleva-Eggleton Ludmila
Kissell Kenneth E.
Klarmann Joseph
Klein Richard I.
Kleinmann Douglas E.
Klemola Arnold R.
Klemperer Wilfred K.
Klepczynski William J.
Klimchuk James A.
Klinglesmith Daniel A.
Klinglesmith III Daniel A.
Kliore Arvydas Joseph
Klock Benny L.
Knacke Roger F.
Knapp Gillian R.
Kniffen Donald A.
Knoelker Michael
Knowles Stephen H.
Ko Hsien C.
Kobayashi Naoto
Koch David G.
Koch Robert H.
Koekemoer Anton M.
Kofman Lev
Kohl John L.
Kolb Edward W.
Kondo Yoji
Kong Albert
Konigl Arieh
Koo David C-Y
Kopp Roger A.
Kopp Greg
Koratkar Anuradha P.
Korchagin Vladimir
Kormendy John
Kosovichev Alexander
Koupelis Theodoros
Kouveliotou Chryssa
Kovalev Yuri Y.
Kovar N. S.
Kovar Robert P.
Kowal Charles Thomas
Kraft Robert P.
Kramer Kh N.
Kramida Alexander
Kraushaar William L.
Kreidl Tobias J. N.
Krieger Allen S.
Kriss Gerard A.
Krogdahl W. S.
Krolik Julian H.
Kron Richard G.

Krucker Sam
Krupp Edwin C.
Kudritzki Rolf-Peter
Kuhi Leonard V.
Kuhn Jeffery Richard
Kuin Nicolaas Paulus M.
Kuiper Thomas B. H.
Kulkarni Shrinivas R.
Kulsrud Russell M.
Kumar C. Krishna
Kumar Shiv S.
Kundu Mukul R.
Kundu Arunav
Kurfess James D.
Kurtz Michael Julian
Kurucz Robert L.
Kutner Marc Leslie
Kutter G. Siegfried
Kwitter Karen Beth
Lacy Claud H.
Lacy John H.
Lada Charles Joseph
Laird John B.
Lamb Richard C.
Lamb Susan Ann
Lamb Frederick K.
Lamb Jr Donald Quincy
Lambert David L.
Lampton Michael
Lande Kenneth
Landecker Peter Bruce
Landman Donald A.
Landolt Arlo U.
Lane Arthur Lonne
Lane Adair P.
Lang Kenneth R.
Langer William David
Langhoff Stephanie R.
Langston Glen I.
Lanning Howard H.
Lanz Thierry
LaRosa Theodore N.
Larsen Jeffrey A.
Larson Stephen M.
Larson Harold P.
Larson Richard B.
Lasala Gerald J.
Lasher Gordon Jewett
Latham David W.
Latter William B.
Lattimer James M.
Lauroesch James T.
Lautman D. A.
Lawlor Timothy M.
Lawrence John K.
Lawrence Charles R.
Lawrence G. M.
Lawrie David G.
Lawson Peter R.
Layden Andrew Choisy
Layzer David
Lazio Joseph
Lea Susan Maureen
Leacock Robert Jay
Lebofsky Larry Allen
Lebovitz Norman R.
Lecar Myron
Leckrone David S.
Lee Paul D.
Leggett Sandy K.
Leibacher John
Leighly Karen Marie
Leisawitz David
Leitherer Claus
Leka Kimberly D.
Lemmon Mark
Leonard Peter James T.
Lepp Stephen H.
Lester Daniel F.
Leung Kam Ching
Leung Chun Ming
Levine Randolph H.
Levine Stephen
Levison Harold F.
Levreault Russell M.
Levy Eugene H.
Lewin Walter H. G.
Lewis Brian Murray
Lewis J. S.
Li Hong-Wei
Li Linghuai
Liang Edison P.
Libbrecht Kenneth G.
Liddell U.
Liebert James W.
Lieske Jay H.
Likkel Lauren Jones
Lilley Edward A.
Lillie Charles F.
Lin Chia C.
Lin Douglas N. C.
Lindsey Charles Allan
Lingenfelter Richard E.
Linke Richard Alan
Linnell Albert P.
Linsky Jeffrey L.
Linsley John
Lippincott Zimmerman Sarah Lee
Lipschutz Michael E.
Lis Dariusz C.
Lissauer Jack J.
Lisse Carey M.
Lister Matthew L.
Liszt Harvey Steven
Little-Marenin Irene R.
Littleton John E.
Litvak Marvin M.
Liu Sou-Yang
Liu Yang
Livingston William C.
Livio Mario
Lo Kwok-Yung
Lochner James Charles
Lockman Felix J.
Lockwood G. Wesley
Lodders Katharina
Long Knox S.
Longmore Andrew J.
Lonsdale Carol J.
Lopes-Gautier Rosaly
Lord Steven Donald
Loren Robert Bruce
Lovas Francis John
Lovelace Richard V. E.
Low Frank J.
Low Boon Chye
Lu Limin
Lu Phillip K.
Lubin Lori M.
Lubowich Donald A.
Luck R. Earle
Lucke Peter B.
Lugger Phyllis M.
Lundquist Charles A.
Luttermoser Donald
Lutz Barry L.
Lutz Julie H.
Luu Jane X.
Luzum Brian J.

Lynch David K.
Lynds Beverly T.
Lynds Roger C.
Ma Chopo
Mac Clain Edward F.
Macalpine Gordon M.
Macchetto Ferdinando
MacConnell Darrell Jack
MacDonald James
Mack Peter
Macquart Jean-Pierre
MacQueen Robert M.
Macri Lucas M.
Macy William Wray
Madau Piero
Madore Barry Francis
Magee-Sauer Karen P.
Magnani Loris Alberto
Makarov Valeri
Malhotra Sageeta
Malhotra Renu
Malina Roger Frank
Malitson Harriet H.
Malkamaeki Lauri J.
Malkan Matthew Arnold
Malville J. Mckim
Mamajek Eric E.
Mangum Jeffrey Gary
Manset Nadine C.
Maran Stephen P.
Marcialis Robert
Marcy Geoffrey W.
Marengo Massimo
Margon Bruce H.
Margot Jean-Luc
Margrave Jr Thomas E.
Mariska John T.
Markworth Norman Lee
Marley Mark S.
Marochnik L. S.
Marr Jonathon M.
Marschall Laurence A.
Marscher Alan Patrick
Marsden Brian G.
Marshall Herman Lee
Marston Anthony Philip
Martin Rene Pierre
Martin William C.
Martin Robert N.
Martin Donn Christopher
Martin Crystal L.
Martin Christopher L.
Martini Paul
Martins Donald Henry
Marvel Kevin B.
Marvin Ursula B.
Mason Brian D.
Mason Glenn M.
Massa Derck Louis
Massey Philip L.
Masson Colin R.
Matese John J.
Mather John Cromwell
Mathews William G.
Mathieu Robert D.
Mathis John S.
Matsakis Demetrios N.
Matson Dennis L.
Matthews Brenda C.
Matthews Lynn E.
Matthews Clifford
Matthews Henry E.
Matthews Thomas A.
Mattox John
Matz Steven Micheal
Matzner Richard A.
Mauche Christopher W.
Max Claire E.
Maxwell Alan
Mayfield Earle B.
Mazurek Thaddeus John
Mazzarella Joseph M.
McAlister Harold A.
McAteer R. T. James
McCabe Marie K.
McCammon Dan
McCarthy Dennis D.
McClintock Jeffrey E.
McCluskey Jr George E.
McCord Thomas B.
McCray Richard
McCrosky Richard E.
McCullough Peter R.
McDavid David A.
McDonald Frank B.
McDonough Thomas R.
McElroy M. B.
McFadden Lucy Ann
McGaugh Stacy Sutton
McGimsey Jr Ben Q.
McGrath Melissa Ann
McGraw John T.
McIntosh Patrick S.
McKee Christopher F.
McKinnon William Beall
McLaren Robert A.
McLean Brian J.
McLean Ian S.
McMahan Robert Kenneth
McMillan Robert S.
McMullin Joseph P.
McNamara Delbert H.
McSwain Mary V.
Mead Jaylee Montague
Meech Karen J.
Meeks M. Littleton
Megeath S. Thomas
Mehringer David Michael
Meier David L.
Meier Robert R.
Meisel David D.
Meixner Margaret
Melbourne William G.
Melia Fulvio
Melnick Gary J.
Melott Adrian L.
Mendis Devamitta Asoka
Merline William J.
Mertz Lawrence N.
Messmer Peter
Meszaros Peter
Metcalfe Travis S.
Metevier Anne
Meyer David M.
Meyers Karie Ann
Michel Raul
Michel F. Curtis
Mickelson Michael E.
Mighell Kenneth John
Mihalas Dimitri
Mihalas Barbara R. Weibel
Mikesell Alfred H.
Milkey Robert W.
Miller Joseph S.
Miller Guy Scott
Miller Michael C.
Miller Hugh R.
Miller Richard H.
Miller Neal A.
Milligan J. E.

Millikan Allan G.
Millis Robert L.
Minchin Robert F.
Mink Douglas J.
Minter Anthony H.
Mintz Blanco Betty
Miralles Mari Paz
Misawa Toru
Misconi Nebil Yousif
Misner Charles W.
Mitchell Kenneth J.
Miyaji Takamitsu
Miyazaki Satoshi
Mo Houjun
Mo Jinger
Modali Sarma B.
Modisette Jerry L.
Moellenbrock III George
Moffett David A.
Moffett Thomas J.
Mohr Joseph J.
Molnar Michael R.
Momjian Emmanuel
Monet David G.
Monet Alice K. B.
Monnier John D.
Moody Joseph Ward
Mook Delo E.
Moore Elliott P.
Moore Ronald L.
Moore Marla H.
Moos Henry Warren
Moran James M.
Morgan Thomas H.
Morgan John Adrian
Moriarty-Schieven Gerald H.
Morino Jun-Ichi
Morris Charles S.
Morris Steven
Morris Mark Root
Morrison Nancy D.
Morrison David
Morton G. A.
Moss Christopher
Motz Lloyd
Mould Jeremy R.
Mouschovias Telemachos Ch.
Moustakas Leonidas A.
Mozurkewich David
Mueller Beatrice E. A.
Mueller Ivan I.
Muench Guido
Mufson Stuart Lee
Mukai Koji
Mukherjee Krishna
Mulchaey John S.
Mullan Dermott J.
Muller Richard A.
Mumford George S.
Mumma Michael Jon
Mundy Lee G.
Munro Richard H.
Murdock Thomas Lee
Murphy Brian William
Murphy Robert E.
Murray Stephen S.
Murray Stephen David
Musen Peter
Mushotzky Richard
Musielak Zdzislaw E.
Musman Steven
Mutel Robert Lucien
Mutschlecner J. Paul
Myers Philip C.
Nacozy Paul E.
Nahar Sultana N.
Narayan Ramesh
Nather R. Edward
Nave Gillian
Neff James Edward
Neff Susan Gale
Neidig Donald F.
Nelson George Driver
Nelson Robert M.
Nelson Jerry E.
Nelson Robert A.
Nelson Burt
Nemiroff Robert
Ness Norman F.
Neugebauer Gerry
Neupert Werner M.
Newburn Jr Ray L.
Newhall X. X.
Newman Michael John
Newsom Gerald H.
Newton Robert R.
Nice David J.
Nichols Joy
Nicolas Kenneth Robert
Nicoll Jeffrey Fancher
Nicollier Claude
Niedner Malcolm B.
Niell Arthur E.
Nilsson Carl
Ninkov Zoran
Nishikawa Ken-Ichi
Nishimura Tetsuo
Noerdlinger Peter D.
Nolan Michael C.
Noll Keith Stephen
Noonan Thomas W.
Noriega-Crespo Alberto
Norman Colin A.
Norman Dara J.
Noumaru Junichi
Novick Robert
Noyes Robert W.
Nulsen Paul E. J.
Nuth Joseph A. III
O'Brian Thomas R.
O'Connell Robert F.
O'Connell Robert West
O'Dea Christopher P.
Odell Andrew P.
O'Dell Stephen L.
O'Dell Charles R.
Odenwald Sten F.
Oegerle William R.
Oemler Jr Augustus
Ofman Leon
Ogelman Hakki B.
O'Handley Douglas A.
Ohashi Nagayoshi
Ohtsuki Keiji
Ojha Roopesh
Oka Takeshi
Oliver John Parker
Olivier Scot Stewart
Olling Robert P.
Olmi Luca
Olowin Ronald Paul
Olsen Knut A. G.
Olsen Kenneth H.
Olson Edward C.
O'Neal Douglas B.
O'Neil Karen L.
Onello Joseph S.
Opendak Michael

Orlin Hyman
Ormes Jonathan F.
Orton Glenn S.
Osborn Wayne
Osmer Patrick S.
Osten Rachel A.
Ostriker Eve C.
Ostriker Jeremiah P.
Ostro Steven J.
Oswalt Terry D.
Ouchi Masami
Owen Tobias C.
Owen Frazer Nelson
Owen Jr William Mann
Owocki Stanley Peter
Ozsvath I.
Pacholczyk Andrzej G.
Paciesas William S.
Paerels Frederik B. S.
Palmer Patrick E.
Pan Xiao-Pei
Panagia Nino
Panek Robert J.
Pang Kevin
Pankonin Vernon Lee
Pannuti Thomas G.
Pap Judit
Papaliolios Costas
Parhi Shyamsundar
Parise Ronald A.
Parker Eugene N.
Parker Robert A. R.
Parkinson Truman
Parkinson William H.
Parrish Allan
Parsons Sidney B.
Partridge Robert B.
Pasachoff Jay M.
Pascu Dan
Patel Nimesh A.
Pauls Thomas Albert
Pavlov George G.
Payne David G.
Peale Stanton J.
Pearson Timothy J.
Pearson Kevin J.
Peck Alison B.
Peebles P. James E.
Peery Benjamin F.
Pellerin Jr Charles J.
Pence William D.
Pendleton Yvonne Jean
Penzias Arno A.
Perkins Francis W.
Perley Richard Alan
Perry Peter M.
Pesch Peter
Peters Geraldine Joan
Peters William L. III
Peterson Charles John
Peterson Ruth Carol
Peterson Laurence E.
Peterson Deane M.
Peterson Bradley Michael
Petre Robert
Petro Larry David
Petrosian Vahe
Pettengill Gordon H.
Petuchowski Samuel J.
Pevtsov Alexei A.
Pfeiffer Raymond J.
Phelps Randy L.
Philip A. G. Davis
Phillips Thomas Gould
Pickles Andrew John
Pier Jeffrey R.
Pierce David Allen
Pilachowski Catherine
Pilcher Carl Bernard
Pines David
Pingree David
Pinsonneault Marc Howard
Pinto Philip Alfred
Pipher Judith L.
Pisano Daniel J.
Platais Imants
Plavec Zdenka
Plavec Mirek J.
Pneuman Gerald W.
Pogge Richard William
Poland Arthur I.
Polidan Ronald S.
Pontoppidan Klaus M.
Potter Andrew E.
Pound Marc W.
Pradhan Anil K.
Prasad Sheo S.
Pravdo Steven H.
Press William H.
Preston George W.
Preston Robert Arthur
Price Michael J.
Price R. M.
Price Stephan Donald
Pritzl Barton J.
Probstein R. F.
Prochaska Jason X.
Proffitt Charles R.
Protheroe William M.
Pryor Carlton Philip
Puche Daniel
Puetter Richard C.
Puschell Jeffery J.
Pyper Smith Diane M.
Quirk William J.
Rabin Douglas Mark
Radford Simon John E.
Radick Richard R.
Rafert James Bruce
Rafferty Theodore J.
Ralchenko Yuri
Ramsey Lawrence W.
Rand Richard J.
Rank David M.
Rankin Joanna M.
Ransom Scott M.
Rasio Frederic A.
Ratcliff Stephen J.
Ray James R.
Raymond John Charles
Reach William
Readhead Anthony C. S.
Reames Donald V.
Reasenberg Robert D.
Rebull Luisa
Rector Travis A.
Reed Bruce Cameron
Reeves Edmond M.
Reichert Gail Anne
Reid Neill
Reid Mark Jonathan
Reipurth Bo
Reitsema Harold J.
Rense William A.
Rettig Terrence W.
Revelle Douglas Orson
Reyes Francisco
Reynolds Stephen P.
Reynolds Ronald J.
Rhoads James

Rhodes Jr Edward J.
Rich Robert M.
Richards Mercedes T.
Richardson R. S.
Richstone Douglas O.
Rickard Lee J.
Rickard James Joseph
Ricker George R.
Rickett Barnaby James
Ricotti Massimo
Riddle Anthony C.
Ridgway Stephen T.
Riegel Kurt W.
Riegler Guenter R.
Rindler Wolfgang
Ringwald Frederick Arthur
Rivolo Arthur Rex
Roark Terry P.
Roberge Wayne G.
Roberts Morton S.
Roberts David Hall
Roberts Jr William W.
Roberts, Jr Lewis C.
Robertson Douglas S.
Robinson Leif J.
Robinson Edward L.
Robinson I.
Robinson Lloyd B.
Robinson Jr Richard D.
Roddier Claude
Roddier Francois
Rodman Richard B.
Roeder Robert
Roellig Thomas L.
Roemer Elizabeth
Rogers Alan E. E.
Rogers Forrest J.
Rogerson John B.
Rogstad David H.
Roman Nancy Grace
Romani Roger William
Romanishin William
Romano-Diaz Emilio
Romer Anita K.
Romney Jonathan D.
Rood Robert T.
Rood Herbert J.
Roosen Robert G.
Rose William K.
Rose James Anthony
Rosen Edward
Rosendhal Jeffrey D.
Rosner Robert
Ross Dennis K.
Rots Arnold H.
Rountree Janet
Rouse Carl A.
Routly Paul M.
Roy Jean-Rene
Rubenstein Eric
Rubin Vera C.
Rubin Robert Howard
Rudnick Lawrence
Rugge Hugo R.
Rule Bruce H.
Russell Jane L.
Russell Christopher T.
Rust David M.
Ruszkowski Mateusz
Rybicki George B.
Sackmann Ingrid Juliana
Sadun Alberto Carlo
Safko John L.
Sage Leslie John
Saha Abhijit
Sahai Raghvendra
Sahu Kailash C.
Sakai Shoko
Salama Farid
Salas Luis
Salisbury J. W.
Salpeter Edwin E.
Salstein David A.
Salter Christopher John
Salzer John Joseph
Samarasinha Nalin H.
Samec Ronald G.
Sandage Allan
Sandell Goran Hans l
Sanders Wilton Turner III
Sanders David B.
Sanders Walt L.
Sandford Maxwell T. II
Sandford Scott Alan
Sandmann William H.
Sanyal Ashit
Sarajedini Ata
Sarazin Craig L.
Sargent Wallace L. W.
Sargent Annelia I.
Sarma Anuj P.
Sartori Leo
Sasaki Toshiyuki
Saslaw William C.
Sasselov Dimitar D.
Savage Blair D.
Savedoff Malcolm P.
Savin Daniel Wolf
Sawyer Constance B.
Scalo John Michael
Scargle Jeffrey D.
Schaefer Bradley E.
Schaefer Martha W.
Schatten Kenneth H.
Schechter Paul L.
Scheeres Daniel J.
Scherb Frank
Scherrer Philip H.
Schild Rudolph E.
Schiller Stephen
Schlegel Eric M.
Schleicher David G.
Schlesinger Barry M.
Schloerb F. Peter
Schmahl Edward J.
Schmalberger Donald C.
Schmelz Joan T.
Schmidt Edward G.
Schmidt Maarten
Schmidtke Paul C.
Schmitt Henrique R.
Schmitz Marion
Schneider Donald P.
Schneider Nicholas M.
Schneider Glenn H.
Schneps Matthew H.
Schnopper Herbert W.
Schoolman Stephen A.
Schou Jesper
Schreier Ethan J.
Schrijver Karel J.
Schroeder Daniel J.
Schucking Engelbert L.
Schulte D. H.
Schulte-Ladbeck Regina E.
Schultz David R.
Schultz Alfred Bernard
Schulz Norbert S.
Schutz Bob Ewald
Schwartz Daniel A.

Schwartz Richard D.
Schwartz Philip R.
Schweizer François
Sconzo Pasquale
Scott John S.
Scott Eugene Howard
Scoville Nicholas Z.
Searle Leonard
Sears Richard Langley
Seeds Michael A.
Seeger Charles Louis III
Seeger Philip A.
Seidelmann P. Kenneth
Seielstad George A.
Seigar Marcus S.
Seitzer Patrick
Sekanina Zdenek
Sekiguchi Kazuhiro
Sellgren Kristen
Sellwood Jerry A.
Sembach Kenneth R.
Serabyn Eugene
Severson Scott
Seward Frederick D.
Shaffer David B.
Shafter Allen W.
Shaham Jacob
Shandarin Sergei F.
Shane William W.
Shank Michael H.
Shao Cheng-yuan
Shapero Donald C.
Shapiro Maurice M.
Shapiro Irwin I.
Shapiro Stuart L.
Shara Michael
Sharma A. Surjalal
Sharp Christopher
Sharpless Stewart
Shaw John H.
Shaw James Scott
Shaw Richard A.
Shawl Stephen J.
Shaya Edward J.
Shea Margaret A.
Sheeley Neil R.
Sheffield Charles
Sheikh Suneel I.
Sheinis Andrew I.
Shelus Peter J.
Shen Benjamin S. P.
Shepherd Debra S.
Shetrone Matthew C.
Shields Gregory A.
Shields Joseph C.
Shine Richard A.
Shinnaga Hiroko
Shipman Harry L.
Shivanandan Kandiah
Shlosman Isaac
Shore Bruce W.
Shostak G. Seth
Shull John Michael
Shull Peter Otto
Sigurdsson Steinn
Silva David Richard
Silverberg Eric C.
Simkin Susan M.
Simon George W.
Simon Michal
Simon Norman R.
Simon Theodore
Simonson S. Christian
Sinha Rameshwar P.
Sion Edward Michael
Sitko Michael L.
Sjogren William L.
Skalafuris Angelo J.
Skillman Evan D.
Skumanich Andrew
Slade Martin A. III
Slane Patrick
Slater Timothy F.
Sloan Gregory Clayton
Slovak Mark Haines
Smale Alan Peter
Smecker-Hane Tammy A.
Smith Harding E.
Smith Howard Alan
Smith Randall K.
Smith Tracy L.
Smith J. Allyn
Smith Horace A.
Smith Graeme H.
Smith Verne V.
Smith Eric Philip
Smith Peter L.
Smith Bradford A.
Smith Haywood C.
Smith Barham W.
Smith Bruce F.
Smith Dean F.
Smith Myron A.
Smith Wm Hayden
Smith Jr Harding E.
Smoot III George F.
Sneden Chris
Snell Ronald L.
Snow Theodore P.
Snyder Lewis E.
Soberman Robert K.
Sobieski Stanley
Soderblom Larry
Soderblom David R.
Sofia Ulysses J.
Sofia Sabatino
Soifer Baruch T.
Solomon Philip M.
Sonett Charles P.
Song Inseok
Sonneborn George
Soon Willie H.
Sovers Ojars J.
Sowell James Robert
Spahr Timothy B.
Sparke Linda
Sparks William Brian
Sparks Warren M.
Spencer John R.
Spencer John Howard
Spergel David N.
Spicer Daniel Shields
Spiegel E.
Spinrad Hyron
Sprague Ann Louise
Squires Gordon K.
Sramek Richard A.
Sridharan Tirupati K.
Stacey Gordon J.
Stachnik Robert V.
Stahler Steven W.
Stahr-Carpenter M.
Stancil Philip C.
Standish E. Myles
Stanford Spencer A.
Stanghellini Letizia
Stanimirovic Snezana
Stanley G. J.
Stapelfeldt Karl R.
Stark Glen

Stark Antony A.
Starrfield Sumner
Statler Thomas S.
Stauffer John Richard
Stebbins Robin
Stecher Theodore P.
Stecker Floyd W.
Stefanik Robert
Steiger W. R.
Steigman Gary
Steiman-Cameron Thomas
Stein Wayne A.
Stein Robert F.
Stein John William
Steinolfson Richard S.
Stellingwerf Robert F.
Stencel Robert Edward
Stepinski Tomasz
Stern Robert Allan
Stern S. Alan
Stiavelli Massimo
Stier Mark T.
Stinebring Daniel R.
Stocke John T.
Stockman Jr Hervey S.
Stockton Alan N.
Stokes Grant H.
Stone Remington P. S.
Stone R. G.
Stone Edward C.
Stone James McLellan
Storrie-Lombardi Lisa
Storrs Alexander D.
Strachan Leonard Jr
Strelnitski Vladimir
Stringfellow Guy S.
Strittmatter Peter A.
Strobel Darrell F.
Strohmayer Tod E.
Strom Karen M.
Strom Robert G.
Strom Stephen E.
Strong Ian B.
Strong Keith T.
Struble Mitchell F.
Struck-Marcell Curtis J.
Stryker Linda L.
Sturrock Peter A.
Suess Steven T.
Sukumar Sundarajan
Sulentic Jack W.
Sullivan Woodruff T.
Sutton Edmund Charles
Svalgaard Leif
Swade Daryl Allen
Swank Jean Hebb
Sweigart Allen V.
Sweitzer James Stuart
Swenson Jr George W.
Swerdlow Noel
Sykes Mark Vincent
Synnott Stephen P.
Szalay Alex
Szkody Paula
Taam Ronald Everett
Tademaru Eugene
Taff Laurence G.
Takata Tadafumi
Talbot Jr Raymond J.
Tandberg-Hanssen Einar A.
Tapley Byron D.
Tarnstrom Guy
Tarter C. Bruce
Tarter Jill C.
Tayal Swaraj S.
Taylor Donald J.
Taylor Joseph H.
Taylor Keith
Taylor Gregory Benjamin
Teays Terry J.
Tedesco Edward F.
Telesco Charles M.
Ten Brummelaar Theo A.
Terrell Dirk C.
Terrell James
Terrile Richard John
Terzian Yervant
Teske Richard G.
Teuben Peter J.
Thaddeus Patrick
Tholen David J.
Thomas Roger J.
Thomas John H.
Thompson A. Richard
Thompson Ian
Thonnard Norbert
Thorne Kip S.
Thornley Michele D.
Thorsett Stephen Erik
Thorstensen John R.
Thronson Jr Harley Andrew
Thuan Trinh Xuan
Tifft William G.
Tilanus Remo P. J.
Timothy J. Gethyn
Tipler Frank Jennings
Title Alan Morton
Tody Douglas C.
Tohline Joel Edward
Tokunaga Alan Takashi
Tolbert Charles R.
Toller Gary N.
Tomasko Martin G.
Toner Clifford George
Tonry John
Toomre Juri
Toomre Alar
Torres Guillermo
Torres Diego F.
Torres Dodgen Ana V.
Tothill Nicholas F. H.
Townes Charles Hard
Trafton Laurence M.
Trammell Susan R.
Traub Wesley Arthur
Treffers Richard R.
Tremaine Scott Duncan
Trimble Virginia L.
Tripicco Michael J.
Troland Thomas Hugh
TruranJr James W.
Tsuruta Sachiko
Tsutsumi Takahiro
Tsvetanov Zlatan I.
Tucker Wallace H.
Tull Robert G.
Tully Richard Brent
Turner Edwin L.
Turner Kenneth C.
Turner Barry E.
Turner Michael S.
Turner Jean L.
Turner Nils H.
Turnshek David A.
Twarog Bruce A.
Tyler Jr G. Leonard
Tylka Allan J.
Tyson John Anthony
Tytler David
Ulich Bobby Lee

Ulmer Melville P.
Ulrich Roger K.
Ulvestad James Scott
Underwood James H.
Unwin Stephen C.
Uomoto Alan
Upgren Arthur R.
Upson Walter L. II
Upton E. K. l.
Urban Sean E.
Urry Claudia Megan
Usher Peter D.
Uson Juan M.
Valenti Jeff A.
van Altena William F.
van Belle Gerard T.
van Breugel Wil
van Citters Gordon W.
van der Marel Roeland P.
van der Veen
Wilhelmus E. C.
van Dorn Bradt Hale
van Dyk Schuyler
van Flandern Tom
van Gorkom Jacqueline H.
van Hamme Walter
van Horn Hugh M.
van Hoven Gerard
van Moorsel Gustaaf
van Putten Maurice
van Riper Kenneth A.
van Speybroeck Leon P.
van Zee Liese E.
VandenBout Paul A.
Vandervoort Peter O.
Vaughan Arthur H.
Veeder Glenn J.
Veillet Christian
Veilleux Sylvain
Velusamy T.
Venugopal V. R.
Verner Ekaterina
Verschuur Gerrit L.
Verter Frances
Vesecky J. F.
Vesperini Enrico
Veverka Joseph
Vijh Uma P.
Vilas Faith
Vishniac Ethan T.
Vogel Stuart Newcombe
Vogt Steven Scott
Volk Kevin
von Hippel Theodore A.
Vorpahl Joan A.
Vrba Frederick J.
Vrtilek Saeqa Dil
Vrtilek Jan M.
Wachter Stefanie
Waddington C. Jake
Wade Richard Alan
Wagner Robert M.
Wagner William J.
Wagner Raymond L.
Wagoner Robert V.
Wainscoat Richard J.
Wakker Bastiaan Pieter
Walborn Nolan R.
Walker Merle F.
Walker Robert M. A.
Walker Alta Sharon
Walker Robert C.
Wallace Richard K.
Wallace Lloyd V.
Waller William H.
Wallerstein George
Wallin John Frederick
Walter Fabian
Walter Frederick M.
Walterbos Rene A. M.
Wampler E. Joseph
Wang Qingde Daniel
Wang Zhong
Wang Yi-ming
Wang Haimin
Wannier Peter Gregory
Ward Richard A.
Ward William R.
Wardle John F. C.
Warner John W.
Warren Jr Wayne H.
Warwick James W.
Wasserman Lawrence H.
Wasson John T.
Watson William D.
Wdowiak Thomas J.
Weaver Harold F.
Weaver Thomas A.
Weaver William Bruce
Weaver Kimberly A.
Webb David F.
Webber John C.
Webbink Ronald F.
Weber Stephen Vance
Webster Zodiac T.
Weedman Daniel W.
Weekes Trevor C.
Wegner Gary Alan
Wehinger Peter A.
Wehrle Ann Elizabeth
Wei Mingzhi
Weidenschilling S. J.
Weiler Kurt W.
Weiler Edward J.
Weill Gilbert M.
Weinberg Jerry L.
Weinberg Steven
Weis Edward W.
Weisberg Joel Mark
Weisheit Jon C.
Weisskopf Martin Ch.
Weissman Paul Robert
Weistrop Donna
Welch William J.
Weller Charles S.
Wells Eddie Neil
Wells Donald C.
Welsh William F.
Wentzel Donat G.
Westerhout Gart
Weymann Ray J.
Wheeler John A.
Wheeler J. Craig
Whipple Arthur L.
Whitaker Ewen A.
White Oran R.
White R. Stephen
White Nathaniel M.
White James Clyde
White Raymond Edwin III
White Richard E.
White Richard L.
White Nicholas Ernest
White Stephen Mark
Whitmore Bradley C.
Whitney Charles A.
Whittet Douglas C. B.
Whittle D. Mark
Widing Kenneth G.
Wiese Wolfgang L.

Wiita Paul Joseph
Wiklind Tommy
Wilcots Eric M.
Wilkening Laurel L.
Wilkes Belinda J.
Will Clifford M.
Williamon Richard M.
Williams Carol A.
Williams Gareth V.
Williams Theodore B.
Williams John A.
Williams James G.
Williams Robert
Williams Barbara A.
Williams Glen A.
Williams Jonathan P.
Williams Thomas R.
Willner Steven Paul
Wills Beverley J.
Wills Derek
Willson Lee Anne M.
Willson Robert Frederick
Wilner David J.
Wilson Robert E.
Wilson Curtis A.
Wilson Andrew S.
Wilson Albert G.
Wilson James R.
Wilson William J.
Wilson Richard
Wilson Robert W.
Wilson Thomas L.
Windhorst Rogier A.
Winebarger Amy R.
Wing Robert F.
Winget Donald E.
Winkler Karl-Heinz A.
Winkler Paul Frank
Winkler Gernot M. R.
Winn Joshua N.
Withbroe George L.
Witt Adolf N.
Witten Louis
Wolfe Arthur M.
Wolff Sidney C.
Wolff Michael J.
Wolfire Mark Guy
Wolfson C. Jacob
Wolfson Richard
Wolszczan Alexander
Wood Matthew Alan
Wood Douglas O. S.
Wood John A.
Wood H. J.
Wooden Diane H.
Wooden William Hugh
Woodward Paul R.
Woolf Neville J.
Woosley Stanley E.
Wootten Henry Alwyn
Worden Simon P.
Wouterloot Jan G. A.
Wray James D.
Wright Melvyn C. H.
Wright Edward L.
Wright James P.
Wrobel Joan Marie
Wu Nailong
Wu Chi Chao
Wu Shi Tsan
Wuelser Jean-Pierre
Wyckoff Susan
Wyse Rosemary F.
Xanthopoulos Emily
Yahil Amos
Yamashita Takuya
Yanamandra-Fisher Padma A.
Yaplee B. S.
Yau Kevin K. C.
Yeh Tyan
Yeomans Donald K.
Yin Qi-Feng
Yoder Charles F.
York Donald G.
Yoshida Shin'ichirou
Yoshino Kouichi
Yoss Kenneth M.
Young Arthur
Young Louise Gray
Young Andrew T.
Young Judith Sharn
Yu Yan
Yusef-Zadeh Farhad
Zabriskie F. R.
Zacharias Norbert
Zare Khalil
Zarro Dominic M.
Zayer Igor
Zeilik Michael Ii
Zellner Benjamin H.
Zepf Stephen Edward
Zezas Andreas
Zhang Shouzhong
Zhang Sheng-Pan
Zhang Cheng-Yue
Zhang Er-Ho
Zhang Xiaolei
Zhang Qizhou
Zhang William W.
Zhao Jun Hui
Zhao Junwei
Zheng Wei
Zinn Robert J.
Zirin Harold
Zirker Jack B.
Ziurys Lucy Marie
Zombeck Martin V.
Zuckerman Ben M.

Uruguay

Fernandez Julio Angel
Gallardo Castro Carlos Tabare
Motta Veronica
Tancredi Gonzalo

Vatican City State

Casanovas Juan
Consolmagno Guy Joseph
Coyne George V.
Omizzolo Alessandro
Stoeger William R.

Venezuela

Abad Hiraldo Carlos
Bautista Manuel A.
Bongiovanni Angel
Briceno César
Bruzual Gustavo R.
Calvet Nuria
Falcon Veloz Nelson L.
Ferrin Ignacio
Fuenmayor Francisco J.
Hernandez Jesús O.
Ibanez S. Miguel H.
Magris Gladis C.
Mendoza Claudio
Mendoza-Briceno César A.
Parravano Antonio
Ramirez Jose M.
Rengel Miriam E.
Rosenzweig-Levy Patrica
Sigalotti Leonardo G.
Vivas Anna Katherina

INDIVIDUAL MEMBERSHIP BY NON-ADHERING COUNTRY

Albania

Hafizi Mimoza

Algeria

Abdelatif Toufik E
Irbah Abdanour
Makhlouf Amar

Azerbaijan (Republic of)

Aslanov I. A.
Asvarov Abdul I.
Babayev Elchin S.
Eminzade T. A.
Guseinov O. H.
Gusejnov Ragim Eh
Ismailov Nariman Z.
Kasumov Fikret K.O.
Sultanov G. F.

Colombia

Brieva Eduardo
de Greiff J. Arias

Ecuador

Lopez Ericson D.

Ethiopia

Kebede Legesse W.

Honduras

Pineda de Carias Maria Cristina
Ponce Gustavo A.

Iraq

Abdulla Shaker Abdul Aziz
Jabbar Sabeh Rhaman
Jabir Niama Lafta
Kadouri Talib Hadi
Mohammed Ali Talib
Sadik Aziz R.
Younis Saad M.

Kazakhstan

Denisyuk Edvard K.
Genkin Igor L.
Karygina Zoya V.
Kharitonov Andrej V.
Omarov Tuken B.
Rozhkovskij Dimitrij A.
Tejfel Viktor G.
Vilkoviskij Emmanuil Y.

Korea (Democratic People's Republic of)

Baek Chang Ryong
Bang Yong Gol
Cha Gi Ung
Cha Du Jin
Chio Chol Zong
Choi Won Chol
Dong Il Zun
Hong Hyon Ik
Kang Jin Sok
Kang Gon Ik
Kim Yul
Kim Zong Dok
Kim Jik Su
Kim Yong Uk
Kim Yong Hyok

Li Son Jae
Li Gi Man
Li Gyong Won
Li Hyok Ho
Li Sin Hyong

Macedonia (Former Yugoslav Republic of)

Apostolovska Gordana

Malta

Mallia Edward A.

Mauritius

Goelbasi Orhan
Golap Kumar

Pakistan

Quamar Jawaid

Singapore

Snowden Michael J.

Slovenia

Cadez Andrej
Dintinjana Bojan
Dominko Fran
Kilar Bogdan
Slosar Anze
Zwitter Tomaz

Sri Lanka

de Silva L. Nalin
Maheswaran Murugesapillai
Ratnatunga Kavan U.

Trinidad and Tobago

Haque Shirin T.

United Arab Emirates

Al-Naimiy Hamid M. K.
Guessoum Nidhal
Mohamed Ali Alaa Eldin Fouad

Uzbekistan

Ehgamberdiev Shuhrat
Hojaev Alisher S.
Ilyasov Sabit
Kalmykov A. M.
Latypov A. A.
Mamadazimov Mamadmuso
Muminov Muydinjon
Nuritdinov Salakhutdin
Sattarov Isroil
Yuldashbaev Taimas S.
Zakirov Mamnum

Viet Nam

Dinh Van M.
Huan Nguyen Dinh
Nguyen Mau Tung

October 2007

MEMBERSHIP BY COMMISSION

Composition of Commission 4

Ephemerides / Ephémérides

President Fukushima Toshio

Vice-President Kaplan George H.

Secretary Hohenkerk Catherine Y.

Organizing Committee

Arlot Jean-Eudes
Bangert John A.
Lara Martin
Pitjeva Elena V.
Urban Sean E.
Vondrak Jan

Members

Abalakin Viktor K.
Ahn Youngsook
Aoki Shinko
Bandyopadhyay A.
Bell Steven A.
Brumberg Victor A.
Capitaine Nicole
Chapront Jean
Chapront-Touze Michelle
Chollet Fernand
Coma Juan Carlos
Cooper Nicholas J.
de Greiff J. Arias
Deprit Andre
Di Xiaohua
Dickey Jean O'Brien
Duncombe Raynor L.
Dunham David W.
Eroshkin Georgij I.
Fiala Alan D.
Fienga Agnès
Fominov Aleksandr M.
Fu Yanning
Glebova Nina I.
Harper David
He Miao-fu
Henrard Jacques
Hilton James Lindsay
Howard Sethanne
Ilyas Mohammad
Janiczek Paul M.
Johnston Kenneth J.
Kinoshita Hiroshi
Klepczynski William J.
Kolaczek Barbara
Krasinsky George A.
Laskar Jacques
Lehmann Marek
Li Gi Man
Li Hyok Ho
Lieske Jay H.
Lopez Moratalla Teodoro
Lukashova Marina V.
Majid Abdul Bin A H
Morrison Leslie V.
Mueller Ivan I.
Newhall X. X.
Nguyen Mau Tung
O'Handley Douglas A.
Reasenberg Robert D.
Rodin Alexander E.
Romero Perez M. Pilar
Rossello Gaspar
Salazar Antonio
Schwan Heiner
Seidelmann P. Kenneth
Shapiro Irwin I.
Shiryaev Alexander A.
Simon Jean-Louis
Soma Mitsuru
Standish E. Myles
Thuillot William
Ting Yeou-Tswen
Van Flandern Tom
Wielen Roland
Wilkins George A.
Williams Carol A.
Williams James G.
Winkler Gernot M. R.
Wytrzyszczak Iwona
Yallop Bernard D.

MEMBERSHIP BY COMMISSION

Composition of Commission 5

Documentation & Astronomical Data

Documentation & Données Astronomiques

President Norris Raymond P.

Vice-President Ohishi Masatoshi

Organizing Committee

Genova Françoise
Grothkopf Uta
Hanisch Robert J.
Malkov Oleg Yu.
Pence William D.
Schmitz Marion
Zhou Xu

Members

Abalakin Viktor K.
Abt Helmut A.
A'Hearn Michael F.
Aizenman Morris L.
Alvarez Pedro
Andernach Heinz J
Banhatti Dilip G.
Benacchio Leopoldo
Benn Chris R.
Berthier Jerôme
Bessell Michael S.
Bond Ian A.
Borde Suzanne
Bouska Jiri
Boyce Peter B.
Brouw Willem N.
Calabretta Mark R.
Chang Hsiang-Kuang
Chapman Jacqueline F.
Cheung Cynthia Y.
Chiappetti Lucio
Chu Yaoquan
Coletti Donna J.
Coluzzi Regina
Corbin Brenda G.
Creze Michel
Cunniffe John
Dalla Silvia C.
Davis Morris S.
Davis Robert J.
de Boer Klaas Sjoerds
Dewhirst David W.
Dickel Helene R.
Dixon Robert S.
Dluzhnevskaya Olga B.
Dobrzycki Adam
Dubois Pascal
Ducati Jorge R.
Duncombe Raynor L.
Durand Daniel
Egret Daniel
Elia Davide
Fyfe Duncan J.
Garaimov Vladimir I.
Garstang Roy H.
Gomez Monique
Green David A
Greisen Eric
Griffin Roger F.
Griffin R. Elizabeth
Grosbol Preben Johnson
Guibert Jean
Guo Hongfang
Harvel Christopher Alvin
Hauck Bernard
Heck Andre
Hefele Herbert
Helou George
Hopkins Andrew M.
Hudkova Ludmila
Jenkner Helmut
Kadla Zdenka I.(†)
Kalberla Peter
Kaplan George H.
Kembhavi Ajit K.
Kharin Arkadiy S.
Kleczek Josip
Kovaleva Dana A.
Kuin Nicolaas Paulus M.
Lantos Pierre(†)
Lequeux James
Lesteven Soizick
Linde Peter
Lonsdale Carol J.
Lortet Marie-Claire
Madore Barry Francis
Mann Robert G.
Matz Steven Micheal
McLean Brian J.
McNally Derek
McNamara Delbert H.
Mead Jaylee Montague
Meadows A. Jack
Mein Pierre
Mermilliod Jean-Claude
Michel Laurent
Mink Douglas J.
Mitton Simon
Morris Rhys H.
Murphy Tara
Murtagh Fionn
Nakajima Koichi
Nishimura Shiro E.

Ochsenbein François
Pakhomov Yuri V.
Pamyatnykh Aleksej A.
Pasian Fabio
Pasinetti Laura E.
Paturel Georges
Pecker Jean-Claude
Philip A. G. Davis
Piskunov Anatolij E.
Pizzichini Graziella
Polechova Pavla
Pucillo Mauro
Quintana Hernan
Raimond Ernst
Ratnatunga Kavan U.
Reardon Kevin
Renson P. F. M.
Riegler Guenter R.
Roman Nancy Grace
Rots Arnold H.
Russo Guido
Samodurov Vladimir A.
Sarasso Maria
Schade David J.
Schilbach Elena
Schlesinger Barry M.
Schlueter A.
Schmadel Lutz D.
Schneider Jean
Schroder Anja C.
Serrano Alfonso
Shakeshaft John R.
Shaw Richard A.
Spite François M.
Tedds Jonathan A.
Terashita Yoichi
Teuben Peter J.
Tody Douglas C.
Tritton Susan Barbara
Tsvetkov Milcho K.
Turner Kenneth C.
Uesugi Akira
Valeev Sultan G.
Wallace Patrick T.
Warren Jr Wayne H.
Weidemann Volker
Wells Donald C.
Wenger Marc
Westerhout Gart
Wielen Roland
Wilkins George A.
Wright Alan E.
Yang Hong-Jin
Zhao Yongheng
Zhao Jun Liang

MEMBERSHIP BY COMMISSION

Composition of Commission 6

Astronomical Telegrams / Télégrammes Astronomiques

President Gilmore Alan C.

Vice-President Samus Nikolaj N.

Organizing Committee

Aksnes Kaare
Green Daniel William Edward
Isobe Syuzo(†)
Marsden Brian G.
Nakano Syuichi
Roemer Elizabeth
Ticha Jana
Yamaoka Hitoshi

Members

Apostolovska Gordana
Coletti Donna J.
Corbin Brenda G.
Filippenko Alexei V.
Grindlay Jonathan E.
Hers Jan
Kastel' Galina R.
Kouveliotou Chryssa
Mattila Seppo K.
Nakamura Tsuko
Phillips Mark M.
Tholen David J.
Tsvetkov Milcho K.
West Richard M.
Williams Gareth V.

MEMBERSHIP BY COMMISSION

Composition of Commission 7

Celestial Mechanics & Dynamical Astronomy

Mécanique Céleste & Astronomie Dynamique

President Burns Joseph A.

Vice-President Knezevic Zoran

Secretary Vokrouhlicky David

Organizing Committee

Athanassoula Evangelia
Beauge Christian
Erdi Bálint
Lemaitre Anne
Maciejewski Andrzej J.
Malhotra Renu
Milani Andrea
Morbidelli Alessandro
Peale Stanton J.
Sidlichovsky Milos
Zhou Ji-Lin

Members

Abad Medina Alberto J.
Abalakin Viktor K.
Ahmed Mostafa Kamal
Akim Efraim L.
Aksnes Kaare
Anosova Joanna
Antonacopoulos Gregory
Aoki Shinko
Archinal Brent A.
Balmino Georges G.
Barabanov Sergey I.
Barberis Bruno
Barbosu Mihail
Barkin Yuri V.
Batrakov Yurij V.
Benest Daniel
Bettis Dale G.
Beutler Gerhard
Bhatnagar K. B.
Boccaletti Dino
Bois Eric
Borderies Nicole
Boss Alan P.
Bozis George
Branham Richard L.
Breiter Slawomir
Brieva Eduardo
Brookes Clive J.
Broucke Roger
Brumberg Victor A.
Brunini Adrian
Cai Michael
Calame Odile
Caranicolas Nicholas
Carpino Mario
Cefola Paul J.
Celletti Alessandra
Chakrabarty Dalia
Chambers John E.
Chapront Jean
Chapront-Touze Michelle
Choi Kyu-Hong
Christou Apostolos
Cionco Rodolfo G.
Colin Jacques
Contopoulos George
Cooper Nicholas J.
Counselman Charles C.
Danby J. M. Anthony
Davis Morris S.
Deleflie Florent
Deprit Andre
Descamps Pascal
Di Sisto Romina P.
Dikova Smilyana D.
Dong Xiaojun
Dormand John Richard
Dourneau Gerard
Drozyner Andrzej
Duncombe Raynor L.
Duriez Luc
Dvorak Rudolf
El Bakkali Larbi
Elipe Sánchez Antonio
Emelianov Nikolaj V.
Fernandez Silvia M.
Ferraz Mello Sylvio
Ferrer Martinez Sebastian
Fiala Alan D.
Floria Peralta Luis Mario
Fong Chugang
Froeschle Claude
Fukushima Toshio
Galletto Dionigi
Gaposchkin Edward M.
Gaska Stanislaw
Giacaglia Giorgio E.
Giordano Claudia M.
Giuliatti Winter Silvia M.
Goldreich Peter
Gonzalez Antonio Camacho(†)
Gomes Rodney D. S.
Goodwin Simon P.
Goudas Constantine L.
Gozdziewski Krzysztof
Grebenikov Evgenij A.

Greenberg Richard
Gronchi Giovanni Federico
Gusev Alexander V.
Hadjidemetriou John D.
Haghighipour Nader
Hallan Prem P.
Hamid S. El Din
Hamilton Douglas P.
Hanslmeier Arnold
He Miao-fu
Heggie Douglas C.
Helali Yhya E.
Henon Michel C.
Henrard Jacques
Hori Genichiro
Hu Xiaogong
Huang Cheng
Huang Tianyi
Hurley Jarrod R.
Hut Piet
Ipatov Sergei I.
Ismail Mohamed Nader
Ito Takashi
Ivanova Violeta
Janiczek Paul M.
Jefferys William H.
Ji Jianghui
Jiang Ing-Guey
Jovanovic Bozidar
Jupp Alan H.
Kalvouridis Tilemahos
Kammeyer Peter C.
Kholshevnikov Konstantin V.
Kim Sungsoo S.
King-Hele Desmond G.
Kinoshita Hiroshi
Klioner Sergei A.
Klocok Lubomir
Klokocnik Jaroslav
Kokubo Eiichiro
Korchagin Vladimir
Kovacevic Andjelka B.
Kovalevsky Jean
Kozai Yoshihide
Krasinsky George A.
Krivov Alexander V.
Kuznetsov Eduard D.
La Spina Alessandra
Lala Petr
Laskar Jacques
Lazovic Jovan P.
Lecavelier des Etangs Alain
Lega Elena
Levine Stephen
Liao Xinhao
Lieske Jay H.
Lin Douglas N. C.
Lissauer Jack J.
Liu Wenzhong
Lu BenKui
Lucchesi David M.
Lundquist Charles A.
Ma Jingyuan
Makhlouf Amar
Marchal Christian
Marsden Brian G.
Martinet Louis
Matas Vladimir R.
Mavraganis Anastasios G.
Melbourne William G.
Merman Hirsh G. A.
Message Philip J.
Metris Gilles
Michel Patrick
Mignard François
Mikkola Seppo
Mioc Vasile
Moreira Morais Maria Helena
Mulholland John Derral
Musen Peter
Muzzio Juan C.
Myachin Vladimir F.
Nacozy Paul E.
Namouni Fathi
Nobili Anna M.
Novoselov Viktor S.
O'Handley Douglas A.
Ollongren A.
Omarov Tuken B.
Orellana Rosa B.
Orlov Victor V.
Osorio Jose Pereira
Osorio Isabel Maria T. V. P.
Pal Arpad(†)
Parv Bazil
Pauwels Thierry
Petit Jean-Marc
Petrovskaya Margarita S.
Pilat-Lohinger Elke E.
Polyakhova Elena N.
Puetzfeld Dirk
Robinson William J.
Rodin Alexander E.
Rodriguez-Eillamil R.
Roig Fernando V.
Roman Rodica
Rossi Alessandro
Roy Archie E.
Ryabov Yurij A.
Saad Abdel-naby S.
Sansaturio Maria Eugenia
Scheeres Daniel J.
Scholl Hans
Schubart Joachim
Sconzo Pasquale
Segan Stevo
Sehnal Ladislav
Seidelmann P. Kenneth
Sein-Echaluce M. Luisa
Shapiro Irwin I.
Sheng Wan Xiao
Shevchenko Ivan I.
Sima Zdislav
Simo Carles
Simon Jean-Louis
Skripnichenko Vladimir I.
Soffel Michael H.
Sokolov Leonid L.
Sokolov Viktor G.
Sorokin Nikolaj A.
Souchay Jean
Standish E. Myles
Stellmacher Irène
Steves Bonita Alice
Sultanov G. F.
Sun Yisui
Sweatman Winston L.
Szenkovits Ferenc
Tao Jin-he
Tatevyan Suriya K.
Tawadros Maher Jacoub
Taylor Donald Boggia
Thiry Yves R.
Thuillot William
Tremaine Scott Duncan
Tsuchida Masayoshi
Valsecchi Giovanni B.
Valtonen Mauri J.

Varvoglis Harry
Vashkovyak Sofja N.
Vassiliev Nikolaj N.
Veillet Christian
Vieira Martins Roberto
Vieira Neto Ernesto
Vienne Alain
Vilhena Rodolpho
Moraes R.
Vinet Jean-Yves
Vondrak Jan
Walch Jean-Jacques
Walker Ian Walter
Watanabe Noriaki
Whipple Arthur L.
Wiegert Paul A.
Williams Carol A.
Winter Othon Cabo
Wnuk Edwin
Wu Lianda
Wytrzyszczak Iwona
Xia Yi
Yano Taihei
Yarov-Yarovoj Mikhail S.
Yi Zhaohua
Yokoyama Tadashi
Yong Zhou Li
Yoshida Haruo
Yoshida Junzo
Yuasa Manabu
Zafiropoulos Basil
Zare Khalil
Zhang Sheng-Pan
Zhao You
Zhao Changyin
Zhdanov Valery
Zheng Jia-Qing
Zheng Xuetang
Zhou Hongnan
Zhu Wenyao

MEMBERSHIP BY COMMISSION

Composition of Commission 8

Astrometry / Astrométrie

President Kumkova Irina I.

Vice-President Evans Dafydd Wyn

Organizing Committee

Andrei Alexandre H.
Fresneau Alain
Platais Imants
Popescu Petre P.
Scholz Ralf-Dieter
Soma Mitsuru
Zacharias Norbert
Zhu Zi

Members

Abad Hiraldo Carlos
Abhyankar Krishna D.
Ahmed Abdel-aziz Bakry
Andronova Anna A.
Arenou Frederic
Argyle Robert William
Arias Elisa Felicitas
Arlot Jean-Eudes
Assafin Marcelo
Babusiaux Carine
Bacchus Pierre
Backer Donald Ch.
Bakhtigaraev Nail S.
Ballabh Goswami Mohan
Bangert John A.
Barkin Yuri V.
Bastian Ulrich
Belizon Fernando
Benedict George F.
Benevides Soares Paulo
Bernstein Hans-Heinrich
Bien Reinhold
Boboltz David A.
Bougeard Mireille L.
Bradley Arthur J.
Branham Richard L.
Brosche Peter
Brouw Willem N.
Brown Anthony G. A.
Bucciarelli Beatrice
Bunclark Peter Stephen
Capitaine Nicole
Carrasco Guillermo
Chakrabarty Dalia
Chen Li
Cioni Maria-Rosa L.
Cooper Nicholas J.
Corbin Thomas Elbert
Costa Edgardo
Creze Michel
Crifo Francoise
Cudworth Kyle McC.
Dahn Conard Curtis
Danylevsky Vassyl
De Bruijne Jos H. J.
Dejaiffe Rene J.
Delmas Christian
Devyatkin Aleksandr V.
Dick Wolfgang R.
Dick Steven J.
Dinescu-Casetti Dana Ioana
Dommanget Jean
Ducourant Christine
Duma Dmitrij P.
Duncombe Raynor L.
Einicke Ole H.
Emilio Marcelo
Fabricius Claus V.
Fan Yu
Fey Alan Lee
Firneis Maria G.
Fomin Valerij A.
Fomin Valery
Franz Otto G.
Fredrick Laurence W.
Froeschle Michel
Fujishita Mitsumi
Fukushima Toshio
Gatewood George
Gaume Ralph A.
Gauss Stephen F.
Geffert Michael
Germain Marvin E.
Goncharov Georgij A.
Gouda Naoteru
Goyal A. N.
Guibert Jean
Guseva Irina S.
Hajian Arsen R.
Hanson Robert B.
Hartkopf William I.
Helmer Leif
Hemenway Paul D.
Hering Roland
Heudier Jean-Louis
Hill Graham
Hoeg Erik
Hong Zhang
Ianna Philip A.
Ilin Alexey E.
Irwin Michael John
Jackson Paul
Jahreiss Hartmut
Jefferys William H.
Jin WenJing
Johnston Kenneth J.
Jones Burton
Jones Derek H. P.
Jordi Carme

Kanayev Ivan I.
Kaplan George H.
Kazantseva Liliya
Kharchenko Nina
Kharin Arkadiy S.
Kislyuk Vitalij S.
Klemola Arnold R.
Klock Benny L.
Kolesnik Yuri B
Kovalevsky Jean
Kuimov Konstantin V.
Kurzynska Krystyna
Kuzmin Andrej V.
Lattanzi Mario G.
Latypov A. A.
Lazorenko Peter F.
Le Poole Rudolf S.
Lenhardt Helmut
Li Qi
Li Zhigang
Lindegren Lennart
Lopez Carlos
Lopez Jose A.
Lu Phillip K.
Lu Chunlin
Ma Wenzhang
MacConnell Darrell Jack
Makarov Valeri
Marschall Laurence A.
Martin Vera A. F.
Mavridis Lyssimachos N.
McAlister Harold A.
McLean Brian J.
Mignard François
Mink Douglas J.
Mioc Vasile
Miyamoto Masanori
Monet David G.
Morbidelli Roberto
Muinos Haro Jose L.
Murray Andrew C.
Nakajima Koichi
Nefedeva Antonina I.
Nikoloff Ivan
Noel Fernando
Nunez Jorge
Ohnishi Kouji
Oja Tarmo
Olsen Fogh H. J.
Osborn Wayne
Osorio Jose Pereira
Pakvor Ivan
Pannunzio Renato
Pascu Dan
Pauwels Thierry
Penna Jucira L.
Perryman Michael A. C.
Pinigin Gennadij I.
Polozhentsev Dmitrij D.
Poma Angelo
Poppe Paulo C. d. R.
Pourbaix Dimitri
Protsyuk Yuri
Proverbio Edoardo
Rafferty Theodore J.
Raimond Ernst
Reynolds John
Rizvanov Naufal G.
Rodin Alexander E.
Roemer Elizabeth
Roeser Siegfried
Russell Jane L.
Sadzakov Sofija
Sanders Walt L.
Sarasso Maria
Sato Koichi
Schilbach Elena
Schildknecht Thomas
Schmeidler F.
Schwan Heiner
Schwekendiek Peter
Segransan Damien
Seidelmann P. Kenneth
Shelus Peter J.
Shen Kaixian
Shokin Yurij A.
Shulga Oleksandr V.
Smart Richard L.
Soderhjelm Staffan
Solaric Nikola
Sovers Ojars J.
Spoljaric Drago
Standish E. Myles
Stein John William
Steinert Klaus Guenter
Steinmetz Matthias
Suganuma Masahiro
Tang Zheng Hong
Tedds Jonathan A.
Teixeira Ramachrisna
ten Brummelaar Theo A.
Thuillot William
Tsujimoto Takuji
Turon Catherine
Unwin Stephen C.
Upgren Arthur R.
Urban Sean E.
Vallejo Miguel
van Altena William F.
van Leeuwen Floor
Vass Gheorghe
Vilkki Erkki U.
Volyanskaya Margarita Yu
Wallace Patrick T.
Walter Hans G.
Wang Jiaji
Wang Zhengming
Wasserman Lawrence H.
Westerhout Gart
White Graeme Lindsay
Wielen Roland
Xia Yifei
Xu Jiayan
Yamada Yoshiyuki
Yang Tinggao
Yano Taihei
Yasuda Haruo
Yatsenko Anatolij I.
Yatskiv Yaroslav S.
Ye Shuhua
Yoshizawa Masanori

MEMBERSHIP BY COMMISSION

Composition of Commission 10

Solar Activity / Activité Solaire

President Klimchuk James A.

Vice-President van Driel-Gesztelyi Lidia

Secretary Schrijver Karel J.

Organizing Committee

Fletcher Lyndsay
Gopalswamy Nat
Harrison Richard A.
Mandrini Cristina H.
Melrose Donald B.
Peter Hardi
Tsuneta Saku
Vrsnak Bojan
Wang Jingxiu

Members

Abbett William P.
Abdelatif Toufik E
Aboudarham Jean
Abraham Peter
Abramenko Valentina
Ahluwalia Harjit S.
Ai Guoxiang
Akita Kyo
Alissandrakis Costas
Almleaky Yasseen
Altrock Richard C.
Altschuler Martin D.
Altyntsev Alexandre T.
Aly Jean-Jacques
Ambastha Ashok K.
Ambroz Pavel
Anastasiadis Anastasios
Andersen Bo Nyborg
Anderson Kinsey A.
Andretta Vincenzo
Antiochos Spiro K.
Antonucci Ester
Anzer Ulrich
Aschwanden Markus
Atac Tamer
Athay R. Grant
Aurass Henry
Avignon Yvette
Babin Arthur
Bagala Liria G.
Bagare S. P.
Balasubramaniam K. S.
Balikhin Michael
Ballester Jose L.
Bao Shudong
Baranovsky Edward A.
Barrow Colin H.
Barta Miroslav
Basu Sarbani
Batchelor David A.
Beckers Jacques M.
Bedding Timothy R.
Beebe Herbert A.
Bell Barbara
Bellot Rubio Luis Ramon
Belvedere Gaetano
Bemporad Alessandro
Benevolenskaya Elena
Benz Arnold O.
Berger Mitchell
Bergeron Jacqueline A.
Berghmans David
Bergman Per-Goeran
Berrilli Francesco
Bewsher Danielle
Bianda Michele
Bingham Robert
Bobylev Vadim V.
Bocchia Romeo
Bogdan Thomas J.
Bommier Veronique
Bondal Krishna Raj
Bornmann Patricia L.
Botha Gert J. J.
Bothmer Volker
Bougeret Jean-Louis
Boyer Rene
Brajsa Roman
Brandenburg Axel
Brandt Peter N.
Braun Douglas C.
Bray Robert J.
Brekke Pål O. L.
Bromage Barbara J. I.
Brooke John M.
Brosius Jeffrey W.
Brown John C.
Browning Philippa
Browning Philippa K.
Bruner Marilyn E.
Bruns Andrey V.
Brynildsen Nils
Buecher Alain
Buechner Joerg
Bumba Vaclav
Busa Innocenza
Cadez Vladimir
Cally Paul S.
Cane Hilary V.
Carbonell Marc
Cargill Peter J.
Cauzzi Gianna
Chae Jongchul

Chambe Gilbert
Chandra Suresh
Chang Heon-Young
Channok Chanruangrit
Chaoudhuri Arnab R.
Chapman Gary A.
Charbonneau Paul
Chen Peng Fei
Chernov Gennadij P.
Chertok Ilya M.
Chertoprud Vadim E.
Chiuderi-Drago Franca Pr.
Chiueh Tzihong
Cho Kyung Suk
Choudhary Debi Prasad
Chupp Edward L.
Cliver Edward W.
Coffey Helen E.
Collados Manuel
Conway Andrew J.
Cook John W.
Correia Emilia
Costa Joaquim E. R.
Craig Ian
Cramer Neil F.
Crannell Carol Jo
Culhane Leonard
Curdt Werner
Dalla Silvia C.
Dasso Sergio
Datlowe Dayton
Davila Joseph M.
De Groof Anik
de Jager Cornelis
Del Toro Iniesta Jose C.
Demoulin Pascal
Deng YuanYong
Dennis Brian R.
Dere Kenneth P.
Deubner Franz-Ludwig
Dialetis Dimitris
Ding Mingde
Dinulescu Simona
Dobler Wolfgang
Dobrzycka Danuta
Dollfus Audouin
Dorch Søren Bertil F.
Dorotovic Ivan
Dryer Murray
Dubau Jacques
Dubois Marc A.
Duchlev Peter I.
Duldig Marcus L.
Dumitrache Cristiana
Dwivedi Bhola N.
Eddy John A.
Efimenko Volodymyr M.
Elste Gunther H.
Emslie A. Gordon
Engvold Oddbjørn
Enome Shinzo
Erdelyi Robert
Ermolli Ilaria
Falchi Ambretta
Falciani Roberto
Falewicz Robert K.
Fang Cheng
Farnik Frantisek
Ferreira Joao M.
Ferriz Mas Antonio
Fisher George H.
Fluri Dominique M.
Foing Bernard H.
Fontenla Juan M.
Forbes Terry G.
Forgacs-Dajka Emese
Fossat Eric G.
Fu Hsieh-Hai
Gabriel Alan H.
Gaizauskas Victor
Galal A. A.
Galloway David
Galsgaard Klaus
Gan Weiqun
Garaimov Vladimir I.
Garcia de la Rosa Ignacio
Garcia Howard A.(†)
Gary Gilmer Allen
Gelfreikh Georgij B.
Gergely Tomas E.
Ghizaru Mihai
Gibson David M.
Gill Peter B. J.
Gilliland Ronald L.
Gilman Peter A.
Gimenez de Castro Carlos Guillermo
Glatzmaier Gary A.
Gleisner Hans
Godoli Giovanni
Goedbloed Johan P.
Gokhale Moreshwar H.
Gomez Daniel O.
Gontikakis Constantin
Goossens Marcel
Gopasyuk Olga S.
Graffagnino Vito G.
Grandpierre Attila
Gray Norman
Grechnev Victor V.
Grib Sergey A.
Gudiksen Boris V.
Guhathakurta Madhulika
Gupta Surendra S.
Gurman Joseph B.
Gyori Lajos
Hagyard Mona J.
Hammer Reiner
Hanaoka Yoichiro
Hanasz Jan
Hansen Richard T.
Hanslmeier Arnold
Hara Hirohisa
Harra Louise K.
Harvey John W.
Hasan S. Sirajul
Hathaway David H.
Haugan Stein Vidar H.
Hayward John
Heinzel Petr
Henoux Jean-Claude
Herdiwijaya Dhani
Hermans Dirk
Hiei Eijiro
Hildebrandt Joachim
Hildner Ernest
Hochedez Jean-François E.
Hoeksema Jon Todd
Hohenkerk Catherine Y.
Hollweg Joseph V.
Holman Gordon D.
Holzer Thomas E.
Hong Hyon Ik
Hood Alan W.
Houdebine Eric
Howard Robert F.
Hoyng Peter
Hudson Hugh S.
Hughes David W.
Hurford Gordon J.

Ioshpa Boris A.
Ireland Jack
Ishii Takako T.
Ishitsuka Mutsumi
Isliker Heinz
Ivanchuk Victor I.
Ivanov Evgenij V.
Ivchenko Vasily
Jackson Bernard V.
Jain Rajmal
Jakimiec Jerzy
Janssen Katja
Jardine Moira M.
Ji Haisheng
Jiang Yun Chun
Jimenez Mancebo A. J.
Jing Hairong
Jockers Klaus
Jones Harrison P.
Jordan Stuart D.
Joselyn Jo Ann c
Jovanovic Bozidar
Kaburaki Osamu
Kahler Stephen W.
Kallenbach Reinald
Kalman Bela
Kaltman Tatyana I.
Kane Sharad R.
Kang Jin Sok
Karlicky Marian
Karpen Judith T.
Kasparova Jana
Katsova Maria M.
Kaufmann Pierre
Khan Josef I.
Kim Kap-sung
Kim Iraida S.
Kiplinger Alan L.
Kitai Reizaburo
Kitchatinov Leonid
Kjeldseth-Moe Olav
Kleczek Josip
Klein Karl L.
Kliem Bernhard
Klvana Miroslav
Kondrashova Nina N.
Kontar Eduard P.
Kopecky Miloslav(†)
Kostik Roman I.
Kotrc Pavel
Koutchmy Serge
Kovacs Agnes
Kozlovsky Ben Z.
Krimigis Stamatios M.
Krivsky Ladislav(†)
Krucker Sam
Kryshtal Alexander N.
Kryvodubskyj Valery N
Kubota Jun
Kucera Ales
Kundu Mukul R.
Kuperus Max
Kurochka Evgenia Vasilevna
Kurokawa Hiroki
Kusano Kanya
Kuznetsov Vladimir D.
Labrosse Nicolas
Landi Simone
Landman Donald A.
Lang Kenneth R.
Lantos Pierre(†)
Lawrence John K.
Lazrek Mohamed
Leibacher John
Leiko Uliana
Leka Kimberly D.
Leroy Bernard
Leroy Jean-Louis
Li Son Jae
Li Wei
Li Hui
Li Kejun
Lie-Svendsen Oystein
Lima Joao J.
Lin Yong
Liritzis Ioannis
Liu Xinping
Liu Yang
Livshits Mikhail A.
Longbottom Aaron
Low Boon Chye
Lozitskij Vsevolod
Lundstedt Henrik
Luo Xianhan
Machado Marcos
Mackay Duncan H.
MacKinnon Alexander L.
MacQueen Robert M.
Makarov Valentin I.
Makita Mitsugu
Malherbe Jean-Marie
Malitson Harriet H.
Malville J. Mckim
Manabe Seiji
Mann Gottfried
Marilena Mierla
Maris Georgeta
Mariska John T.
Markova Eva
Martens Petrus C.
Mason Glenn M.
Masuda Satoshi
Matsuura Oscar T.
Mattig W.
Maxwell Alan
McAteer R. T. James
McCabe Marie K.
McIntosh Patrick S.
McKenna Lawlor S.
McLean Donald J.
Mein Pierre
Melnik Valentin N.
Mendes Da Costa Aracy
Mendoza-Briceno César A.
Messerotti Mauro
Messmer Peter
Michalek Grzegorz
Miletsky Eugeny V.
Miralles Mari Paz
Mogilevskij Emmanuil I.
Mohan Anita
Moreno Insertis Fernando
Moriyama Fumio
Motta Santo
Muller Richard
Musielak Zdzislaw E.
Nakajima Hiroshi
Nakariakov Valery M.
Namba Osamu
Narain Udit
Neidig Donald F.
Neukirch Thomas
Neupert Werner M.
Nickeler Dieter H.
Nishi Keizo
Nocera Luigi
Noens Jacques-Clair
Noyes Robert W.
Nussbaumer Harry

Obridko Vladimir N.
Oekten Adnan
Ofman Leon
Ohki Kenichiro
Oliver Ramón
Oraevsky Victor N.
Orlando Salvatore
O'Shea Eoghan
Ozguc Atila
Padmanabhan Janardhan
Paletou Frederic
Pallavicini Roberto
Palle Pere Lluís
Palus Pavel
Pan Liande
Pap Judit
Parfinenko Leonid D.
Park Young-Deuk
Parkinson John H.
Parkinson William H.
Parnell Clare E.
Pasachoff Jay M.
Paterno Lucio
Peres Giovanni
Petrosian Vahe
Petrovay Kristof
Pevtsov Alexei A.
Pflug Klaus
Phillips Kenneth J. H.
Pick Monique
Pneuman Gerald W.
Poedts Stefaan
Pohjolainen Silja H.
Poland Arthur I.
Poquerusse Michel
Preka-Papadema Panagiota
Pres Pawek T.
Priest Eric R.
Proctor Michael R. E.
Prokakis Theodore J.
Raadu Michael A.
Ramelli Renzo G. E.
Rao A. Pramesh
Raoult Antoinette
Raulin Jean-Pierre
Rayrole Jean R.
Reale Fabio
Rees David E.
Reeves Hubert
Reeves Edmond M.
Regnier Stephane
Regulo Clara R.
Rendtel Juergen
Rengel Miriam E.
Riehokainen Aleksandr
Rijnbeek Richard Philip
Robinson Jr Richard D.
Roca Cortes Teodoro
Roemer Max
Romanchuk Pavel R.
Romoli Marco
Rompolt Bogdan
Rosa Dragan
Roudier Thierry
Rovira Marta G.
Roxburgh Ian W.
Rozelot Jean-Pierre
Rudawy Pawel
Ruderman Michael S.
Ruediger Guenther
Ruffolo David
Ruiz Cobo Basilio
Rusin Vojtech
Rust David M.
Ruzdjak Vladimir
Ruzickova-Topolova B.
Rybak Jan
Rybansky Milan
Sahal-Brechot Sylvie
Saito Kuniji
Saiz Alejandro
Sakao Taro
Sakurai Kunitomo
Sanchez Almeida J.
Saniga Metod
Sattarov Isroil
Sawyer Constance B.
Schindler Karl
Schlichenmaier Roef
Schmahl Edward J.
Schmelz Joan T.
Schmidt H. U.
Schmieder Brigitte
Schober Hans J.
Schuessler Manfred
Schwartz Pavol
Schwenn Rainer W.
Semel Meir
Shapley Alan H.(†)
Shea Margaret A.
Sheeley Neil R.
Shibasaki Kiyoto
Shimojo Masumi
Shine Richard A.
Sigalotti Leonardo G.
Silva Adriana V. R.
Simnett George M.
Simon Guy
Sinha Krishnanand
Smaldone Luigi A.
Smith Dean F.
Smol'kov Gennadij Ya
Snegirev Sergey D.
Sobotka Michal
Solanki Sami K.
Soliman Mohamed Ahmed
Soloviev Alexandr A.
Somov Boris V.
Spadaro Daniele
Spicer Daniel Shields
Spruit Henk C.
Steiner Oskar
Stellmacher Götz
Stenflo Jan O.
Stepanian Natali N.
Stepanov Aleksander V.
Steshenko N. V.
Stewart Ronald T.
Stix Michael
Stoker Pieter H.
Strong Keith T.
Struminsky Alexei
Sturrock Peter A.
Subramanian Prasad
Subramanian K. R.
Sukartadiredja Darsa
Suzuki Takeru
Svestka Zdenek
Sykora Julius
Sylwester Barbara
Sylwester Janusz A.
Szalay Alex
Takano Toshiaki
Talon Raoul
Tamenaga Tatsuo
Tandberg-Hanssen Einar A.
Tang Yuhua
Tapping Kenneth F.
Ternullo Maurizio
Teske Richard G.

Thomas John H.
Thomas Roger J.
Tifrea Emilia
Tikhomolov Evgeniy
Tlamicha Antonin
Tlatov Andrej G.
Tobias Steven M.
Tomczak Michal
Treumann Rudolf A.
Tripathy Sushanta C.
Tritakis Basil P.
Trottet Gerard
Tsap Yuri
Tsinganos Kanaris
Tuominen Ilkka V.
Uddin Wahab
Underwood James H.
Valnicek Boris
van Allen James A(†)
van den Oord Bert H. J.
Van der Linden Ronald
van Hoven Gerard
van't-Veer Frans
Vaughan Arthur H.
Veck Nicholas
Vekstein Gregory
Velli Marco
Venkatakrishnan P.
Ventura Rita
Vergez Madeleine
Verheest Frank
Verma V. K.
Verwichte Erwin A. O.
Vial Jean-Claude
Vilmer Nicole
Vinod S. Krishan
Voitenko Yuriy M.
Walker Simon N.
Walsh Robert
Wang Yi-ming
Wang Haimin
Wang Min
Wang Huaning
Webb David F.
Wentzel Donat G.
White Stephen Mark
Wiehr Eberhard
Wikstol Oivind
Wild John Paul
Wilson Peter R.
Winebarger Amy R.
Wittmann Axel D.
Woehl Hubertus
Wolfson Richard
Woltjer Lodewijk
Wu De Jin
Wu Shi Tsan
Xu Aoao
Xu Jun
Yan Yihua
Yang Zhiliang
Yang Hong-Jin
Yeh Tyan
Yi Yu
Yoshimura Hirokazu
Yu Dai
Yun Hong-Sik
Zachariadis Theodosios
Zappala Rosario Aldo
Zelenka Antoine
Zhang Mei
Zhitnik Igor A.
Zhou Daoqi
Zhugzhda Yuzef D.
Zhukov Vladimir I.
Zirin Harold
Zlobec Paolo

MEMBERSHIP BY COMMISSION

Composition of Commission 12

Solar Radiation & Structure / Radiation & Structure Solaires

President	Martinez Pillet Valentin
Vice-President	Kosovichev Alexander
Secretary	Mariska John T.

Organizing Committee

Asplund Martin
Bogdan Thomas J.
Cauzzi Gianna
Christensen-Dalsgaard Jørgen
Cram Lawrence Edward
Gan Weiqun
Gizon Laurent
Heinzel Petr
Rovira Marta G.
Venkatakrishnan P.

Members

Abbett William P.
Aboudarham Jean
Acton Loren W.
Ai Guoxiang
Aime Claude
Alissandrakis Costas
Altrock Richard C.
Altschuler Martin D.
Andersen Bo Nyborg
Ando Hiroyasu
Andretta Vincenzo
Ansari S. M. Razaullah
Antia H. M.
Arnaud Jean-Paul
Artzner Guy
Asai Ayumi
Athay R. Grant
Ayres Thomas R.
Babayev Elchin S.
Baliunas Sallie L.
Balthasar Horst
Barta Miroslav
Basu Sarbani
Baturin Vladimir A.
Beckers Jacques M.
Beckman John E.
Beebe Herbert A.
Beiersdorfer Peter
Bemporad Alessandro
Benford Gregory
Bhattacharyya J. C.
Bi Shao Lan
Bianda Michele
Bingham Robert
Blackwell Donald E.
Blamont Jacques-Emile
Bocchia Romeo
Bommier Veronique
Bonnet Roger M.
Book David L.
Bornmann Patricia L.
Borovik Valery N.
Bougeret Jean-Louis
Brandt Peter N.
Brault James W.
Bray Robert J.
Breckinridge James B.
Brosius Jeffrey W.
Bruls Jo H.
Bruner Marilyn E.
Bruning David H.
Bumba Vaclav
Cadez Vladimir
Cavallini Fabio
Ceppatelli Guido
Chambe Gilbert
Chan Kwing Lam
Chapman Gary A.
Chertok Ilya M.
Clark Thomas Alan
Clette Frederic
Collados Manuel
Cook John W.
Cox Arthur N.
Craig Ian
Cramer Neil F.
Dara Helen
Dasso Sergio
de Jager Cornelis
Degenhardt Detlev
Delbouille Luc(†)
Del Toro Iniesta Jose C.
Deliyannis Jean
Demarque Pierre
Deming Leo Drake
Deubner Franz-Ludwig
Di Mauro Maria Pia
Ding Mingde
Diver Declan Andrew
Dogan Nadir
Donea Alina C.
Dravins Dainis
Dumont Simone
Duvall Jr Thomas L.
Ehgamberdiev Shuhrat
Einaudi Giorgio
Elliott Ian
Elste Gunther H.
Epstein Gabriel Leo
Ermolli Ilaria

Esser Ruth
Evans J. V.
Falciani Roberto
Falewicz Robert K.
Fang Cheng
Feldman Uri
Fiala Alan D.
Fisher George H.
Fleck Bernhard
Fluri Dominique M.
Fofi Massimo
Fomichev Valerij V.
Fontenla Juan M.
Forgacs-Dajka Emese
Fossat Eric G.
Foukal Peter V.
Frazier Edward N.
Froehlich Claus
Gabriel Alan H.
Gaizauskas Victor
Garcia Howard A.(†)
Garcia Rafael A.
Garcia-Berro Enrique
Glatzmaier Gary A.
Godoli Giovanni
Goldman Martin V.
Gomez Maria Teresa
Gopalswamy Nat
Grevesse Nicolas
Guhathakurta Madhulika
Hagyard Mona J.
Hamedivafa Hashem
Hammer Reiner
Harvey John W.
Hein Righini Giovanna
Hejna Ladislav
Hiei Eijiro
Hildner Ernest
Hill Frank
Hoang Binh Dy
Hotinli Metin
House Lewis L.
Howard Robert F.
Hoyng Peter
Huang Guangli
Illing Rainer M. E.
Ivanov Evgenij V.
Jabbar Sabeh Rhaman
Jackson Bernard V.
Janssen Katja
Jefferies Stuart M
Jones Harrison P.
Jordan Stuart D.
Jordan Carole
Kalkofen Wolfgang
Kalman Bela
Kaltman Tatyana I.
Karlicky Marian
Karpen Judith T.
Kaufmann Pierre
Keil Stephen L.
Khan Josef I.
Khetsuriani Tsiala S.
Kim Iraida S.
Kim Yong-Cheol
Klein Karl L.
Kneer Franz
Knoelker Michael
Kononovich Edvard V.
Kopecky Miloslav(†)
Kostik Roman I.
Kotov Valery
Kotrc Pavel
Koutchmy Serge
Krivova Natalie A.
Kryvodubskyj Valery N
Kubicela Aleksandar
Kucera Ales
Kundu Mukul R.
Kuperus Max
Labrosse Nicolas
Labs Dietrich
Landi Degli Innocenti
 Egidio
Landman Donald A.
Landolfi Marco
Lantos Pierre(†)
Lanzafame Alessandro C.
Leibacher John
Leroy Jean-Louis
Li Linghuai
Linsky Jeffrey L.
Livingston William C.
Locke Jack L.
Lopez Arroyo M.
Luest Reimar
Makarov Valentin I.
Makita Mitsugu
Mandrini Cristina H.
Marilli Ettore
Marmolino Ciro
Mattig W.
McAteer R. T. James
McKenna Lawlor S.
Mein Pierre
Melrose Donald B.
Mendoza-Briceno César A.
Meyer Friedrich
Michard Raymond
Mihalas Dimitri
Milkey Robert W.
Monteiro Mario
 Joao P. F. G.
Moore Ronald L.
Moreno Insertis Fernando
Moriyama Fumio
Mouradian Zadig M.
Muller Richard
Munro Richard H.
Namba Osamu
Neckel Heinz
Nesis Anastasios
New Roger
Nicolas Kenneth Robert
Nishi Keizo
Nordlund Aake
Noyes Robert W.
O'Shea Eoghan
Ossendrijver Mathieu
Owocki Stanley Peter
Padmanabhan Janardhan
Palle Pere Lluís
Palus Pavel
Papathanasoglou Dimitrios
Parkinson William H.
Pasachoff Jay M.
Pauluhn Anuschka
Pecker Jean-Claude
Petrovay Kristof
Pflug Klaus
Phillips Kenneth J. H.
Poquerusse Michel
Povel Hanspeter
Priest Eric R.
Prokakis Theodore J.
Qu Zhong Quan
Radick Richard R.
Ramelli Renzo G. E.
Raoult Antoinette
Reardon Kevin

Rees David E.
Reeves Edmond M.
Regulo Clara R.
Rengel Miriam E.
Riehokainen Aleksandr
Roca Cortes Teodoro
Roddier Francois
Rodriguez Hidalgo Inés
Roudier Thierry
Rouppe van der Voort
 Luc H. M.
Rusin Vojtech
Rutten Robert J.
Ryabov Boris I.
Rybansky Milan
Sakai Junichi
Samain Denys
Sanchez Almeida J.
Saniga Metod
Sauval A. Jacques
Schleicher Helmold
Schmahl Edward J.
Schmidt Wolfgang
Schmieder Brigitte
Schmitt Dieter
Schober Hans J.
Schou Jesper
Schuessler Manfred
Schwartz Steven Jay
Schwartz Pavol
Schwenn Rainer W.
Seaton Michael J.(†)
Semel Meir
Severino Giuseppe
Shchukina Nataliya
Sheeley Neil R.
Sheminova Valentina A.
Shine Richard A.
Sigalotti Leonardo G.
Sigwarth Michael
Simon Guy
Simon George W.
Singh Jagdev
Sinha Krishnanand
Sivaraman K. R.
Skumanich Andrew
Smith Peter L.
Solanki Sami K.
Soloviev Alexandr A.
Spicer Daniel Shields
Stathopoulou Maria
Staude Juergen
Stebbins Robin
Steffen Matthias
Steiner Oskar
Stenflo Jan O.
Stix Michael
Stodilka Myroslav
Straus Thomas
Suematsu Yoshinori
Svestka Zdenek
Tandberg-Hanssen Einar A.
Teplitskaya Raisa B.
Thomas John H.
Torelli M.
Trujillo Bueno Javier
Tsap Teodor T.
Tsiklauri David
Tsiropoula Georgia
van Hoven Gerard
Vaughan Arthur H.
Ventura Paolo
Vial Jean-Claude
Vilmer Nicole
Voitsekhovska Anna
Volonte Sergio
von der Luehe Oskar
Vukicevic-Karabin
 Mirjana M.
Wang Jingxiu
Warwick James W.
Weiss Nigel O.
Wentzel Donat G.
Wiik Toutain Jun E.
Wilson Peter R.
Wittmann Axel D.
Woehl Hubertus
Worden Simon P.
Wu Hsin-Heng
Yoshimura Hirokazu
Youssef Nahed H.
Yun Hong-Sik
Zampieri Luca
Zarro Dominic M.
Zelenka Antoine
Zhao Junwei
Zhou Daoqi
Zhugzhda Yuzef D.
Zirin Harold
Zirker Jack B.

MEMBERSHIP BY COMMISSION

Composition of Commission 14

Atomic & Molecular Data / Données Atomiques & Moléculaires

President Federman Steven Robert

Vice-President Wahlgren Glenn Michael

Organizing Committee

Dimitrijevic Milan
Johansson Sveneric
Jorissen Alain
Mashonkina Lyudmila I.
Salama Farid
Tennyson Jonathan
van Dishoeck Ewine F.

Members

Adelman Saul J.
Aggarwal Kanti Mal
Allard Nicole
Allen Jr John E.
Allende Prieto Carlos
Arduini-Malinovsky Monique
Artru Marie-Christine
Balanca Christian
Barklem Paul S
Barnbaum Cecilia
Bartaya R. A.
Bautista Manuel A.
Beiersdorfer Peter
Bely-Dubau Francoise
Berrington Keith Adrian
Biemont Emile
Black John Harry
Boechat-Roberty Heloisa M.
Bommier Veronique
Borysow Aleksandra
Branscomb L. M.
Brault James W.
Bromage Gordon E.
Burgess Alan
Carbon Duane F.
Carroll P. Kevin
Chance Kelly V.
Corliss C. H.
Cornille Marguerite
Czyzak Stanley J.
Dalgarno Alexander
Davis Sumner P.
de Frees Douglas J.
Delsemme Armand H.
Desesquelles Jean
d'Hendecourt Louis
Diercksen Geerd H. F.
Dubau Jacques
Dufay Maurice
Dulieu Francois
Eidelsberg Michele
Epstein Gabriel Leo
Feautrier Nicole
Federici Luciana
Fillion Jean-Hugues
Fink Uwe
Flower David R.
Fluri Dominique M.
Fraser Helen J.
Fuhr Jeffrey Robert
Gabriel Alan H.
Gallagher III John S.
Gargaud Muriel
Garstang Roy H.
Garton W. R. S.
Glagolevskij Yurij V.
Glinski Robert J.
Goldbach Claudine
Grant Ian P.
Grevesse Nicolas
Hartman Henrik
Hesser James E.
Hoang Binh Dy
Horacek Jiri
House Lewis L.
Huber Martin C. E.
Huebner Walter F.
Ignjatovic Ljubinko M.
Iliev Ilian
Irwin Alan W.
Irwin Patrick G. J.
Jamar Claude A. J.
Johnson Fred M.
Johnson Donald R.
Joly Francois
Jordan Carole
Jorgensen Henning E.
Jorgensen Uffe Graae
Kanekar Nissim
Kato Takako
Kennedy Eugene T.
Kerber Florian
Kielkopf John F.
Kim Zong Dok
Kingston Arthur E.
Kipper Tonu
Kirby Kate P.
Kohl John L.
Kramida Alexander
Kroto Harold
Kuan Yi-Jehng
Kupka Friedrich
Kurucz Robert L.
Lambert David L.
Landman Donald A.
Lang James
Langhoff Stephanie R.

Launay Françoise
Launay Jean-Michel
Lawrence G. M.
Layzer David
Le Bourlot Jacques
Le Floch André
Leach Sydney
Leger Alain
Lemaire Jean-louis
Linnartz Harold
Loulergue Michelle
Lovas Francis John
Lutz Barry L.
Maillard Jean-Pierre
Martin William C.
Mason Helen E.
McWhirter R. W. Peter
Mickelson Michael E.
Mihajlov Anatolij A.
Morton Donald C.
Mumma Michael Jon
Nahar Sultana N.
Nave Gillian
Newsom Gerald H.
Nicholls Ralph W.
Nilsson Hampus
Nollez Gerard
Nussbaumer Harry
Obi Shinya
O'Brian Thomas R.
Oetken L.
Oka Takeshi
Omont Alain
Orton Glenn S.
Ozeki Hiroyuki
Palmeri Patrick
Parkinson William H.
Peach Gillian
Pei Chunchuan
Petrini Daniel
Petropoulos Basil Ch.
Pettini Marco
Piacentini Ruben
Pradhan Anil K.
Querci Francois R.
Quinet Pascal
Ralchenko Yuri
Ramirez Jose M.
Redman Matthew P.
Rogers Forrest J.
Rostas Francois
Roueff Evelyne M. A.
Ruder Hanns
Rudzikas Zenonas R.
Ryabchikova Tatiana A.
Sahal-Brechot Sylvie
Sarre Peter J.
Savanov Igor S.
Savin Daniel Wolf
Schrijver Johannes
Schultz David R.
Seaton Michael J.(†)
Sharp Christopher
Shore Bruce W.
Sinha Krishnanand
Smith Wm Hayden
Smith Peter L.
Somerville William B.
Song In-Ok
Spielfiedel Annie
Stancil Philip C.
Stark Glen
Stehle Chantal
Strachan Leonard Jr
Strelnitski Vladimir
Summers Hugh P.
Sutherland Ralph S.
Swings Jean-Pierre
Takayanagi Kazuo
Tatum Jeremy B.
Tayal Swaraj S.
Tchang-Brillet Lydia
Thorne Anne P.
Tozzi Gian Paolo
TranMinh Nguyet
Trefftz Eleonore E.
van Rensbergen Walter
Varshalovich Dmitrij A.
Voelk Heinrich J.
Volonte Sergio
Vujnovic Vladis
Wiese Wolfgang L.
Winnewisser Gisbert
Wunner Guenter
Yoshino Kouichi
Young Louise Gray
Yu Yan
Zeippen Claude
Zeng Qin
Zhao Gang
Zirin Harold

MEMBERSHIP BY COMMISSION

Composition of Commission 15

Physical Studies of Comets & Minor Planets

Etude Physique des Comètes & Petites Planètes

President Huebner Walter F.

Vice-President Cellino Alberto

Secretary Boice Daniel Craig

Organizing Committee

Bockelee-Morvan Dominique
Jenniskens Petrus Matheus Marie
Lupishko Dmitrij F.
Ma Yuehua
Reitsema Harold J.
Schulz Rita M.
Tancredi Gonzalo

Members

Agata Hidehiko
A'Hearn Michael F.
Allegre Claude
Altwegg Kathrin
Andrienko Dmitry A.
Angeli Claudia A.
Archinal Brent A.
Arnold James R.
Arpigny Claude
Axford W. Ian
Babadzhanov Pulat B.
Bailey Mark Edward
Barabanov Sergey I.
Barker Edwin S.
Bar-Nun Akiva
Barriot Jean-Pierre
Barucci Maria A.
Bell Jeffrey F.
Belskaya Irina N.
Belton Michael J. S.
Bemporad Alessandro
Bingham Robert
Binzel Richard P.
Birch Peter Vaughan
Birlan Mirel I.
Biver Nicolas
Blamont Jacques-Emile
Blanco Armando
Boehnhardt Hermann
Bonev Tanyu
Borysenko Sergiy
Bouska Jiri
Bowell Edward L. G.
Brandt John C.
Brecher Aviva
Britt Daniel T.
Brown Robert Hamilton
Brownlee Donald E.
Brunk William E.
Buie Marc W.
Buratti Bonnie J.
Burlaga Leonard F.
Burns Joseph A.
Butler Bryan J.
Campins Humberto
Capaccioni Fabrizio
Capria Maria Teresa
Carruthers George R.
Carsenty Uri
Carusi Andrea
Carvano Jorge M. F.
Ceplecha Zdenek
Cerroni Priscilla
Chandrasekhar Thyagarajan
Chapman Clark R.
Chapman Robert D.
Chen Daohan
Clairemidi Jacques
Clayton Geoffrey C.
Clayton Donald D.
Clube S. V. M.
Cochran William David
Cochran Anita L.
Colom Pierre
Combi Michael R.
Connors Martin G.
Consolmagno Guy Joseph
Cosmovici Cristiano Batalli
Cremonese Gabriele
Cristescu Cornelia G.(†)
Crovisier Jacques
Cruikshank Dale P.
Cuypers Jan
Danks Anthony C.
Davidsson Björn J.
Davies John K.
de Almeida Amaury A.
de Pater Imke
de Sanctis Giovanni
De Sanctis M. Cristina
Debehogne Henri Sc.
Delbo Marco
Dell' Oro Aldo
Delsemme Armand H.
Dermott Stanley F.
Deutschman William A.

Di Martino Mario
Donn Bertram D.
Dotto Elisabetta
Dryer Murray
Duncan Martin J.
Durech Josef
Dzhapiashvili Victor P.
Encrenaz Therese
Erard Stéphane
Ershkovich Alexander
Eviatar Aharon
Farnham Tony L.
Feldman Paul Donald
Fernandez Julio Angel
Ferrin Ignacio
Fitzsimmons Alan
Forti Giuseppe
Foryta Dietmar William
Fraser Helen J.
Froeschle Christiane D.
Fujiwara Akira
Fulchignoni Marcello
Furusho Reiko
Galad Adrian
Gammelgaard Peter
Gehrels Tom
Geiss Johannes
Gerakines Perry A.
Gerard Eric
Gibson James
Gil-Hutton Ricardo A.
Giovane Frank
Gradie Jonathan Carey
Grady Monica M.
Green Simon F.
Green Daniel William
 Edward
Greenberg Richard
Gronkowski Piotr M.
Grossman Lawrence
Gruen Eberhard
Grundy William M.
Gustafson Bo A. S.
Hadamcik Edith
Halliday Ian
Hanner Martha S.
Hapke Bruce W.
Harris Alan William
Hartmann William K.
Harwit Martin
Haupt Hermann F.
Helin Eleanor Francis
Hestroffer Daniel
Howell Ellen S.
Hughes David W.
Huntress Wesley T.
Ibadinov Khursand I.
Ip Wing-Huen
Irvine William M.
Irwin Patrick G. J.
Isobe Syuzo(†)
Israelevich Peter
Ivanova Violeta
Ivanova Aleksandra V.
Ivezic Zeljko
Jackson William M.
Jedicke Robert
Jockers Klaus
Johnson Torrence V.
Jorda Laurent
Kaasalainen Mikko K.
Kaeufl Hans Ulrich
Kavelaars JJ. Matthew
Kawakita Hideyo
Keay Colin S. l.
Keil Klaus
Keller Horst U.
Kidger Mark R.
Kim Bong Gyu
Kiselev Nikolai N.
Klacka Jozef
Kliem Bernhard
Knacke Roger F.
Knezevic Zoran
Koeberl Christian
Kohoutek Lubos
Korsun Pavlo P.
Kowal Charles Thomas
Kozasa Takashi
Krimigis Stamatios M.
Krishna Swamy K. S.
Kristensen Leif Kahl
Krugly Yurij N.
Kryszczynska Agnieszka
Kuan Yi-Jehng
La Spina Alessandra
Lagage Pierre-Olivier
Lagerkvist Claes-Ingvar
Lamy Philippe
Lancaster Brown Peter
Lane Arthur Lonne
Larson Stephen M.
Larson Harold P.
Lazzarin Monica
Lebofsky Larry Allen
Lee Thyphoon
Levasseur-Regourd
 Anny-Chantal
Liller William
Lillie Charles F.
Lindsey Charles Allan
Lipschutz Michael E.
Lissauer Jack J.
Lisse Carey M.
Lodders Katharina
Lopes-Gautier Rosaly
Lukyanyk Igor V.
Lumme Kari A.
Lutz Barry L.
Luu Jane X.
Lyon Ian C.
Magee-Sauer Karen P.
Magnusson Per
Makalkin Andrei B.
Maran Stephen P.
Marcialis Robert
Maris Michele
Marsden Brian G.
Marzari Francesco
Matson Dennis L.
Matsuura Oscar T.
McCord Thomas B.
McCrosky Richard E.
McDonnell J. A. M.
McFadden Lucy Ann
McKenna Lawlor S.
Meech Karen J.
Meisel David D.
Mendis Devamitta Asoka
Merline William J.
Michalowski Tadeusz
Milani Andrea
Millis Robert L.
Moehlmann Diedrich
Moore Elliott P.
Morrison David
Mothe-Diniz Thais
Mueller Thomas G.
Muinonen Karri
Mukai Tadashi

Mumma Michael Jon
Nakamura Tsuko
Nakamura Akiko M.
Napier William M.
Neukum G.
Newburn Jr Ray L.
Niedner Malcolm B.
Ninkov Zoran
Nolan Michael C.
Noll Keith Stephen
O'Dell Charles R.
Paolicchi Paolo
Parisot Jean-Paul
Pellas Paul
Pendleton Yvonne Jean
Perez de Tejada Hector A.
Piironen Jukka O.
Pilcher Carl Bernard
Pillinger Colin
Pittich Eduard M.
Pittichova Jana
Prialnik-Kovetz Dina
Proisy Paul E.
Revelle Douglas Orson
Rickman Hans
Roemer Elizabeth
Rossi Alessandro
Rousselot Philippe
Russel Sara S.
Russell Kenneth S.
Sagdeev Roald Z.
Saito Takao
Salitis Antonijs
Samarasinha Nalin H.
Schleicher David G.
Schloerb F. Peter
Schmidt Maarten
Schmidt H. U.
Schober Hans J.
Scholl Hans
Sekanina Zdenek
Sekiguchi Tomohiko
Serra Ricart Miquel
Shanklin Jonathan D.
Sharma A. Surjalal
Sharp Christopher
Shevchenko Vasilij
Shkodrov Vladimir G.
Shor Viktor A.
Shulman Leonid M.
Sims Mark R.
Sivaraman K. R.
Sizonenko Yuri V.
Smith Bradford A.
Snyder Lewis E.
Solc Martin
Spinrad Hyron
Steel Duncan I.
Stern S. Alan
Subhon Ibadov
Surdej Jean M. G.
Svoren Jan
Swade Daryl Allen
Sykes Mark Vincent
Szego Karoly
Szutowicz Slawomira E.
Tacconi-Garman Lowell E.
Takeda Hidenori
Tanabe Hiroyoshi
Tanga Paolo
Tao Jun
Tatum Jeremy B.
Tedesco Edward F.
Terentjeva Aleksandra K.
Tholen David J.
Thomas Nicolas
Tomita Koichiro
Toth Imre
Tozzi Gian Paolo
Valdes-Sada Pedro A.
Van Flandern Tom
Veeder Glenn J.
Veverka Joseph
Vilas Faith
Walker Alistair Robin
Wallis Max K.
Wasson John T.
Watanabe Jun-ichi
Wdowiak Thomas J.
Weaver Harold F.
Wehinger Peter A.
Weidenschilling S. J.
Weissman Paul Robert
Wells Eddie Neil
West Richard M.
Wetherhill George W.(†)
Wilkening Laurel L.
Williams Iwan P.
Wood John A.
Wooden Diane H.
Woolfson Michael M.
Woszczyk Andrzej
Wyckoff Susan
Yabushita Shin A.
Yanagisawa Masahisa
Yang Jongmann
Yeomans Donald K.
Yi Yu
Yoshida Fumi
Zappala Vincenzo
Zarnecki John Charles
Zellner Benjamin H.
Zhu Jin

MEMBERSHIP BY COMMISSION

Composition of Commission 16

Physical Study of Planets & Satellites

Etude Physique des Planètes & Satellites

President Courtin Régis
Vice-President McGrath Melissa Ann
Secretary Lara Luisa M.

Organizing Committee

Blanco Carlo
Consolmagno Guy Joseph
Ksanfomality Leonid V.
Morrison David
Spencer John R.
Tejfel Viktor G.

Members

Akimov Leonid
Alexandrov Yuri V.
Appleby John F.
Archinal Brent A.
Arthur David W. G.
Atkinson David H.
Atreya Sushil K.
Balikhin Michael
Barkin Yuri V.
Barrow Colin H.
Batson Raymond Milner
Battaner Eduardo
Baum William A.
Beebe Reta Faye
Beer Reinhard
Bell III James F.
Belton Michael J. S.
Bender Peter L.
Ben-Jaffel Lofti
Berge Glenn L.
Bergstralh Jay T.
Bertaux Jean-Loup
Beurle Kevin
Bezard Bruno G.
Billebaud Francoise
Binzel Richard P.
Blamont Jacques-Emile
Blanco Armando
Bondarenko Lyudmila N.
Bosma Pieter B.
Boss Alan P.
Boyce Peter B.
Brahic André
Brecher Aviva
Broadfoot A. Lyle
Brown Robert Hamilton
Brunk William E.
Buie Marc W.
Buratti Bonnie J.
Burba George A.
Burns Joseph A.
Calame Odile
Caldwell John James
Cameron Winifred S.
Campbell Donald B.
Capria Maria Teresa
Carsmaru Maria M.
Catalano Santo
Chapman Clark R.
Chen Daohan
Chevrel Serge
Clairemidi Jacques
Cochran Anita L.
Combi Michael R.
Connes Janine
Coradini Angioletta
Counselman Charles C.
Cruikshank Dale P.
Davies Ashley Gerard
Davis Gary R.
de Bergh Catherine
de Pater Imke
Dermott Stanley F.
Dickel John R.
Dickey Jean O'Brien
Dlugach Zhanna M.
Dollfus Audouin
Drake Frank D.
Drossart Pierre
Dunkin Sarah K.
Durrance Samuel T.
Dzhapiashvili Victor P.
El Baz Farouk
Elliot James L.
Elston Wolfgang E.
Encrenaz Therese
Epishev Vitali P.
Eshleman Von R.
Esposito Larry W.
Evans Michael W.
Ferrari Cecile
Fielder Gilbert
Fink Uwe
Fox W. E.
Fox Kenneth
Fujiwara Akira
Gautier Daniel
Gehrels Tom
Geiss Johannes
Gerard Jean-Claude M. C.
Giclas Henry L.
Gierasch Peter J.
Goldreich Peter

Goldstein Richard M.
Goody R. M.
Gorenstein Paul
Gor'kavyi Nikolai
Goudas Constantine L.
Grav Tommy
Green Jack
Grieger Bjoern
Grossman Lawrence
Guest John E.
Gulkis Samuel
Gurshtein Alexander A.
Hagfors Tor(†)
Halliday Ian
Hammel Heidi B.
Hanninen Jyrki
Harris Alan William
Harris Alan William
Hasegawa Ichiro
Hide Raymond
Holberg Jay B.
Horedt Georg Paul
Hovenier J. W.
Hubbard William B.
Hunt G. E.
Hunten Donald M.
Irvine William M.
Irwin Patrick G. J.
Iwasaki Kyosuke
Johnson Torrence V.
Jurgens Raymond F.
Kaeufl Hans Ulrich
Kascheev Rafael A.
Kiladze R. I.
Killen Rosemary M.
Kim Yongha
Kislyuk Vitalij S.
Kley Wilhelm
Kowal Charles Thomas
Krimigis Stamatios M.
Kumar Shiv S.
Kurt Vladimir G.
Kuzmin Arkadij D.
Lane Arthur Lonne
Larson Stephen M.
Larson Harold P.
Lemmon Mark
Lewis J. S.
Lineweaver Charles H.
Lissauer Jack J.
Lockwood G. Wesley
Lodders Katharina
Lopes-Gautier Rosaly
Lopez Moreno Jose Juan
Lopez Puertas Manuel
Lopez Valverde M. A.
Lumme Kari A.
Lutz Barry L.
Luz David
Lyon Ian C.
Mahra H. S.
Makalkin Andrei B.
Marcialis Robert
Margot Jean-Luc
Marov Mikhail Ya
Marzo Giuseppe A.
Matson Dennis L.
Matsui Takafumi
McCord Thomas B.
McCullough Peter R.
McElroy M. B.
McKinnon William Beall
Meadows A. Jack
Mickelson Michael E.
Mikhail Joseph Sidky
Millis Robert L.
Mills Franklin P.
Moehlmann Diedrich
Molina Antonio
Montmessin Franck
Moore Patrick
Moreno Fernando
Morozhenko A. V.
Mosser Benoît
Mulholland John Derral
Mumma Michael Jon
Murphy Robert E.
Nakagawa Yoshitsugu
Nelson Richard P.
Ness Norman F.
Neukum G.
Noll Keith Stephen
Ohtsuki Keiji
Owen Tobias C.
Pang Kevin
Paolicchi Paolo
Petit Jean-Marc
Petropoulos Basil Ch.
Pettengill Gordon H.
Pillinger Colin
Pokorny Zdenek
Potter Andrew E.
Predeanu Irina
Psaryov Volodymyr A.
Rao M. N.
Rodionova Zhanna F.
Rodrigo Rafael
Roos-Serote Maarten C.
Roques Françoise
Rossi Alessandro
Ruskol Evgeniya L.
Saissac Joseph
Sanchez-Lavega Agustin
Schleicher David G.
Schloerb F. Peter
Schneider Nicholas M.
Shapiro Irwin I.
Shevchenko Vladislav V.
Shkuratov Yurii
Sicardy Bruno
Sims Mark R.
Sjogren William L.
Smith Bradford A.
Snellen Ignas A. G.
Soderblom Larry
Sonett Charles P.
Sprague Ann Louise
Stern S. Alan
Stoev Alexey D.
Stone Edward C.
Strobel Darrell F.
Strom Robert G.
Synnott Stephen P.
Tanga Paolo
Taylor Fredric W.
Tchouikova Nadezhda A.
Tedds Jonathan A.
Terrile Richard John
Tholen David J.
Thomas Nicolas
Trafton Laurence M.
Tran-Minh Francoise
Tyler Jr G. Leonard
van Allen James A(†)
Van Flandern Tom
Veiga Carlos Henrique
Veverka Joseph
Vidmachenko Anatoliy P.
Walker Simon N.
Walker Alta Sharon

MEMBERSHIP BY COMMISSION

Composition of Commission 19

Rotation of the Earth / Rotation de la Terre

President Brzezinski Aleksander

Vice-President Ma Chopo

Organizing Committee

Charlot Patrick
Defraigne Pascale
Dehant Véronique
Dickey Jean O'Brien
Souchay Jean
Vondrak Jan

Members

Arabelos Dimitrios
Archinal Brent A.
Arias Elisa Felicitas
Bang Yong Gol
Banni Aldo
Barkin Yuri V.
Barlier Francois E.
Beutler Gerhard
Bizouard Christian
Bolotin Sergei
Bolotina Olga
Boucher Claude
Bougeard Mireille L.
Boytel Jorge del Pino
Brosche Peter
Capitaine Nicole
Cazenave Anny
Chao Benjamin F.
De Biasi Maria S.
De Viron Olivier
Debarbat Suzanne V.
Dejaiffe Rene J.
Deleflie Florent
Dick Wolfgang R.
Dickman Steven R.
Djurovic Dragutin M.
El Shahawy Mohamad
Eppelbaum Lev V.
Fernandez Laura I.
Ferrandiz Jose Manuel
Fliegel Henry F.
Folgueira Marta
Fong Chugang
Fujishita Mitsumi
Fukushima Toshio
Gambis Daniel
Gao Buxi
Gaposchkin Edward M.
Gayazov Iskander S.
Gontier Anne-Marie
Gozhy Adam
Gross Richard Sewart
Groten Erwin
Guinot Bernard R.
Han Tianqi
Han Yanben
Hefty Jan
Huang Cheng
Huang Cheng-Li
Hugentobler Urs
Iijima Shigetaka
Jin WenJing
Johnson Thomas James
Kakuta Chuichi
Kameya Osamu
Khoda Oleg
Klepczynski William J.
Knowles Stephen H.
Kolaczek Barbara
Korsun Alla
Kosek Wieslaw
Kostelecky Jan
Kouba Jan
Lehmann Marek
Li Jinling
Liao Dechun
Lieske Jay H.
Liu Ciyuan
Luzum Brian J.
Malkin Zinovy M.
Manabe Seiji
McCarthy Dennis D.
Meinig Manfred
Melbourne William G.
Merriam James B.
Monet Alice K. B.
Morgan Peter
Morrison Leslie V.
Mueller Ivan I.
Nastula Jolanta
Naumov Vitalij A.
Newhall X. X.
Niemi Aimo
Nothnagel Axel
Paquet Paul
Park Pil-Ho P.
Pejovic Nadezda R.
Pesek Ivan
Petit Gérard
Petrov Sergei D.
Picca Domenico
Pilkington John D. H.
Poma Angelo
Popelar Josef
Proverbio Edoardo
Ray James R.
Richter Bernd
Robertson Douglas S.
Ron Cyril
Roosbeek Fabian
Rothacher Markus
Ruder Hanns

Rusu I.
Rykhlova Lidiya V.
Sadzakov Sofija
Salstein David A.
Sasao Tetsuo
Sato Koichi
Schillak Stanislaw
Schuh Harald
Schutz Bob Ewald
Sekiguchi Naosuke
Sevilla Miguel J.
Shapiro Irwin I.
Shelus Peter J.
Sidorenkov Nikolaj S.
Soffel Michael H.
Stanila George
Stephenson F. Richard
Sugawa Chikara
Tapley Byron D.
Tarady Vladimir K.
Titov Oleg A.
Tsao Mo
Veillet Christian
Vicente Raimundo O.
Wallace Patrick T.
Wang Zhengming
Wang Kemin
Weber Robert
Williams James G.
Wilson P.
Wooden William Hugh
Wu Bin
Wu Shouxian
Wuensch Johann Jakob
Xiao Naiyuan
Xu Jiayan
Yang Fumin
Yatskiv Yaroslav S.
Ye Shuhua
Yokoyama Koichi
Yu Nanhua
Zhang Zhongping
Zharov Vladimir E.
Zhong Min
Zhou Yonghong
Zhu Yaozhong

MEMBERSHIP BY COMMISSION

Composition of Commission 20

Positions & Motions of Minor Planets, Comets & Satellites

Positions & Mouvements des Petites Planètes, Comètes & Satellites

President Fernandez Julio Angel

Vice-President Yoshikawa Makoto

Secretary Chesley Steven R.

Organizing Committee

Chernetenko Yulia A.
Gilmore Alan C.
Lazzaro Daniela
Muinonen Karri
Pravec Petr
Spahr Timothy B.
Tholen David J.
Ticha Jana
Valsecchi Giovanni B.
Zhu Jin

Members

Abalakin Viktor K.
A'Hearn Michael F.
Aikman G. Chris L.
Aksnes Kaare
Arlot Jean-Eudes
Babadzhanov Pulat B.
Baggaley William J.
Bailey Mark Edward
Batrakov Yurij V.
Behrend Raoul
Benest Daniel
Berthier Jerôme
Bien Reinhold
Blanco Carlo
Blow Graham L.
Boerngen Freimut
Bowell Edward L. G.
Branham Richard L.
Burns Joseph A.
Calame Odile
Carpino Mario
Carusi Andrea
Chapront-Touze Michelle
Chio Chol Zong
Chodas Paul Winchester
Cooper Nicholas J.
Cristescu Cornelia G.(†)
de Sanctis Giovanni
Debehogne Henri Sc.
Delbo Marco
Delsemme Armand H.
Di Sisto Romina P.
Dollfus Audouin
Donnison John Richard
Dourneau Gerard
Doval Jorge M. Pérez
Dunham David W.
Dvorak Rudolf
Dybczynski Piotr A.
Edmondson Frank K.
Elliot James L.
Elst Eric Walter
Emelianov Nikolaj V.
Epishev Vitali P.
Evans Michael W.
Ferraz Mello Sylvio
Ferreri Walter
Forti Giuseppe
Franklin Fred A.
Fraser Brian D.
Freitas Mourao R.
Froeschle Claude
Fuse Tetsuharu
Gaizauskas Victor
Gehrels Tom
Gibson James
Giclas Henry L.
Green Daniel William Edward
Greenberg Richard
Hahn Gerhard J.
Harper David
Harris Alan William
Hasegawa Ichiro
Haupt Hermann F.
He Miao-fu
Helin Eleanor Francis
Hemenway Paul D.
Henrard Jacques
Hers Jan
Heudier Jean-Louis
hol Pedro E.
Hudkova Ludmila
Hurnik Hieronim
Hurukawa Kiitiro
Ianna Philip A.
Isobe Syuzo(†)
Ivanova Violeta
Jacobson Robert A.
Kazantsev Anatolii M.
Khatisashvili Alfez Sh.
Kiang Tao
Kilmartin Pamela
Kinoshita Hiroshi
Kisseleva Tamara P.
Klemola Arnold R.

Knezevic Zoran
Kohoutek Lubos
Kosai Hiroki
Kowal Charles Thomas
Kozai Yoshihide
Krasinsky George A.
Kristensen Leif Kahl
Krolikowska-Soltan Malgorzata
Krugly Yurij N.
Kulikova Nelli V.
Lagerkvist Claes-Ingvar
Larsen Jeffrey A.
Lemaitre Anne
Li Guangyu
Lieske Jay H.
Lomb Nicholas Ralph
Lovas Miklos
Mahra H. S.
Manara Alessandro A.
Marsden Brian G.
Matese John J.
Maury Alain J.
McCrosky Richard E.
McMillan Robert S.
McNaught Robert H.
Medvedev Yurij D.
Melita Mario Daniel
Message Philip J.
Milani Andrea
Millis Robert L.
Mintz Blanco Betty
Monet Alice K. B.
Moravec Zdenek
Morris Charles S.
Mulholland John Derral
Murray Carl D.
Nacozy Paul E.
Nakamura Tsuko
Nakano Syuichi
Neslusan Lubos
Nobili Anna M.
Owen Jr William Mann
Pandey A. K.
Pascu Dan
Pauwels Thierry
Pierce David Allen
Pittich Eduard M.
Polyakhova Elena N.
Porubcan Vladimir
Pozhalova Zhanna
Qiao Rongchuan
Rajamohan R.
Raju Vasundhara
Rapaport Michel
Reitsema Harold J.
Rickman Hans
Roemer Elizabeth
Roeser Siegfried
Rossi Alessandro
Rui Qi
Russell Kenneth S.
Sato Isao
Schmadel Lutz D.
Schober Hans J.
Scholl Hans
Schubart Joachim
Schuster William John
Seidelmann P. Kenneth
Sekanina Zdenek
Shanklin Jonathan D.
Shelus Peter J.
Shen Kaixian
Shkodrov Vladimir G.
Shor Viktor A.
Sitarski Grzegorz
Solovaya Nina A.
Soma Mitsuru
Standish E. Myles
Steel Duncan I.
Stellmacher Irène
Stokes Grant H.
Sultanov G. F.
Svoren Jan
Synnott Stephen P.
Szutowicz Slawomira E.
Tancredi Gonzalo
Tatum Jeremy B.
Taylor Donald Boggia
Thuillot William
Tomita Koichiro
Torkelsson Ulf J.
Torres Carlos
Tsuchida Masayoshi
Tuccari Gino
Van Flandern Tom
van Houten-Groeneveld Ingrid
Veillet Christian
Vieira Martins Roberto
Vienne Alain
Wasserman Lawrence H.
Weissman Paul Robert
West Richard M.
Whipple Arthur L.
Wild Paul
Williams Iwan P.
Williams James G.
Williams Gareth V.
Yabushita Shin A.
Yeomans Donald K.
Yim Hong-Suh
Yuasa Manabu
Zagretdinov Renat V.
Zappala Vincenzo
Zhang Jiaxiang
Zhang Qiang
Ziolkowski Krzysztof

MEMBERSHIP BY COMMISSION

Composition of Commission 21

Light of the Night Sky / Lumière du Ciel Nocturne

President Witt Adolf N.

Vice-President Murthy Jayant

Organizing Committee

Baggaley William J.
Dwek Eli
Gustafson Bo A. S.
Levasseur-Regourd Anny-Chantal
Mann Ingrid
Mattila Kalevi
Watanabe Jun-ichi

Members

Angione Ronald J.
Belkovich Oleg I.
Blamont Jacques-Emile
Bowyer C. Stuart
Broadfoot A. Lyle
Clairemidi Jacques
Dermott Stanley F.
d'Hendecourt Louis
Dodonov Sergej N.
Dubin Maurice
Dufay Maurice
Dumont Rene
Feldman Paul Donald
Fujiwara Akira
Gadsden Michael
Giovane Frank
Gruen Eberhard
Hanner Martha S.
Harwit Martin
Hauser Michael G.
Hecht James H.
Henry Richard Conn
Hofmann Wilfried
Hong Seung Soo
Hurwitz Mark V.
Ivanov-Kholodny Gor S.
Jackson Bernard V.
James John F.
Joubert Martine
Karygina Zoya V.
Kopylov Aleksandr I.
Koutchmy Serge
Kramer Busaba H.
Kulkarni Prabhakar V.
Lamy Philippe
Leger Alain
Leinert Christoph
Lemke Dietrich
Lillie Charles F.
Lopez Gonzalez Maria J.
Lopez Moreno Jose Juan
Lopez Puertas Manuel
Lumme Kari A.
Maihara Toshinori
Martin Donn Christopher
Mather John Cromwell
Matsumoto Toshio
Maucherat J.
McDonnell J. A. M.
Mikhail Joseph Sidky
Misconi Nebil Yousif
Morgan David H.
Muinonen Karri
Mukai Tadashi
Nakamura Akiko M.
Nawar Samir
Nishimura Tetsuo
Paresce Francesco
Perrin Jean-Marie
Pfleiderer Jorg
Reach William
Renard Jean-Baptiste
Robley Robert
Rodrigo Rafael
Rozhkovskij Dimitrij A.
Sanchez Francisco M.
Sanchez-Saavedra M. Luisa
Saxena P. P.
Schlosser Wolfhard
Schuh Harald
Schwehm Gerhard
Shefov Nikolaj N.
Soberman Robert K.
Sparrow James G.
Staude Hans Jakob
Sykes Mark Vincent
Tanabe Hiroyoshi
Toller Gary N.
Toroshlidze Teimuraz I.
Tyson John Anthony
Ueno Munetaka
Vrtilek Jan M.
Wallis Max K.
Weinberg Jerry L.
Wesson Paul S.
Wheatley Peter J.
Wilson P.
Wolstencroft Ramon D.
Woolfson Michael M.
Yamamoto Tetsuo
Yamashita Kojun
Zerull Reiner H.

MEMBERSHIP BY COMMISSION

Composition of Commission 22

Meteors, Meteorites & Interplanetary Dust

Météores, Météorites & Poussière Interplanétaire

President Spurny Pavel

Vice-President Watanabe Jun-ichi

Secretary Borovicka Jiri

Organizing Committee

Baggaley William J.
Brown Peter Gordon
Consolmagno Guy Joseph
Jenniskens Petrus Matheus Marie
Pellinen-Wannberg Asta K.
Porubcan Vladimir
Williams Iwan P.
Yano Hajime

Members

Alexandrov Alexander N.
Asher David J.
Babadzhanov Pulat B.
Belkovich Oleg I.
Bhandari N.
Brownlee Donald E.
Campbell-Brown Margaret D.
Carusi Andrea
Ceplecha Zdenek
Cevolani Giordano
Clifton Kenneth St.
Clube S. V. M.
Cooper Timothy
Djorgovski Stanislav
Dubin Maurice
Elford William Graham
Forti Giuseppe
Glass Billy Price
Gorbanev Jury
Goswami J. N.
Grady Monica M.
Gruen Eberhard
Gustafson Bo A. S.
Hajdukova Maria
Halliday Ian
Hanner Martha S.
Harvey Gale A.
Hasegawa Ichiro
Hawkes Robert Lewis
Helin Eleanor Francis
Hey James Stanley
Hodge Paul W.
Hong Seung Soo
Hughes David W.
Jennison Roger C.(†)
Jones James
Jopek Tadeusz Jan
Kalenichenko Valentin
Kapisinsky Igor
Keay Colin S. l.
Koeberl Christian
Kokhirova Gulchehra I.
Koten Pavel
Kramer Kh N.
Kruchinenko Vitaliy G.
Lamy Philippe
Lemaire Joseph F.
Levasseur-Regourd Anny-Chantal
Lindblad Bertil A.(†)
Lodders Katharina
Lovell Sir Bernard
Lyon Ian C.
Makalkin Andrei B.
Mann Ingrid
Maris Michele
Marvin Ursula B.
Mason John William
McCrosky Richard E.
McDonnell J. A. M.
McIntosh Bruce A.
Meisel David D.
Miles Howard G.
Misconi Nebil Yousif
Murray Andrew C.
Murray Carl D.
Nakamura Takuji
Nakazawa Kiyoshi
Napier William M.
Newburn Jr Ray L.
Nuth Joseph A. III
Pecina Petr
Pillinger Colin
Plavec Zdenka
Poole Graham
Rendtel Juergen
Revelle Douglas Orson
Rickman Hans
Ripken Hartmut W.
Ryabova Galina O.
Sekanina Zdenek
Shao Cheng-yuan
Simek Milos
Soberman Robert K.
Steel Duncan I.
Svestka Jiri
Svoren Jan
Tatum Jeremy B.
Taylor Andrew

Tedesco Edward F.
Terentjeva Aleksandra K.
Tomita Koichiro
Trigo-Rodriguez Josep M.
Valsecchi Giovanni B.
Vaubaillon Jérémie J.
Voloschuk Yuri I.
Webster Alan R.
Weinberg Jerry L.
Wetherhill George W.(†)
Wood John A.
Woolfson Michael M.
Yamamoto Masayuki
Yeomans Donald K.
Zhu Jin

MEMBERSHIP BY COMMISSION

Composition of Commission 25

Stellar Photometry & Polarimetry

Photométrie & Polarimétrie Stellaire

President Martinez Peter

Vice-President Milone Eugene F.

Organizing Committee

Jordi Carme
Landolt Arlo U.
Mironov Aleksey V.
Qian Shengbang
Schmidt Edward G.
Sterken Christiaan L.

Members

Ables Harold D.
Adelman Saul J.
Ahumada Javier Alejandro
Albrecht Rudolf
Anandaram Mandayam N.
Andreuzzi Gloria
Angel J. Roger P.
Angione Ronald J.
Anthony-Twarog Barbara J.
Arnaud Jean-Paul
Arsenijevic Jelisaveta
Ashok N. M.
Aspin Colin
Axon David
Baldinelli Luigi
Baliyan Kiran S.
Balona Luis A.
Barnes III Thomas G.
Barrett Paul Everett
Bastien Pierre
Baume Gustavo L.
Behr Alfred
Bellazzini Michele
Berdyugin Andrei V.
Bessell Michael S.
Bjorkman Jon E.
Blanco Victor M.
Blecha Andre Boris G.
Bookmyer Beverly B.
Borgman Jan
Borra Ermanno F.
Breger Michel
Brown Douglas N.
Brown Thomas M.
Buser Roland
Carney Bruce William
Carter Brian
Castelaz Micheal W.
Celis Leopoldo
Chen Wen Ping
Cioni Maria-Rosa L.
Clem James L.
Connolly Leo P.
Coyne George V.
Cramer Noel
Crawford David L.
Cuypers Jan
Dachs Joachim
Dahn Conard Curtis
Danford Stephen C.
Deshpande M. R.
Dolan Joseph F.
Dubout Renee
Ducati Jorge R.
Ducourant Christine
Dzervitis Uldis
Edwards Paul J.
Efimov Yuri
Elkin Vladimir
Fabregat Juan
Fabrika Sergei Nikolaevich
Feinstein Alejandro
Fernie J. Donald
Fluri Dominique M.
Forte Juan C.
Freyhammer Lars Michael
Galadi Enriquez David
Gallouet Louis
Gehrz Robert Douglas
Genet Russel M.
Gerbaldi Michèle
Ghosh S. K.
Gilliland Ronald L.
Glass Ian S.
Golay Marcel
Goy Gerald
Graham John A.
Grauer Albert D.
Grenon Michel
Grewing Michael
Grundahl Frank
Guetter Harry Hendrik
Gutierrez-Moreno A.
Hall Douglas S.
Hauck Bernard
Hayes Donald S.
Heck Andre
Hensberge Herman
Hilditch Ronald W.
Hubrig Swetlana
Huovelin Juhani
Hyland Harry R. Harry
Irwin Alan W.
Ivezic Zeljko
Jerzykiewicz Mikolaj

Joshi Umesh C.
Kawara Kimiaki
Kazlauskas Algirdas S.
Kebede Legesse W.
Keller Stefan C.
Kepler S. O.
Kilkenny David
Kim Seung-Lee
King Ivan R.
Knude Jens Kirkeskov
Koch Robert H.
Kornilov Viktor G.
Kulkarni Prabhakar V.
Kunkel William E.
Kurtz Donald W.
Labhardt Lukas
Landstreet John D.
Laskarides Paul G.
Lazauskaite Romualda
Lemke Michael
Lenzen Rainer
Leroy Jean-Louis
Li Sin Hyong
Li Li Qingkang
Linde Peter
Lockwood G. Wesley
Lub Jan
Luna Homero G.
Maitzen Hans M.
Manfroid Jean
Manset Nadine C.
Markkanen Tapio
Marraco Hugo G.
Marsden Stephen C.
Martinez Roger Carlos
Masani A.
Maslennikov Kirill L.
Mathys Gautier
Mayer Pavel
McDavid David A.
McLean Ian S.
Mendoza V. Eugenio E.
Menzies John W.
Metcalfe Travis S.
Mianes Pierre
Miller Joseph S.
Mintz Blanco Betty
Moffett Thomas J.
Moitinho André
Mourard Denis
Mumford George S.
Munari Ulisse
Naylor Tim
Neiner Coralie
Nicolet Bernard
Noguchi Kunio
Notni Peter
Oblak Edouard
Oestreicher Roland
Orsatti Ana María
Page Arthur
Pedreros Mario
Pel Jan Willem
Penny Alan John
Petit Pascal
Pfeiffer Raymond J.
Philip A. G. Davis
Piirola Vilppu E.
Platais Imants
Pokrzywka Bartlomiej
Rao Pasagada Vivekananda
Raveendran A. V.
Reglero-Velasco Victor
Robb Russell M.
Robinson Edward L.
Romanyuk Yaroslav O.
Roslund Curt
Rostopchina Alla
Rufener Fredy G.
Santos Agostinho Rui J.
Schuster William John
Sekiguchi Kazuhiro
Shakhovskoj Nikolay M.
Shawl Stephen J.
Smith J. Allyn
Smyth Michael J.
Snowden Michael
Steinlin Uli
Stetson Peter B.
Stockman Jr Hervey S.
Stone Remington P. S.
Straizys Vytautas P.
Stritzinger Maximilian D.
Sudzius Jokubas
Sullivan Denis John
Szkody Paula
Szymanski Michal
Tandon S. N.
Tapia-Perez Santiago
Taranova Olga G.
Tedds Jonathan A.
Tinbergen Jaap
Todoran Ioan
Tokunaga Alan Takashi
Tolbert Charles R.
Umeda Hideyuki
Ureche Vasile
Vaughan Arthur H.
Verma R. P.
Voloshina Irina B.
Vrba Frederick J.
Walker Alistair Robin
Walker William S. G.
Warren Jr Wayne H.
Weiss Werner W.
Weistrop Donna
Wesselius Paul R.
Wheatley Peter J.
White Nathaniel M.
Wielebinski Richard
Willstrop Roderick V.
Winiarski Maciej
Wramdemark Stig S.
Yamashita Yasumasa
Yao Yongqiang
Young Andrew T.
Yudin Ruslan V.
Ziznovsky Jozef

MEMBERSHIP BY COMMISSION

Composition of Commission 26

Double & Multiple Stars / Etoiles Doubles & Multiples

President Allen Christine

Vice-President Docobo Jose A. Durantez

Organizing Committee

Balega Yurij Yu
Hartkopf William I.
Mason Brian D.
Oblak Edouard
Oswalt Terry D.
Pourbaix Dimitri
Scarfe Colin David

Members

Abt Helmut A.
Ahumada Javier Alejandro
Anosova Joanna
Arenou Frederic
Argyle Robert William
Armstrong John Thomas
Bacchus Pierre
Bagnuolo Jr William G.
Bailyn Charles D.
Batten Alan H.
Beavers Willet I.
Bernacca Pierluigi
Boden Andrew F.
Bonneau Daniel
Brosche Peter
Budaj Jan
Cester Bruno
Chen Wen Ping
Clarke Catherine
Couteau Paul
Culver Roger Bruce
Cvetkovic Zorica D.
Dadaev Aleksandr N.
Davis John
De Cat Peter
Dommanget Jean
Dukes Jr. Robert
Dunham David W.
Elkin Vladimir
Fekel Francis C.
Fernandes Joao
Ferrer Osvaldo E.
Fletcher J. Murray
Franz Otto G.
Fredrick Laurence W.
Freitas Mourao R.
Freyhammer Lars Michael
Gatewood George
Gaudenzi Silvia
Geyer Edward H.
Ghez Andrea
Goodwin Simon P.
Hakkila Jon E.
Halbwachs Jean-Louis
Hartigan Patrick M.
Heacox William D.
Hershey John L.
Hidayat Bambang
Hill Graham
Hillwig Todd C.
Hindsley Robert Bruce
Horch Elliott P.
Hummel Christian Aurel
Hummel Wolfgang
Ianna Philip A.
Jahreiss Hartmut
Jassur Davoud MZ
Jurdana-Sepic Rajka
Kazantseva Liliya
Kiselev Aleksej A.
Kisseleva-Eggleton Ludmila
Kitsionas Spyridon
Kley Wilhelm
Kroupa Pavel
Kubat Jiri
Lampens Patricia
Latham David W.
Lattanzi Mario G.
Leinert Christoph
Lim Jeremy
Ling Josefina F.
Lippincott Zimmerman Sarah Lee
Loden Kerstin R.
Lyubchik Yuri
Maddison Sarah T.
Malkov Oleg Yu.
Marsakova Vladislava
Martin Eduardo L.
Mathieu Robert D.
McAlister Harold A.
McDavid David A.
Mikkola Seppo
Mikolajewski Maciej P.
Mohan Chander
Morbey Christopher L.
Morbidelli Roberto
Morel Pierre-Jacques
Negueruela Ignacio
Neuhaeuser Ralph
Nurnberger Dieter E. A.
Orlov Victor V.
Pannunzio Renato
Pauls Thomas Albert
Pereira Claudio B.
Peterson Deane M.
Petr-Gotzens Monika G.
Pluzhnik Eugene
Pollacco Don

Popovic Georgije
Poveda Arcadio
Prieto Cristina
Prieur Jean-Louis
Rakos Karl D.
Reipurth Bo
Roberts, Jr Lewis C.
Russell Jane L.
Sagar Ram
Salukvadze G. N.
Scardia Marco
Schmidtke Paul C.
Schoeller Markus
Shakht Natalia A.
Simon Michal
Sinachopoulos Dimitris
Skokos Charalambos
Smak Jozef I.
Smith J. Allyn
Soderhjelm Staffan
Sowell James Robert
Stein John William
Szabados Laszlo
Tamazian Vakhtang S.
Tango William J.
Tarasov Anatolii E.
ten Brummelaar Theo A.
Terquem Caroline E.
Tokovinin Andrei A.
Torres Guillermo
Trimble Virginia L.
Tsay Wean-Shun
Turner Nils H.
Upgren Arthur R.
Valtonen Mauri J.
van Altena William F.
van der Hucht Karel A.
van Dessel Edwin Ludo
Vaz Luiz Paulo Ribeiro
Wang Jiaji
Weis Edward W.
Zheleznyak Alexander P.
Zinnecker Hans

MEMBERSHIP BY COMMISSION

Composition of Commission 27

Variable Stars / Etoiles Variables

President Kawaler Steven D.

Vice-President Handler Gerald

Organizing Committee

Aerts Conny
Bedding Timothy R.
Catelan Márcio
Cunha Margarida
Eyer Laurent
Martinez Peter
Olah Katalin
Pollard Karen
Somasundaram Seetha

Members

Abada-Simon Meil
Aizenman Morris L.
Albinson James
Albrow Michael
Alfaro Emilio Javier
Allan David W.
Alpar Ali
Amado Gonzalez Pedro
Ando Hiroyasu
Andrievsky Sergei
Antipova Lyudmila I.
Antonello Elio
Antonyuk Kirill
Antov Alexandar
Arellano Ferro Armando
Arentoft Torben
Arkhipova Vera P.
Arsenijevic Jelisaveta
Asteriadis Georgios
Avgoloupis Stavros
Baade Dietrich
Baglin Annie
Balona Luis A.
Barnes III Thomas G.
Bartolini Corrado
Barwig Heinz
Baskill Darren S.
Bastien Pierre
Bateson Frank M. O.(†)
Bath Geoffrey T.
Bauer Wendy H.
Bazot Michael
Beaulieu Jean-Philippe R.
Bedogni Roberto
Belmonte Aviles Juan Antonio
Belserene Emilia P.
Belvedere Gaetano
Benkoe Jozsef M.
Benson Priscilla J.
Berdnikov Leonid N.
Bersier David
Berthomieu Gabrielle
Bessell Michael S.
Bianchini Antonio
Bjorkman Karen S.
Bochonko D. Richard
Bolton Charles Thomas
Bond Howard E.
Bopp Bernard W.
Boulon Jacques J.
Bowen George H.
Boyd David R.
Bradley Paul A.
Breger Michel
Brown Douglas N.
Bruntt Hans
Buchler J. R.
Burki Gilbert
Burwitz Vadim
Busa Innocenza
Busko Ivo C.
Butler Dennis
Butler Christopher John
Buzasi Derek
Cacciari Carla
Caldwell John A. R.
Cameron Andrew Collier
Cao Huilai
Carrier Fabien
Casares Jorge
Catchpole Robin M.
Chadid-Vernin Merieme
Cherepashchuk Anatolij M.
Chou Yi
Christensen-Dalsgaard Jørgen
Christie Grant W.
Christy Robert F.
Cioni Maria-Rosa L.
Clementini Gisella
Cohen Martin
Connolly Leo P.
Contadakis Michael E.
Cook Kem H.
Costa Vitor
Cottrell Peter L.
Coulson Iain M.
Coutts-Clement Christine
Cox Arthur N.
Cutispoto Giuseppe
Cuypers Jan
Dall'Ora Massimo
D'Amico Nicolo'
Danford Stephen C.
Daszynska-Daszkiewicz Jadwiga
De Cat Peter
de Groot Mart

De Ridder Joris
Delgado Antonio J.
Demers Serge
Deng LiCai
Deupree Robert G.
Di Mauro Maria Pia
Dickens Robert J.
Diethelm Roger
Donahue Robert A.
Dorokhova Tetyana
Downes Ronald A.
Dukes Jr. Robert
Dunlop Storm
Dupuy David L.
Dziembowski Wojciech A
Edwards Paul J.
Efremov Yurij N.
El Basuny Ahmed Alawy
Elkin Vladimir
Eskioglu A. Nihat
Evans Nancy R.
Evans Aneurin
Evren Serdar
Fadeyev Yurij A.
Feast Michael W.
Ferland Gary J.
Fernie J. Donald
Fitch Walter S.
Fokin Andrei
Formiggini Lilliana
Freyhammer Lars Michael
Friedjung Michael
Fu Hsieh-Hai
Fu Jian-Ning
Fu Jianning
Fujiwara Tomoko
Gahm Goesta F.
Galis Rudolf
Gameiro Jorge
Garrido Rafael
Gascoigne S. C. B.
Genet Russel M.
Gershberg R. E.
Geyer Edward H.
Gieren Wolfgang P.
Gies Douglas R.
Gillet Denis
Glagolevskij Yurij V.
Godoli Giovanni
Gondoin Philippe A. D.
Gosset Eric
Gough Douglas O.
Goupil Marie-Jose
Graham John A.
Grasberg Ernest K.
Grasdalen Gary L.
Green Daniel William Edward
Grinin Vladimir P.
Groenewegen Martin
Grygar Jiri
Guerrero Gianantonio
Guinan Edward F.
Gursky Herbert
Guzik Joyce A.
Hackwell John A.
Haefner Reinhold
Haisch Bernhard Michael
Halbwachs Jean-Louis
Hall Douglas S.
Hamdy M. A. M.
Hansen Carl J.
Hao Jinxin
Harmanec Petr
Hawley Suzanne Louise
Heiser Arnold M.
Hempelmann Alexander M.
Henden Arne A.
Herbig George H.
Hers Jan
Hesser James E.
Hill Henry A.
Hintz Eric G.
Hoffleit E. Dorrit(†)
Hojaev Alisher S.
Horner Scott D.
Houdek Gunter
Houk Nancy
Howell Steve B.
Huenemoerder David P.
Hutchings John B.
Iben Jr Icko
Iijima Takashi
Ishida Toshihito
Ismailov Nariman Z.
Ita Yoshifusa
Ivezic Zeljko
Jablonski Francisco J.
Jankov Slobodan S.
Jeffery Christopher S.
Jerzykiewicz Mikolaj
Jetsu Lauri J.
Jewell Philip R.
Jiang Biwei
Joner Michael D.
Jones Albert F.
Jurcsik Johanna
Kadouri Talib Hadi
Kaeufl Hans Ulrich
Kambe Eiji
Kanamitsu Osamu
Kanbur Shashi
Kanyo Sandor
Karitskaya Evgeniya A.
Karovska Margarita
Karp Alan H.
Katsova Maria M.
Kaufer Andreas
Kaye Anthony B.
Kazarovets Elena V.
Keller Stefan C.
Kepler S. O.
Khaliullin Khabibrachman F.
Kilkenny David
Kim Tu-Whan
Kim Chulhee
Kim Seung-Lee
Kiplinger Alan L.
Kippenhahn Rudolf
Kiss Laszlo L.
Kjeldsen Hans
Kjurkchieva Diana
Kochukhov Oleg
Koen Marthinus
Koevari Zsolt
Kollath Zoltan
Komzik Richard
Konstantinova-Antova Renada K.
Kopacki Grzegorz
Korhonen Heidi H.
Kraft Robert P.
Krautter Joachim
Kreiner Jerzy M.
Krisciunas Kevin
Krzeminski Wojciech
Krzesinski Jerzy H.
Kubiak Marcin A.
Kuhi Leonard V.

Kunjaya Chatief
Kunkel William E.
Kurtz Donald W.
Lago Maria T. V. T.
Lampens Patricia
Landolt Arlo U.
Laney Clifton D.
Lanning Howard H.
Lanza Antonino F.
Larionov Valeri M.
Laskarides Paul G.
Lawlor Timothy M.
Lawson Warrick
Lazaro Carlos
Le Bertre Thibaut R.
Lebzelter Thomas
Lee Jae-Woo
Leite Scheid P.
Leung Kam Ching
Li Yan
Li Zhiping
Little-Marenin Irene R.
Lloyd Christopher
Lockwood G. Wesley
Longmore Andrew J.
Lopez De Coca M. D. P.
Lorenz-Martins Silvia
Lub Jan
Machado Folha Daniel F.
Macri Lucas M.
Madore Barry Francis
Maeder Andre
Maffei Paolo
Mahmoud Farouk M. A. B.
Mahra H. S.
Makarenko Ekaterina N.
Mantegazza Luciano
Marchev Dragomir V.
Marconi Marcella
Margrave Jr Thomas E.
Markoff Sera B.
Marsakova Vladislava
Masani A.
Mathias Philippe
Matsumoto Katsura
Matthews Jaymie
Mauche Christopher W.
Mavridis Lyssimachos N.
McGraw John T.
McNamara Delbert H.
McSaveney Jennifer A.
Melikian Norair D.
Messina Sergio
Michel Eric
Mikolajewski Maciej P.
Milone Eugene F.
Milone Luis A.
Minnikulov Nasriddin K.
Mkrtichian David E.
Moffett Thomas J.
Mohan Chander
Monteiro Mario
Joao P. F. G.
Morales Rueda Luisa
Morrison Nancy D.
Moskalik Pawel
Mukai Koji
Mumford George S.
Murdin Paul G.
Nather R. Edward
Naylor Tim
Neiner Coralie
Niarchos Panayiotis
Nikolov Nikola S.
Nikolov Andrej
Nugis Tiit
Odgers Graham J.
O'Donoghue Darragh
Ogloza Waldemar
Opolski Antoni
Ostensen Roy H.
Oswalt Terry D.
O'Toole Simon J.
Papaloizou John C. B.
Paparo Margit
Parsamyan Elma S.
Parthasarathy Mudumba
Patat Ferdinando
Paterno Lucio
Pavlovski Kresimir
Pearson Kevin J.
Percy John R.
Perez Hernandez
Fernando J.
Petersen J. Otzen
Petit Pascal
Petrov Peter P.
Pettersen Bjørn R.
Piirola Vilppu E.
Pijpers Frank Peter
Plachinda Sergei I.
Pollacco Don
Pont Frederic
Pop Alexandru V.
Pop Vasile
Pringle James E.
Pritzl Barton J.
Provost Janine
Pugach Alexander F.
Rakos Karl D.
Ransom Scott M.
Rao N. Kameswara
Ratcliff Stephen J.
Reale Fabio
Reinsch Klaus
Renson P. F. M.
Rey Soo-Chang
Robinson Edward L.
Rodriguez Eloy
Romano Giuliano
Romanov Yuri S.
Rosenbush Alexander E.
Rountree Janet
Russev Ruscho Minchev
Sachkov Mikhail E.
Sadik Aziz R.
Saha Abhijit
Samus Nikolaj N.
Sandmann William H.
Sanyal Ashit
Sareyan Jean-Pierre
Sasselov Dimitar D.
Schaefer Bradley E.
Schlegel Eric M.
Schmidt Edward G.
Schmidtobreick Linda
Schuh Sonja
Schwartz Philip R.
Schwarzenberg-Czerny A.
Schwope Axel
Scuflaire Richard
Seeds Michael A.
Shahul Hameed Mohin
Shakhovskaya Nadejda I.
Shara Michael
Sharma Dharma P.
Shenavrin Victor I.
Sherwood William A.
Shobbrook Robert R.
Silvotti Roberto

Smak Jozef I.
Smeyers Paul
Smit Jan A.
Smith Myron A.
Soliman Mohamed Ahmed
Soszynski Igor
Srivastava Ram K.
Starrfield Sumner
Stellingwerf Robert F.
Stepien Kazimierz
Sterken Christiaan L.
Strassmeier Klaus G.
Stringfellow Guy S.
Strom Stephen E.
Strom Karen M.
Szabados Laszlo
Szatmary Karoly
Szecsenyi-Nagy Gábor A.
Szeidl Bela
Szkody Paula
Takata Masao
Takeuti Mine
Tammann Gustav A.
Tamura Shin'ichi
Tarasova Taya
Taylor John K.
Teixeira Teresa C. V. S.
Tempesti Piero
Terzan Agop
Tjin-a-Djie Herman R. E.
Tremko Jozef
Tsvetkov Milcho K.
Tsvetkova Katja
Turner David G.
Tutukov Aleksandr V.
Tylenda Romuald
Udovichenko Sergei N.
Uemura Makoto
Usher Peter D.
Utrobin Victor P.
Valtier Jean-Claude
van Genderen Arnoud M.
van Hoolst Tim
Ventura Rita
Verheest Frank
Viotti Roberto
Vivas Anna K.
Vogt Nikolaus
Voloshina Irina B.
Waelkens Christoffel
Walker Edward N.
Walker William S. G.
Walker Merle F.
Wallerstein George
Warner Brian
Watson Robert
Webbink Ronald F.
Wehlau Amelia F.
Weis Kerstin
Weiss Werner W.
Welch Douglas L.
Wheatley Peter J.
Whitelock Patricia Ann
Williamon Richard M.
Willson Lee Anne M.
Wilson Lionel
Wing Robert F.
Wittkowski Markus
Wood Peter R.
Xiong Da Run
Zijlstra Albert
Zola Stanislaw
Zsoldos Endre
Zuckerman Ben M.

MEMBERSHIP BY COMMISSION

Composition of Commission 28

Galaxies / Galaxies

President Combes Françoise

Vice-President Davies Roger L.

Organizing Committee

Dekel Avishai
Franx Marijn
Gallagher III John S.
Karachentseva Valentina
Knapp Gillian R.
Kraan-Korteweg Renée C.
Leibundgut Bruno
Nakai Naomasa
Narlikar Jayant V.
Rubio Monica
Sadler Elaine M.

Members

Aalto Suzanne E.
Ables Harold D.
Abrahamian Hamlet V.
Adler David Scott
Afanas'ev Viktor L.
Aguero Estela L.
Aguilar Luis A. C.
Ahmad Farooq
Akiyama Masayuki
Alcaino Gonzalo
Aldaya Victor
Alexander Tal E.
Alladin Saleh Mohamed
Allen Ronald J.
Allington-Smith Jeremy R.
Alloin Danielle
Almaini Omar
Aloisi Alessandra
Alonso Maria V.
Amram Philippe
Andernach Heinz J
Andrillat Yvette
Ann Hong Bae
Anosova Joanna
Anton Sonia
Aoki Kentaro
Aparicio Antonio
Aragon-Salamanca Alfonso
Ardeberg Arne L.
Aretxaga Itziar
Arkhipova Vera P.
Arnaboldi Magda
Artamonov Boris P.
Athanassoula Evangelia
Avila-Reese Vladimir
Ayani Kazuya
Azzopardi Marc
Bachev Rumen S.
Baes Maarten
Bailey Mark Edward
Bajaja Esteban
Baker Andrew J.
Baldwin Jack A.
Balkowski-Mauger Chantal
Ballabh Goswami Mohan
Balogh Michael L.
Banhatti Dilip G.
Barbon Roberto
Barcons Xavier
Barnes David G.
Barr Jordi McGregor
Barth Aaron J.
Barthel Peter
Barton Elizabeth J.
Bassino Lilia P.
Basu Baidyanath
Battaner Eduardo
Battinelli Paolo
Baum William A.
Baum Stefi Alison
Bautista Manuel A.
Beaulieu Sylvie F.
Beck Rainer
Begeman Kor G.
Bender Ralf
Benedict George F.
Benetti Stefano
Bensby Thomas L.
Berczik Peter
Bergeron Jacqueline A.
Berkhuijsen Elly M.
Berman Vladimir
Berta Stefano
Bertola Francesco
Bettoni Daniela
Bian Yulin
Bianchi Simone
Biermann Peter L.
Bijaoui Albert
Binette Luc
Binggeli Bruno
Binney James J.
Biretta John Anthony
Birkinshaw Mark
Bjornsson Claes-Ingvar
Blakeslee John P.
Bland-Hawthorn Jonathan
Blitz Leo
Block David Lazar
Blumenthal George R.
Boeker Torsten
Boissier Samuel
Boisson Catherine
Boksenberg Alec
Boles Thomas
Bolzonella Micol

Bomans Dominik J.
Bongiovanni Angel
Borchkhadze Tengiz M.
Borne Kirk D.
Bosma Albert
Bottinelli Lucette
Bower Gary Allen
Braccesi Alessandro
Braine Jonathan
Braun Robert
Bravo-Alfaro Hector
Brecher Kenneth
Bressan Alessandro
Bridges Terry J.
Briggs Franklin
Brinkmann Wolfgang
Brinks Elias
Brodie Jean P.
Brosch Noah
Brouillet Nathalie
Brown Thomas M.
Bruzual Gustavo R.
Buat Véronique
Buote David A.
Burbidge Eleanor Margaret
Burbidge Geoffrey R.
Bureau Martin G.
Burgarella Denis
Burkert Andreas M.
Burns Jr Jack O'Neal
Burstein David
Busarello Giovanni
Buta Ronald J.
Butcher Harvey R.
Byrd Gene G.
Byun Yong-Ik
Cai Michael
Calderon Jesus
Calura Francesco
Calzetti Daniela
Campusano Luis E.
Cannon Russell D.
Cannon John M.
Canzian Blaise
Cao Xinwu
Capaccioli Massimo
Cappellari Michele
Carigi Leticia
Carollo Marcella
Carrillo Moreno Rene
Carswell Robert F.
Carter David
Casoli Fabienne
Cayatte Veronique
Cellone Sergio Aldo
Cepa Jordi
Cha Seung-Hoon
Chakrabarti Sandip Kumar
Chakrabarty Dalia
Chamaraux Pierre
Chang Ruixiag
Charmandaris Vassilis
Chatterjee Tapan Kumar
Chatzichristou Eleni T.
Chavushyan Vahram
Chen Yang
Chen Jiansheng
Chiappini Moraes Leite Cristina
Chiba Masashi
Chincarini Guido L.
Chou Chih-Kang
Chu Yaoquan
Chugai Nikolaj N.
Chun Sun Y.
Cinzano Pierantonio
Cioni Maria-Rosa L.
Ciotti Luca
Clavel Jean
Clementini Gisella
Cohen Ross D.
Colbert Edward J. M.
Colina Luis
Comte Georges
Conselice Christopher J.
Contopoulos George
Cook Kem H.
Corbett Elizabeth A.
Corbin Michael R.
Corsini Enrico M.
Corwin Jr Harold G.
Cote Patrick
Cote Stéphanie
Couch Warrick
Courbin Frederic Y. M.
Courtes Georges
Courvoisier Thierry J.-L.
Cowsik Ramanath
Coziol Roger
Crane Philippe
Crawford Carolin
Cress Catherine M.
Cunniffe John
Cunow Barbara
da Costa Luiz A. N.
Dallacasa Daniele
Danks Anthony C.
Dantas Christine C.
Davidge Timothy J.
Davies Rodney D.
Davies Jonathan Ivor
Davis Marc
De Blok Erwin
de Boer Klaas Sjoerds
de Bruyn A. Ger
de Carvalho Reinaldo
de Grijs Richard
de Jong Roelof S.
De Mello Duilia F.
De Propris Roberto
De Rijcke Sven
de Zeeuw Pieter T.
Dejonghe Herwig B.
Demers Serge
Deng Zugan
Dennefeld Michel
Dessauges-Zavadsky Miroslava
Dettmar Ralf-Juergen
Diaferio Antonaldo
Diaz Angeles Isabel
Dickey John M.
Dietrich Matthias
d'Odorico Sandro
Doi Mamoru
Dokuchaev Vyacheslav I.
Donas Jose
Donea Alina C.
Donner Karl Johan
D'Onofrio Mauro
Donzelli Carlos J.
Dopita Michael A.
Doroshenko Valentina T.
Dottori Horacio A.
Dovciak Michal
Doyon Rene
Dressel Linda L.
Dressler Alan
Drinkwater Michael J.
Driver Simon P.

Duc Pierre-Alain
Dufour Reginald James
Dultzin-Hacyan Deborah
Dumont Anne-Marie
Durret Florence
Duval Marie-France
Eales Stephen A.
Edelson Rick
Edmunds Michael Geoffrey
Efstathiou George
Einasto Jaan
Ekers Ronald D.
Ellis Simon C.
Elmegreen Debra M.
Elvis Martin S.
Elvius Aina M.
English Jayanne
Espey Brian Russell
Evans Robert
Fabbiano Giuseppina
Faber Sandra M.
Fabricant Daniel G.
Fairall Anthony P.
Falco Emilio E.
Fall S. Michael
Fan Junhui
Fasano Giovanni
Feast Michael W.
Feinstein Carlos
Feitzinger Johannes
Ferguson Annette M.
Ferland Gary J.
Ferrarese Laura
Ferreras Ignacio P.
Ferrini Federico
Fharo Toshihiro
Field George B.
Filippenko Alexei V.
Flin Piotr
Florsch Alphonse
Foltz Craig B.
Forbes Duncan Alan
Ford Holland C.
Ford Jr W. Kent
Fouque Pascal
Fraix-Burnet Didier
Francis Paul
Freedman Wendy L.
Freeman Kenneth C.
Fricke Klaus
Fried Josef Wilhelm
Fritze Klaus
Fritze-von Alvensleben Uta
Frogel Jay Albert
Fuchs Burkhard
Fujita Yutaka
Fukugita Masataka
Funato Yoko
Funes José G.
Gallart Carme
Gallego Jesús
Galletta Giuseppe
Gallimore Jack F.
Gamaleldin Abdulla I.
Gardner Jonathan P.
Garilli Bianca
Gascoigne S. C. B.
Gelderman Richard
Geller Margaret Joan
Georgiev Tsvetan
Gerhard Ortwin
Ghigo Francis D.
Ghosh P.
Giacani Elsa B.
Gibson Brad K.
Gigoyan Kamo S.
Giovanardi Carlo
Giovanelli Riccardo
Glass Ian S.
Godlowski Wlodzimierz
Gonzalez Serrano J. I.
Gonzalez Delgado Rosa M.
Gonzalez-Solares Eduardo
Goodrich Robert W.
Gorgas Garcia Javier
Goss W. Miller
Goto Tomotsugu
Gottesman Stephen T.
Gouguenheim Lucienne
Graham John A.
Graham Alister W. McK.
Granato Gian Luigi
Gray Meghan
Grebel Eva K.
Gregg Michael David
Greve Thomas R.
Griffiths Richard E.
Griv Evgeny
Gronwall Caryl A.
Grupe Dirk
Gu Qiusheng
Gunn James E.
Gurzadyan Grigor A.
Guseva Natalia G.
Gyulbudaghian Armen L.
Hagen-Thorn Vladimir A.
Hamabe Masaru
Hambaryan Valeri V.
Hammer Francois
Han Cheongho
Hanami Hitoshi
Hara Tetsuya
Hardy Eduardo
Harms Richard James
Harnett Julienne
Hasan Hashima
Hashimoto Yasuhiro
Hattori Makoto
He XiangTao
Heckman Timothy M.
Heidt Jochen
Held Enrico V.
Helou George
Henning Patricia A.
Henry Richard B. C.
Hensler Gerhard
Hewitt Adelaide
Hewitt Anthony V.
Hickson Paul
Hintzen Paul Michael N.
Hirashita Hiroyuki
Hjalmarson Ake G.
Hjorth Jens
Ho Luis Chi
Hodge Paul W.
Hong Wu
Hopkins Andrew M.
Hopp Ulrich
Horellou Cathy
Hornstrup Allan
Hou Jinliang
Houdashelt Mark L.
Hough James
Hu Fuxing
Hua Chon Trung
Huang Keliang
Huang Jiasheng
Huchra John Peter
Huchtmeier Walter K.
Huettemeister Susanne

Hughes David H.
Humphreys Roberta M.
Hunstead Richard W.
Hunter James H.
Hunter Christopher
Hwang Chorng-Yuan
Ichikawa Takashi
Ichikawa Shin-ichi
Idiart Thais E.
Illingworth Garth D.
Im Myungshin
Imanishi Masatoshi
Impey Christopher D.
Infante Leopoldo
Irwin Judith
Isaak Kate G.
Ishimaru Yuhri
Israel Frank P.
Issa Issa Aly
Ivezic Zeljko
Ivison Robert J.
Iwamuro Fumihide
Iye Masanori
Izotov Yuri
Izotova Iryna Yu.
Jafelice Luiz C.
Jaffe Walter Joseph
Jang Minwhan
Jarrett Thomas H.
Jerjen Helmut
Jiang Ing-Guey
Jog Chanda J.
Joly Monique
Jones Thomas Walter
Jones Paul
Jones Christine
Jorgensen Inger
Joshi Umesh C.
Jovanovic Predrag
Joy Marshall J.
Jugaku Jun
Jungwiert Bruno
Junkes Norbert
Junkkarinen Vesa T.
Junor William
Kalinkov Marin P.
Kalloglian Arsen T.
Kandalyan Rafik A.
Kanekar Nissim
Kaneko Noboru
Karachentsev Igor D.
Karoji Hiroshi
Kashikawa Nobunari
Kaspi Shai
Kassim Namir E.
Katgert Peter
Kauffmann Guinevere A. M.
Kaufman Michele
Kawakatu Nozomu
Keel William C.
Kellermann Kenneth I.
Kemp Simon N.
Kennicutt Robert C.
Khachikian Edward Yerem
Khanna Ramon
Khare Pushpa
Khosroshahi Habib G.
Kilborn Virginia A.
Kim Sungsoo S.
Kim Dong-Woo
King Ivan R.
Kinman Thomas D.
Kirshner Robert Paul
Kissler-Patig Markus
Klein Ulrich
Knapen Johan Hendrik
Kniazev Alexei
Knudsen Kirsten K.
Ko Chung-Ming
Kobayashi Chiaki
Kochhar Rajesh K.
Kodaira Keiichi
Kodama Tadayuki
Kogoshvili Natela G.
Kollatschny Wolfram
Komiyama Yutaka
Kontizas Evangelos
Kontorovich Victor M.
Koo David C-Y
Koopmans Leon V. E.
Koratkar Anuradha P.
Koribalski Baerbel Silvia
Kormendy John
Kotilainen Jari
Krause Marita
Krishna Gopal
Kron Richard G.
Kumai Yasuki
Kunchev Peter
Kuno Nario
Kunth Daniel
Kuntschner Harald
La Barbera Francesco
Lancon Ariane
Larsen Soeren S.
Larson Richard B.
Laurikainen Eija
Layzer David
Leacock Robert Jay
Leeuw Lerothodi L.
Lefevre Olivier
Lehto Harry J.
Lequeux James
Li Qibin
Li Li-Xin
Lilly Simon J.
Lim Jeremy
Lima Neto Gastao B.
Lin Chia C.
Lindblad Per O.
Linden-Voernle Michael J. D.
Lo Kwok-Yung
Lobo Catarina
Londrillo Pasquale
Lopez Ericson D.
Lopez Cruz Omar
Lopez Hermoso Maria Rosario
Lord Steven Donald
Loup Cecile
Low Frank J.
Lu Limin
Lugger Phyllis M.
Luminet Jean-Pierre
Lutz Dieter
Lynden-Bell Donald
Lynds Beverly T.
Lynds Roger C.
Macalpine Gordon M.
Maccagni Dario
Maccarone Thomas J.
Macchetto Ferdinando
Maciejewski Witold B.
Mackie Glen
Macquart Jean-Pierre
Madden Suzanne
Madore Barry Francis
Magorrian Stephen J.

Magris Gladis C.
Mahtessian Abraham P.
Maiolino Roberto
Makarov Dmitry I.
Makarova Lidia N.
Malagnini Maria Lucia
Malhotra Sageeta
Mann Robert G.
Mannucci Filippo
Marcelin Michel
Marco Olivier
Marconi Alessandro
Markoff Sera B.
Marquez Isabel
Marr Jonathon M.
Marston Anthony Philip
Martin Maria C.
Martin Rene Pierre
Martin Crystal L.
Martinet Louis
Martinez Garcia Vicent J.
Martini Paul
Marziani Paola
Masegosa Gallego J.
Mathewson Donald S.
Matthews Lynn E.
Mattila Seppo K.
Mauersberger Rainer
Maurice Eric N.
Mayya Divakara
Mazzarella Joseph M.
McBreen Brian Philip
McGaugh Stacy Sutton
Mediavilla Evencio
Mehlert Dörte
Meier David L.
Meikle William P. S.
Meisenheimer Klaus
Mendes de Oliveira Cláudia L.
Menon T. K.
Mercurio Amata
Merluzzi Paola
Merrifield Michael R.
Metevier Anne
Meusinger Helmut
Mihov Boyko M.
Miley George K.
Miller Joseph S.
Miller Hugh R.
Miller Richard H.
Miller Neal A.
Milvang-Jensen Bo
Ming Bai Jin
Mirabel Igor Felix
Misawa Toru
Mizuno Takao
Moiseev Alexei V.
Moles Mariano J.
Molinari Emilio
Monaco Pierluigi
Moody Joseph Ward
Mori Masao
Moss Christopher
Motohara Kentaro
Mould Jeremy R.
Mueller Volker
Mulchaey John S.
Muller Erik M.
Munoz Tunon Casiana
Muratorio Gerard
Murayama Takashi
Murphy Michael T.
Murray Stephen S.
Mushotzky Richard
Muzzio Juan C.
Nagashima Masahiro
Nair Sunita
Nakanishi Kouichiro
Nakanishi Hiroyuki
Namboodiri P. M. S.
Napolitano Nicola R.
Navarro Julio Fernando
Nedialkov Petko L.
Ninkovic Slobodan
Nipoti Carlo
Nishikawa Ken-Ichi
Nityananda Ram
Noguchi Masafumi
Noonan Thomas W.
Norman Colin A.
Nucita Achille A.
Nulsen Paul E. J.
O'Connell Robert West
O'Dea Christopher P.
Oemler Jr Augustus
Ohta Kouji
Okamoto Takashi
Okamura Sadanori
Olling Robert P.
Olofsson Kjell
Olsen Lisbeth F.
Omizzolo Alessandro
Oosterloo Thomas
Origlia Livia
Osman Anas Mohamed
Osterbrock Donald E.(†)
Ostlin Göran T.
Ostriker Eve C.
Ott Juergen A.
Ouchi Masami
Oyabu Shinki
Pacholczyk Andrzej G.
Page Mathew J.
Palmer Philip
Palumbo Giorgio G. C.
Pannuti Thomas G.
Papayannopoulos Theodoros
Park Jang-Hyun
Parker Quentin A.
Pastoriza Miriani G.
Paturel Georges
Pearce Frazer R.
Peimbert Manuel
Pellegrini Silvia
Pello Roser Descayre
Perea-Duarte Jaime D.
Perez Fournon Ismael
Perry Judith J.
Peters William L. III
Peterson Charles John
Petit Jean-Marc
Petrosian Artaches R.
Petrov Georgi Trendafilov
Petuchowski Samuel J.
Pfenniger Daniel
Philipp Sabine D.
Phillipps Steven
Phillips Mark M.
Pikichian Hovhannes Vahram
Pipino Antonio
Pisano Daniel J.
Pizzella Alessandro
Plana Henri M.
Pogge Richard William
Poggianti Bianca M.
Polyachenko Evgeny V.
Popescu Cristina Carmen

Popovic Luka C.
Portinari Laura
Poveda Arcadio
Prabhu Tushar P.
Pracy Michael B.
Prandoni Isabella
Press William H.
Prevot-Burnichon Marie-Louise
Pritchet Christopher J.
Proctor Robert N.
Pronik Iraida I.
Pronik Vladimir I.
Proust Dominique
Puerari Ivanio
Pustilnik Simon A.
Qin Yi-Ping
Quinn Peter
Quintana Hernan
Rafanelli Piero
Raiteri Claudia M.
Rampazzo Roberto
Rand Richard J.
Rasmussen Jesper
Raychaudhury Somak
Raychaudhury Somak
Read Andrew M.
Rector Travis A.
Reichert Gail Anne
Rejkuba Marina
Rephaeli Yoel
Reshetnikov Vladimir P.
Revaz Yves
Revnivtsev Mikhail G.
Rey Soo-Chang
Richer Harvey B.
Richstone Douglas O.
Richter Gotthard
Rix Hans-Walter
Robert Carmelle
Roberts Timothy P.
Roberts Morton S.
Roberts Jr William W.
Roeser Hermann-Josef
Romano Patrizia
Romeo Alessandro B.
Romero-Colmenero Encarnacion
Roos Nicolaas
Rosa Michael Richard
Rosado Margarita
Rose James Anthony
Rots Arnold H.
Rozas Maite
Rubin Vera C.
Rudnicki Konrad
Ryder Stuart
Sackett Penny
Sadat Rachida
Sadun Alberto Carlo
Sahibov Firuz H.
Saitoh Takayuki
Saiz Alejandro
Sakai Shoko
Sala Ferran
Salvador-Sole Eduardo
Samland Markus
Samurovic Srdjan S.
Sanahuja Blai
Sancisi Renzo
Sanders Robert
Sanders David B.
Sanroma Manuel
Sansom Anne E.
Sapre Ashok Kumar
Saracco Paolo
Sarazin Craig L.
Sargent Wallace L. W.
Sasaki Toshiyuki
Sasaki Minoru
Saslaw William C.
Sastry Shankara K.
Savage Ann
Sawa Takeyasu
Scaramella Roberto
Schaye Joop
Schechter Paul L.
Schmidt Maarten
Schmitt Henrique R.
Schmitz Marion
Schroder Anja C.
Schucking Engelbert L.
Schwarz Ulrich J.
Schweizer François
Scodeggio Marco
Scorza de Appl Cecilia
Scoville Nicholas Z.
Searle Leonard
Seigar Marcus S.
Sellwood Jerry A.
Semelin Benoit
Sempere Maria J.
Sergeev Sergey G.
Serjeant Stephen
Serote Roos Margarida
Seshadri Sridhar
Setti Giancarlo
Shapovalova Alla I.
Sharples Ray
Shaver Peter A.
Shaya Edward J.
Sherwood William A.
Shields Gregory A.
Shields Joseph C.
Shimasaku Kazuhiro
Shostak G. Seth
Shukurov Anvar Muhamadovich
Siebenmorgen Ralf
Sigurdsson Steinn
Sil'chenko Olga K.
Sillanpaa Aimo Kalevi
Silva David Richard
Silva Laura
Simkin Susan M.
Simone Zaggia
Singh Kulinder Pal
Siopis Christos
Skillman Evan D.
Slezak Eric
Smail Ian
Smecker-Hane Tammy A.
Smith Eric Philip
Smith Haywood C.
Smith Malcolm G.
Smith Harding E.(†)
Smith Jr Harding E.
Soares Domingos S. L.
Sobouti Yousef
Sohn Young-Jong
Soltan Andrzej Maria
Sorai Kazuo
Sparks William Brian
Spinoglio Luigi
Spinrad Hyron
Srinivasan G.
Statler Thomas S.
Staveley-Smith Lister
Stavrev Konstantin Y.
Steiman-Cameron Thomas

Steinbring Eric
Stiavelli Massimo
Stirpe Giovanna M.
Stone Remington P. S.
Storchi-Bergmann Thaisa
Strom Robert G.
Strom Richard G.
Sugai Hajime
Sulentic Jack W.
Sullivan Woodruff T.
Sundin Maria I.
Sutherland Ralph S.
Tacconi Linda J.
Tacconi-Garman Lowell E.
Tagger Michel
Takata Tadafumi
Takato Naruhisa
Takeuchi Tsutomu T.
Takizawa Motokazu
Tammann Gustav A.
Tanaka Yutaka D.
Taniguchi Yoshiaki
Telles Eduardo
Tenjes Peeter
Terlevich Roberto Juan
Terzian Yervant
Theis Christian
Thomasson Magnus
Thonnard Norbert
Thornley Michele D.
Thuan Trinh Xuan
Tiersch Heinz
Tifft William G.
Tilanus Remo P. J.
Tissera Patricia B.
Tolstoy Eline
Tomita Akihiko
Toomre Alar
Tovmassian Hrant Mushegi
Toyama Kiyotaka
Traat Peeter
Trager Scott C.
Tremaine Scott Duncan
Trimble Virginia L.
Trinchieri Ginevra
Tsuchiya Toshio
Tsvetkov Dmitry Yu.
Tuffs Richard J.
Tully Richard Brent
Turner Edwin L.
Tyson John Anthony
Tyul'bashev Sergei A.
Ulrich Marie-Helene D.
Urbanik Marek
Utrobin Victor P.
Valentijn Edwin A.
Vallenari Antonella
Valotto Carlos A.
Valtonen Mauri J.
van Albada Tjeerd S.
Van den Bergh Sidney
van der Hulst Jan M.
van der Kruit Pieter C.
van der Laan Harry
van der Marel Roeland P.
van Driel Wim
van Gorkom Jacqueline H.
van Moorsel Gustaaf
van Woerden Hugo
Van Zee Liese E.
Vansevicius Vladas
Varma Ram Kumar
Vaughan Simon A.
Vauglin Isabelle
Vavilova Iryna B.
Vazdekis Alexandre
Veilleux Sylvain
Vercellone Stefano
Verdes-Montenegro Lourdes
Vermeulen Rene Cornelis
Veron Marie-Paule
Veron Philippe
Vigroux Laurent
Villata Massimo
Vivas Anna K.
Vlasyuk Valerij
Voglis Nikos
Vollmer Bernd
Volonteri Marta
Vrtilek Jan M.
Wada Keiichi
Wagner Stefan
Wakamatsu Ken-ichi
Walker Mark Andrew
Walter Fabian
Walterbos Rene A. M.
Wanajo Shinya
Wang Yiping
Wang Tinggui
Ward Martin John
Weedman Daniel W.
Wei Jianyan
Weiler Kurt W.
Welch Gary A.
Westerlund Bengt E.
White Simon David Manion
Whitmore Bradley C.
Wielebinski Richard
Wielen Roland
Wiita Paul Joseph
Wilcots Eric M.
Wild Wolfgang
Williams Theodore B.
Williams Barbara A.
Williams Robert
Wills Beverley J.
Wills Derek
Wilson Andrew S.
Windhorst Rogier A.
Winkler Hartmut
Wise Michael W.
Wisotzki Lutz
Wlerick Gerard
Woosley Stanley E.
Worrall Diana Mary
Woudt Patrick A.
Wozniak Herve
Wrobel Joan Marie
Wu Xuebing
Wunsch Richard
Wynn-Williams C. G.
Xanthopoulos Emily
Xia Xiao-Yang
Xilouris Emmanouel
Xue Suijian
Yagi Masafumi
Yahagi Hideki
Yakovleva Valerija A.
Yamada Yoshiyuki
Yamada Toru
Yamagata Tomohiko
Yi Sukyoung Ken
Yonehara Atsunori
Yoshida Michitoshi
Young Judith Sharn
Zamorano Jaime
Zaroubi Saleem
Zasov Anatolij V.
Zavatti Franco
Zeilinger Werner W.

MEMBERSHIP BY COMMISSION

Composition of Commission 29

Stellar Spectra / Spectre Stellaire

President Parthasarathy Mudumba

Vice-President Piskunov Nikolai E.

Organizing Committee

Carpenter Kenneth G.
Castelli Fiorella
Cunha Katia
Eenens Philippe
Hubeny Ivan
Rossi Silvia C. F.
Sneden Chris
Takada-Hidai Masahide
Wahlgren Glenn Michael
Weiss Werner W.

Members

Abhyankar Krishna D.
Abia Carlos A.
Abt Helmut A.
Adelman Saul J.
Aikman G. Chris L.
Ake III Thomas Bellis
Alcala Juan Manuel
Alecian Georges
Allende Prieto Carlos
Ambruster Carol
Andretta Vincenzo
Andreuzzi Gloria
Andrillat Yvette
Annuk Kalju
Aoki Wako
Appenzeller Immo
Arkharov Arkadij A.
Artru Marie-Christine
Atac Tamer
Audard Marc
Baade Dietrich
Bagnulo Stefano
Baliunas Sallie L.
Ballereau Dominique
Banerjee Dipankar P. K.
Baratta Giovanni Battista
Barbuy Beatriz
Baron Edward
Basri Gibor B.
Batalha Celso Correa
Bauer Wendy H.
Beckman John E.
Beiersdorfer Peter
Bellas-Velidis Ioannis
Bensby Thomas L.
Berger Jacques G.
Bessell Michael S.
Bikmaev Ilfan F.
Boehm Torsten C.
Boesgaard Ann M.
Boggess Albert
Bohlender David
Bond Howard E.
Bonifacio Piercarlo
Bonsack Walter K.
Bopp Bernard W.
Bouvier Jerôme
Bragaglia Angela
Brandi Elisande E.
Breysacher Jacques
Brickhouse Nancy S.
Briot Danielle
Brown Douglas N.
Brown Paul James Frank
Bruhweiler Frederick C.
Bruning David H.
Bruntt Hans
Bues Irmela D.
Burkhart Claude
Busa Innocenza
Butler Keith
Carney Bruce William
Carretta Eugenio
Carter Bradley Darren
Catala Claude
Catalano Santo
Catanzaro Giovanni
Catchpole Robin M.
Cayrel Roger
Cayrel de Strobel Giusa
Chavez-Dagostino Miguel
Cidale Lydia S.
Claudi Riccardo U.
Climenhaga John L.
Coluzzi Regina
Conti Peter S.
Corbally Christopher
Cornide Manuel
Cottrell Peter L.
Cowley Anne P.
Cowley Charles R.
Crowther Paul
da Silva Licio
Dacic Miodrag D.
Daflon Simone
Damineli Neto Augusto
Dawanas Djoni N.
de Araujo Francisco X.
de Castro Elisa
de Groot Mart
de Laverny Patrick
del Peloso Eduardo F.
Divan Lucienne
Doazan Vera
Dolidze Madona V.
Doppmann Gregory W.

Dragunova Alina V.
Drake Natalia
Duncan Douglas Kevin
Dworetsky Michael M.
Elkin Vladimir
Faraggiana Rosanna
Feast Michael W.
Felenbok Paul
Fernandez-Figueroa M. J.
Fitzpatrick Edward L.
Floquet Michele
Foing Bernard H.
Foy Renaud
Franchini Mariagrazia
Francois Patrick
Frandsen Soeren
Freire Ferrero Rubens G.
Freyhammer Lars Michael
Friedjung Michael
Friel Eileen D.
Fujita Yoshio
Fulbright Jon P.
Fullerton Alexander W.
Garcia Lopez Ramon J.
Garmany Catherine D.
Garrison Robert F.
Gautier Daniel
Gehren Thomas
Gerbaldi Michèle
Gershberg R. E.
Gesicki Krzysztof
Ghosh Kajal Kumar
Giampapa Mark S.
Gilra Daya P.
Giovannelli Franco
Glagolevskij Yurij V.
Glazunova Ljudmila
Glushneva Irina N.
Goebel John H.
Gonzalez Guillermo
Gopka Vera F.
Grady Carol Anne
Gratton Raffaele G.
Gray David F.
Griffin R. Elizabeth
Griffin Roger F.
Gu Sheng-hong
Gustafsson Bengt
Guthrie Bruce N. G.
Hack Margherita
Hanuschik Reinhard
Harmer Charles F. W.
Harmer Dianne L.
Hartman Henrik
Hartmann Lee William
Hashimoto Osamu
Hearnshaw John B.
Heber Ulrich
Heintze J. R. W.
Heiter Ulrike
Henrichs Hubertus F.
Herbig George H.
Heske Astrid
Hessman Frederic Victor
Hill Grant
Hill Vanessa M.
Hinkle Kenneth H.
Hirai Masanori
Hirata Ryuko
Hoeflich Peter
Horaguchi Toshihiro
Houk Nancy
Houziaux Leo
Hron Josef
Hubert-Delplace Anne-Marie
Hubrig Swetlana
Huenemoerder David P.
Hyland Harry R. Harry
Israelian Garik
Ivans Inese
Izumiura Hideyuki
Jankov Slobodan S.
Jehin Emmanuel
Johnson Jennifer A.
Johnson Hollis R.
Jordan Carole
Jugaku Jun
Kaeufl Hans Ulrich
Kipper Tonu
Kitchin Christopher R.
Klochkova Valentina G.
Kochukhov Oleg
Kodaira Keiichi
Kogure Tomokazu
Kolka Indrek
Kordi Ayman S.
Korn Andreas J.
Korotin Sergey
Kotnik-Karuza Dubravka
Koubsky Pavel
Kovachev Bogomil Jivkov
Kovtyukh Valery V.
Kraft Robert P.
Krempec-Krygier Janina
Kwok Sun
Lago Maria T. V. T.
Lagrange Anne-Marie
Laird John B.
Lambert David L.
Lamers Henny J. G. L. M.
Lamontagne Robert
Landstreet John D.
Lanz Thierry
Le Contel Jean-Michel
Lebre Agnes
Leckrone David S.
Lee Jae-Woo
Leedjarv Laurits
Lester John B.
Leushin Valerij V.
Levato Hugo
Liebert James W.
Little-Marenin Irene R.
Lodders Katharina
Lopes Dalton De faria
Lubowich Donald A.
Luck R. Earle
Lundstrom Ingemar
Lyubimkov Leonid S.
Magain Pierre
Magazzu Antonio
Maillard Jean-Pierre
Maitzen Hans M.
Malaroda Stella M.
Manteiga Outeiro Minia
Marilli Ettore
Marsden Stephen C.
Massey Philip L.
Mathys Gautier
Matsuura Mikako
Mazzali Paolo A.
McDavid David A.
McGregor Peter John
McNamara Delbert H.
McSaveney Jennifer A.
McSwain Mary V.
Megessier Claude
Melendez Jorge
Melo Claudio H.

Mickaelian Areg M.
Mikulasek Zdenek
Moffat Anthony F. J.
Molaro Paolo
Monin Dmitry
Montes David
Moos Henry Warren
Morossi Carlo
Morrison Nancy D.
Napiwotzki Ralf
Nazarenko Victor
Neckel Heinz
Neiner Coralie
Nicholls Ralph W.
Niedzielski Andrzej
Niemela Virpi S.(†)
Nilsson Hampus
Nishimura Shiro E.
Norris John
North Pierre
Nugis Tiit
Okazaki Atsuo T.
O'Neal Douglas B.
O'Toole Simon J.
Owocki Stanley Peter
Pagel Bernard E. J(†)
Pakhomov Yuri V.
Pallavicini Roberto
Parsons Sidney B.
Pasinetti Laura E.
Pavlenko Yakov V.
Pedoussaut Andre
Peery Benjamin F.
Perrin Marie-Noel
Peters Geraldine Joan
Peterson Ruth Carol
Petit Pascal
Pilachowski Catherine
Pintado Olga I.
Plavec Mirek J.
Plez Bertrand
Polcaro V. F.
Polidan Ronald S.
Polosukhina-Chuvaeva Nina
Pompeia Luciana
Porto de Mello Gustavo F.
Praderie Françoise
Primas Francesca
Prinja Raman
Querci Francois R.
Querci Monique
Raassen Ion
Rao N. Kameswara
Rashkovskij Sergey L.
Rautela B. S.
Rauw Gregor
Rebolo Rafael
Rego Fernandez M.
Reimers Dieter
Rettig Terrence W.
Ringuelet Adela E.
Romanyuk Iosif I.
Rose James Anthony
Rossi Lucio
Rossi Corinne
Rutten Robert J.
Ryan Sean Gerard
Ryde Nils A. E.
Sachkov Mikhail E.
Sadakane Kozo
Sahade Jorge
Sanchez Almeida J.
Sanwal Basant Ballabh
Sareyan Jean-Pierre
Sarre Peter J.
Sbordone Luca
Schild Rudolph E.
Schroeder Klaus Peter
Schuh Sonja
Seggewiss Wilhelm
Shetrone Matthew C.
Sholukhova Olga
Shore Steven N.
Simon Theodore
Simone Zaggia
Singh Mahendra
Sinnerstad Ulf E.
Smalley Barry
Smith Myron A.
Smith Graeme H.
Smith Verne V.
Snow Theodore P.
Soderblom David R.
Sonneborn George
Sonti Sreedhar Rao
Spite François M.
Spite Monique
Stalio Roberto
Stateva Ivanka K.
Stathakis Raylee A.
Stawikowski Antoni
Stecher Theodore P.
Steffen Matthias
Stefl Stanislav
Stencel Robert Edward
St-Louis Nicole
Suntzeff Nicholas B.
Svolopoulos Sotirios
Swings Jean-Pierre
Szeifert Thomas
Talavera Antonio
Tautvaisiene Grazina
Thevenin Frederic
Torrejon Jose Miguel
Tuominen Ilkka V.
Usenko Igor
Utrobin Victor P.
Utsumi Kazuhiko
Valenti Jeff A.
Valtier Jean-Claude
Valyavin Gennady G.
van der Hucht Karel A.
Van Eck Sophie
van Winckel Hans
van't Veer-Menneret Claude
Vasu-Mallik Sushma
Verdugo Eva
Verheijen Marc A. W.
Vilhu Osmi
Viotti Roberto
Vladilo Giovanni
Vogt Steven Scott
Vogt Nikolaus
Vreux Jean Marie
Wade Gregg A.
Wallerstein George
Waterworth Michael
Wegner Gary Alan
Wehinger Peter A.
Williams Peredur M.
Wing Robert F.
Wolf Bernhard
Wolff Sidney C.
Wood H. J.
Wyckoff Susan
Yamashita Yasumasa
Yoshioka Kazuo
Yushkin Maxim V.
Zhang Huawei
Zhu Zhenxi
Zorec Juan
Zverko Juraj

MEMBERSHIP BY COMMISSION

Composition of Commission 30

Radial Velocities / Vitesses Radiales

President Udry Stephane

Vice-President Torres Guillermo

Organizing Committee

Fekel Francis C.
Freeman Kenneth C.
Glushkova Elena V.
Marcy Geoffrey W.
Mathieu Robert D.
Nordstrom Birgitta
Pourbaix Dimitri
Turon Catherine
Zwitter Tomaz

Members

Abt Helmut A.
Al-Malki M. B.
Andersen Johannes
Arnold Richard A.
Balona Luis A.
Batten Alan H.
Beavers Willet I.
Beers Timothy C.
Bernstein Hans-Heinrich
Beuzit Jean-Luc
Boulon Jacques J.
Breger Michel
Burki Gilbert
Carney Bruce William
Cochran William David
Crampton David
Crifo Francoise
da Costa Luiz A. N.
Davis Marc
Davis Robert J.
de Jonge J. K.
de Medeiros Jose-Renan
Dravins Dainis
Dubath Pierre
Edmondson Frank K.
Elkin Vladimir
Fairall Anthony P.
Fehrenbach Charles
Fletcher J. Murray
Florsch Alphonse
Foltz Craig B.
Forveille Thierry
Garcia Beatriz E.
Georgelin Yvon P.
Gilmore Gerard F.
Giovanelli Riccardo
Gnedin Yurij N.
Gonzalez Jorge F.
Gouguenheim Lucienne
Gray David F.
Griffin Roger F.
Halbwachs Jean-Louis
Hearnshaw John B.
Heintze J. R. W.
Hewett Paul
Hilditch Ronald W.
Hill Graham
Holmberg Johan
Hrivnak Bruce J.
Hube Douglas P.
Hubrig Swetlana
Huchra John Peter
Imbert Maurice
Irwin Alan W.
Jorissen Alain
Kadouri Talib Hadi
Karachentsev Igor D.
Katz David A.
Khalesseh Bahram
Kraft Robert P.
Latham David W.
Levato Hugo
Lewis Brian Murray
Lindgren Harri
Marschall Laurence A.
Maurice Eric N.
Mayor Michel
Mazeh Tsevi
McMillan Robert S.
Melnick Gary J.
Mermilliod Jean-Claude
Meylan Georges
Mink Douglas J.
Missana Marco
Mkrtichian David E.
Morbey Christopher L.
Morrell Nidia
Naef Dominique
Napolitano Nicola R.
Oetken L.
Pedoussaut Andre
Pellegrini Paulo S. S.
Pepe Francesco A.
Perrier-Bellet Christian
Peterson Ruth Carol
Philip A. G. Davis
Popov Viktor S.
Preston George W.
Quintana Hernan
Rastorguev Aleksej S.
Ratnatunga Kavan U.
Romanov Yuri S.
Royer Frédéric
Rubenstein Eric
Rubin Vera C.
Sachkov Mikhail E.
Samus Nikolaj N.
Santos Nuno Miguel C.
Scarfe Colin David

Schroder Anja C.
Simone Zaggia
Sivan Jean-Pierre
Skuljan Jovan
Smith Myron A.
Solivella Gladys R.
Stefanik Robert
Steinmetz Matthias
Stickland David J.
Suntzeff Nicholas B.
Szabados Laszlo
Szecsenyi-Nagy Gábor A.
Tokovinin Andrei A.
Tonry John
van Dessel Edwin Ludo
Verschueren Werner
Vinko Jozsef
Walker Gordon A. H.
Wegner Gary Alan
Willstrop Roderick V.
Yang Stephenson L. S.
Yoss Kenneth M.

MEMBERSHIP BY COMMISSION

Composition of Commission 31

Time / Temps (L'Heure)

President Defraigne Pascale

Vice-President Manchester Richard N.

Organizing Committee

Hosokawa Mizuhiko
Leschiutta S.
Matsakis Demetrios N.
Petit Gérard

Members

Abele Maris K.
Afanasjeva Praskovja M.
Ahn Youngsook
Allan David W.
Alley Carrol O.
Aoki Shinko
Archinal Brent A.
Arias Elisa Felicitas
Bender Peter L.
Breakiron Lee Allen
Brumberg Victor A.
Bruyninx Carine
Carter William Eugene
Chou Yi
Dehant Véronique
Dick Wolfgang R.
Dickey Jean O'Brien
Douglas R. J.
Fallon Frederick W.
Fliegel Henry F.
Fujimoto Masa-Katsu
Fukushima Toshio
Gambis Daniel
Gang Zhang Shou
Gao Yuping
Granveaud Michel
Guinot Bernard R.
Han Tianqi
Hers Jan
Hu Yonghui
Iijima Shigetaka
Ilyasov Yuri P.
Ivanov Dmitrij V.
Jin WenJing
Kakuta Chuichi
Klepczynski William J.
Kolaczek Barbara
Koshelyaevsky Nikolay B.
Kovalevsky Jean
Kwok Sun
Lieske Jay H.
Lu BenKui
Luck John M.
Luo Dingchang
Ma Zhenguo
McCarthy Dennis D.
Meinig Manfred
Melbourne William G.
Mendes Virgilio B.
Millar Thomas J.
Morgan Peter
Mueller Ivan I.
Naumov Vitalij A.
Nelson Robert A.
Newhall X. X.
Noel Fernando
Paquet Paul
Pilkington John D. H.
Pineau des Forets Guillaume
Popelar Josef
Pushkin Sergej B.
Ray James R.
Robertson Douglas S.
Rodin Alexander E.
Sheikh Suneel I.
Smylie Douglas E.
Song Jinan
Stanila George
Stappers Benjamin W.
Thomas Claudine
Tuckey Philip A.
Vernotte François
Vicente Raimundo O.
Wilkins George A.
Wu Shouxian
Wu Dong Shao
Wu Haitao
Wu Guichen
Yatskiv Yaroslav S.
Ye Shuhua

MEMBERSHIP BY COMMISSION

Composition of Commission 33

Structure & Dynamics of the Galactic System

Structure & Dynamique du Système Galactique

President Gerhard Ortwin

Vice-President Wyse Rosemary F.

Organizing Committee

Efremov Yurij N.
Evans Wyn
Flynn Chris
Grindlay Jonathan E.
Lazio Joseph
Nordstrom Birgitta
Whitelock Patricia Ann
Yuan Chi

Members

Aarseth Sverre J.
Acosta Pulido Jose A.
Adamson Andrew
Afanas'ev Viktor L.
Aguilar Luis A. C.
Alcobe Santiago
Allende Prieto Carlos
Altenhoff Wilhelm J.
Ambastha Ashok K.
Andersen Johannes
Antonov Vadim A.
Aoki Shinko
Ardeberg Arne L.
Ardi Eliani
Arnold Richard A.
Asteriadis Georgios
Athanassoula Evangelia
Babusiaux Carine
Baier Frank W.
Balazs Lajos G.
Balbus Steven A.
Balcells Marc
Baldwin John E.
Banhatti Dilip G.
Baranov Alexander S.
Barberis Bruno
Bartasiute Stanislava
Bash Frank N.
Basu Baidyanath
Baud Boudewijn
Bellazzini Michele
Bensby Thomas L.
Berkhuijsen Elly M.
Bienayme Olivier
Binney James J.
Blaauw Adriaan
Blanco Victor M.
Blitz Leo
Bloemen Hans (J. B. G. M.)
Blommaert Joris A. D. L.
Bobylev Vadim V.
Boulon Jacques J.
Brand Jan
Bronfman Leonardo
Brown Warren R.
Burke Bernard F.
Burton W. Butler
Butler Raymond F.
Caldwell John A. R.
Cane Hilary V.
Carollo Daniela
Carpintero Daniel Diego
Carrasco Luis
Caswell James L.
Cesarsky Diego A.
Cesarsky Catherine J.
Cha Seung-Hoon
Chakrabarty Dalia
Chapman Jessica
Chen Li
Christodoulou Dmitris
Churchwell Edward B.
Cincotta Pablo M.
Cioni Maria-Rosa L.
Ciurla Tadeusz
Clemens Dan P.
Clube S. V. M.
Cohen Richard S.
Comins Neil Francis
Contopoulos George
Costa Edgardo
Courtes Georges
Crampton David
Crawford David L.
Creze Michel
Cropper Mark
Cubarsi Rafael
Cudworth Kyle McC.
Cuisinier Francois C.
Cuperman Sami
Dambis Andrei K.
Dauphole Bertrand
Davies Rodney D.
Dawson Peter
de Jong Teije
Dejonghe Herwig B.
Dekel Avishai
Diaferio Antonaldo
Dickel Helene R.
Dickel John R.
Dickman Robert L.
Dieter Conklin Nannielou H.

Djorgovski Stanislav
Do Nascimento Jose D.
Downes Dennis
Drilling John S.
Drimmel Ronald E.
Ducati Jorge R.
Ducourant Christine
Dzigvashvili R. M.
Edmondson Frank K.
Egret Daniel
Einasto Jaan
Elmegreen Debra M.
Evangelidis E. A.
Faber Sandra M.
Feast Michael W.
Fehrenbach Charles
Feitzinger Johannes
Fenkart Rolf P.
Ferguson Annette M.
Fernandez David
Figueras Francesca
Freeman Kenneth C.
Fridman Aleksej M.
Fuchs Burkhard
Fujimoto Masa-Katsu
Fujiwara Takao
Fux Roger M.
Galletto Dionigi
Garzon Francisco L.
Gemmo Alessandra
Genkin Igor L.
Genzel Reinhard
Georgelin Yvon P.
Gilmore Gerard F.
Goldreich Peter
Gomez Ana E.
Gordon Mark A.
Gottesman Stephen T.
Grayzeck Edwin J.
Green Anne M.
Grenon Michel
Gupta Sunil K.
Habe Asao
Habing Harm J.
Hakkila Jon E.
Hamajima Kiyotoshi
Hanami Hitoshi
Hartkopf William I.
Hawkins Michael R. S.
Hayli Abraham
Heiles Carl
Helmi Amina
Herbst William
Hernandez-Pajares Manuel
Hetem Jr. Annibal
Hobbs Robert W.
Holmberg Johan
Honma Mareki
Hori Genichiro
Hozumi Shunsuke
Hron Josef
Hulsbosch A. N. M.
Humphreys Roberta M.
Hunter Christopher
Iguchi Osamu
Ikeuchi Satoru
Inagaki Shogo
Innanen Kimmo A.
Isobe Syuzo(†)
Israel Frank P.
Ivezic Zeljko
Iwaniszewska Cecylia
Iye Masanori
Jablonka Pascale
Jackson Peter D.
Jahreiss Hartmut
Jalali Mir Abbas
Jasniewicz Gerard
Jiang Dongrong
Jiang Ing-Guey
Jog Chanda J.
Johnson Hugh M.
Jones Derek H. P.
Kalandadze N. B.
Kalnajs Agris J.
Kang Yong Hee
Kasumov Fikret K. O.
Kato Shoji
Kim Sungsoo S.
King Ivan R.
Kinman Thomas D.
Klare Gerhard
Knapp Gillian R.
Korchagin Vladimir
Kormendy John
Kulsrud Russell M.
Kutuzov Sergej A.
Lafon Jean-Pierre J.
Larson Richard B.
Latham David W.
Lecar Myron
Lee Sang Gak
Lee Hyung Mok
Lee Kang Hwan
Liebert James W.
Lin Qing
Lin Chia C.
Lindblad Per O.
Lockman Felix J.
Loden Kerstin R.
MacConnell Darrell Jack
Macrae Donald A.(†)
Manchester Richard N.
Marochnik L. S.
Martin Christopher L.
Martinet Louis
Martos Marco A.
Mathewson Donald S.
Matteucci Francesca
Mavridis Lyssimachos N.
Mayor Michel
McClure-Griffiths Naomi M.
McGregor Peter John
Méndez René A.
Merrifield Michael R.
Mezger Peter G.
Mikkola Seppo
Miller Richard H.
Mirabel Igor Felix
Mishurov Yuri N.
Miyamoto Masanori
Moffat Anthony F. J.
Mohammed Ali Talib
Moitinho André
Monet David G.
Monnet Guy J.
Morales Rueda Luisa
Moreno Lupianez Manuel
Morris Rhys H.
Morris Mark Root
Muench Guido
Nakasato Naohito
Namboodiri P. M. S.
Napolitano Nicola R.
Neckel Th.
Nelemans Gijs
Nelson Alistair H.
Nikiforov Igor I.
Ninkovic Slobodan

Nishida Mitsugu
Nishida Minoru
Norman Colin A.
Nuritdinov Salakhutdin
Oblak Edouard
Odenkirchen Michael
Oh Kap-Soo
Oja Tarmo
Ojha Devendra K.
Okuda Haruyuki
Olano Carlos A.
Ollongren A.
Orlov Victor V.
Ortiz Roberto
Ostriker Eve C.
Ostriker Jeremiah P.
Palmer Patrick E.
Palous Jan
Pandey A. K.
Pandey Bierndra P.
Papayannopoulos Theodoros
Parmentier Genevieve A.
Patsis Panos
Pauls Thomas Albert
Peimbert Manuel
Perek Lubos
Perryman Michael A. C.
Pesch Peter
Philip A. G. Davis
Pier Jeffrey R.
Polyachenko Evgeny V.
Portinari Laura
Price R. M.
Rabolli Monica
Raharto Moedji
Ratnatunga Kavan U.
Reid Neill
Reif Klaus
Reyle Céline
Rich Robert M.
Riegel Kurt W.
Roberts Morton S.
Roberts Jr William W.
Robin Annie C. R
Rocha-Pinto Helio J.
Rohlfs Kristen
Rong Jianxiang
Rubin Vera C.
Ruelas-Mayorga R. A.
Ruiz Maria Teresa
Rybicki George B.
Saar Enn
Sakano Masaaki
Sala Ferran
Sanchez-Saavedra M. Luisa
Sandqvist Aage
Santillan Alfredo J.
Sanz Jaume
Sargent Annelia I.
Schechter Paul L.
Schmidt Maarten
Schmidt-Kaler Theodor
Seggewiss Wilhelm
Seimenis John
Sellwood Jerry A.
Serabyn Eugene
Shane William W.
Shu Frank H.
Sigalotti Leonardo G.
Simone Zaggia
Simonson S. Christian
Sobouti Yousef
Solomon Philip M.
Sotnikova Natalia Y.
Soubiran Caroline
Sparke Linda
Spergel David N.
Spiegel E.
Stecker Floyd W.
Steinlin Uli
Stibbs Douglas W. N.
Strobel Andrzej
Surdin Vladimir G.
Svolopoulos Sotirios
Sygnet Jean-Francois
Tammann Gustav A.
Terzides Charalambos
The Pik-Sin
Thielheim Klaus O.
Thomas Claudine
Tinney Christopher G.
Tobin William
Tomisaka Kohji
Toomre Juri
Toomre Alar
Torra Jordi
Tosa Makoto
Trefzger Charles F.
Tsujimoto Takuji
Turon Catherine
Upgren Arthur R.
Valtonen Mauri J.
van der Kruit Pieter C.
van Woerden Hugo
Vandervoort Peter O.
Varela Perez Antonia M.
Vega E. I.
Venugopal V. R.
Vergne María Marcela
Verschuur Gerrit L.
Vetesnik Miroslav
Villas da Rocha Jaime F.
Vivas Anna K.
Volkov Evgeni V.
Volonteri Marta
Voroshilov Volodymyr I.
Wachlin Felipe C.
Weaver Harold F.
Weistrop Donna
Westerhout Gart
Westerlund Bengt E.
Whiteoak John B.
Whittet Douglas C. B.
Wielebinski Richard
Wielen Roland
Woltjer Lodewijk
Woodward Paul R.
Wouterloot Jan G. A.
Wramdemark Stig S.
Wunsch Richard
Yamagata Tomohiko
Yim Hong-Suh
Yoshii Yuzuru
Younis Saad M.
Zachilas Loukas

MEMBERSHIP BY COMMISSION

Composition of Commission 34

Interstellar Matter / Matière Interstellaire

President Millar Thomas J.

Vice-President Chu You-Hua

Organizing Committee

Breitschwerdt Dieter
Burton Michael G.
Cabrit Sylvie
Caselli Paola
De Gouveia Dal Pino Elisabete M.
Dyson John E.
Ferland Gary J.
Juvela Mika J.
Koo Bon-Chul
Kwok Sun
Lizano Susana
Rozyczka Michal
Toth Laszlo V.
Tsuboi Masato
Yang Ji

Members

Aannestad Per Arne
Abgrall Herve
Acker Agnes
Adams Fred
Aiad A. Zaki
Aikawa Yuri
Aitken David K.
Akabane Kenji A.
Alcolea Javier
Al-Mostafa Zaki A.
Altenhoff Wilhelm J.
Alves Joao F.
Anantharamaiah Kuduvalli R.(†)
Andersen Anja C.
Andersen Morten
Andersson Bengt Goeran
Andrillat Yvette
Andronov Ivan
Anglada Guillem
Arkhipova Vera P.
Arny Thomas T.
Arthur Jane
Audard Marc
Avery Lorne W.
Axford W. Ian
Azcarate Diana E.
Baars Jacob W. M.
Baart Edward E.
Bachiller Rafael
Baker Andrew J.
Baldwin John E.
Ballesteros-Paredes Javier
Balser Dana S.
Baluteau Jean-Paul
Bania Thomas Michael
Barlow Michael J.
Barnes Aaron
Bash Frank N.
Basu Shantanu
Baudry Alain
Bautista Manuel A.
Becklin Eric E.
Beckman John E.
Beckwith Steven V. W.
Bedogni Roberto
Benaydoun Jean-Jacques
Bergeron Jacqueline A.
Bergin Edwin A.
Bergstrom Lars
Berkhuijsen Elly M.
Bernat Andrew Plous
Bertout Claude
Bhat Ramesh N D
Bhatt H. C.
Bianchi Luciana
Bieging John Harold
Bignell R. Carl
Binette Luc
Birkle Kurt
Black John Harry
Blades John Chris
Blair Guy Norman
Blair William P.
Bless Robert C.
Blitz Leo
Bloemen Hans (J. B. G. M.)
Bobrowsky Matthew
Bocchino Fabrizio
Bochkarev Nikolai G.
Bode Michael F.
Bodenheimer Peter
Boeshaar Gregory Orth
Boggess Albert
Bohlin Ralph C.
Boisse Patrick
Boland Wilfried
Bontemps Sylvain
Bordbar Gholam H.
Borgman Jan
Borkowski Kazimierz J.
Boulanger Francois
Boumis Panayotis
Bourke Tyler L.
Bouvier Jerôme
Brand Peter W. J.l.
Brand Jan
Briceno César
Brinkmann Wolfgang
Bromage Gordon E.

Brooks Kate J.
Brouillet Nathalie
Brown Ronald D.
Bruhweiler Frederick C.
Bryce Myfanwy
Bujarrabal Valentin
Burgess Alan
Burke Bernard F.
Burton W. Butler
Bykov Andrei M.
Bystrova Natalja V.
Bzowski Maciej
Cambresy Laurent
Canto Jorge
Caplan James
Cappa de Nicolau Cristina
Capriotti Eugene R.
Capuzzo Dolcetta Roberto
Carretti Ettore
Carruthers George R.
Castaneda Héctor F.
Castelletti Gabriela
Caswell James L.
Cecchi-Pellini Cesare
Centurion Martin Miriam
Cernicharo Jose
Cerruti Sola Monica
Cersosimo Juan Carlos
Cesarsky Diego A.
Cesarsky Catherine J.
Cha Seung-Hoon
Chandra Suresh
Chen Yang
Chen Yafeng
Cheng Kwang Ping
Chevalier Roger A.
Chini Rolf
Chopinet Marguerite
Churchwell Edward B.
Ciardullo Robin
Clark Frank Oliver
Clarke David A.
Clegg Robin E. S.
Code Arthur D.
Codella Claudio
Cohen Marshall H.
Colangeli Luigi
Collin Suzy
Colomb Fernando R.
Combes Françoise
Corbelli Edvige
Corradi Wagner J. B.
Costero Rafael
Courtes Georges
Cowie Lennox Lauchlan
Cox Donald P.
Cox Pierre
Coyne George V.
Crane Philippe
Crawford Ian Andrew
Crovisier Jacques
Cruvellier Paul E.
Cudaback David D.
Cunningham Maria R.
Czyzak Stanley J.
Dahn Conard Curtis
Dalgarno Alexander
Danks Anthony C.
Danly Laura
Davies Rodney D.
Davies Jonathan Ivor
Davis Christopher J.
de Almeida Amaury A.
De Avillez Miguel P.
de Boer Klaas Sjoerds
De Buizer James M.
de Jong Teije
de la Noe Jerome
De Marco Orsola
Decourchelle Anne C.
Deguchi Shuji
Deharveng Lise
Deiss Bruno M.
Dennefeld Michel
Dent William
Dewdney Peter E. F.
d'Hendecourt Louis
Di Fazio Alberto
Dias da Costa Roberto D.
Dickel Helene R.
Dickel John R.
Dickey John M.
Dinerstein Harriet L.
Dinh Van M.
Disney Michael J.
Djamaluddin Thomas
d'Odorico Sandro
Dokuchaev Vyacheslav I.
Dokuchaeva Olga D.
Dominik Carsten
Donn Bertram D.
Dopita Michael A.
Dorschner Johann
Dottori Horacio A.
Downes Dennis
Draine Bruce T.
Dreher John W.
Dubner Gloria M.
Dubout Renee
Dudorov Alexander E.
Dufour Reginald James
Duley Walter W.
Dupree Andrea K.
Duvert Gilles
Dwarkadas Vikram V.
Dwek Eli
Egan Michael P.
Ehlerova Sona
Eisloeffel Jochen
Elia Davide
Elitzur Moshe
Elliott Kenneth Harrison
Elmegreen Bruce Gordon
Elmegreen Debra M.
Elvius Aina M.
Emerson James P.
Encrenaz Pierre J.
Escalante Vladimir
Esipov Valentin F.
Esteban César
Evans Neal J.
Evans Aneurin
Falgarone Edith
Falk Jr Sydney W.
Falle Samuel A.
Federman Steven Robert
Feitzinger Johannes
Felli Marcello
Felten James E.
Fendt Christian
Ferlet Roger
Fernandes Amadeu
Ferriere Katia M.
Ferrini Federico
Fesen Robert A.
Fiebig Dirk
Field David
Field George B.
Fierro Julieta
Fischer Jacqueline

Kamijo Fumio
Kanekar Nissim
Kantharia Nimisha G.
Kassim Namir E.
Kazes Ilya
Keene Jocelyn Betty
Kegel Wilhelm H.
Kennicutt Robert C.
Khesali Ali R.
Khromov Gavriil S.
Kim Jongsoo
Kimura Toshiya
King David Leonard
Kirkpatrick Ronald C.
Kirshner Robert Paul
Klessen Ralf S.
Knacke Roger F.
Knapp Gillian R.
Knude Jens Kirkeskov
Ko Chung-Ming
Kobayashi Naoto
Kohoutek Lubos
Koike Chiyoe
Kondo Yoji
Koornneef Jan F.
Korpi Maarit J.
Kostyakova Elena B.
Kozasa Takashi
Kramer Busaba H.
Krautter Joachim
Kravchuk Sergei
Kreysa Ernst
Krishna Swamy K. S.
Kuan Yi-Jehng
Kudoh Takahiro
Kuiper Thomas B. H.
Kumar C. Krishna
Kundu Mukul R.
Kunth Daniel
Kutner Marc Leslie
Kwitter Karen Beth
Kylafis Nikolaos D.
Lada Charles Joseph
Lafon Jean-Pierre J.
Langer William David
Latter William B.
Laureijs Rene J.
Laurent Claudine
Lauroesch James T.
Lazio Joseph
Le Squeren Anne-Marie
Lee Terence J.
Lee Hee-Won
Lee Dae-Hee
Leger Alain
Lehtinen Kimmo K.
Leisawitz David
Lepine Jacques R. D.
Lequeux James
Leto Giuseppe
Leung Chun Ming
Ligori Sebastiano
Likkel Lauren Jones
Liller William
Limongi Marco
Lin Chia C.
Linke Richard Alan
Linnartz Harold
Lis Dariusz C.
Liseau Rene
Liszt Harvey Steven
Lo Kwok-Yung
Lockman Felix J.
Lodders Katharina
Loinard Laurent R.
Lopez Jose Alberto
Loren Robert Bruce
Lortet Marie-Claire
Louise Raymond
Lovas Francis John
Low Frank J.
Lozinskaya Tatjana A.
Lucas Robert
Luo Shaoguang
Lynds Beverly T.
Lyon Ian C.
Mac Low Mordecai-Mark
Maciel Walter J.
MacLeod John M.
Madsen Gregory J.
Maihara Toshinori
Makiuti Sin'itirou
Malbet Fabien
Mampaso Antonio
Manchado Arturo T.
Manchester Richard N.
Manfroid Jean
Marston Anthony Philip
Martin Peter G.
Martin Robert N.
Martin Christopher L.
Martin-Pintado Jesus
Masson Colin R.
Mather John Cromwell
Mathews William G.
Mathewson Donald S.
Mathis John S.
Matsuhara Hideo
Matsumoto Tomoaki
Matsumura Masafumi
Matthews Brenda C.
Mattila Kalevi
Mauersberger Rainer
McCall Marshall Lester
McClure-Griffiths Naomi M.
Mccombie June
McCray Richard
McGee Richard Xavier
McGregor Peter John
McKee Christopher F.
McNally Derek
Meaburn John
Mebold Ulrich
Mehringer David Michael
Meier Robert R.
Meixner Margaret
Mellema Garrelt
Melnick Gary J.
Mennella Vito
Menon T. K.
Menzies John W.
Meszaros Peter
Mezger Peter G.
Miller Joseph S.
Milne Douglas K.
Minn Young Key
Minter Anthony H.
Mitchell George F.
Miyama Syoken
Mizuno Shun
Mo Jinger
Monin Jean-Louis
Montmerle Thierry
Moore Marla H.
Moreno-Corral Marco A.
Morgan David H.
Moriarty-Schieven Gerald H.
Morimoto Masaki

Morris Mark Root
Morton Donald C.
Mouschovias Telemachos Ch.
Muench Guido
Mufson Stuart Lee
Mulas Giacomo
Muller Erik M.
Murthy Jayant
Myers Philip C.
Nagata Tetsuya
Nakada Yoshikazu
Nakagawa Takao
Nakamoto Taishi
Nakamura Fumitaka
Nakano Makoto
Nakano Takenori
Natta Antonella
Neugebauer Gerry
Nguyen-Quang Rieu
Nikolic Silvana
Nishi Ryoichi
Nordh Lennart H.
Norman Colin A.
Nulsen Paul E. J.
Nurnberger Dieter E. A.
Nussbaumer Harry
Nuth Joseph A. III
O'Dell Charles R.
O'Dell Stephen L.
Ohtani Hiroshi
Okuda Haruyuki
Okumura Shin-ichiro
Olofsson Hans
Omont Alain
Omukai Kazuyuki
Onaka Takashi
Onello Joseph S.
Opendak Michael
Orlando Salvatore
Osborne John L.
Osterbrock Donald E.(†)
Ostriker Eve C.
Ott Juergen A.
Pagani Laurent P.
Pagano Isabella
Pagel Bernard E. J.(†)
Palla Francesco
Palmer Patrick E.
Palumbo Maria Elisabetta
Panagia Nino
Pandey Bierndra P.
Pankonin Vernon Lee
Parker Eugene N.
Parthasarathy Mudumba
Pauls Thomas Albert
Pecker Jean-Claude
Peimbert Manuel
Pellegrini Silvia
Pena Miriam
Pendleton Yvonne Jean
Penzias Arno A.
Pequignot Daniel
Perault Michel
Perinotto Mario(†)
Persi Paolo
Peters William L. III
Petrosian Vahe
Petuchowski Samuel J.
Philipp Sabine D.
Phillips Thomas Gould
Phillips John Peter
Pineau des Forets Guillaume
Pittard Julian M.
Plume Rene
Poeppel Wolfgang G. L.
Pollacco Don
Pongracic Helen
Pontoppidan Klaus M.
Porceddu Ignazio E. P.
Pottasch Stuart R.
Pound Marc W.
Pouquet Annick
Prasad Sheo S.
Preite Martinez Andrea
Price R. M.
Prochaska Jason X.
Pronik Iraida I.
Prusti Timo
Puget Jean-Loup
Qin Zhihai
Radhakrishnan V.
Raimond Ernst
Ramirez Jose M.
Ratag Mezak Arnold
Rawlings Jonathan
Raymond John Charles
Redman Matthew P.
Reipurth Bo
Rengarajan T. N.
Rengel Miriam E.
Reyes Rafael Carlos
Reynolds Ronald J.
Reynoso Estela M.
Rickard Lee J.
Roberge Wayne G.
Roberts Jr William W.
Robinson Garry
Roche Patrick F.
Rodriguez Monica
Rodriguez Luis F.
Roelfsema Peter
Roeser Hans-peter
Roger Robert S.
Rogers Alan E. E.
Rohlfs Kristen
Rosa Michael Richard
Rosado Margarita
Rose William K.
Rouan Daniel
Roxburgh Ian W.
Rozhkovskij Dimitrij A.
Rubin Robert Howard
Ryabov Michael I.
Sabano Yutaka
Sabbadin Franco
Sahu Kailash C.
Saigo Kazuya
Sakano Masaaki
Salama Farid
Salinari Piero
Salpeter Edwin E.
Salter Christopher John
Samodurov Vladimir A.
Sanchez-Saavedra M. Luisa
Sancisi Renzo
Sandell Goran Hans l
Sandqvist Aage
Sarazin Craig L.
Sargent Annelia I.
Sarma N. V. G.
Sarre Peter J.
Sato Shuji
Sato Fumio
Savage Blair D.
Savedoff Malcolm P.
Scalo John Michael
Scappini Flavio
Scarrott Stanley M.

Schatzman Evry
Scherb Frank
Schilke Peter
Schlemmer Stephan
Schmid-Burgk J.
Schmidt-Kaler Theodor
Schroder Anja C.
Schulz R. Andreas
Schwartz Philip R.
Schwartz Richard D.
Schwarz Ulrich J.
Scott Eugene Howard
Scoville Nicholas Z.
Seaton Michael J.(†)
Seki Munezo
Sellgren Kristen
Sembach Kenneth R.
Shadmehri Mohsen
Shane William W.
Shao Cheng-yuan
Shapiro Stuart L.
Sharpless Stewart
Shaver Peter A.
Shawl Stephen J.
Shchekinov Yuri A.
Shematovich Valerij I.
Sherwood William A.
Shields Gregory A.
Shipman Russell F.
Shmeld Ivar
Shu Frank H.
Shull John Michael
Shull Peter Otto
Shustov Boris M.
Siebenmorgen Ralf
Sigalotti Leonardo G.
Silich Sergey
Silk Joseph I.
Silva Laura
Silvestro Giovanni
Simons Stuart
Sitko Michael L.
Sivan Jean-Pierre
Skilling John
Skulskyj Mychajlo Y.
Slane Patrick
Sloan Gregory Clayton
Smith Michael
Smith Randall K.
Smith Tracy L.
Smith Barham W.
Smith Peter L.
Smith Craig H.
Smith Robert G.
Snell Ronald L.
Snow Theodore P.
Sobolev Andrej M.
Sofia Sabatino
Sofia Ulysses J.
Sofue Yoshiaki
Solc Martin
Solomon Philip M.
Somerville William B.
Song In-Ok
Spaans Marco
Stahler Steven W.
Stanga Ruggero
Stanghellini Letizia
Stanimirovic Snezana
Stapelfeldt Karl R.
Stark Ronald
Stasinska Grazyna
Stecher Theodore P.
Stecklum Bringfried
Stenholm Lars
Stenholm Björn
Stone James McLellan
Strom Richard G.
Suh Kyung-Won
Sun Jin
Sutherland Ralph S.
Suzuki Tomoharu
Swade Daryl Allen
Sylvester Roger
Szczerba Ryszard
Tachihara Kengo
Tafalla Mario
Takahashi Junko
Takakubo Keiya
Takano Toshiaki
Tamura Shin'ichi
Tamura Motohide
Tanaka Masuo
Taylor Kenneth N. R.
Tenorio-Tagle Guillermo
Terzian Yervant
Testi Leonardo
Thaddeus Patrick
The Pik-Sin
Thompson A. Richard
Thonnard Norbert
Thronson Jr Harley Andrew
Tilanus Remo P. J.
Tokarev Yurij V.
Torrelles Jose M.
Torres-Peimbert Silvia
Tosi Monica
Tothill Nicholas F. H.
Townes Charles Hard
Trammell Susan R.
Treffers Richard R.
Turner Barry E.
Turner Kenneth C.
Tyul'bashev Sergei A.
Ulrich Marie-Helene D.
Urosevic Dejan V.
Van de Steene Griet C.
van den Ancker Mario E.
van der Hulst Jan M.
van der Laan Harry
van der Tak Floris F. S.
van Dishoeck Ewine F.
van Gorkom Jacqueline H.
van Woerden Hugo
VandenBout Paul A.
Varshalovich Dmitrij A.
Velazquez Pablo F.
Verheijen Marc A. W.
Verner Ekaterina
Verschuur Gerrit L.
Viala Yves
Viallefond Francois
Vidal Jean-Louis
Vidal-Madjar Alfred
Viegas Sueli M. M.
Vijh Uma P.
Vilchez Jose M.
Vink Jacco
Viti Serena
Volk Kevin
Vorobyov Eduard I.
Voshchinnikov Nikolai V.
Vrba Frederick J.
Wakker Bastiaan Pieter
Walker Gordon A. H.
Walmsley C. Malcolm
Walsh Wilfred M.
Walsh Andrew J.
Walton Nicholas A.
Wang Qingde Daniel

MEMBERSHIP BY COMMISSION

Composition of Commission 35

Stellar Constitution / Constitution des Etoiles

President D'Antona Francesca

Vice-President Charbonnel Corinne

Organizing Committee

Fontaine Gilles
Larson Richard B.
Lattanzio John
Liebert James W.
Mueller Ewald
Weiss Achim
Yungelson Lev R.

Members

Adams Mark T.
Aiad A. Zaki
Aizenman Morris L.
Anand S. P. S.
Angelov Trajko
Antia H. M.
Appenzeller Immo
Arai Kenzo
Arentoft Torben
Argast Dominik
Arimoto Nobuo
Arnett W. David
Arnould Marcel L.
Audouze Jean
Baglin Annie
Barnes Sydney A.
Basu Sarbani
Baym Gordon Alan
Bazot Michael
Beaudet Gilles
Becker Stephen A.
Belmonte Aviles Juan Antonio
Benz Willy
Bergeron Pierre
Bertelli Gianpaolo
Berthomieu Gabrielle
Bisnovatyi-Kogan Gennadij S.
Blaga Christina O.
Bludman Sidney A.
Bocchia Romeo
Bodenheimer Peter
Bombaci Ignazio
Bono Giuseppe
Boss Alan P.
Brassard Pierre
Bravo Eduardo
Bressan Alessandro
Brownlee Robert R.
Bruenn Stephen W.
Buchler J. R.
Burbidge Geoffrey R.
Busso Maurizio
Callebaut Dirk K.
Caloi Vittoria
Canal Ramon M.
Caputo Filippina
Carson T. R.
Castellani Vittorio(†)
Castor John I.
Chaboyer Brian C.
Chabrier Gilles
Chan Roberto
Chan Kwing Lam
Charpinet Stéphane
Chechetkin Valerij M.
Chiosi Cesare S.
Chitre Shashikumar M.
Chkhikvadze Iakob N.
Christensen-Dalsgaard Jørgen
Christy Robert F.
Cohen Jeffrey M.
Connolly Leo P.
Cowan John J.
Das Mrinal Kanti
Daszynska-Daszkiewicz Jadwiga
Davis Jr Cecil G.
de Greve Jean-Pierre
de Jager Cornelis
de Loore Camiel
de Medeiros Jose Renan
Dearborn David Paul K.
Deinzer W.
Deliyannis Constantine P.
Demarque Pierre
Despain Keith Howard
Deupree Robert G.
Di Mauro Maria Pia
Dluzhnevskaya Olga B.
Dominguez Inma
Dupuis Jean
Durisen Richard H.
Dziembowski Wojciech A
Edwards Alan Ch.
Eggleton Peter P.
Eminzade T. A.
Endal Andrew S.
Eriguchi Yoshiharu
Ezer-Eryurt Dilhan
Fadeyev Yurij A.
Faulkner John
Flannery Brian Paul
Forbes J. E.
Forestini Manuel
Fossat Eric G.
Foukal Peter V.

Fujimoto Masayuki
Gabriel Maurice R.
Gallino Roberto
Garcia Domingo
Gautschy Alfred
Geroyannis Vassilis S.
Giannone Pietro
Gimenez Alvaro
Girardi Leo A.
Giridhar Sunetra
Glatzmaier Gary A.
Goriely Stephane
Gough Douglas O.
Goupil Marie-Jose
Graham Eric
Greggio Laura
Guenther David Bruce
Guzik Joyce A.
Hachisu Izumi
Hammond Gordon L.
Han Zhanwen
Hashimoto Masa-aki
Hayashi Chushiro
Henry Richard B. C.
Hernanz Margarita
Hirschi Raphael
Hollowell David Earl
Huang Runqian
Huggins Patrick J.
Humphreys Roberta M.
Iben Jr Icko
Iliev Ilian
Imbroane Alexandru
Imshennik Vladimir S.
Isern Jorge
Ishizuka Toshihisa
Itoh Naoki
James Richard A.
Jørgensen Jes K.
Jose Jordi
Kaehler Helmuth
Kato Mariko
Kiguchi Masayoshi
King David S.
Kippenhahn Rudolf
Kiziloglu Nilguen
Knoelker Michael
Kochhar Rajesh K.
Koester Detlev
Konar Sushan
Kosovichev Alexander
Kovetz Attay
Kozlowski Maciej
Kumar Shiv S.
Kwok Sun
Labay Javier
Lamb Susan Ann
Lamb Jr Donald Quincy
Lamzin Sergei A.
Langer Norbert
Laskarides Paul G.
Lasota Jean-Pierre
Lebovitz Norman R.
Lebreton Yveline
Lee Thyphoon
Leitherer Claus
Lepine Jacques R. D.
Li Zongwei
Li Li Qingkang
Liebendoerfer Matthias
Lignieres François
Limongi Marco
Linnell Albert P.
Littleton John E.
Livio Mario
Maeda Keiichi
Maeder Andre
Maheswaran Murugesapillai
Masani A.
Massevich Alla G.
Matteucci Francesca
Mazurek Thaddeus John
Mazzitelli Italo
McDavid David A.
Mendes Luiz T. S.
Mestel Leon
Meyer-Hofmeister Eva
Meynet Georges
Michaud Georges J.
Mitalas Romas Assoc
Miyaji Shigeki
Moellenhoff Claus
Mohan Chander
Moiseenko Sergey G.
Monaghan Joseph J.
Monier Richard
Monteiro Mario
Joao P. F. G.
Moore Daniel R.
Morgan John Adrian
Moskalik Pawel
Moss David L.
Mowlavi Nami
Nadyozhin Dmitrij K.
Nakamura Takashi
Nakano Takenori
Nakazawa Kiyoshi
Narasimha Delampady
Narita Shinji
Nelemans Gijs
Newman Michael John
Nishida Minoru
Noels Arlette
Nomoto Ken'ichi
Odell Andrew P.
Ohyama Noboru
Okamoto Isao
Oliveira Joana M.
Osaki Yoji
Ostriker Jeremiah P.
Oswalt Terry D.
O'Toole Simon J.
Paczynski Bohdan(†)
Pamyatnykh Aleksej A.
Pande Girish Chandra
Papaloizou John C. B.
Pearce Gillian
Phillips Mark M.
Pines David
Pinotsis Antonis D.
Plavec Mirek J.
Pongracic Helen
Pontoppidan Klaus M.
Porfir'ev Vladimir V.
Poveda Arcadio
Prentice Andrew J. R.
Prialnik-Kovetz Dina
Proffitt Charles R.
Provost Janine
Qu Qinyue
Raedler K. H.
Ramadurai Souriraja
Rauscher Thomas
Ray Alak
Rayet Marc
Reeves Hubert
Renzini Alvio
Reyniers Maarten
Ritter Hans
Roca Cortes Teodoro

Rood Robert T.
Rouse Carl A.
Roxburgh Ian W.
Ruiz-Lapuente María P.
Sackmann Ingrid Juliana
Saio Hideyuki
Sakashita Shiro
Salpeter Edwin E.
Santos Filipe D.
Sarna Marek Jacek
Sato Katsuhiko
Savedoff Malcolm P.
Savonije Gerrit Jan
Scalo John Michael
Schatten Kenneth H.
Schatzman Evry
Schild Hansruedi
Schoenberner Detlef
Schutz Bernard F.
Scuflaire Richard
Sears Richard Langley
Seidov Zakir F.
Sengbusch Kurt V.
Shaviv Giora
Shibahashi Hiromoto
Shibata Yukio
Shustov Boris M.
Sienkiewicz Ryszard
Sigalotti Leonardo G.
Signore Monique
Sills Alison I.
Silvestro Giovanni
Sion Edward Michael
Smeyers Paul
Smith Robert Connon
Sobouti Yousef
Sofia Sabatino
Sparks Warren M.
Spiegel E.
Sreenivasan S. Ranga
Starrfield Sumner
Stellingwerf Robert F.
Stibbs Douglas W. N.
Stringfellow Guy S.
Strittmatter Peter A.
Suda Kazuo
Suda Takuma
Sugimoto Daiichiro
Sweigart Allen V.
Taam Ronald Everett
Takahara Mariko
Thielemann Friedrich-Karl
Thomas Hans Christoph
Tjin-a-Djie Herman R. E.
Tohline Joel Edward
Toomre Juri
Tornambe Amedeo
Trimble Virginia L.
TruranJr James W.
Tscharnuter Werner M.
Tuominen Ilkka V.
Turck-Chieze Sylvaine
Tutukov Aleksandr V.
Uchida Juichi
Ulrich Roger K.
Unno Wasaburo
Utrobin Victor P.
van den Heuvel
Edward P. J.
van der Borght Rene
van der Raay Herman B.
van Horn Hugh M.
van Loon Jacobus Th.
van Riper Kenneth A.
VandenBerg Don
Vardya M. S.
Vauclair Gérard P.
Ventura Paolo
Vila Samuel C.
Vilhu Osmi
Vilkoviskij Emmanuil Y.
Vink Jorick S.
Ward Richard A.
Weaver Thomas A.
Webbink Ronald F.
Weiss Nigel O.
Wheeler J. Craig
Willson Lee Anne M.
Wilson Robert E.
Winkler Karl-Heinz A.
Wood Matthew Alan
Wood Peter R.
Woosley Stanley E.
Xiong Da Run
Yamaoka Hitoshi
Yi Sukyoung Ken
Yorke Harold W.
Yoshida Shin'ichirou
Yoshida Takashi
Yushkin Maxim V.
Zahn Jean-Paul
Ziolkowski Janusz

MEMBERSHIP BY COMMISSION

Composition of Commission 36

Theory of Stellar Atmospheres / Théorie des Atmosphères Stellaires

President Landstreet John D.

Vice-President Asplund Martin

Organizing Committee

Balachandran Suchitra C.
Berdyugina Svetlana V.
Hauschildt Peter H.
Ludwig Hans G.
Mashonkina Lyudmila I.
Nagendra K. N.
Puls Joachim
Randich Sofia
Spite Monique
Tautvaisiene Grazina

Members

Abbott David C.
Abhyankar Krishna D.
Allende Prieto Carlos
Altrock Richard C.
Andretta Vincenzo
Arpigny Claude
Atanackovic-Vukmanovic Olga M.
Athay R. Grant
Auer Lawrence H.
Auman Jason R.
Avrett Eugene H.
Ayres Thomas R.
Baade Dietrich
Baird Scott R.
Baliunas Sallie L.
Balona Luis A.
Barbuy Beatriz
Baschek Bodo
Basri Gibor B.
Bell Roger A.
Bennett Philip D.
Bernat Andrew Plous
Bertout Claude
Bingham Robert
Blanco Carlo
Bless Robert C.
Blomme Ronny
Bodo Gianluigi
Boesgaard Ann M.
Bonifacio Piercarlo
Bopp Bernard W.
Bowen George H.
Brown Douglas N.
Brown Alexander
Bues Irmela D.
Busa Innocenza
Cameron Andrew Collier
Carbon Duane F.
Carlsson Mats
Carson T. R.
Cassinelli Joseph P.
Castelli Fiorella
Castor John I.
Catala Claude
Catalano Franco A.
Catalano Santo
Cayrel Roger
Cayrel de Strobel Giusa
Chan Kwing Lam
Chen Peisheng
Chugai Nikolaj N.
Cidale Lydia S.
Conti Peter S.
Cowley Charles R.
Cram Lawrence Edward
Crivellari Lucio
Cruzado Alicia
Cugier Henryk
Cuntz Manfred
Cuny Yvette J.
Daszynska-Daszkiewicz Jadwiga
Davis Jr Cecil G.
de Koter Alex
Decin Leen K. E.
Deliyannis Constantine P.
Dimitrijevic Milan
Doazan Vera
Donati Jean-Francois
Doyle John Gerard
Drake Stephen A.
Dravins Dainis
Dreizler Stefan
Duari Debiprosad
Dufton Philip L.
Dupree Andrea K.
Edvardsson Bengt
Elste Gunther H.
Eriksson Kjell
Evangelidis E. A.
Faraggiana Rosanna
Faurobert-Scholl Marianne
Feigelson Eric D.
Ferreira Joao M.
Fitzpatrick Edward L.
Fluri Dominique M.
Fontaine Gilles
Fontenla Juan M.
Forveille Thierry
Foy Renaud
Freire Ferrero Rubens G.
Fremat Yves
Freytag Bernd
Friend David B.
Frisch Helene

Frisch Uriel
Froeschle Christiane D.
Gail Hans-Peter
Gallino Roberto
Garcia Lopez Ramon J.
Gebbie Katharine B.
Gesicki Krzysztof
Giampapa Mark S.
Gigas Detlef
Gonzalez Jean-Francois
Gordon Charlotte
Gough Douglas O.
Grant Ian P.
Gratton Raffaele G.
Gray David F.
Grevesse Nicolas
Grinin Vladimir P.
Guedel Manuel
Gussmann Ernst August
Gustafsson Bengt
Hack Margherita
Haisch Bernhard Michael
Hall Douglas S.
Hamann Wolf-Rainer
Harper Graham M.
Hartmann Lee William
Harutyunian Haik A.
Heasley James Norton
Heber Ulrich
Heiter Ulrike
Hill Vanessa M.
Hoare Melvin G.
Hoeflich Peter
Hoefner Susanne
Holzer Thomas E.
Hotinli Metin
House Lewis L.
Huang He
Hubeny Ivan
Hui bon Hoa Alain
Hutchings John B.
Ignace Richard
Ignjatovic Ljubinko M.
Ito Yutaka
Ivanov Vsevolod V.
Jahn Krzysztof
Jankov Slobodan S.
Jatenco-Pereira Vera
Jevremovic Darko
Johnson Hollis R.
Jordan Stefan
Judge Philip
Kadouri Talib Hadi
Kalkofen Wolfgang
Kamp Lucas Willem
Kandel Robert S.
Karp Alan H.
Kasparova Jana
Katsova Maria M.
Kiselman Dan
Klein Richard I.
Kochukhov Oleg
Kodaira Keiichi
Koester Detlev
Kolesov Aleksandr K.
Kondo Yoji
Kontizas Evangelos
Korcakova Daniela
Korn Andreas J.
Kraus Michaela
Krikorian Ralph
Krishna Swamy K. S.
Krticka Jiri
Kubat Jiri
Kudritzki Rolf-Peter
Kuhi Leonard V.
Kumar Shiv S.
Kupka Friedrich
Kurucz Robert L.
Lambert David L.
Lamers Henny J. G. L. M.
Lanz Thierry
Lee Jae-Woo
Leibacher John
Leitherer Claus
Liebert James W.
Linnell Albert P.
Linsky Jeffrey L.
Liu Caipin
Loskutov Viktor M.
Luck R. Earle
Luo Qinghuan
Luttermoser Donald
Lyubimkov Leonid S.
Machado Maria A. D.
Madej Jerzy
Magazzu Antonio
Magnan Christian
Marlborough J. Michael
Marley Mark S.
Massaglia Silvano
Mathys Gautier
Mauas Pablo
Medupe Rodney T.
Michaud Georges J.
Mihajlov Anatolij A.
Mihalas Dimitri
Molaro Paolo
Muench Guido
Musielak Zdzislaw E.
Mutschlecner J. Paul
Nagirner Dmitrij I.
Najarro de la Parra Francisco
Narasimha Delampady
Nariai Kyoji
Nikoghossian Arthur G.
Nishimura Masayoshi
Nordlund Aake
Owocki Stanley Peter
Pacharin-Tanakun P.
Pagel Bernard E. J.(†)
Pakhomov Yuri V.
Pallavicini Roberto
Pandey Bierndra P.
Panek Robert J.
Pasinetti Laura E.
Pavlenko Yakov V.
Pecker Jean-Claude
Peraiah Annamaneni
Perinotto Mario(†)
Peters Geraldine Joan
Pinsonneault Marc Howard
Pintado Olga I.
Pinto Philip Alfred
Piskunov Nikolai E.
Plez Bertrand
Pogodin Mikhail A.
Pottasch Stuart R.
Praderie Françoise
Przybilla Norbert
Querci Francois R.
Querci Monique
Rachkovsky D. N.
Ramsey Lawrence W.
Rangarajan K. E.
Rao D. Mohan
Rauch Thomas
Reale Fabio
Reimers Dieter

Rostas Francois
Rovira Marta G.
Rucinski Slavek M.
Rutten Robert J.
Ryabchikova Tatiana A.
Rybicki George B.
Sachkov Mikhail E.
Saio Hideyuki
Saito Kuniji
Sakhibullin Nail A.
Sapar Lili
Sapar Arved
Sarre Peter J.
Sasselov Dimitar D.
Sauty Christophe
Savanov Igor S.
Sbordone Luca
Schaerer Daniel
Scharmer Goeran Bjarne
Schmalberger Donald C.
Schmid-Burgk J.
Schmutz Werner
Schoenberner Detlef
Scholz M.
Schrijver Karel J.
Seaton Michael J(†)
Sedlmayer Erwin
Sengupta Sujan K.
Shine Richard A.
Shipman Harry L.
Sigut Aaron T. A.
Simon Klaus Peter
Simon Theodore
Simonneau Eduardo
Skumanich Andrew
Sneden Chris
Snezhko Leonid I.
Soderblom David R.
Spiegel E.
Spite François M.
Spruit Henk C.
Stalio Roberto
Stauffer John Richard
Stee Philippe
Steffen Matthias
Stein Robert F.
Stepien Kazimierz
Stern Robert Allan
Stibbs Douglas W. N.
Strom Stephen E.
Szecsenyi-Nagy Gábor A.
Takeda Yoichi
Thejll Peter Andreas
Toomre Juri
Traving Gerhard
Tsuji Takashi
Tuominen Ilkka V.
Ueno Sueo
Uesugi Akira
Ulmschneider Peter
Unno Wasaburo
Utrobin Victor P.
Vakili Farrokh
van't Veer-Menneret Claude
van't-Veer Frans
Vardavas Ilias Mihail
Vardya M. S.
Vasu-Mallik Sushma
Vaughan Arthur H.
Velusamy T.
Viik Tõnu
Vilhu Osmi
Vink Jorick S.
Walter Frederick M.
Watanabe Tetsuya
Waters Laurens B. F. M.
Weber Stephen Vance
Wehrse Rainer
Weidemann Volker
Werner Klaus
White Richard L.
Wickramasinghe N. Chandra
Willson Lee Anne M.
Wilson Peter R.
Wilson S. J.
Woehl Hubertus
Wolff Sidney C.
Yanovitskij Edgard G.
Yengibarian Norair
Yorke Harold W.
Zacs Laimons
Zahn Jean-Paul

MEMBERSHIP BY COMMISSION

Composition of Commission 37

Star Clusters & Associations / Amas Stellaires & Associations

President Hatzidimitriou Despina

Vice-President Lada Charles Joseph

Organizing Committee

Cannon Russell D.
Castellani Vittorio(†)
Cudworth Kyle McC.
Da Costa Gary Stewart
Deng LiCai
Lee Young Wook
Sarajedini Ata
Tosi Monica

Members

Aarseth Sverre J.
Abou'el-ella Mohamed S.
Ahumada Javier Alejandro
Ahumada Andrea V.
Aiad A. Zaki
Akeson Rachel
Alcaino Gonzalo
Alfaro Emilio Javier
Alksnis Andrejs
Allen Lori E.
Allen Christine
Andreuzzi Gloria
Aparicio Antonio
Armandroff Taft E.
Auriere Michel
Bailyn Charles D.
Balazs Bela A.
Barrado Navascues David
Bastian Nathan J.
Baume Gustavo L.
Bell Roger A.
Bellazzini Michele
Bijaoui Albert
Blum Robert D.
Boily Christian M.
Bonatto Charles J.
Bosch Guillermo L.
Bragaglia Angela
Brown Anthony G. A.
Buonanno Roberto
Burderi Luciano
Burkhead Martin S.
Butler Dennis
Butler Raymond F.
Buzzoni Alberto
Byrd Gene G.
Calamida Annalisa
Callebaut Dirk K.
Caloi Vittoria
Caputo Filippina
Capuzzo Dolcetta Roberto
Carney Bruce William
Carraro Giovanni
Chaboyer Brian C.
Chavarria-K Carlos
Cheng Kwang Ping
Chiosi Cesare S.
Christian Carol Ann
Chryssovergis Michael
Chun Mun-suk
Claria Juan
Clementini Gisella
Colin Jacques
Covino Elvira
Cropper Mark
D'Amico Nicolo'
Danford Stephen C.
D'Antona Francesca
Dapergolas Anastasios
Daube-Kurzemniece Ilga A.
Davies Melvyn B.
de Grijs Richard
De Marchi Guido
DeGioia-Eastwood Kathleen
Dehghani Mohammad Hossein
Demarque Pierre
Demers Serge
Di Fazio Alberto
Dickens Robert J.
Djupvik Anlaug A.
Dluzhnevskaya Olga B.
Drissen Laurent
Durrell Patrick R.
El Basuny Ahmed Alawy
Elmegreen Bruce Gordon
Fall S. Michael
Feinstein Alejandro
Forbes Douglas
Forte Juan C.
Freyhammer Lars Michael
Friel Eileen D.
Fukushige Toshiyuki
Fusi-Pecci Flavio
Garcia Beatriz E.
Gascoigne S. C. B.
Geffert Michael
Geisler Douglas P.
Giersz Miroslav
Glushkova Elena V.
Golay Marcel
Goodwin Simon P.
Gouliermis Dimitrios
Gratton Raffaele G.
Green Elizabeth M.
Griffiths William K.
Grundahl Frank

Guetter Harry Hendrik
Haisch Jr Karl E.
Hanes David A.
Harris Hugh C.
Harris Gretchen L. H.
Harvel Christopher Alvin
Hassan S. M.(†)
Hawarden Timothy G.
Hazen Martha L.
Heggie Douglas C.
Henon Michel C.
Herbst William
Hesser James E.
Heudier Jean-Louis
Hilker Michael
Hillenbrand Lynne A.
Hills Jack G.
Hodapp Klaus-Werner
Huensch Matthias
Hut Piet
Iben Jr Icko
Illingworth Garth D.
Ishida Keiichi
Janes Kenneth A.
Joshi Umesh C.
Kadla Zdenka I.(†)
Kamp Lucas Willem
Kilambi G. C.
Kim Sungsoo S.
Kim Seung-Lee
King Ivan R.
Kitsionas Spyridon
Ko Chung-Ming
Kontizas Evangelos
Kontizas Mary
Kraft Robert P.
Kroupa Pavel
Kun Maria
Kundu Arunav
Kurtev Radostin G.
Landolt Arlo U.
Lapasset Emilio
Larsson-Leander Gunnar
Laval Annie
Lee Kang Hwan
Lee Jae-Woo
Leisawitz David
Leonard Peter James T.
Lloyd Evans Thomas Harry
Loktin Alexhander V.
Lu Phillip K.
Lynden-Bell Donald
Maccarone Thomas J.
Maeder Andre
Makalkin Andrei B.
Makino Junichiro
Mamajek Eric E.
Marco Amparo
Mardling Rosemary A.
Markkanen Tapio
Markov Haralambi S.
Marraco Hugo G.
Marsden Stephen C.
Marshall Kevin P.
Martinez Roger Carlos
Martins Donald Henry
Menon T. K.
Menzies John W.
Mermilliod Jean-Claude
Meylan Georges
Milone Eugene F.
Minniti Dante
Moehler Sabine
Mohan Vijay
Moitinho André
Mould Jeremy R.
Muminov Muydinjon
Murray Andrew C.
Muzzio Juan C.
Naylor Tim
Nemec James
Nesci Roberto
Neuhaeuser Ralph
Newell Edward B.
Ninkov Zoran
Nurnberger Dieter E. A.
Ogura Katsuo
Oliveira Joana M.
Origlia Livia
Ortolani Sergio
Osman Anas Mohamed
Pandey A. K.
Parmentier Genevieve A.
Parsamyan Elma S.
Paunzen Ernst
Pedreros Mario
Penny Alan John
Peterson Charles John
Petrovskaya Margarita S.
Peykov Zvezdelin I.
Phelps Randy L.
Piatti Andrés E.
Pilachowski Catherine
Piskunov Anatolij E.
Platais Imants
Portegies Zwart Simon F.
Poveda Arcadio
Pritchet Christopher J.
Raimondo Gabriella
Rebull Luisa
Renzini Alvio
Rey Soo-Chang
Richer Harvey B.
Richtler Tom
Rountree Janet
Royer Pierre
Russeva Tatjana
Sagar Ram
Salukvadze G. N.
Samus Nikolaj N.
Sanders Walt L.
Santos Jr Joao F.
Schild Hansruedi
Schweizer François
Seitzer Patrick
Semkov Evgeni Hristov
Shawl Stephen J.
Sher David
Shobbrook Robert R.
Shu Chenggang
Simoda Mahiro
Simone Zaggia
Smith J. Allyn
Smith Graeme H.
Song Inseok
Spurzem Rainer
Stauffer John Richard
Stetson Peter B.
Stringfellow Guy S.
Sugimoto Daiichiro
Sung Hwankyung
Suntzeff Nicholas B.
Szecsenyi-Nagy Gábor A.
Tadross Ashraf Latif
Takahashi Koji
Taylor John K.
Terranegra Luciano
Terzan Agop
Thoul Anne A.
Tornambe Amedeo

Tripicco Michael J.
Trullols I. Farreny Enric
Tsvetkov Milcho K.
Tsvetkova Katja
Turner David G.
Twarog Bruce A.
Upgren Arthur R.
van Altena William F.
Van den Bergh Sidney
VandenBerg Don
Vazquez Ruben A.
Ventura Paolo
Verschueren Werner
Vesperini Enrico
von Hippel Theodore A.
Walker Merle F.
Walker Gordon A. H.
Warren Jr Wayne H.
Weaver Harold F.
Wehlau Amelia F.
Wielen Roland
Wramdemark Stig S.
Wu Hsin-Heng
Xiradaki Evangelia
Yi Sukyoung Ken
Yim Hong-Suh
Zakharova Polina E.
Zhao Jun Liang
Zinn Robert J.

MEMBERSHIP BY COMMISSION

Composition of Commission 40

Radio Astronomy / Radioastronomie

President Nan Ren-Dong

Vice-President Taylor Russell A.

Organizing Committee

Anantharamaiah Kuduvalli R.(†)
Carilli Christopher L.
Chapman Jessica
Dubner Gloria M.
Garrett Michael
Goss W. Miller
Hills Richard E.
Hirabayashi Hisashi
Rodriguez Luis F.
Shastri Prajval
Torrelles Jose M.

Members

Abdulla Shaker Abdul Aziz
Abraham Peter
Ade Peter A. R.
Akabane Kenji A.
Akujor Chidi E.
Alberdi Antonio
Alexander Paul
Alexander Joseph K.
Allen Ronald J.
Aller Margo F.
Aller Hugh D.
Altenhoff Wilhelm J.
Altunin Valery I.
Ambrosini Roberto
Andernach Heinz J
Aparici Juan
Arnal Edmundo Marcelo
Asareh Habibolah
Aschwanden Markus
Assousa George Elias
Aubier Monique G.
Augusto Pedro
Aurass Henry
Avery Lorne W.
Avignon Yvette
Axon David
Baan Willem A.
Baars Jacob W. M.
Baart Edward E.
Baath Lars B.
Bachiller Rafael
Backer Donald Ch.
Bagri Durgadas S.
Bailes Matthew
Bajaja Esteban
Bajkova Anisa T.
Baker Joanne
Baker Andrew J.
Balasubramanian V.
Balasubramanyam Ramesh
Baldwin John E.
Balklavs-Grinhofs Arturs E.
Ball Lewis
Bally John
Balonek Thomas J.
Banhatti Dilip G.
Barrow Colin H.
Bartel Norbert Harald
Barthel Peter
Barvainis Richard
Bash Frank N.
Basu Dipak
Baudry Alain
Baum Stefi Alison
Beasley Anthony James
Beck Rainer
Benaglia Paula
Benn Chris R.
Bennett Charles L.
Benson Priscilla J.
Benz Arnold O.
Berge Glenn L.
Berkhuijsen Elly M.
Bhandari Rajendra
Bhat Ramesh N D
Bieging John Harold
Biermann Peter L.
Biggs James
Bignell R. Carl
Biraud François
Biretta John Anthony
Birkinshaw Mark
Blair David Gerald
Blandford Roger David
Bloemhof Eric E.
Blundell Katherine M.
Boboltz David A.
Bock Douglas C.-J.
Bockelee-Morvan Dominique
Bolatto Alberto D.
Bondi Marco
Bos Albert
Bottinelli Lucette
Bower Geoffrey C.
Bowers Phillip F.
Bracewell Ronald N.(†)
Breahna Iulian
Bregman Jacob D. Ir.
Bridle Alan H.
Brinks Elias
Britzen Silke H.
Broderick John
Bronfman Leonardo
Brooks Kate J.
Broten Norman W.

Brouw Willem N.
Browne Ian W. A.
Brunetti Gianfranco
Bujarrabal Valentin
Burbidge Geoffrey R.
Burderi Luciano
Burke Bernard F.
Campbell Robert M.
Campbell-Wilson Duncan
Carlqvist Per A.
Caroubalos Constantinos A.
Carr Thomas D.
Carretti Ettore
Carvalho Joel C.
Casoli Fabienne
Castelletti Gabriela
Castets Alain
Caswell James L.
Cawthorne Timothy V.
Cernicharo Jose
Chan Kwing Lam
Chandler Claire
Charlot Patrick
Chen Yongjun
Chengalur Jayaram N.
Chikada Yoshihiro
Chin Yi-nan
Chini Rolf
Cho Se-Hyung
Christiansen Wayne A.
Christiansen Wilbur(†)
Chung Hyun Soo
Chyzy Krzysztof Tadeusz
Clark David H.
Clark Barry G.
Clark Frank Oliver
Clemens Dan P.
Cohen Raymond James(†)
Cohen Richard S.
Cohen Marshall H.
Coleman Paul Henry
Colomb Fernando R.
Colomer Francisco
Combes Françoise
Combi Jorge A.
Condon James J.
Conklin Edward K.
Contreras Maria E.
Conway John
Conway Robin G.
Cordes James M.
Costa Marco E.
Cotton Jr William D.
Crane Patrick C.
Crawford Fronefield
Crovisier Jacques
Crutcher Richard M.
Cudaback David D.
Cunningham Maria R.
Dagkesamansky Rustam D.
Daintree Edward J.
Daishido Tsuneaki
Dallacasa Daniele
D'Amico Nicolo'
Davies Rodney D.
Davis Michael M.
Davis Robert J.
Davis Richard J.
de Bergh Catherine
de Jager Cornelis
de la Noe Jerome
de Ruiter Hans Rudolf
de Vicente Pablo
de Young David S.
Degaonkar S. S.
Delannoy Jean
Denisse Jean-Francois
Dent William A.
Deshpande Avinash
Despois Didier
Dewdney Peter E. F.
Dhawan Vivek
Diamond Philip John
Dickel Helene R.
Dickel John R.
Dickey John M.
Dickman Robert L.
Dieter Conklin Nannielou H.
Dixon Robert S.
Dobashi Kazuhito
Doubinskij Boris A.
Dougherty Sean M.
Douglas James N.
Downes Dennis
Downs George S.
Drake Stephen A.
Drake Frank D.
Dravskikh Aleksandr F.
Dreher John W.
Duffett-Smith Peter James
Dulk George A.
Dwarakanath K. S.
Dyson Freeman J.
Eales Stephen A.
Edelson Rick
Ekers Ronald D.
Elia Davide
Ellingsen Simon P.
Ellis G. R. A.
Emerson Darrel Trevor
Enome Shinzo
Epstein Eugene E.
Erickson William C.
Eshleman Von R.
Evans Kenton Dower
Ewing Martin S.
Ezawa Hajime
Facondi Silvia Rosa
Falcke Heino D.
Fanaroff Bernard L.
Fanti Roberto
Faulkner Andrew J.
Fedotov Leonid V.
Feigelson Eric D.
Feldman Paul A.
Felli Marcello
Felten James E.
Feretti Luigina
Ferrari Attilio
Fey Alan Lee
Fharo Toshihiro
Field George B.
Filipovic Miroslav D.
Finkelstein Andrej M.
Fleischer Robert
Florkowski David R.
Foley Anthony
Fomalont Edward B.
Fort David N.
Forveille Thierry
Fouque Pascal
Frail Dale Andrew
Frater Robert H.
Frey Sandor
Friberg Per
Fuerst Ernst
Fukui Yasuo
Gabuzda Denise C.
Gaensler Bryan M.

Gallego Juan Daniel
Gallimore Jack F.
Galt John A.
Garaimov Vladimir I.
Garay Guido
Garrington Simon T.
Gasiprong Nipon
Gaume Ralph A.
Gaylard Michael John
Geldzahler Bernard J.
Gelfreikh Georgij B.
Genzel Reinhard
Gerard Eric
Gergely Tomas E.
Gervasi Massimo
Ghigo Francis D.
Ghosh Tapashi
Gil Janusz A.
Gimenez Alvaro
Ginzburg Vitalij L.
Gioia Isabella M.
Giovannini Gabriele
Goldwire Jr Henry C.
Gomez Gonzalez Jesus
Gopalswamy Nat
Gordon Mark A.
Gorgolewski Stanislaw Pr.
Gorschkov Aleksandr G.
Gosachinskij Igor V.
Gottesman Stephen T.
Gower Ann C.
Graham David A.
Green David A
Green Anne
Gregorini Loretta
Gregorio-Hetem Jane C.
Gregory Philip C.
Grewing Michael
Gubchenko Vladimir M.
Guelin Michel
Guesten Rolf
Guidice Donald A.
Gulkis Samuel
Gull Stephen F.
Gupta Yashwant
Gurvits Leonid I.
Gwinn Carl R.
Haddock Fred T.
Hall Peter J.
Han JinLin
Hanasz Jan
Handa Toshihiro
Hanisch Robert J.
Hankins Timothy Hamilton
Hardee Philip
Harnett Julienne
Harris Daniel E.
Harten Ronald H.
Haschick Aubrey
Hasegawa Tetsuo
Hayashi Masahiko
Haynes Martha P.
Haynes Raymond F.
Hazard Cyril
Heeschen David S.
Heiles Carl
Helou George
Henkel Christian
Herpin Fabrice
Heske Astrid
Hewish Antony
Hey James Stanley
Hibbard John E.
Higgs Lloyd A.
Hirota Tomoya
Hjalmarson Ake G.
Ho Paul T. P.
Hoang Binh Dy
Hobbs Robert W.
Hoeglund Bertil
Hofner Peter
Hogbom Jan A.
Hogg David E.
Hollis Jan Michael
Hong Xiaoyu
Hopkins Andrew M.
Horiuchi Shinji
Howard William E.
Huchtmeier Walter K.
Hughes Philip
Hughes David H.
Hulsbosch A. N. M.
Hunstead Richard W.
Hwang Chorng-Yuan
Iguchi Satoru
Ikhsanov Robert N.
Ikhsanova Vera N.
Imai Hiroshi
Inatani Junji
Inoue Makoto
Ipatov Aleksandr V.
Irvine William M.
Ishiguro Masato
Israel Frank P.
Ivanov Dmitrij V.
Iwata Takahiro
Jackson Neal J.
Jackson Carole A.
Jacq Thierry
Jaffe Walter Joseph
Janssen Michael Allen
Jauncey David L.
Jenkins Charles R.
Jennison Roger C(†)
Jewell Philip R.
Jin Shenzeng
Johansson Lars Erik B.
Johnson Donald R.
Johnston Kenneth J.
Joly Francois
Jones Dayton L.
Jones Paul
Jung Jae Hoon
Kaftan May A.
Kahlmann Hans C.
Kaidanovski Mikhail N.
Kaifu Norio
Kakinuma Takakiyo T.
Kalberla Peter
Kaltman Tatyana I.
Kameya Osamu
Kandalyan Rafik A.
Kanekar Nissim
Kang Gon Ik
Kardashev Nicolay S.
Kassim Namir E.
Kasuga Takashi
Kaufmann Pierre
Kawabata Kinaki
Kawabe Ryohei
Kawaguchi Kentarou
Kawamura Akiko
Kazes Ilya
Kedziora-Chudozer Lucyna L.
Kellermann Kenneth I.
Kesteven Michael J. l.
Khaikin Vladimir B.
Kijak Jaroslaw
Kilborn Virginia A.

Killeen Neil
Kim Kwang-tae
Kim Tu-Whan
Kim Bong Gyu
Kim Hyun-Goo
Kislyakov Albert G.
Kitaeff Vyacheslav V.
Klein Ulrich
Klein Karl L.
Knudsen Kirsten K.
Ko Hsien C.
Kobayashi Hideyuki
Kocharovsky Vitalij V.
Koda Jin
Kohno Kotaro
Kojima Masayoshi
Kondratiev Vladislav I.
Konovalenko Alexander A.
Korzhavin Anatolij N.
Kovalev Yuri Y.
Kovalev Yuri A.
Kramer Busaba H.
Kramer Michael
Kreysa Ernst
Krichbaum Thomas P.
Krishna Gopal
Krishnan Thiruvenkata
Kronberg Philipp
Krugel Endrik
Krygier Bernard
Kuan Yi-Jehng
Kuijpers H. Jan M. E.
Kuiper Thomas B. H.
Kulkarni Prabhakar V.
Kulkarni Shrinivas R.
Kulkarni Vasant K.
Kumkova Irina I.
Kundt Wolfgang
Kundu Mukul R.
Kuril'chik Vladimir N.
Kus Andrzej Jan
Kutner Marc Leslie
Kuzmin Arkadij D.
Kwok Sun
Lada Charles Joseph
Laing Robert
Landecker Thomas L.
Lang Kenneth R.
Langer William David
Langston Glen I.
Lantos Pierre(†)
LaRosa Theodore N.
Lasenby Anthony
Lawrence Charles R.
Le Squeren Anne-Marie
Leahy J. Patrick
Lee Chang-Won
Lee Yong Bok
Lee Youngung
Legg Thomas H.
Lepine Jacques R. D.
Lequeux James
Lesch Harald
Lestrade Jean-Francois
Leung Chun Ming
Levreault Russell M.
Li Gyong Won
Li Hong-Wei
Liang Shiguang
Likkel Lauren Jones
Lilley Edward A.
Lim Jeremy
Lindqvist Michael
Linke Richard Alan
Lis Dariusz C.
Liseau Rene
Lister Matthew L.
Little Leslie T.(†)
Litvinenko Leonid N.
Liu Yuying
Liu Xiang
Lo Kwok-Yung
Locke Jack L.
Lockman Felix J.
Loiseau Nora
Longair Malcolm S.
Loren Robert Bruce
Lovell Sir Bernard
Lovell James E.
Lozinskaya Tatjana A.
Lubowich Donald A.
Luks Thomas
Luo Xianhan
Lyne Andrew G.
Macchetto Ferdinando
MacDonald Geoffrey H.
MacDonald James
Machalski Jerzy
Mack Karl-Heinz
MacLeod John M.
Macrae Donald A.(†)
Maehara Hideo
Malofeev Valery Mikhailovich
Malumian Vigen H.
Manchester Richard N.
Mandolesi Nazzareno
Mantovani Franco
Mao Rui-Qing
Maran Stephen P.
Marcaide Juan-Maria
Mardyshkin Vyacheslav V.
Marecki Andrzej
Markoff Sera B.
Marques Dos Santos P.
Marscher Alan Patrick
Marti Josep
Martin Christopher L.
Martin Robert N.
Martin-Pintado Jesus
Marvel Kevin B.
Masheder Michael
Maslowski Jozef
Masson Colin R.
Matheson David Nicholas
Matsakis Demetrios N.
Matsuo Hiroshi
Matsushita Satoki
Matthews Brenda C.
Matthews Henry E.
Mattila Kalevi
Matveenko Leonid I.
Mauersberger Rainer
Maxwell Alan
May J.
McAdam Bruce W. B.
McConnell David
McCulloch Peter M.
McKenna Lawlor S.
McLean Donald J.
McMullin Joseph P.
Mebold Ulrich
Meeks M. Littleton
Meier David L.
Menon T. K.
Menten Karl M.
Mezger Peter G.
Michalec Adam
Miley George K.
Miller Neal A.

Mills Bernard Y.
Milne Douglas K.
Milogradov-Turin Jelena
Mirabel Igor Felix
Mitchell Kenneth J.
Miyawaki Ryosuke
Miyazaki Atsushi
Miyoshi Makoto
Mizuno Norikazu
Mizuno Akira
Moellenbrock III George
Moffett David A.
Momjian Emmanuel
Momose Muntake
Montmerle Thierry
Moran James M.
Morganti Raffaella
Morimoto Masaki
Morison Ian
Morita Koh-ichiro
Morita Kazuhiko
Moriyama Fumio
Morras Ricardo
Morris Mark Root
Morris David
Moscadelli Luca
Muller Erik M.
Mundy Lee G.
Murdoch Hugh S.
Murphy Tara
Mutel Robert Lucien
Muxlow Thomas
Myers Philip C.
Nadeau Daniel
Nagnibeda Valerij G.
Nakajima Junichi
Nakano Takenori
Nakashima Jun-ichi
Neeser Mark J.
Nguyen-Quang Rieu
Nicastro Luciano
Nice David J.
Nicolson George D.
Nikolic Silvana
Nishio Masanori
Norris Raymond P.
Nurnberger Dieter E. A.
O'Dea Christopher P.
Oezel Mehmet Emin
Ogawa Hideo
Ohashi Nagayoshi
Ohishi Masatoshi
Ojha Roopesh
Okoye Samuel E.
Okumura Sachiko
Olberg Michael
Onishi Toshikazu
Onuora Lesley Irene
Orchiston Wayne
Osterbrock Donald E(†)
O'Sullivan John David
Otmianowska-Mazur Katarzyna
Ott Juergen A.
Owen Frazer Nelson
Pacholczyk Andrzej G.
Padman Rachael
Palmer Patrick E.
Pankonin Vernon Lee
Paredes Poy Josep M.
Parijskij Yurij N.
Parker Edward A.
Parma Paola
Parrish Allan
Pasachoff Jay M.
Pashchenko Mikhail I.
Patel Nimesh A.
Pauls Thomas Albert
Payne David G.
Pearson Timothy J.
Peck Alison B.
Pedersen Holger
Pedlar Alan
Peng Bo
Penzias Arno A.
Perez Fournon Ismael
Perley Richard Alan
Peters William L. III
Petrova Svetlana
Pettengill Gordon H.
Philipp Sabine D.
Phillips Christopher J.
Phillips Thomas Gould
Pick Monique
Pisano Daniel J.
Planesas Pere
Pogrebenko Sergei
Polatidis Antonios
Ponsonby John E. B.
Pooley Guy
Porcas Richard
Prandoni Isabella
Preston Robert Arthur
Preuss Eugen
Price R. M.
Puschell Jeffery J.
Radford Simon John E.
Radhakrishnan V.
Rahimov Ismail A.
Raimond Ernst
Ransom Scott M.
Rao A. Pramesh
Raoult Antoinette
Ray Thomas P.
Razin Vladimir A.
Readhead Anthony C. S.
Redman Matthew P.
Reich Wolfgang
Reid Mark Jonathan
Reif Klaus
Reyes Francisco
Reynolds John
Rhee Myung-Hyun
Ribo Marc
Richer John
Rickard Lee J.
Rickett Barnaby James
Riihimaa Jorma J.
Riley Julia M.
Rioja Maria J.
Rizzo Jose Ricardo
Roberts Morton S.
Roberts David Hall
Robertson James Gordon
Robertson Douglas S.
Robinson Jr Richard D.
Roeder Robert
Roelfsema Peter
Roennaeng Bernt O.
Roeser Hans-peter
Roger Robert S.
Rogers Alan E. E.
Rogstad David H.
Rohlfs Kristen
Romanov Andrey M.
Romero Gustavo E.
Romney Jonathan D.
Rowson Barrie
Rubin Robert Howard
Rubio Monica

Rudnick Lawrence
Rudnitskij Georgij M.
Russell Jane L.
Rydbeck Gustaf H. B.
Rys Stanislaw
Sadler Elaine M.
Saikia Dhruba Jyoti
Sakamoto Seiichi
Salpeter Edwin E.
Salter Christopher John
Samodurov Vladimir A.
Sanamian V. A.
Sandell Goran Hans l
Sanders David B.
Sargent Annelia I.
Sarma Anuj P.
Sarma N. V. G.
Sastry Ch. V.
Sato Fumio
Saunders Richard D. E.
Savage Ann
Sawada-Satoh Satoko
Sawant Hanumant S.
Scalise Jr Eugenio
Schaal Ricardo E.
Schilizzi Richard T.
Schilke Peter
Schlickeiser Reinhard
Schmidt Maarten
Schroder Anja C.
Schuch Nelson Jorge
Schulz R. Andreas
Schwartz Philip R.
Schwarz Ulrich J.
Scott Paul F.
Scott John S.
Seaquist Ernest R.
Seielstad George A.
Sekido Mamoru
Sekimoto Yutaro
Setti Giancarlo
Shaffer David B.
Shakeshaft John R.
Shaposhnikov Vladimir E.
Shaver Peter A.
Shepherd Debra S.
Sheridan K. V.
Shevgaonkar R. K.
Shibata Katsunori M.
Shimmins Albert John
Shinnaga Hiroko
Shitov Yurij P.(†)
Shmeld Ivar
Sholomitsky Gennady B.
Shone David
Shulga Valery
Sieber Wolfgang
Singal Ashok K.
Sinha Rameshwar P.
Skillman Evan D.
Slade Martin A. III
Slee O. B.
Slysh Vyacheslav I.
Smith Dean F.
Smith Francis Graham
Smith Niall
Smol'kov Gennadij Ya
Smolentsev Sergej G.
Snellen Ignas A. G.
Sobolev Yakov M.
Soboleva Natalja S.
Sodin Leonid
Sofue Yoshiaki
Sokolov Konstantin
Sorochenko Roman L.
Spencer John Howard
Spencer Ralph E.
Spoelstra T. A. Th.
Sramek Richard A.
Sridharan Tirupati K.
Stahr-Carpenter M.
Stairs Ingrid H.
Stanghellini Carlo
Stanley G. J.
Stannard David
Stappers Benjamin W.
Steffen Matthias
Steinberg Jean-Louis
Stewart Paul
Stewart Ronald T.
Stone R. G.
Storey Michelle C.
Strom Richard G.
Strukov Igor A.
Subrahmanya C. R.
Subrahmanyan Ravi
Sugitani Koji
Sukumar Sundarajan
Sullivan Woodruff T.
Sunada Kazuyoshi
Swarup Govind
Swenson Jr George W.
Szymczak Marian
Takaba Hiroshi
Takakubo Keiya
Takano Toshiaki
Takano Shuro
Tanaka Riichiro
Tapping Kenneth F.
Tarter Jill C.
Tatematsu Ken'ichi
te Lintel Hekkert Peter
Tello Bohorquez Camilo
Terasranta Harri T.
Terzian Yervant
Theureau Gilles
Thomasson Peter
Thompson A. Richard
Thum Clemens
Tingay Steven J.
Tlamicha Antonin
Tofani Gianni
Tolbert Charles R.
Tomasi Paolo
Tornikoski Merja T.
Tosaki Tomoka
Tovmassian Hrant Mushegi
Townes Charles Hard
Trigilio Corrado
Tritton Keith Peter
Troland Thomas Hugh
Truong Bach
Trushkin Sergej A.
Tsuboi Masato
Tsutsumi Takahiro
Tuccari Gino
Turlo Zygmunt
Turner Kenneth C.
Turner Barry E.
Turner Jean L.
Turtle A. J.
Tyul'bashev Sergei A.
Tzioumis Anastasios
Udal'tsov Vyacheslav A.
Udaya Shankar N.
Ukita Nobuharu
Ulrich Marie-Helene D.
Ulvestad James Scott
Umana Grazia
Umemoto Tomofumi

Unger Stephen
Unwin Stephen C.
Urama Johnson O.
Urosevic Dejan V.
Uson Juan M.
Vallee Jacques P.
Valtaoja Esko
Valtonen Mauri J.
Val'tts Irina E.
van der Hulst Jan M.
van der Kruit Pieter C.
van der Laan Harry
van der Tak Floris F. S.
van Driel Wim
van Gorkom Jacqueline H.
van Langevelde Huib Jan
van Woerden Hugo
VandenBout Paul A.
Vats Hari OM
Vaughan Alan
Velusamy T.
Venturi Tiziana
Venugopal V. R.
Verheijen Marc A. W.
Verkhodanov Oleg V.
Vermeulen Rene Cornelis
Veron Philippe
Verschuur Gerrit L.
Verter Frances
Vilas Faith
Vilas-Boas José W. S.
Vivekanand M.
Vogel Stuart Newcombe
Volvach Alexander
Walker Robert C.
Wall Jasper V.
Wall William F.
Walmsley C. Malcolm
Walsh Wilfred M.
Walsh Andrew J.
Wang Na
Wang Shouguan
Wannier Peter Gregory
Wardle John F. C.
Ward-Thompson Derek
Warmels Rein Herm
Warner Peter J.
Warwick James W.
Watson Robert A.
Wehrle Ann Elizabeth
Wei Mingzhi
Weigelt Gerd
Weiler Edward J.
Weiler Kurt W.
Welch William J.
Wellington Kelvin
Wendker Heinrich J.
Wenlei Shan
Westerhout Gart
Whiteoak John B.
Wickramasinghe N. Chandra
Wielebinski Richard
Wiik Kaj J.
Wiklind Tommy
Wild John Paul
Wild Wolfgang
Wilkinson Peter N.
Willis Anthony G.
Wills Beverley J.
Wills Derek
Willson Robert Frederick
Wilner David J.
Wilson Andrew S.
Wilson William J.
Wilson Robert W.
Wilson Thomas L.
Windhorst Rogier A.
Wink Joern Erhard(†)
Winnberg Anders
Winnewisser Gisbert
Witzel Arno
Wolszczan Alexander
Woltjer Lodewijk
Wood Douglas O. S.
Woodsworth Andrew W.
Wootten Henry Alwyn
Wright Alan E.
Wrobel Joan Marie
Wu Nailong
Wu Xinji
Wu Yuefang
Yang Ji
Yang Zhigen
Yao Qijun
Ye Shuhua
Yin Qi-Feng
Yonekura Yoshinori
Younis Saad M.
Yu Zhiyao
Yusef-Zadeh Farhad
Zabolotny Vladimir F.
Zaitsev Valerij V.
Zanichelli Alessandra
Zannoni Mario
Zensus J-Anton
Zhang Hongbo
Zhang Qizhou
Zhang Xizhen
Zhang Jin
Zhang Jian
Zhao Jun Hui
Zheleznyak Alexander P.
Zheleznyakov Vladimir V.
Zheng Xinwu
Zieba Stanislaw
Zinchenko Igor
Zlobec Paolo
Zlotnik Elena Ya
Zuckerman Ben M.
Zylka Robert

MEMBERSHIP BY COMMISSION

Composition of Commission 41

History of Astronomy / Histoire de l'Astronomie

President Nha Il-Seong

Vice-President Ruggles Clive L. N.

Secretary Kochhar Rajesh K.

Organizing Committee

de Jong Teije
Gurshtein Alexander A.
Nakamura Tsuko
Orchiston Wayne
Videira Antonio A.
Warner Brian

Members

Ahn Youngsook
Ansari S. M. Razaullah
Aoki Shinko
Bandyopadhyay A.
Batten Alan H.
Belmonte Aviles Juan Antonio
Bennett Jim A.
Benson Priscilla J.
Berendzen Richard
Bessell Michael S.
Bishop Roy L.
Boccaletti Dino
Bonoli Fabrizio
Botez Elvira
Brooks Randall C.
Brosche Peter
Bruck Mary T.
Brunet Jean-Pierre
Burman Ronald R.
Carlson John B.
Chapman Allan
Chen Kwan-yu
Chen Meidong
Chin Yi-nan
Chinnici Ileana
Clifton Gloria
Corbin Brenda G.
Cornejo Alejandro A.
Cui Shizhu
Cui Zhenhua
Dadic Zarko
Danezis Emmanuel
de Freitas Mourao Ronaldo R.
Debarbat Suzanne V.
Deeming Terence J.
Dekker E.
Devorkin David H.
Dewhirst David W.
Dick Steven J.
Dick Wolfgang R.
Dorokhova Tetyana
Duerbeck Hilmar W.
Dumont Simone
Eddy John A.
Edmondson Frank K.
Ehgamberdiev Shuhrat
Esteban César
Fernie J. Donald
Firneis Maria G.
Flin Piotr
Florides Petros S.
Fluke Christopher J.
Freeman Kenneth C.
Freitas Mourao R.
Gingerich Owen
Glass Ian S.
Green Daniel William Edward
Han Wonyong
Hasegawa Ichiro
Hayli Abraham
Haynes Raymond F.
Haynes Roslynn
Hearnshaw John B.
Heck Andre
Hemenway Mary Kay M.
Herrmann Dieter
Hidayat Bambang
Hingley Peter D.
Hirai Masanori
Hockey Thomas Arnold
Holmberg Gustav
Hopkins Andrew M.
Hu Tiezhu
Huan Nguyen Dinh
Hurukawa Kiitiro
Hwang Chorng-Yuan
Hysom Edmund J.
Hyung Siek
Idlis Grigorij M.
Jarrell Richard A.
Jauncey David L.
Jeong Jang Hae
Jiang Xiaoyuan
Jovanovic Bozidar
Khromov Gavriil S.
Kiang Tao
Kim Chun Hwey
Kim Yonggi
King David S.
King Henry C.(†)
Kollerstrom Nicholas
Krisciunas Kevin
Krupp Edwin C.

Lang Kenneth R.
Launay Françoise
Le Guet Tully Françoise
Lee Yong Bok
Lee Eun-Hee
Lee Woo-baik
Lee Yong-Sam
Lerner Michel-Pierre
Levy Eugene H.
Liu Ciyuan
Locher Kurt H.
Lopes-Gautier Rosaly
Lopez Carlos
Marco Olivier
Mathewson Donald S.
McAdam Bruce W. B.
McKenna Lawlor S.
Mickelson Michael E.
Mikhail Joseph Sidky
Mikisha Anatoly
Min Wang Yu
Moesgaard Kristian P.
Nadal Robert
Nguyen Mau Tung
Nicolaidis Efthymios
Norris Raymond P.
North John David
Oh Kyu Dong
Ohashi Yukio
Ohashi Nagayoshi
Oproiu Tiberiu
Osterbrock Donald E.(†)
Papathanasoglou Dimitrios
Peterson Charles John
Pettersen Bjørn R.
Pigatto Luisa
Pingree David
Polozhentsev Dmitrij D.
Polyakhova Elena N.
Poulle Emmanuel
Pozhalova Zhanna
Prokakis Theodore J.
Proverbio Edoardo
Pustylnik Izold
Rafferty Theodore J.
Satterthwaite Gilbert E.
Sbirkova-Natcheva T.
Schaefer Bradley E.
Schmadel Lutz D.
Seconds Alain-Philippe
Shank Michael H.
Shukla K.
Signore Monique
Sima Zdislav
Simpson Allen D. C.
Sobouti Yousef
Solc Martin
Soonthornthum Boonrucksar
Stathopoulou Maria
Steele John
Steinle Helmut R.
Stephenson F. Richard
Sterken Christiaan L.
Stoev Alexey D.
Sullivan Woodruff T.
Sun Xiaochun
Sundman Anita
Svolopoulos Sotirios
Swerdlow Noel
Taton Rene
Taub Liba
Theodossiou Efstratios
Tobin William
Trimble Virginia L.
Van Gent Robert H.
Vargha Magda
Vass Gheorghe
Verdet Jean-Pierre
Verdun Andreas K.
Volyanskaya Margarita Yu
Wang Rongbin
Whitaker Ewen A.
White Graeme Lindsay
Whiteoak John B.
Whitrow Gerald James
Wilkins George A.
Williams Thomas R.
Wilson Curtis A.
Wolfschmidt Gudrun
Xi Zezong
Yang Hong-Jin
Yau Kevin K. C.
Yeomans Donald K.
Zanini Valeria
Zhang Shouzhong
Zhou Yonghong
Zsoldos Endre

MEMBERSHIP BY COMMISSION

Composition of Commission 42

Close Binary Stars / Etoiles Doubles Serrées

President Rucinski Slavek M.

Vice-President Ribas Ignasi

Organizing Committee

Gimenez Alvaro
Harmanec Petr
Hilditch Ronald W.
Kaluzny Janusz
Niarchos Panayiotis
Nordstrom Birgitta
Olah Katalin
Richards Mercedes T.
Scarfe Colin David
Sion Edward Michael
Torres Guillermo
Vrielmann Sonja

Members

Abhyankar Krishna D.
Al-Naimiy Hamid M. K.
Andersen Johannes
Antipova Lyudmila I.
Antokhin Igor I.
Antonopoulou Evgenia
Anupama G. C.
Aquilano Roberto Oscar
Awadalla Nabil Shoukry
Baba Hajime
Bailyn Charles D.
Baptista Raymundo
Barkin Yuri V.
Barone Fabrizio
Bartolini Corrado
Bateson Frank M. O.(†)
Bath Geoffrey T.
Batten Alan H.
Bell Steven A.
Bianchi Luciana
Blair William P.
Blundell Katherine M.
Boffin Henri M. J.
Bolton Charles Thomas
Bonazzola Silvano
Bookmyer Beverly B.
Bopp Bernard W.
Borisov Nikolay V.
Boyd David R.
Boyle Stephen
Bozic Hrvoje
Bradstreet David H.
Brandi Elisande E.
Broglia Pietro
Brownlee Robert R.
Bruch Albert
Bruhweiler Frederick C.
Budding Edwin
Bunner Alan N.
Burderi Luciano
Busa Innocenza
Busso Maurizio
Callanan Paul
Canalle Joao B. G.
Catalano Santo
Cester Bruno
Chambliss Carlson R.
Chapman Robert D.
Chaty Sylvain
Chaubey Uma Shankar
Chen Kwan-yu
Cherepashchuk Anatolij M.
Chochol Drahomir
Choi Kyu-Hong
Choi Chul-Sung
Chou Yi
Ciardi David R.
Claria Juan
Clausen Jens Viggo
Collins George W. II
Cowley Anne P.
Cropper Mark
Cutispoto Giuseppe
Dadaev Aleksandr N.
D'Amico Nicolo'
D'Antona Francesca
de Greve Jean-Pierre
de Groot Mart
de Loore Camiel
Delgado Antonio J.
Demircan Osman
Diaz Marcos P.
Dobrzycka Danuta
Dorfi Ernst Anton
Dougherty Sean M.
Drechsel Horst
Duemmler Rudolf
Duerbeck Hilmar W.
Dupree Andrea K.
Durisen Richard H.
Duschl Wolfgang J.
Eaton Joel A.
Edalati Sharbaf Mohammad Taghi
Eggleton Peter P.
Elias Nicholas
Etzel Paul B.
Eyres Stewart P. S.
Fabrika Sergei Nikolaevich
Fang Li Li
Faulkner John
Fekel Francis C.
Ferluga Steno
Ferrario Lilia
Ferrer Osvaldo E.
Flannery Brian Paul

Frank Juhan
Fredrick Laurence W.
Friedjung Michael
Gaensicke Boris T.
Gallagher III John S.
Garcia Lia G.
Garmany Catherine D.
Gasiprong Nipon
Geldzahler Bernard J.
Geyer Edward H.
Giannone Pietro
Gies Douglas R.
Giovannelli Franco
Goldman Itzhak
Gonzalez Martinez Pais Ignacio
Gosset Eric
Graffagnino Vito G.
Groot Paul J.
Grygar Jiri
Guinan Edward F.
Gulliver Austin Fraser
Gunn Alastair G.
Gursky Herbert
Guseinov O. H.
Hadrava Petr
Hakala Pasi J.
Hall Douglas S.
Hammerschlag-Hensberge G.
Hanawa Tomoyuki
Hantzios Panayiotis
Hassall Barbara J. M.
Haswell Carole A.
Hayasaki Kimitake
Hazlehurst John
Hegedues Tibor
Hellier Coel
Helt Bodil E.
Hempelmann Alexander M.
Hensler Gerhard
Herczeg Tibor J.
Hill Graham
Hills Jack G.
Hillwig Todd C.
Hoard Donald W.
Holmgren David E.
Holt Stephen S.
Honeycutt R. Kent
Horiuchi Ritoku
Hric Ladislav
Hrivnak Bruce J.
Huang Runqian
Hube Douglas P.
Hutchings John B.
Ibanoglu Cafir
Imamura James
Imbert Maurice
Jabbar Sabeh Rhaman
Jasniewicz Gerard
Jeong Jang Hae
Jonker Peter G.
Joss Paul Christopher
Kadouri Talib Hadi
Kaitchuck Ronald H.
Kang Young-Woon
Karetnikov Valentin G. R.
Kato Taichi
Kawabata Shusaku
Kenny Harold
Kenyon Scott J.
Khalesseh Bahram
Kim Chun Hwey
Kim Ho-il
King Andrew R.
Kitamura Masatoshi
Kjurkchieva Diana
Kley Wilhelm
Koch Robert H.
Kolb Ulrich
Kolesnikov Sergey
Kondo Yoji
Koubsky Pavel
Kraft Robert P.
Kraicheva Zdravka
Krautter Joachim
Kreiner Jerzy M.
Kruchinenko Vitaliy G.
Kruszewski Andrzej
Krzeminski Wojciech
Kudashkina Larisa S.
Kumsiashvily Mzia I.
Kurpinska-Winiarska M.
Kuznetsov Oleg A.(†)
Kwee K. K.
Lacy Claud H.
Lamb Jr Donald Quincy
Landolt Arlo U.
Lanning Howard H.
Lapasset Emilio
Larsson Stefan
Larsson-Leander Gunnar
Lavrov Mikhail I.
Lee Woo-baik
Lee Yong-Sam
Leedjarv Laurits
Leung Kam Ching
Li Zhongyuan
Lim Jeremy
Linnell Albert P.
Linsky Jeffrey L.
Liu Qingzhong
Livio Mario
Lloyd Huw
Lucy Leon B.
Lyuty Victor M.
MacDonald James
Maceroni Carla
Malasan Hakim Luthfi
Manimanis Vassilios
Mardirossian Fabio
Maria de Garcia J. M.
Marilli Ettore
Markoff Sera B.
Markworth Norman Lee
Marsh Thomas R.
Mathieu Robert D.
Mauder Horst
Mayer Pavel
Mazeh Tsevi
McCluskey Jr George E.
Meintjes Petrus J.
Melia Fulvio
Meliani Mara T.
Mereghetti Sandro
Meyer-Hofmeister Eva
Mezzetti Marino
Mikolajewska Joanna
Mikulasek Zdenek
Milano Leopoldo
Milone Eugene F.
Mineshige Shin
Miyaji Shigeki
Mochnacki Stephan W.
Morales Rueda Luisa
Morgan Thomas H.
Morrell Nidia
Mouchet Martine
Mumford George S.
Munari Ulisse

Murray James R.
Mutel Robert Lucien
Nakamura Yasuhisa
Nakao Yasushi
Nariai Kyoji
Nather R. Edward
Naylor Tim
Neff James Edward
Nelemans Gijs
Nelson Burt
Newsom Gerald H.
Nha Il-Seong
Niemela Virpi S.(†)
Norton Andrew J.
Oezkan Mustafa Tuerker
Ogloza Waldemar
Oh Kyu Dong
Okazaki Akira
Oliver John Parker
Olson Edward C.
Osaki Yoji
Paczynski Bohdan(†)
Padalia T. D.
Pandey Uma Shankar
Park Hong Suh
Parthasarathy Mudumba
Patkos Laszlo
Pavlenko Elena
Pavlovski Kresimir
Pearson Kevin J.
Peters Geraldine Joan
Piccioni Adalberto
Piirola Vilppu E.
Plavec Mirek J.
Pojmanski Grzegorz
Polidan Ronald S.
Pollacco Don
Postnov Konstantin A.
Potter Stephen B.
Pribulla Theodor
Pringle James E.
Prokhorov Mikhail E.
Pustylnik Izold
Qiao Guojun
Rafert James Bruce
Rahunen Timo
Rakos Karl D.
Ramsey Lawrence W.
Ransom Scott M.
Rao Pasagada Vivekananda
Rasio Frederic A.
Refsdal Sjur
Reglero-Velasco Victor
Rey Soo-Chang
Ringwald Frederick Arthur
Ritter Hans
Robb Russell M.
Robertson John Alistair
Robinson Edward L.
Rovithis Peter
Rovithis-Livaniou Helen
Roxburgh Ian W.
Ruffert Maximilian
Russo Guido
Sadik Aziz R.
Sahade Jorge
Saijo Keiichi
Samec Ronald G.
Sanyal Ashit
Savonije Gerrit Jan
Schiller Stephen
Schmid Hans Martin
Schmidtke Paul C.
Schmidtobreick Linda
Schober Hans J.
Seggewiss Wilhelm
Semeniuk Irena
Shafter Allen W.
Shakura Nikolaj I.
Shaviv Giora
Shaw Simon E.
Shu Frank H.
Sima Zdislav
Simmons John Francis l
Sistero Roberto F.
Skopal Augustin
Slovak Mark Haines
Smak Jozef I.
Smith Robert Connon
Sobieski Stanley
Soderhjelm Staffan
Solheim Jan Erik
Sonti Sreedhar Rao
Sowell James Robert
Sparks Warren M.
Srivastava J. B.
Srivastava Ram K.
Stagg Christopher
Stanishev Vallery D.
Starrfield Sumner
Steiman-Cameron Thomas
Steiner Joao E.
Stencel Robert Edward
Sterken Christiaan L.
Stringfellow Guy S.
Sugimoto Daiichiro
Sundman Anita
Svechnikov Marij A.
Szkody Paula
Taam Ronald Everett
Tan Huisong
Tauris Thomas M.
Taylor John K.
Teays Terry J.
Terrell Dirk C.
Todoran Ioan
Tout Christopher
Tremko Jozef
Trimble Virginia L.
Turolla Roberto
Tutukov Aleksandr V.
Ureche Vasile
van den Heuvel Edward P. J.
van Hamme Walter
van Kerkwijk Marten H.
van't-Veer Frans
Vaz Luiz Paulo Ribeiro
Vetesnik Miroslav
Vilhu Osmi
Voloshina Irina B.
Wachter Stefanie
Wade Richard Alan
Walder Rolf
Walker William S. G.
Ward Martin John
Warner Brian
Webbink Ronald F.
Weiler Edward J.
Wheatley Peter J.
Wheeler J. Craig
White James Clyde
Williamon Richard M.
Williams Robert
Williams Glen A.
Wilson Robert E.
Wood Janet H.
Yamaoka Hitoshi
Yamasaki Atsuma
Yoon Tae S.

MEMBERSHIP BY COMMISSION

Composition of Commission 44

Space & High Energy Astrophysics

Astrophysique Spatiales & des Hautes Energies

President Hasinger Günther

Vice-President Jones Christine

Organizing Committee

Braga João
Brosch Noah
de Graauw Thijs
Gurvits Leonid I.
Helou George
Howarth Ian Donald
Kunieda Hideyo
Montmerle Thierry
Okuda Haruyuki
Salvati Marco
Singh Kulinder Pal

Members

Abramowicz Marek
Acharya Bannanje S.
Acton Loren W.
Agrawal P. C.
Ahluwalia Harjit S.
Ahmad Imad Aldean
Alexander Joseph K.
Allington-Smith Jeremy R.
Almleaky Yasseen
Andersen Bo Nyborg
Apparao K. M. V.
Arafune Jiro
Arnaud Monique D.
Arnould Marcel L.
Arons Jonathan
Aschenbach Bernd
Asseo Estelle
Asvarov Abdul I.
Audard Marc
Audouze Jean
Awaki Hisamitsu
Axford W. Ian
Ayres Thomas R.
Baan Willem A.
Badiali Massimo
Bailyn Charles D.
Balikhin Michael
Baliunas Sallie L.
Baring Matthew G.
Barstow Martin Adrian
Baskill Darren S.
Basu Dipak
Baym Gordon Alan
Becker Robert Howard
Becker Werner
Begelman Mitchell Craig
Beiersdorfer Peter
Belloni Tomaso
Benedict George F.
Benford Gregory
Bennett Charles L.
Bennett Kevin
Benvenuto Omar
Bergeron Jacqueline A.
Bernacca Pierluigi
Berta Stefano
Beskin Vasily S.
Beskin Gregory M.
Bhattacharjee Pijushpani
Bhattacharya Dipankar
Bianchi Luciana
Bicknell Geoffrey V.
Biermann Peter L.
Bignami Giovanni F.
Bingham Robert
Biswas Sukumar
Blamont Jacques-Emile
Blandford Roger David
Bleeker Johan A. M. Ir.
Bless Robert C.
Blinnikov Sergey I.
Bloemen Hans (J. B. G. M.)
Blondin John M.
Bludman Sidney A.
Bocchino Fabrizio
Boer Michel
Boggess Nancy W.
Boggess Albert
Bohlin Ralph C.
Boksenberg Alec
Bonazzola Silvano
Bonnet Roger M.
Bonnet-Bidaud Jean-Marc
Bonometto Silvio A.
Borozdin Konstantin N.
Bougeret Jean-Louis
Bowyer C. Stuart
Boyd Robert L. F.
Bradley Arthur J.
Brandt John C.
Brandt William N.
Brandt Soeren K.
Brecher Kenneth
Breslin Ann
Brinkman Bert C.
Brown Alexander
Bruhweiler Frederick C.
Bruner Marilyn E.
Brunetti Gianfranco
Bumba Vaclav

Bunner Alan N.
Buote David A.
Burbidge Geoffrey R.
Burderi Luciano
Burenin Rodion A.
Burger Marijke
Burke Bernard F.
Burrows David Nelson
Burrows Adam Seth
Burton William M.
Butler Christopher John
Butterworth Paul S
Caccianiga Alessandro
Cai Michael
Camenzind Max
Campbell Murray F.
Cappi Massimo
Caraveo Patrizia
Cardenas Rolando P.
Cardini Daniela
Carlson Per
Carpenter Kenneth G.
Carroll P. Kevin
Casandjian Jean-Marc
Cash Jr Webster C.
Casse Michel
Castro-Tirado Alberto J.
Catura Richard C.
Cavaliere Alfonso G.
Cesarsky Catherine J.
Chakrabarti Sandip Kumar
Chakraborty Deo K.
Chang Hsiang-Kuang
Chang Heon-Young
Channok Chanruangrit
Chapman Sandra C.
Chapman Robert D.
Charles Philip Allan
Chartas George
Chechetkin Valerij M.
Chenevez Jerome
Cheng Kwongsang
Cheung Cynthia Y.
Chian Abraham Chian-Long
Chiappetti Lucio
Chikawa Michiyuki
Chitre Shashikumar M.
Chochol Drahomir
Choi Chul-Sung
Chou Yi
Chubb Talbot A.
Chupp Edward L.
Churazov Evgenij M.
Ciotti Luca
Clark Thomas Alan
Clark George W.
Clay Roger
Code Arthur D.
Cohen Jeffrey M.
Collin Suzy
Comastri Andrea
Condon James J.
Contopoulos Ioannis
Corbet Robin Henry D.
Corbett Ian F.
Corcoran Michael Francis
Cordova France A. D.
Courtes Georges
Courvoisier Thierry J.-L.
Cowie Lennox Lauchlan
Cowsik Ramanath
Crannell Carol Jo
Crocker Roland M.
Cropper Mark
Cruise Adrian Michael
Culhane Leonard
Cunniffe John
Curir Anna
Cusumano Giancarlo
da Costa Jose Marques
da Costa Antonio A.
Dadhich Naresh
Dai Zigao
Damian Audley M.
D'Amico Flavio
Damle S. V.
Davidson William
Davis Robert J.
Davis Michael M.
Dawson Bruce
de Aguiar Odylio Denys
de Felice Fernando
de Jager Cornelis
de Martino Domitilla
de Young David S.
Della Ceca Roberto
Dempsey Robert C.
den Herder Jan-Willem
Dennerl Konrad
Dennis Brian R.
Dermer Charles Dennison
Dewitt Bryce S.
Di Cocco Guido
Digel Seth William
Disney Michael J.
Dokuchaev Vyacheslav I.
Dolan Joseph F.
Domingo Vicente
Donea Alina C.
Dotani Tadayasu
Dovciak Michal
Downes Turlough
Drake Frank D.
Drury Luke O'Connor
Duorah Hira Lal
Dupree Andrea K.
Durouchoux Philippe
Duthie Joseph G.
Edelson Rick
Edwards Paul J.
Eichler David
Eilek Jean
El Raey Mohamed E.
Elvis Martin S.
Emanuele Alessandro
Ensslin Torsten A.
Ettori Stefano
Eungwanichayapant Anant
Evans W. Doyle
Fabian Andrew C.
Fabricant Daniel G.
Fang Li-zhi
Fang Liu Bi
Faraggiana Rosanna
Fazio Giovanni G.
Feldman Paul Donald
Felten James E.
Fender Robert P.
Fendt Christian
Fenton K. B.
Ferrari Attilio
Fichtel Carl E.
Field George B.
Fisher Philip C.
Fishman Gerald J.
Fitton Brian
Foing Bernard H.
Fomin Valery

Fonseca Gonzalez Maria Victoria
Forman William Richard
Franceschini Alberto
Frandsen Soeren
Frank Juhan
Fransson Claes
Fredga Kerstin
Fujimoto Shin-ichiro
Fujita Mitsutaka
Furniss Ian
Fyfe Duncan J.
Gabriel Alan H.
Gaisser Thomas K.
Galeotti Piero
Galloway Duncan K.
Garcia Howard A.(†)
Garmire Gordon P.
Gaskell C. Martin
Gathier Roel
Gehrels Neil
Georgantopoulos Ioannis
Geppert Ulrich R. M. E.
Gezari Daniel Ysa
Ghia Piera Luisa
Ghirlanda Giancarlo
Ghisellini Gabriele
Giacconi Riccardo
Gilra Daya P.
Ginzburg Vitalij L.
Gioia Isabella M.
Glaser Harold
Goldsmith Donald W.
Goldwurm Andrea
Gomez de Castro Ana I.
Gondhalekar Prabhakar
Gonzales'a Walter D.
Gotthelf Eric
Graffagnino Vito G.
Grebenev Sergej A.
Greenhill John
Greisen Kenneth I.(†)
Grenier Isabelle
Grewing Michael
Greyber Howard D.
Griffiths Richard E.
Grindlay Jonathan E.
Gull Theodore R.
Gunn James E.
Gursky Herbert
Guseinov O. H.
Hack Margherita
Haddock Fred T.
Hakkila Jon E.
Halevin Alexander V.
Hall Andrew Norman
Hallam Kenneth L.
Hameury Jean-Marie
Hannikainen Diana C.
Hardcastle Martin J.
Harms Richard James
Harris Daniel E.
Harvey Paul Michael
Harvey Christopher C.
Harwit Martin
Hasan Hashima
Hatsukade Isamu
Haubold Hans J.
Hauser Michael G.
Hawkes Robert Lewis
Hawking Stephen W.
Hawkins Isabel
Hayama Kazuhiro
Haymes Robert C.
Heckathorn Harry M.
Hein Righini Giovanna
Heise John
Helfand David John
Helmken Henry F.
Henoux Jean-Claude
Henriksen Richard N.
Henry Richard Conn
Hensberge Herman
Heske Astrid
Hoffman Jeffrey Alan
Holberg Jay B.
Holloway Nigel J.
Holt Stephen S.
Hornstrup Allan
Houziaux Leo
Hoyng Peter
Hu Wenrui
Huang Jiasheng
Huang YongFeng
Huber Martin C. E.
Hulth Per Olof
Hunt Leslie
Hurley Kevin C.
Hutchings John B.
Hwang Chorng-Yuan
Ichimaru Setsuo
Illarionov Andrei F.
Imamura James
Imhoff Catherine L.
Inoue Makoto
Inoue Hajime
in't Zand Johannes J. M.
Ipser James R.
Ishida Manabu
Israel Werner
Ito Kensai A.
Itoh Masayuki
Jackson John Charles
Jaffe Walter Joseph
Jakobsson Pall
Jamar Claude A. J.
Janka Hans Thomas
Jaranowski Piotr
Jenkins Edward B.
Jokipii J. R.
Jones Frank Culver
Jones Thomas Walter
Jonker Peter G.
Jordan Carole
Jordan Stuart D.
Joss Paul Christopher
Juliusson Einar
Kafatos Minas
Kahabka Peter
Kaneda Hidehiro
Kaper Lex
Kapoor Ramesh Chander
Kaspi Victoria M.
Kasturirangan K.
Kato Tsunehiko
Kato Yoshiaki
Katsova Maria M.
Katz Jonathan I.
Kawai Nobuyuki
Kellermann Kenneth I.
Kellogg Edwin M.
Kembhavi Ajit K.
Kessler Martin F.
Killeen Neil
Kim Yonggi
Kimble Randy A.
Kinugasa Kenzo
Kirk John
Klinkhamer Frans
Klose Sylvio

Knapp Johannes
Ko Chung-Ming
Kobayashi Shiho
Kocharov Grant E.
Koch-Miramond Lydie
Koide Shinji
Kojima Yasufumi
Kolb Edward W.
Kondo Yoji
Kondo Masaaki
Kong Albert
Koshiba Masatoshi
Koupelis Theodoros
Koyama Katsuji
Kozlowski Maciej
Kozma Cecilia A.
Kraushaar William L.
Kristiansson Krister
Kryvdyk Volodymyr G.
Kuiper Lucien
Kulsrud Russell M.
Kumagai Shiomi
Kuncic Zdenka
Kundt Wolfgang
Kurt Vladimir G.
Kusunose Masaaki
Lamb Susan Ann
Lamb Frederick K.
Lamb Jr Donald Quincy
Lamers Henny J. G. L. M.
Lampton Michael
Lapington Jonathan S.
Lasher Gordon Jewett
Lattimer James M.
Lea Susan Maureen
Leckrone David S.
Leighly Karen Marie
Lemaire Philippe
Levin Yuri
Lewin Walter H. G.
Li Yuanjie
Li Tipei
Li Zongwei
Li Zhongyuan
Li Li-Xin
Li Xiangdong
Liang Edison P.
Lin Xuan-bin
Lindblad Bertil A.(†)
Linsky Jeffrey L.
Linsley John
Lochner James Charles
Long Knox S.
Longair Malcolm S.
Lovelace Richard V. E.
Lovell Sir Bernard
Lu Tan
Lu Jufu
Luest Reimar
Luminet Jean-Pierre
Luo Qinghuan
Lutovinov Alexander A.
Lynden-Bell Donald
Lyubarsky Yury E.
Ma YuQian
Maccacaro Tommaso
Maccarone Thomas J.
Macchetto Ferdinando
Machida Mami
Maggio Antonio
Makarov Valeri
Malitson Harriet H.
Malkan Matthew Arnold
Manara Alessandro A.
Mandolesi Nazzareno
Maran Stephen P.
Marar T. M. k.
Markoff Sera B.
Marov Mikhail Ya
Martin Inacio Malmonge
Martinis Mladen E.
Masai Kuniaki
Mason Keith Owen
Mason Glenn M.
Mather John Cromwell
Matsumoto Ryoji
Matsumoto Hironori
Matsuoka Masaru
Matt Giorgio
Matz Steven Micheal
Mazurek Thaddeus John
McBreen Brian Philip
McCluskey Jr George E.
McCray Richard
McWhirter R. W. Peter
Mead Jaylee Montague
Medina Jose
Medina Tanco Gustavo A.
Meier David L.
Meiksin Avery Abraham
Melatos Andrew
Melia Fulvio
Melnick Gary J.
Melrose Donald B.
Mendez Mariano
Mereghetti Sandro
Mestel Leon
Meszaros Peter
Meyer Jean-Paul
Meyer Friedrich
Micela Giuseppina
Michel F. Curtis
Miller Michael C.
Miller Guy Scott
Miller John C.
Min Yuan Wei
Mineo Teresa
Miyaji Takamitsu
Miyaji Shigeki
Miyata Emi
Mizumoto Yoshihiko
Mizutani Kohei
Moderski Rafal
Modisette Jerry L.
Monet David G.
Moon Shin Haeng
Moos Henry Warren
Morgan Thomas H.
Morton Donald C.
Motch Christian
Motizuki Yuko
Mucciarelli Paola
Mulchaey John S.
Murakami Hiroshi
Murakami Toshio
Murdock Thomas Lee
Murtagh Fionn
Murthy Jayant
Nagataki Shigehiro
Nakayama Kunji
Neeman Yuval
Neff Susan Gale
Ness Norman F.
Neuhaeuser Ralph
Neupert Werner M.
Nichols Joy
Nicollier Claude
Nishimura Osamu
Nitta Shin-ya
Nityananda Ram

Nomoto Ken'ichi
Norci Laura
Nordh Lennart H.
Norman Colin A.
Novick Robert
Noyes Robert W.
Nulsen Paul E. J.
O'Brien Paul Thomas
O'Connell Robert F.
Oezel Mehmet Emin
Ogawara Yoshiaki
Ogelman Hakki B.
Okeke Pius N.
Okoye Samuel E.
Okuda Toru
Olthof Henk
O'Mongain Eon
Oohara Ken-ichi
Orford Keith J.
Orio Marina
Orlandini Mauro
Orlando Salvatore
Osborne Julian P.
Osten Rachel A.
Ostriker Jeremiah P.
Ostrowski Michal
O'Sullivan Denis F.
Ott Juergen A.
Owen Tobias C.
Ozaki Masanobu
Pacholczyk Andrzej G.
Paciesas William S.
Pacini Franco
Page Clive G.
Page Mathew J.
Paltani Stéphane
Palumbo Giorgio G. C.
Pandey Uma Shankar
Papadakis Iossif
Park Myeong-gu
Parker Eugene N.
Parkinson William H.
Parkinson John H.
Paul Biswajit
Pavlov George G.
Peacock Anthony
Pearce Mark
Pearson Kevin J.
Pellegrini Silvia
Peng Qiuhe
Perola Giuseppe C.
Perry Peter M.
Peters Geraldine Joan
Peterson Bruce A.
Peterson Laurence E.
Pethick Christopher J.
Petkaki Panagiota
Petro Larry David
Petrosian Vahe
Phillips Kenneth J. H.
Pian Elena
Pinkau K.
Pinto Philip Alfred
Pipher Judith L.
Piran Tsvi
Piro Luigi
Polidan Ronald S.
Pounds Kenneth A.
Poutanen Juri
Prasanna A. R.
Preuss Eugen
Price Stephan Donald
Protheroe Raymond J.
Prusti Timo
Qiu Yulei
Qu Qinyue
Quintana Hernan
Quiros Israel
Radhakrishnan V.
Raiteri Claudia M.
Ramadurai Souriraja
Ramirez Jose M.
Rao Ramachandra V.
Rao Arikkala Raghurama
Rasmussen Ib L.
Rasmussen Jesper
Raubenheimer Barend C.
Raychaudhury Somak
Reale Fabio
Rees Martin J.
Reeves Hubert
Reeves Edmond M.
Reichert Gail Anne
Reig Pablo
Reiprich Thomas H.
Rengarajan T. N.
Rense William A.
Revnivtsev Mikhail G.
Rhoads James
Riegler Guenter R.
Roberts Timothy P.
Roman Nancy Grace
Romano Patrizia
Rosendhal Jeffrey D.
Rosner Robert
Rovero Adrián C.
Ruder Hanns
Ruffini Remo
Ruffolo David
Ruszkowski Mateusz
Rutledge Robert E.
Sabau-Graziati Lola
Safi-Harb Samar
Sagdeev Roald Z.
Sahade Jorge
Saiz Alejandro
Sakano Masaaki
Sakelliou Irini
Salpeter Edwin E.
Samimi Jalal
Sanchez Norma G.
Sanders Wilton Turner III
Santos Nilton Oscar
Sartori Leo
Saslaw William C.
Sato Katsuhiko
Savage Blair D.
Savedoff Malcolm P.
Sazonov Sergey
Scargle Jeffrey D.
Schaefer Gerhard
Schatten Kenneth H.
Schatzman Evry
Schilizzi Richard T.
Schmitt Juergen H. M. M.
Schnopper Herbert W.
Schreier Ethan J.
Schulz Norbert S.
Schwartz Steven Jay
Schwartz Daniel A.
Schwehm Gerhard
Sciortino Salvatore
Scott John S.
Seielstad George A.
Selvelli Pierluigi
Semerak Oldrich
Sequeiros Juan
Setti Giancarlo
Seward Frederick D.
Shaham Jacob

Wells Donald C.
Wentzel Donat G.
Wesselius Paul R.
Wheatley Peter J.
Wheeler John A.
Wheeler J. Craig
White Nicholas Ernest
Wijers Ralph A. M. J.
Wijnands Rudy
Will Clifford M.
Willis Allan J.
Willner Steven Paul
Wilms Jörn
Wilson James R.
Wilson Andrew S.
Winkler Christoph
Wise Michael W.
Wolfendale Arnold W.
Wolstencroft Ramon D.
Wolter Anna
Woltjer Lodewijk
Worrall Diana Mary
Wray James D.
Wu Chi Chao
Wu Xuejun
Wu Shaoping
Wunner Guenter
Xu Renxin
Yadav Jagdish Singh
Yamada Shoichi
Yamamoto Yoshiaki
Yamasaki Tatsuya
Yamasaki Noriko Y.
Yamashita Kojun
Yamauchi Makoto
Yamauchi Shigeo
Yock Philip
Yoshida Atsumasa
You Junhan
Yu Wang Xiang
Yuan Ye-fei
Zamorani Giovanni
Zane Silvia
Zannoni Mario
Zarnecki John Charles
Zdziarski Andrzej
Zezas Andreas
Zhang Shuang Nan
Zhang Li
Zhang William W.
Zhang Jialu
Zheng Wei
Zheng Xiaoping
Zombeck Martin V.

MEMBERSHIP BY COMMISSION

Composition of Commission 45

Stellar Classification / Classification Stellaire

President Giridhar Sunetra

Vice-President Gray Richard O.

Organizing Committee

Bailer-Jones Coryn A. L.
Corbally Christopher
Eyer Laurent
Irwin Michael John
Kirkpatrick Joseph D.
Minniti Dante
Nordström Birgitta

Members

Albers Henry
Allende Prieto Carlos
Ardeberg Arne L.
Arellano Ferro Armando
Babu G. S. D.
Baglin Annie
Barbosa Cassio L.
Bartaya R. A.
Bartkevicius Antanas
Bell Roger A.
Bidelman William P.
Blanco Victor M.
Buser Roland
Celis Leopoldo
Cester Bruno
Cherepashchuk Anatolij M.
Christy James Walter
Claria Juan
Coluzzi Regina
Cowley Anne P.
Crawford David L.
Divan Lucienne
Drilling John S.
Eglitis Ilgmars
Egret Daniel
Faraggiana Rosanna
Feast Michael W.
Fehrenbach Charles
Feltzing Sofia
Fitzpatrick Edward L.
Fukuda Ichiro
Garmany Catherine D.
Garrison Robert F.
Gerbaldi Michèle
Geyer Edward H.
Gizis John E.
Glagolevskij Yurij V.
Golay Marcel
Grenon Michel
Grosso Monica Gladys
Guetter Harry Hendrik
Gupta Ranjan
Gurzadyan Grigor A.
Hack Margherita
Hallam Kenneth L.
Hauck Bernard
Hayes Donald S.
Holmberg Johan
Houk Nancy
Humphreys Roberta M.
Kato Ken-ichi
Kurtanidze Omar M.
Kurtz Michael Julian
Kurtz Donald W.
Labhardt Lukas
Lasala Gerald J.
Lattanzio John
Lee Sang Gak
Leggett Sandy K.
Levato Hugo
Lloyd Evans Thomas Harry
Loden Kerstin R.
Low Frank J.
Lu Phillip K.
Luri Xavier
Lutz Julie H.
MacConnell Darrell Jack
Maehara Hideo
Malagnini Maria Lucia
Malaroda Stella M.
McNamara Delbert H.
Mead Jaylee Montague
Mendoza V. Eugenio E.
Morossi Carlo
Morrell Nidia
Nicolet Bernard
North Pierre
Notni Peter
Oja Tarmo
Olsen Erik H.
Osborn Wayne
Oswalt Terry D.
Pakhomov Yuri V.
Parsons Sidney B.
Pasinetti Laura E.
Philip A. G. Davis
Pizzichini Graziella
Preston George W.
Rautela B. S.
Roman Nancy Grace
Rountree Janet
Rudkjobing Mogens(†)
Schild Rudolph E.
Schmidt-Kaler Theodor
Seitter Waltraut C.(†)
Sharpless Stewart
Shore Steven N.
Shvelidze Teimuraz D.
Sinnerstad Ulf E.

Sion Edward Michael
Smith J. Allyn
Sonti Sreedhar Rao
Soubiran Caroline
Steinlin Uli
Straizys Vytautas P.
Strobel Andrzej
Upgren Arthur R.
von Hippel Theodore A.
Walborn Nolan R.
Walker Gordon A. H.
Warren Jr Wayne H.
Weaver William Bruce
Weiss Werner W.
Wesselius Paul R.
Westerlund Bengt E.
Williams John A.
Wing Robert F.
Wu Hsin-Heng
Wyckoff Susan
Yamashita Yasumasa
Yoss Kenneth M.
Zdanavicius Kazimeras

MEMBERSHIP BY COMMISSION

Composition of Commission 46

Astronomy Education & Development

Education & Développement en Astronomie

President Stavinschi Magda G.

Vice-President Ros Rosa M.

Organizing Committee

Gerbaldi Michèle
Guinan Edward F.
Hearnshaw John B.
Isobe Syuzo(†)
Metaxa Margarita
Morrell Nidia
Othman Mazlan
Pasachoff Jay M.
Percy John R.
Tolbert Charles R.
Torres-Peimbert Silvia
White James Clyde

Members

Acker Agnes
Aguilar Maria Luisa
Aiad A. Zaki
Albanese Lara
Alexandrov Yuri V.
Al-Naimiy Hamid M. K.
Alsabti Abdul Athem
Alvarez Rodrigo
Alvarez-Pomares Oscar
Anandaram Mandayam N.
Andersen Johannes
Andrews Frank
Ansari S. M. Razaullah
Arellano Ferro Armando
Aubier Monique G.
Bajaja Esteban
Barclay Charles E.
Barthel Peter
Baskill Darren S.
Batten Alan H.
Benson Priscilla J.
Bernabeu Guillermo
Birlan Mirel I.
Bittar Jamal
Black Adam Robert S.
Bobrowsky Matthew
Bochonko D. Richard
Booth Roy S.
Borchkhadze Tengiz M.
Botez Elvira
Bottinelli Lucette
Brieva Eduardo
Brosch Noah
Bruck Mary T.
Budding Edwin
Cai Michael
Calvet Nuria
Cannon Wayne H.
Capaccioli Massimo
Carter Brian
Catala Poch M. A.
Celebre Cynthia P.
Chamcham Khalil
Chitre Dattakumar M.
Christensen Lars Lindberg
Clarke David
Codina Ladanberry Sayd J.(†)
Colafrancesco Sergio
Corbally Christopher
Cottrell Peter L.
Couper Heather
Covone Giovanni
Crawford David L.
Cui Zhenhua
Cunningham Maria R.
Dall'Ora Massimo
Daniel Jean-Yves
Danner Rolf
de Greve Jean-Pierre
DeGioia-Eastwood Kathleen
Demircan Osman
Devaney Martin N.
Diego Francisco
Ducati Jorge R.
Dukes Jr. Robert
Dupuy David L.
Duval Marie-France
Dworetsky Michael M.
El Eid Mounib
Eze Romanus N.
Fairall Anthony P.
Fernandez Julio Angel
Fernandez-Figueroa M. J.
Fienberg Richard T.
Fierro Julieta
Fleck Robert Charles
Florsch Alphonse
Forbes Douglas
Fu Hsieh-Hai
Gallino Roberto
Gasiprong Nipon
Germany Lisa M.
Ghobros Roshdy Azer
Gill Peter B. J.
Gimenez Alvaro
Gingerich Owen
Gouguenheim Lucienne
Gray Richard O.

Gregorio-Hetem Jane C.
Hafizi Mimoza
Haque Shirin T.
Haubold Hans J.
Haupt Hermann F.
Havlen Robert J.
Hawkins Isabel
Haywood J.
Hemenway Mary Kay M.
Heudier Jean-Louis
Hidayat Bambang
Hockey Thomas Arnold
Hoff Darrel Barton
Houziaux Leo
Huan Nguyen Dinh
Huang Tianyi
Huettemeister Susanne
Hughes Stephen W.
Ilyas Mohammad
Impey Christopher D.
Inglis Michael
Isaak Kate G.
Ishizaka Chiharu
Iwaniszewska Cecylia
Jafelice Luiz C.
Jarrett Alan H.(†)
Jones Barrie W.
Jorgensen Henning E.
Karetnikov Valentin G. R.
Karttunen Hannu
Keller Hans-Ulrich F.
Khan Josef I.
Kiasatpour Ahmad
Kitchin Christopher R.
Klinglesmith III Daniel A.
Koechlin Laurent
Kolka Indrek
Kononovich Edvard V.
Kourganoff Vladimir(†)
Kozai Yoshihide
Kramer Busaba H.
Kreiner Jerzy M.
Krishna Gopal
Krupp Edwin C.
Kuan Yi-Jehng
Lago Maria T. V. T.
Lai Sebastiana
Lanciano Nicoletta
Lee Kang Hwan
Lee Yong Bok
Leung Chun Ming
Leung Kam Ching
Li Zongwei
Linden-Voernle Michael J. D.
Little-Marenin Irene R.
Lomb Nicholas Ralph
Luck John M.
Ma Xingyuan
Maciel Walter J.
Macrae Donald A.(†)
Maddison Ronald Ch.
Mahoney Terence J.
Malasan Hakim Luthfi
Mamadazimov Mamadmuso
Marco Olivier
Marsh Julian C. D.
Martinet Louis
Martinez Peter
Massey Robert M.
Mavridis Lyssimachos N.
Maza Jose
McKinnon David H.
McNally Derek
Meidav Meir
Milogradov-Turin Jelena
Mizuno Takao
Moreels Guy
Morimoto Masaki
Murphy John A.
Muzzio Juan C.
Najid Nour-Eddine
Narlikar Jayant V.
Nayar S. R. Prabhakaran
Nguyen-Quang Rieu
Nha Il-Seong
Nicolson Iain
Nikolov Nikola S.
Ninkovic Slobodan
Noels Arlette
Norton Andrew J.
Oja Heikki
Okeke Pius N.
Okoye Samuel E.
Olsen Fogh H. J.
Onuora Lesley Irene
Oowaki Naoaki
Orchiston Wayne
Osborn Wayne
Osorio Jose Pereira
Oswalt Terry D.
Pandey Uma Shankar
Parisot Jean-Paul
Penston Margaret
Pokorny Zdenek
Ponce Gustavo A.
Proverbio Edoardo
Quamar Jawaid
Querci Francois R.
Quiros Israel
Raboud Didier
Raghavan Nirupama
Ramadurai Souriraja
Rijsdijk Case
Roberts Morton S.
Robinson Leif J.
Roca Cortes Teodoro
Rosenzweig-Levy Patrica
Roslund Curt
Routly Paul M.
Roy Archie E.
Sabra Bassem Mohamad
Sadat Rachida
Safko John L.
Sahade Jorge
Samodurov Vladimir A.
Sanahuja Blai
Sandqvist Aage
Sattarov Isroil
Satterthwaite Gilbert E.
Saxena P. P.
Sbirkova-Natcheva T.
Schleicher David G.
Schlosser Wolfhard
Schmitter Edward F.
Schroeder Daniel J.
Seeds Michael A.
Shen Chun-Shan
Shipman Harry L.
Slater Timothy F.
Smail Ian
Smith Francis Graham
Solheim Jan Erik
Soriano Bernardo M.
Stefl Vladimir
Stenholm Björn
Stoev Alexey D.
Straizys Vytautas P.
Sukartadiredja Darsa

Svestka Jiri
Swarup Govind
Szecsenyi-Nagy Gábor A.
Szostak Roland
Torres Jesus Rodrigo F.
Touma Jihad Rachid
Urama Johnson O.
van den Heuvel
Edward P. J.
van Santvoort Jacques
Vauclair Sylvie D.
Vilks Ilgonis
Vinuales Gavin Ederlinda
Vujnovic Vladis
Walsh Wilfred M.
Wang Shouguan
Ward Richard A.
Wentzel Donat G.
West Richard M.
Whitelock Patricia Ann
Williamon Richard M.
Willmore A. Peter
Wolfschmidt Gudrun
Ye Shuhua
Yim Hong-Suh
Zakirov Mamnum
Zealey William J.
Zeilik Michael Ii
Zhao Jun Liang
Zimmermann Helmut

MEMBERSHIP BY COMMISSION

Composition of Commission 47

Cosmology / Cosmologie

President Webster Rachel L.

Vice-President Padmanabhan Thanu

Organizing Committee

Campusano Luis E.
Charlot Stephane
da Costa Luiz A. N.
Koo David C-Y
Lahav Ofer
Lefevre Olivier
Peacock John Andrew
Scott Douglas
Suto Yasushi

Members

Abu Kassim Hasan B.
Adami Christophe
Adams Jenni
Alard Christophe L.
Alcaniz Jailson S.
Alimi Jean-Michel A.
Allan Peter M.
Allington-Smith Jeremy R.
Almaini Omar
Amendola Luca
Andersen Michael I.
Andreani Paola Michela
Aretxaga Itziar
Argueso Francisco
Atrio Barandela Fernando
Audouze Jean
Avelino Pedro
Azuma Takahiro
Bagla Jasjeet S.
Bahcall Neta A.
Bajtlik Stanislaw
Baker Andrew J.
Baldwin John E.
Banday Anthony J.
Banerji Sriranjan
Banhatti Dilip G.
Barberis Bruno
Barbuy Beatriz
Bardeen James M.
Bardelli Sandro
Barger Amy J.
Barkana Rennan
Barr Jordi McGregor
Barrow John David
Bartelmann Matthias
Barthel Peter
Barton Elizabeth J.
Basa Stephane
Bassett Bruce A.
Basu Dipak
Battye Richard A.
Bechtold Jill
Beckman John E.
Beesham Aroonkumar
Belinski Vladimir A.
Bennett Charles L.
Bergeron Jacqueline A.
Bergvall Nils Ake Sigvard
Berman Marcelo S.
Berta Stefano
Bertola Francesco
Bertschinger Edmund
Betancor Rijo Juan
Bharadwaj Somnath
Bhavsar Suketu P.
Bianchi Simone
Bicknell Geoffrey V.
Bignami Giovanni F.
Binetruy Pierre
Birkinshaw Mark
Biviano Andrea
Blakeslee John P.
Blanchard Alain
Bleyer Ulrich
Bludman Sidney A.
Blundell Katherine M.
Boehringer Hans
Boksenberg Alec
Bolzonella Micol
Bond John Richard
Bongiovanni Angel
Bonnor W. B.
Borgani Stefano
Bouchet François R.
Boyle Brian J.
Brecher Kenneth
Bridle Sarah L.
Bunker Andrew J.
Buote David A.
Burbidge Geoffrey R.
Burns Jr Jack O'Neal
Calvani Massimo
Canavezes Alexandre G. R.
Cappi Alberto
Cardenas Rolando P.
Carr Bernard John
Carretti Ettore
Castagnino Mario
Cavaliere Alfonso G.
Cesarsky Diego A.
Chang Kyongae
Chang Heon-Young
Chen Hsiao-Wen
Chen DaMing
Chen Jiansheng
Cheng Fuzhen
Chiba Takeshi
Chincarini Guido L.
Chitre Dattakumar M.

Chodorowski Michal
Chu Yaoquan
Claeskens Jean-François
Claria Juan
Clarke Tracy E.
Clowe Douglas I.
Clowes Roger G.
Cocke William John
Cohen Jeffrey M.
Cohen Ross D.
Colafrancesco Sergio
Cole Shaun M.
Coles Peter
Colless Matthew
Colombi Stephane
Condon James J.
Cooray Asantha R.
Cora Sofia A.
Corsini Enrico M.
Courbin Frederic Y. M.
Courteau Stephane J.
Covone Giovanni
Crane Philippe
Crane Patrick C.
Cristiani Stefano V.
Croom Scott M.
Curran Stephen J.
D' Odorico Valentina
Da Costa Gary Stewart
Dadhich Naresh
Dahle Haakon
Daigne Frederic
Danese Luigi
Das P. K.
Davidson William
Davies Paul Charles W.
Davies Roger L.
Davis Tamara M
Davis Marc
Davis Michael M.
de Lapparent-Gurriet Valérie
De Petris Marco
de Ruiter Hans Rudolf
de Silva L. Nalin
de Zotti Gianfranco
Dekel Avishai
Dell'Antonio P. Ian
Demianski Marek
Deustua Susana E.
Dhurandhar Sanjeev
Diaferio Antonaldo
Dionysiou Demetrios
Djorgovski Stanislav
Dobbs Matt A.
Dobrzycki Adam
Dressler Alan
Drinkwater Michael J.
Dultzin-Hacyan Deborah
Dunlop James
Dunsby Peter
Dyer Charles Chester
Eales Stephen A.
Edsjo Joakim
Efstathiou George
Ehlers Jürgen
Einasto Jaan
Elgaroy Oystein
Elizalde Emilio
Ellis George F. R.
Ellis Richard S.
Elvis Martin S.
Enginol Turan B.
Ettori Stefano
Eungwanichayapant Anant
Faber Sandra M.
Fairall Anthony P.
Falk Jr Sydney W.
Fall S. Michael
Fan Zuhui
Fang Li-zhi
Fassnacht Christopher D.
Fedorova Elena
Felten James E.
Feng Long Long
Field George B.
Filippenko Alexei V.
Florides Petros S.
Focardi Paola
Fomin Piotr Ivanovich
Fong Richard
Ford Holland C.
Forman William Richard
Fouque Pascal
Fox Andrew J.
Franceschini Alberto
Frenk Carlos S.
Friaca Amancio C. S.
Fukugita Masataka
Fukui Takao
Fynbo Johan P. U.
Galletto Dionigi
Garilli Bianca
Garrison Robert F.
Geller Margaret Joan
Germany Lisa M.
Ghirlanda Giancarlo
Giallongo Emanuele
Gioia Isabella M.
Goldsmith Donald W.
Gonzalez Alejando
Gonzalez-Solares Eduardo
Goobar Ariel M.
Goret Philippe
Gosset Eric
Gottloeber Stefan
Gouda Naoteru
Govinder Keshlan S.
Goyal Ashok Kumar
Grainge Keith J. B.
Granato Gian Luigi
Gray Richard O.
Gray Meghan
Green Anne M.
Gregory Stephen Albert
Greve Thomas R.
Greyber Howard D.
Griest Kim
Grishchuk Leonid P.
Gudmundsson Einar H.
Gunn James E.
Guzzo Luigi
Haehnelt Martin G.
Hagen Hans-Juergen
Hall Patrick B.
Hamilton Andrew J. S.
Hannestad Steen
Hardy Eduardo
Harms Richard James
Harrison Edward R.(†)
Hawking Stephen W.
Hayashi Chushiro
He XiangTao
Heavens Alan
Heinamaki Pekka S.
Hellaby Charles William
Heller Michael
Hendry Martin A.
Henriksen Mark Jeffrey
Hewett Paul

Hewitt Adelaide
Hnatyk Bohdan Ivanovych
Hu Esther M.
Huang Jiasheng
Huchra John Peter
Hudson Michael J.
Hughes David H.
Hwang Chorng-Yuan
Hwang Jai-chan
Icke Vincent
Ikeuchi Satoru
Im Myungshin
Impey Christopher D.
Inada Naohisa
Iovino Angela
Ishihara Hideki
Iyer B. R.
Jakobsson Pall
Jannuzi Buell Tomasson
Jaroszynski Michal
Jauncey David L.
Jaunsen Andreas O.
Jedamzik Karsten
Jensen Brian L.
Jetzer Philippe F.
Jing Yipeng
Jones Heath D.
Jones Laurence R.
Jones Bernard J. T.
Jones Christine
Jovanovic Predrag
Junkkarinen Vesa T.
Juszkiewicz Roman
Kajino Toshitaka
Kanekar Nissim
Kang Hyesung
Kapoor Ramesh Chander
Karachentsev Igor D.
Kato Shoji
Kauffmann Guinevere A. M.
Kaul Chaman
Kawabata Kinaki
Kawabata Kiyoshi
Kawasaki Masahiro
Kellermann Kenneth I.
Kembhavi Ajit K.
Khare Pushpa
Khmil Sergiy V.
Kim Jik Su
King Lindsay J.
Kirilova Daniela
Kneib Jean-Paul
Kodama Hideo
Koekemoer Anton M.
Kokkotas Konstantinos
Kolb Edward W.
Kompaneets Dmitrij A.
Koopmans Leon V. E.
Kormendy John
Kovalev Yuri A.
Kovetz Attay
Kozai Yoshihide
Kozlovsky Ben Z.
Krasinski Andrzej
Kriss Gerard A.
Kudrya Yury M.
Kunth Daniel
La Barbera Francesco
Lacey Cédric
Lachieze-Rey Marc
Lake Kayll William
Larionov Mikhail G.
Lasota Jean-Pierre
Layzer David
Lequeux James
Leubner Manfred P.
Levin Yuri
Lewis Geraint F.
Li Li-Xin
Lian Luo Xin
Liddle Andrew
Liebscher Dierck-E
Lilje Per Vidar Barth
Lilly Simon J.
Lima Jose A. S.
Liu Yongzhen
Lombardi Marco
Longair Malcolm S.
Lonsdale Carol J.
Lopez Corredoira Martin
Loveday Jon N.
Lu Tan
Lubin Lori M.
Lukash Vladimir N.
Luminet Jean-Pierre
Lynden-Bell Donald
Maartens Roy
Maccagni Dario
MacCallum Malcolm A. H.
Maddox Stephen
Maeda Kei-ichi
Maharaj Sunil Dutt
Maia Marcio R. G. S.
Mamon Gary A.
Mandolesi Nazzareno
Mangalam Arun
Mann Robert G.
Manrique Alberto
Mansouri Reza
Mao Shude
Maoz Dan
Marano Bruno
Mardirossian Fabio
Marek John
Maris Michele
Martinez-Gonzalez Enrique
Martinis Mladen E.
Martins Carlos J. A. P.
Mather John Cromwell
Mathez Guy
Matsumoto Toshio
Matzner Richard A.
Mavrides Stamatia
Mellier Yannick
Melott Adrian L.
Meneghetti Massimo
Merighi Roberto
Meszaros Attila
Meszaros Peter
Meyer David M.
Meylan Georges
Mezzetti Marino
Miralda-Escude Jordi
Miralles Joan-Marc
Miranda Oswaldo D.
Misner Charles W.
Miyazaki Satoshi
Miyoshi Shigeru
Mo Houjun
Mohr Joseph J.
Monaco Pierluigi
Moore Ben
Moreau Olivier
Moscardini Lauro
Motta Veronica
Muanwong Orrarujee
Muecket Jan P.
Muller Richard A.
Murakami Izumi

Murphy Michael T.
Murphy John A.
Nakamichi Akika
Nambu Yasusada
Narasimha Delampady
Narlikar Jayant V.
Naselsky Pavel D.
Nasr-Esfahani Bahram
Neeman Yuval
Neves de Araujo Jose Carlos
Nicoll Jeffrey Fancher
Nishida Minoru
Noerdlinger Peter D.
Noh Hyerim
Noonan Thomas W.
Norman Colin A.
Norman Dara J.
Nottale Laurent
Novello Mario
Novikov Igor D.
Novosyadlyj Bohdan
Novotny Jan
Nozari Kourosh
O'Connell Robert West
Oemler Jr Augustus
Okoye Samuel E.
Oliver Sebastian J.
Olowin Ronald Paul
Omnes Roland
Onuora Lesley Irene
Oscoz Alejandro
Ott Heinz-Albert
Ozsvath I.
Page Don Nelson
Parnovsky Sergei
Partridge Robert B.
Pecker Jean-Claude
Pedersen Kristian
Peebles P. James E.
Pello Roser Descayre
Pen Ue-Li
Penzias Arno A.
Perryman Michael A. C.
Persides Sotirios C.
Peterson Bruce A.
Petitjean Patrick
Petrosian Vahe
Pimbblet Kevin A.
Plionis Manolis
Popescu Nedelia A.
Portinari Laura
Power Chris B.
Prandoni Isabella
Premadi Premana W.
Press William H.
Puetzfeld Dirk
Puget Jean-Loup
Puy Denis
Qin Bo
Qu Qinyue
Quiros Israel
Rahvar Sohrab
Ramella Massimo
Rawlings Steven
Raychaudhuri Amalkumar
Rebolo Rafael
Rees Martin J.
Reeves Hubert
Refsdal Sjur
Reiprich Thomas H.
Revnivtsev Mikhail G.
Rey Soo-Chang
Riazi Nematollah
Riazuelo Alain
Ribeiro Marcelo B.
Ricotti Massimo
Rindler Wolfgang
Rivolo Arthur Rex
Roberts David Hall
Robinson I.
Rocca-Volmerange Brigitte
Roeder Robert
Roettgering Huub
Romano-Diaz Emilio
Romeo Alessandro B.
Romer Anita K.
Rosquist Kjell
Rowan-Robinson Michael
Roxburgh Ian W.
Rubin Vera C.
Rudnick Lawrence
Rudnicki Konrad
Ruffini Remo
Ruszkowski Mateusz
Saar Enn
Sadat Rachida
Sahni Varun
Saiz Alejandro
Salvador-Sole Eduardo
Salzer John Joseph
Sanz Jose L.
Sapar Arved
Saracco Paolo
Sargent Wallace L. W.
Sasaki Misao
Sasaki Shin
Sato Humitaka
Sato Katsuhiko
Sato Shinji
Sato Jun'ichi
Savage Ann
Sazhin Mikhail V.
Scaramella Roberto
Schatzman Evry
Schaye Joop
Schechter Paul L.
Schindler Sabine
Schmidt Maarten
Schmidt Brian P.
Schneider Donald P.
Schneider Jean
Schneider Peter
Schneider Raffaella
Schramm Thomas
Schuch Nelson Jorge
Schucking Engelbert L.
Schuecker Peter
Schumacher Gerard
Scodeggio Marco
Seielstad George A.
Semerak Oldrich
Serjeant Stephen
Setti Giancarlo
Shandarin Sergei F.
Shanks Thomas
Shao Zhengyi
Shaver Peter A.
Shaviv Giora
Shaya Edward J.
Shibata Masaru
Shivanandan Kandiah
Siebenmorgen Ralf
Signore Monique
Silk Joseph I.
Silva Laura
Sironi Giorgio
Sistero Roberto F.
Slosar Anze
Smail Ian
Smette Alain

Smith Rodney M.
Smith Nigel J. T.
Smith Jr Harding E.
Smoot III George F.
Sokolowski Lech
Sollerman Jesper
Song Doo Jong
Souriau Jean-Marie
Spinoglio Luigi
Spyrou Nicolaos
Squires Gordon K.
Srianand Roghunathan
Stavrev Konstantin Y.
Stecker Floyd W.
Steigman Gary
Stewart John Malcolm
Stoeger William R.
Stolyarov Vladislav A.
Storrie-Lombardi Lisa
Straumann Norbert
Stritzinger Maximilian D.
Struble Mitchell F.
Strukov Igor A.
Stuchlik Zdenek
Subrahmanya C. R.
Suginohara Tatsushi
Sugiyama Naoshi
Suhhonenko Ivan
Sunyaev Rashid A.
Surdej Jean M. G.
Sutherland William
Szalay Alex
Szydlowski Marek W.
Tagoshi Hideyuki
Takahara Fumio
Tammann Gustav A.
Tanabe Kenji
Tarter Jill C.
Taruya Atsushi
Tatekawa Takayuki
Taylor Angela C.
Thuan Trinh Xuan
Tifft William G.
Tipler Frank Jennings
Toffolatti Luigi
Tomimatsu Akira
Tomita Kenji
Tonry John
Tormen Giuseppe
Totani Tomonori
Tozzi Paolo
Treder H. J.
Tremaine Scott Duncan
Trevese Dario
Trimble Virginia L.
Trotta Roberto
Tsamparlis Michael
Tully Richard Brent
Turner Michael S.
Turner Edwin L.
Turnshek David A.
Tyson John Anthony
Tytler David
Tyul'bashev Sergei A.
Umemura Masayuki
Uson Juan M.
Vagnetti Fausto
Vaidya P. C.
Vaisanen Petri S. M.
Valls-Gabaud David
van der Laan Harry
van Haarlem Michiel
Vedel Henrik
Vettolani Giampaolo
Viana Pedro
Vishniac Ethan T.
Vishveshwara C. V.
Voglis Nikos
Volonteri Marta
von Borzeszkowski H. H.
Waddington Ian
Wagoner Robert V.
Wainwright John
Wambsganss Joachim
Wanas Mamdouh I.
Watson Darach J.
Webb Tracy M. A.
Webster Adrian S.
Weinberg Steven
Wesson Paul S.
West Michael J.
Wheeler John A.
White Simon David Manion
Whiting Alan B.
Whitrow Gerald James
Widrow Larry M.
Will Clifford M.
Wilson Albert G.
Wilson Andrew S.
Windhorst Rogier A.
Wise Michael W.
Wolfe Arthur M.
Woltjer Lodewijk
Woszczyna Andrzej
Wright Edward L.
Wu Xiangping
Wyithe Stuart
Xiang Shouping
Xu Chongming
Yasuda Naoki
Yi Sukyoung Ken
Yokoyama Jun-ichi
Yoshii Yuzuru
Yoshioka Satoshi
Yushchenko Alexander V.
Zamorani Giovanni
Zanichelli Alessandra
Zannoni Mario
Zaroubi Saleem
Zhang Jialu
Zhang Tongjie
Zhao Donghai
Zhou Youyuan
Zhu Xingfeng
Zieba Stanislaw
Zou Zhenlong
Zucca Elena
Zuiderwijk Edwardus J.

MEMBERSHIP BY COMMISSION

Composition of Commission 49

Interplanetary Plasma & Heliosphere

Plasma Interplanétaire & Héliosphère

President Bougeret Jean-Louis

Vice-President von Steiger Rudolf

Organizing Committee

Ananthakrishnan Subramanian
Cane Hilary V.
Gopalswamy Nat
Kahler Stephen W.
Lallement Rosine
Sanahuja Blai
Shibata Kazunari
Vandas Marek
Verheest Frank
Webb David F.

Members

Ahluwalia Harjit S.
Anderson Kinsey A.
Andretta Vincenzo
Balikhin Michael
Barnes Aaron
Barrow Colin H.
Barta Miroslav
Barth Charles A.
Benz Arnold O.
Bertaux Jean-Loup
Blackwell Donald E.
Blandford Roger David
Bochsler Peter
Bonnet Roger M.
Brandt John C.
Browning Philippa K.
Burlaga Leonard F.
Buti Bimla
Cairns Iver H.
Channok Chanruangrit
Chapman Sandra C.
Chashei Igor V.
Chassefiere Eric
Chitre Shashikumar M.
Chou Chih-Kang
Couturier Pierre A
Cramer Neil F.
Cuperman Sami
Daglis Ioannis A.
Dalla Silvia C.
Dasso Sergio
de Jager Cornelis
De Keyser Johan
Dinulescu Simona
Dorotovic Ivan
Dryer Murray
Duldig Marcus L.
Durney Bernard
Dyson John E.
Eshleman Von R.
Eviatar Aharon
Fahr Hans Joerg
Feynman Joan
Fichtner Horst
Field George B.
Fraenz Markus
Galvin Antoinette B.
Gedalin Michael
Gleisner Hans
Goldman Martin V.
Gosling John T.
Grzedzielski Stanislaw Pr.
Habbal Shadia Rifai
Harvey Christopher C.
Heras Ana M.
Heynderickx Daniel
Heyvaerts Jean
Hollweg Joseph V.
Holzer Thomas E.
Huber Martin C. E.
Humble John Edmund
Inagaki Shogo
Ivanov Evgenij V.
Jokipii J. R.
Joselyn Jo Ann c
Kakinuma Takakiyo T.
Keller Horst U.
Khan Josef I.
Ko Chung-Ming
Kojima Masayoshi
Lafon Jean-Pierre J.
Lai Sebastiana
Landi Simone
Levy Eugene H.
Li Bo
Lotova Natalja A.
Luest Reimar
Lundstedt Henrik
MacQueen Robert M.
Mangeney André
Manoharan P. K.
Marsch Eckart
Mason Glenn M.
Matsuura Oscar T.
Mavromichalaki Helen
Meister Claudia Veronika
Melrose Donald B.
Mendis Devamitta Asoka
Mestel Leon
Michel F. Curtis

Moussas Xenophon
Nickeler Dieter H.
Pandey Bierndra P.
Paresce Francesco
Parhi Shyamsundar
Parker Eugene N.
Perkins Francis W.
Pflug Klaus
Pneuman Gerald W.
Quemerais Eric
Raadu Michael A.
Readhead Anthony C. S.
Rickett Barnaby James
Riddle Anthony C.
Ripken Hartmut W.
Robinson Peter A.
Rosa Reinaldo R.
Rosner Robert
Roth Michel A.
Roxburgh Ian W.
Ruffolo David
Russell Christopher T.
Sagdeev Roald Z.
Saiz Alejandro
Sarris Emmanuel T.
Sastri Hanumath J.
Sawyer Constance B.
Schatzman Evry
Scherb Frank
Schindler Karl
Schmidt H. U.
Schreiber Roman
Schwartz Steven Jay
Schwenn Rainer W.
Setti Giancarlo
Shea Margaret A.
Smith Dean F.
Sonett Charles P.
Stone R. G.
Struminsky Alexei
Sturrock Peter A.
Suess Steven T.
Tritakis Basil P.
Tyul'bashev Sergei A.
Vainshtein Leonid A.
van Allen James A.(†)
Vinod S. Krishan
Voitsekhovska Anna
Vucetich Hector
Walker Simon N.
Wang Yi-ming
Watanabe Takashi
Watari Shinichi
Weller Charles S.
Wild John Paul
Willes Andrew J.
Wu Shi Tsan
Yeh Tyan
Yi Yu

MEMBERSHIP BY COMMISSION

Composition of Commission 50

Protection of Existing & Potential Observatory Sites

Protection des Sites d'Observatoires Existants & Potentiels

President Wainscoat Richard J. (*)

* **Incoming President, following the death of Schwarz Hugo who died on October 20, 2006, and following elections by the OC of Commission 50 and approval by Division XII.**

Organizing Committee

Blanco Carlo
Cohen Raymond James(†)
Crawford David L.
Isobe Syuzo(†)
Metaxa Margarita
Sullivan Woodruff T.

Members

Alvarez del Castillo Elizabeth M.
Ardeberg Arne L.
Arsenijevic Jelisaveta
Baan Willem A.
Barreto Luiz Muniz(†)
Baskill Darren S.
Benkhaldoun Zouhair
Bensammar Slimane
Bhattacharyya J. C.
Blanco Victor M.
Brown Robert Hamilton
Burstein David
Carraminana Alberto
Carrasco Bertha Esperanza
Cayrel Roger
Cinzano Pierantonio
Colas François
Costero Rafael
Coyne George V.
Davis Donald R.
Davis John
de Greiff J. Arias
Dommanget Jean
Dukes Jr. Robert
Edwards Paul J.
Galan Maximino J.
Garcia Beatriz E.
Gergely Tomas E.
Gibson David M.
Goebel Ernst
Green Richard F.
Haenel Andreus
Heck Andre
Helmer Leif
Hidayat Bambang
Ilyasov Sabit
Kadiri Samir
Kahlmann Hans C.
Kontizas Evangelos
Kontizas Mary
Kovalevsky Jean
Kozai Yoshihide
Kramer Busaba H.
Leibowitz Elia M.
Lewis Brian Murray
Lomb Nicholas Ralph
Mahra H. S.
Malin David F.
Markkanen Tapio
Mattig W.
McNally Derek
Mendoza-Torress Jose-Eduardo
Menzies John W.
Mitton Jacqueline
Murdin Paul G.
Nelson Burt
Oezel Mehmet Emin
Osorio Jose Pereira
Owen Frazer Nelson
Pankonin Vernon Lee
Percy John R.
Pound Marc W.
Sanchez Francisco M.
Schilizzi Richard T.
Shetrone Matthew C.
Siebenmorgen Ralf
Smith Malcolm G.
Smith Francis Graham
Spoelstra T. A. Th.
Stencel Robert Edward
Storey Michelle C.
Suntzeff Nicholas B.
Torres Carlos
Tremko Jozef
Tzioumis Anastasios
Upgren Arthur R.
Van den Bergh Sidney
van Driel Wim
Vernin Jean
Vetesnik Miroslav
Walker Merle F.
Whiteoak John B.
Woolf Neville J.
Woszczyk Andrzej
Yano Hajime
Ziznovsky Jozef

MEMBERSHIP BY COMMISSION

Composition of Commission 51

Bio-Astronomy / Bio-Astronomie

President Boss Alan P.

Vice-President Irvine William M.

Organizing Committee

Cosmovici Cristiano Batalli
Ehrenfreund Pascale
Latham David W.
Meech Karen J.
Morrison David
Udry Stephane

Members

Almar Ivan
Al-Naimiy Hamid M. K.
Alsabti Abdul Athem
Ando Hiroyasu
Balazs Bela A.
Ball John A.
Bania Thomas Michael
Barbieri Cesare
Basu Baidyanath
Basu Dipak
Baum William A.
Beaudet Gilles
Beckman John E.
Beckwith Steven V. W.
Beebe Reta Faye
Benest Daniel
Berendzen Richard
Bernacca Pierluigi
Billingham John
Biraud François
Bless Robert C.
Bond Ian A.
Bowyer C. Stuart
Boyce Peter B.
Bracewell Ronald N.
Broderick John
Brown Ronald D.
Burke Bernard F.
Calvin William H.
Campusano Luis E.
Cardenas Rolando P.
Carlson John B.
Carr Thomas D.
Chaisson Eric J.
Chou Kyong Chol
Cirkovic Milan M.
Clark Thomas A.
Colomb Fernando R.
Connes Pierre
Coude du Foresto Vincent
Couper Heather
Cunningham Maria R.
Daigne Gerard
Davis Michael M.
de Jager Cornelis
de Jonge J. K.
de Loore Camiel
de Vincenzi Donald
Deeg Hans
Delsemme Armand H.
Dent William
Dick Steven J.
Dixon Robert S.
Dorschner Johann
Doubinskij Boris A.
Downs George S.
Drake Frank D.
Dutil Yvan
Dyson Freeman J.
Eccles Michael J.
Ellis George F. R.
Epstein Eugene E.
Evans Neal J.
Fazio Giovanni G.
Fejes Istvan
Feldman Paul A.
Field George B.
Firneis Maria G.
Firneis Friedrich J.
Fisher Philip C.
Fraser Helen J.
Fredrick Laurence W.
Freire Ferrero Rubens G.
Fujimoto Masa-Katsu
Gatewood George
Gehrels Tom
Ghigo Francis D.
Ginzburg Vitalij L.
Giovannelli Franco
Godoli Giovanni
Golden Aaron
Goldsmith Donald W.
Gott J. Richard
Goudis Christos D.
Gregory Philip C.
Gulkis Samuel
Gunn James E.
Haddock Fred T.
Haisch Bernhard Michael
Hale Alan
Harrison Edward R.(†)
Hart Michael H.
Heck Andre
Heeschen David S.
Herczeg Tibor J.
Hershey John L.
Heudier Jean-Louis
Hinners Noel W.
Hirabayashi Hisashi
Hoang Binh Dy
Hogbom Jan A.
Hollis Jan Michael

Holm Nils G.
Horowitz Paul
Hunten Donald M.
Hunter James H.
Hysom Edmund J.
Idlis Grigorij M.
Israel Frank P.
Jastrow Robert
Jayawardhana Ray
Jeffers Stanley
Jennison Roger C.(†)
Jones Eric M.
Jugaku Jun
Kafatos Minas
Kapisinsky Igor
Kardashev Nicolay S.
Kaufmann Pierre
Keay Colin S. l.
Keller Hans-Ulrich F.
Kellermann Kenneth I.
Kilston Steven D.
Knowles Stephen H.
Kocer Durcun
Koch Robert H.
Koeberl Christian
Ksanfomality Leonid V.
Kuiper Thomas B. H.
Kuzmin Arkadij D.
Lafon Jean-Pierre J.
Laques Pierre
Lee Sang Gak
Leger Alain
Levasseur-Regourd
Anny-Chantal
Lilley Edward A.
Lineweaver Charles H.
Lippincott Zimmerman
Sarah Lee
Lovell Sir Bernard
Lyon Ian C.
Maffei Paolo
Margrave Jr Thomas E.
Marov Mikhail Ya
Martin Maria C.
Martin Anthony R.
Marzo Giuseppe A.
Matsakis Demetrios N.
Matsuda Takuya
Matthews Clifford
Mavridis Lyssimachos N.
Mayor Michel
McAlister Harold A.
McDonough Thomas R.
Mendoza V. Eugenio E.
Minn Young Key
Minniti Dante
Mirabel Igor Felix
Mokhele Khotso D.
Moore Marla H.
Morimoto Masaki
Morris Mark Root
Muller Richard A.
Naef Dominique
Nakagawa Yoshinari
Nelson Robert M.
Neuhaeuser Ralph
Niarchos Panayiotis
Norris Raymond P.
Ollongren A.
Ostriker Jeremiah P.
Owen Tobias C.
Pacini Franco
Parijskij Yurij N.
Pasinetti Laura E.
Perek Lubos
Pollacco Don
Ponsonby John E. B.
Qiu Yaohui
Quintana Jose M.
Quintana Hernan
Quirrenbach Andreas
Raghavan Nirupama
Rajamohan R.
Reay Newrick K.
Rees Martin J.
Riihimaa Jorma J.
Robinson Leif J.
Rodriguez Luis F.
Rood Robert T.
Rowan-Robinson Michael
Rubin Robert Howard
Russell Jane L.
Sakurai Kunitomo
Sancisi Renzo
Santos Nuno Miguel C.
Sarre Peter J.
Scargle Jeffrey D.
Schatzman Evry
Schild Rudolph E.
Schneider Jean
Schober Hans J.
Schuch Nelson Jorge
Seeger Charles Louis III
Seielstad George A.
Shapiro Maurice M.
Shen Chun-Shan
Shostak G. Seth
Sims Mark R.
Singh Harinder P.
Sivaram C.
Slysh Vyacheslav I.
Snyder Lewis E.
Sofue Yoshiaki
Song In-Ok
Stalio Roberto
Stein John William
Straizys Vytautas P.
Sturrock Peter A.
Sullivan Woodruff T.
Tahtinen Leena
Takaba Hiroshi
Takada-Hidai Masahide
Tarter Jill C.
Tavakol Reza
Tedesco Edward F.
Tejfel Viktor G.
Terzian Yervant
Thaddeus Patrick
Tolbert Charles R.
Toro Tibor
Tovmassian Hrant Mushegi
Townes Charles Hard
Trimble Virginia L.
Turner Kenneth C.
Turner Edwin L.
Vallee Jacques P.
Van Flandern Tom
Varshalovich Dmitrij A.
Vauclair Gérard P.
Vazquez Manuel
Venugopal V. R.
Verschuur Gerrit L.
Vogt Nikolaus
Wallis Max K.
Walsh Wilfred M.
Walsh Andrew J.
Watson Frederick Garnett
Welch William J.
Wellington Kelvin
Wesson Paul S.

Wetherhill George W.(†)
Wielebinski Richard
Williams Iwan P.
Willson Robert Frederick
Wilson Thomas L.
Wolstencroft Ramon D.
Wright Ian P.
Wright Alan E.
Xu Weibiao
Ye Shuhua
Zuckerman Ben M.

MEMBERSHIP BY COMMISSION

Composition of Commission 52

Relativity in Fundamental Astronomy

Relativité en Astronomie Fondamentale

President Klioner Sergei A.

Vice-President Petit Gérard

Organizing Committee

Brumberg Victor A.
Capitaine Nicole
Fienga Agnès
Fukushima Toshio
Guinot Bernard R.
Huang Cheng
Mignard François
Seidelmann P. Kenneth
Soffel Michael H.
Wallace Patrick T.

Members

Boucher Claude
Calabretta Mark R.
de Felice Fernando
Gray Norman
Hestroffer Daniel
Hilton James Lindsay
Huang Tianyi
Kaplan George H.
Luzum Brian J.
Manchester Richard N.
McCarthy Dennis D.
Nelson Robert A.
Osorio Jose Pereira
Pitjeva Elena V.
Ray James R.
Standish E. Myles
Vityazev Veniamin V.

MEMBERSHIP BY COMMISSION

Composition of Commission 53

Extrasolar Planets / Planètes Extrasolaires

President Mayor Michel

Vice-President Boss Alan P.

Organizing Committee

Butler Paul R.
Hubbard William B.
Ianna Philip A.
Kürster Martin
Lissauer Jack J.
Meech Karen J.
Mignard François
Penny Alan John
Quirrenbach Andreas
Tarter Jill C.
Vidal-Madjar Alfred

Composition of Commission 54

Optical & Infrared Interferometry

Interférométrie Optique & Infrarouge

President	Perrin Guy S.
Vice-President	Ridgway Stephen T.
Secretary	van Belle Gerard T.

Organizing Committee

Duvert Gilles	Hummel Christian Aurel	Queloz Didier
Genzel Reinhard	Lawson Peter R.	Tuthill Peter G.
Haniff Christopher	Monnier John D.	Vakili Farrokh

Members

Leisawitz David	Malbet Fabien

MEMBERSHIP BY COMMISSION

Composition of Commission 55

Communicating Astronomy with the Public

Communiquer l'Astronomie au Public

President	Robson Ian E.
Vice-President	Crabtree Dennis
Secretary	Christensen Lars Lindberg

Organizing Committee

Alvarez-Pomares Oscar
Damineli Neto Augusto
Fienberg Richard T.
Green Anne
Kembhavi Ajit K.
Nordström Birgitta
Sekiguchi Kazuhiro
Whitelock Patricia Ann
Zhu Jin

Members

de Grijs Richard

Transactions IAU, Volume XXVIB
Proc. IAU XXVI General Assembly, August 2006
Karel A. van der Hucht, ed.

doi:10.1017/S1743921308024435

ALPHABETICAL LIST OF IAU MEMBERS

A full alphabetical list of IAU members is available online as "Supplementary data" on Cambridge Journals Online: http://www.journals.cup.org/S1743921308024435